Lecture Notes in Networks and Systems

Volume 1010

The series "Lecture Notes in Networks and Systems" publishes the latest developments in Networks and Systems—quickly, informally and with high quality. Original research reported in proceedings and post-proceedings represents the core of LNNS.

Volumes published in LNNS embrace all aspects and subfields of, as well as new challenges in, Networks and Systems.

The series contains proceedings and edited volumes in systems and networks, spanning the areas of Cyber-Physical Systems, Autonomous Systems, Sensor Networks, Control Systems, Energy Systems, Automotive Systems, Biological Systems, Vehicular Networking and Connected Vehicles, Aerospace Systems, Automation, Manufacturing, Smart Grids, Nonlinear Systems, Power Systems, Robotics, Social Systems, Economic Systems and other. Of particular value to both the contributors and the readership are the short publication timeframe and the world-wide distribution and exposure which enable both a wide and rapid dissemination of research output.

The series covers the theory, applications, and perspectives on the state of the art and future developments relevant to systems and networks, decision making, control, complex processes and related areas, as embedded in the fields of interdisciplinary and applied sciences, engineering, computer science, physics, economics, social, and life sciences, as well as the paradigms and methodologies behind them.

Indexed by SCOPUS, EI Compendex, INSPEC, WTI Frankfurt eG, zbMATH, SCImago.

All books published in the series are submitted for consideration in Web of Science.

For proposals from Asia please contact Aninda Bose (aninda.bose@springer.com).

Sivaji Bandyopadhyay · Valentina Emilia Balas ·
Saroj Kumar Biswas · Anish Kumar Saha ·
Dalton Meitei Thounaojam
Editors

Intelligent Computing Systems and Applications

Proceedings of the 2nd International Conference, ICICSA 2023

Springer

Editors
Sivaji Bandyopadhyay
Department of Computer Science
and Engineering
Jadavpur University
Kolkata, West Bengal, India

Valentina Emilia Balas
Department of Automatics and Applied
Software
Aurel Vlaicu University of Arad
Arad, Romania

Saroj Kumar Biswas
Department of Computer Science
and Engineering
National Institute of Technology Silchar
Silchar, Assam, India

Anish Kumar Saha
Department of Computer Science
and Engineering
National Institute of Technology Silchar
Silchar, Assam, India

Dalton Meitei Thounaojam
Department of Computer Science
and Engineering
National Institute of Technology Silchar
Silchar, Assam, India

ISSN 2367-3370 ISSN 2367-3389 (electronic)
Lecture Notes in Networks and Systems
ISBN 978-981-97-5411-3 ISBN 978-981-97-5412-0 (eBook)
https://doi.org/10.1007/978-981-97-5412-0

This Springer imprint is published by the registered company Springer Nature Singapore Pte Ltd.
The registered company address is: 152 Beach Road, #21-01/04 Gateway East, Singapore 189721, Singapore

Contents

Editors and Contributors

About the Editors

Prof. Sivaji Bandyopadhyay is a Professor in the Department of Computer Science and Engineering, at Jadavpur University, India where he has been serving since 1989. He served as a Director of the National Institute of Technology Silchar from December 2017 to November 2022. He has more than 300 publications in reputed journals and conferences. He has edited two books so far. His research interests are in the areas of Natural Language Processing, Machine Translation, Sentiment Analysis, and Medical Imaging among others. He has organized several conferences and has been the Program Committee Member and Area Chair in several reputed conferences.

Prof. Valentina Emilia Balas is currently a Full Professor in the Department of Automatics and Applied Software at the Faculty of Engineering, "Aurel Vlaicu" University of Arad, Romania. She holds a Ph.D. Cum Laude, in Applied Electronics and Telecommunications from Polytechnic University of Timisoara. Dr. Balas is the author of more than 400 research papers in refereed journals and International Conferences. Her research interests are in Intelligent Systems, Fuzzy Control, Soft Computing, Smart Sensors, Information Fusion, Modeling and Simulation.

Dr. Saroj Kumar Biswas is currently working as an Associate Professor and HoD of the Computer Science and Engineering Department, at NIT Silchar. His research interest includes Machine Learning, Artificial Intelligence, Pattern Recognition, Data Mining, Image Processing and Video Processing. 15 Ph.D. scholars are presently doing their Ph.D. work under him and 05 have been completed. He has completed one DEITY-sponsored R&D Project and one NPIU (World Bank) sponsored R&D Project. He is currently running one high-value MEITY-sponsored R&D Project. He has published 56 international journals including 35 unpaid SCI-indexed journals (regular issues) in various reputed publication houses worldwide including Elsevier, IET, Springer, World Scientific, etc., and 68 conference papers.

Dr. Anish Kumar Saha working as an Assistant Professor in the Department of Computer Science and Engineering, National Institute of Technology Silchar since 2018. Prior to that, he served as an Assistant Professor in the Department of Computer Science and Engineering, at the National Institute of Technology Arunachal Pradesh and at the National Institute of Technology Agartala. He has been in the field of teaching and research for the last 12 years. His research interests are Quantum computing, Optimization, and Networking. He had served in different conferences and workshops in various assigned chairs. Currently, he is serving as the Principle Investigator of the MeitY-sponsored project in the field of graph theory and quantum computing. He had served as a co-coordinator for a PM's Gati Shakti-funded workshop "Digital Supply Chain" in 2022. He has published in several journals and conferences of repute. He also obtained a Best Paper award at an IEEE conference at The Seventh ICIC-APTIKOM 2022, Bali, Indonesia.

Dr. Dalton Meitei Thounaojam has been working as an Assistant Professor in the Department of Computer Science and Engineering, National Institute of Technology Silchar since 2012. Prior to that, he served as an Assistant Professor in the Department of Computer Science and Engineering at, the National Institute of Technology Manipur and Manipur Institute of Technology respectively. He has been in the field of teaching and research for the last 13 years. His research interests are Image processing, Video processing, etc. He had served in different conferences and workshops in various assigned chairs. He has published in several journals and conferences of repute.

Contributors

A. M. Abirami Department of Information Technology, Thiagarajar College of Engineering, Madurai, Tamil Nadu, India

Shiv Prasad Aggarwal North Eastern Space Applications Centre, Umiam, Meghalaya, India

Maroi Agrebi Department of Computer Science, LAMIH UMR CNRS 8201, Université Polytechnique Hauts-de-France, Valenciennes, France

Sony K. Ahuja Department of Computer Science and Engineering, Visvesvaraya National Institute of Technology (VNIT), Nagpur, India

B. Akoramurthy Department of CSE, National Institute of Technology Puducherry, Karaikal, India

Azad Ali Department of Computer Science and Engineering, National Institute of Technology, Silchar, Silchar, India

A. Anandkumar Department of Electronics and Communication Engineering, Jai Shriram Engineering College, Tirupur, Tamil Nadu, India

J. Akil Arsath Thiagarajar College of Engineering, Madurai, India

Md. Arshad Ahmed Department of Computer Science and Engineering, National Institute of Technology, Silchar, Silchar, India

T. S. Arulananth Department of Electronics and Communication Engineering, MLR Institute of Technology, Hyderabad, India

M. Ashwin Velammal Institute of Technology Panchetti, Chennai, India

Arti Bahuguna Department of Computer Science and Engineering, National Institute of Technology Sikkim, Sikkim, India

A. S. Bharath Kumar Velammal College of Engineering and Technology, Madurai, India

Gopa Bhaumik Department of Computer Science and Engineering, National Institute of Technology Sikkim, Sikkim, India;
Department of Computer Science and Engineering, National Institute of Technology Jamshedpur, Jharkhand, India

Seema Gupta Bhol Kalinga Institute of Industrial Technology, Bhubaneswar, India

S. Bhuvaneswari Department of Computer Science, School of Engineering and Technology, Pondicherry University, Karaikal, Pondicherry, India

Anupam Biswas National Institute of Technology, Department of Computer Science and Engineering, Silchar, India

K. R. M. Vijaya Chandrakala Department of Electrical and Electronics Engineering, Amrita School of Engineering, Amrita Vishwa Vidyapeetham, Coimbatore, India

T. Chandrakumar Thiagarajar College of Engineering, Madurai, India

Supriya Maganahalli Chandramouli Sri Siddhartha Institute of Technology (SSIT), SSAHE, Tumkur, India

Pranali Chanduka Computer Science and Engineering Department, Motilal Nehru National Institute of Technology, Allahabad, Prayagraj, India

Ashok Chikaraddi School of Computer Science and Engineering Department, KLE Technological University, Hubli, India

Avinash Chouhan North Eastern Space Applications Centre, Umiam, Meghalaya, India

Dibyajyoti Chutia North Eastern Space Applications Centre, Umiam, Meghalaya, India

Sreayan De Department of Electronics and Communication Engineering, B. P. Poddar Institute of Management and Technology, Kolkata, India

N. Dineshwar Velammal Institute of Technology Panchetti, Chennai, India

Ambati Divya School of Electronics Engineering, Vellore Institute of Technology, Chennai, India

Aditi R. Durge Department of Computer Science and Engineering, Visvesvaraya National Institute of Technology (VNIT), Nagpur, India

M. Ganesh Madhan Department of Electronics Engineering, Madras Institute of Technology, Anna University, Chennai, India

K. S. Gayathri Department of Information Technology, Sri Sivasubramaniya Nadar College of Engineering, Chennai, Tamil Nadu, India

Shantala Giraddi School of Computer Science and Engineering Department, KLE Technological University, Hubli, India

N. Gomathi Department of Artificial Intelligence and Data Science, Karpagam Institute of Technology, Coimbatore, Tamil Nadu, India

Ariyaman Goswami Department of Information Technology, Jadavpur University, Kolkata, West Bengal, India

Mahesh Chandra Govil Department of Computer Science and Engineering, National Institute of Technology Sikkim, Sikkim, India

Vidya S. Handur School of Computer Science and Engineering Department, KLE Technological University, Hubli, India

J. V. Hari Harra Sudhan Velammal College of Engineering and Technology, Madurai, India

K. Haripriya Department of Computer Science and Engineering, Anna University, Chennai, India

K. Harshini Department of CSE, VFSTR, Deemed to be University, Guntur, Andhra Pradesh, India

B. Himaja Department of CSE, VFSTR, Deemed to be University, Guntur, Andhra Pradesh, India

Bhavana Inaganti Computer Science and Engineering, SRM University—AP, Amaravathi, India

Kailash Jagadeesh Mechanical Engineering Department, NIT Trichy, Tiruchirapalli, India

W. Jaisingh School of Information Science, Presidency University, Bengaluru, Karnataka, India

Anjana Jalnnavar School of Computer Science and Engineering Department, KLE Technological University, Hubli, India

S. Jayanthi Department of Electronics and Communication Engineering, R.M.D Engineering College, Chennai, India

B. Johnson Thiagarajar College of Engineering, Madurai, India

A. Jose Anand Department of ECE, KCG College of Technology, Chennai, India

N. Jyothi Prakash Velammal College of Engineering and Technology, Madurai, India

R. Kalpana Department of Computer Science and Engineering, Puducherry Technological University, Puducherry, India

Surekha Kamath Department of Instrumentation and Control Engineering, Manipal Institute of Technology, Manipal Academy of Higher Education, Manipal, India

Suvarna G. Kanakareddi School of Computer Science and Engineering Department, KLE Technological University, Hubli, India

Ananya Kanjilal B. P. Poddar Institute of Management & Technology, Kolkata, India

Ramdas Kapila Department of Computer Science and Engineering, Narsimha Reddy Engineering College (Autonomous), Maisammaguda, Secunderabad, Telangana, India

P. Karthikeyan Thiagarajar College of Engineering, Madurai, India

Adil Mustafa Khokhawala National College of Ireland, Dublin, Ireland

P. Jagdish Kumar Department of CSE, IcfaiTech, IFHE, Hyderabad, India

Sushil Kumar KIET Group of Institutions, Delhi-NCR, Ghaziabad, Uttar Pradesh, India

P. Meena Kumari Department of CSE, IcfaiTech, IFHE, Hyderabad, India

Royyuru Manasa Lakshmi Department of Electrical and Electronics Engineering, Amrita School of Engineering, Amrita Vishwa Vidyapeetham, Coimbatore, India

K. Lakshmi Narayanan Department of Computer Science and Engineering, SRM Institute of Science and Technology, Kattankulathur, Chennai, India

Challa Madhavi Latha Department of Computer Science, School of Engineering and Technology, Pondicherry University, Karaikal, Pondicherry, India

Teresa Louis Department of Computer Science and Engineering, National Institute of Technology, Silchar, Silchar, India

Saurav Mallik Department of Pharmacology and Toxicology, R. Ken Coit College of Pharmacy, The University of Arizona, Tucson, AZ, USA

P. Mamatha Department of Computer Science and Engineering, AVN Institute of Engineering and Technology, Hyderabad, Telangana, India

Petla Manaswini Department of Electronics and Communication Engineering, MLR Institute of Technology, Hyderabad, India

Saksham Sneh Mandal B. P. Poddar Institute of Management & Technology, Kolkata, India

S. Menaga Department of Electronics and Communication Engineering, Jai Shriram Engineering College, Tirupur, Tamil Nadu, India

R. Mohana Department of Information Technology, Sri Sivasubramaniya Nadar College of Engineering, Chennai, Tamil Nadu, India

J. R. Mohanty Kalinga Institute of Industrial Technology, Bhubaneswar, India

Sujit Kumar Mondal Department of Information Technology, Jadavpur University, Kolkata, West Bengal, India

Thangavel Murugan Department of Information Systems and Security, College of Information Technology, United Arab Emirates University, Al Ain, UAE

S. Muthulakshmi School of Electronics Engineering, Vellore Institute of Technology, Chennai, India

Sankayya Muthuramalingam Department of Information Technology, Thiagarajar College of Engineering, Madurai, India

K. Naga Poojitha Department of CSE, VFSTR, Deemed to be University, Guntur, Andhra Pradesh, India

Uttej Kumar Nannapaneni Department of CSE, VFSTR, Deemed to be University, Guntur, Andhra Pradesh, India

N. Pathri Narayanan Thiagarajar College of Engineering, Madurai, India

R. Naresh Department of Networking and Communications, SRM Institute of Science and Technology, Kattankulathur, Chennai, India

M. Nirmala Devi Department of Computer Science and Engineering, Thiagarajar College of Engineering, Madurai, Tamil Nadu, India

Kiran Kumar Jakkur Patalappa Sri Siddhartha Institute of Technology (SSIT), SSAHE, Tumkur, India

Het Bhavinkumar Patel University of Guelph, Guelph, ON, Canada

Ripon Patgiri Department of Computer Science and Engineering, National Institute of Technology, Silchar, Silchar, India

P. K. Pattnaik Kalinga Institute of Industrial Technology, Bhubaneswar, India

Rajdeep Paul National Institute of Technology, Department of Computer Science and Engineering, Silchar, India

Vibhav Prakash Singh Computer Science and Engineering Department, Motilal Nehru National Institute of Technology, Allahabad, Prayagraj, India

Nalla Krishna Prasad Department of Electronics and Communication Engineering, MLR Institute of Technology, Hyderabad, India

P. Priyadharshini Department of Computer Science and Engineering, Anna University, Chennai, India

A. M. Rajeswari Velammal College of Engineering and Technology, Madurai, India

Rajesh Ramasamy Department of Electrical and Electronics Engineering, Amrita School of Engineering, Coimbatore, India

Balakrishnan Ramprakash Department of Applied Mathematics and Computational Science, Thiagarajar College of Engineering, Madurai, India

Mitu Ranjan KIET Group of Institutions, Delhi-NCR, Ghaziabad, Uttar Pradesh, India

G. Redyy Taruni Department of Electronics and Communication Engineering, MLR Institute of Technology, Hyderabad, India

M. Sagana Department of Information Technology, Sri Sivasubramaniya Nadar College of Engineering, Chennai, Tamil Nadu, India

Sumeet Sagar Thapar Institute of Engineering and Technology, Patiala, India

Mridu Sahu Department of Information Technology, National Institute of Technology, Raipur, India

Sumalatha Saleti Data Science Laboratory, Computer Science and Engineering, SRM University—AP, Amaravathi, India

Anu Sam Department of Electronics Engineering, Madras Institute of Technology, Anna University, Chennai, India

Satish Kumar Satti Department of CSE, VFSTR, Deemed to be University, Guntur, Andhra Pradesh, India

S. Selva Kumar Department of Electrical and Electronics Engineering, Amrita School of Engineering, Coimbatore, India

Deepti D. Shrimankar Department of Computer Science and Engineering, Visvesvaraya National Institute of Technology (VNIT), Nagpur, India

B. Shruti School of Electronics Engineering, Vellore Institute of Technology, Chennai, India

Khoirom Motilal Singh Department of CSE, Indian Institute of Information Technology, Senapati, Manipur, India

Laiphrakpam Dolendro Singh Department of Computer Science and Engineering, National Institute of Technology, Silchar, Silchar, India

Om Singh National Institute of Technology, Department of Mathematics, Silchar, India

Pawan Kumar Singh Department of Information Technology, Jadavpur University, Kolkata, West Bengal, India

M. Sivakumar Department of Computer Science and Engineering, Thiagarajar College of Engineering, Madurai, Tamil Nadu, India

K. L. S. Soujanya Department of Information Technology, CMR College of Engineering and Technology, Kandlakoya, Medchal, Hyderabad, India

V. Sreejith Velammal College of Engineering and Technology, Madurai, India

A. Sreenidhi School of Electronics Engineering, Vellore Institute of Technology, Chennai, India

Gadicharla Sahiti Sriya Bridgetown International University, Bridgetown, India

B. Subbulakshmi Department of Computer Science and Engineering, Thiagarajar College of Engineering, Madurai, Tamil Nadu, India

N. Subhashini School of Electronics Engineering, Vellore Institute of Technology, Chennai, India

D. Suganya Department of Computer Science and Engineering, Puducherry Technological University, Puducherry, India

S. Sumitra Department of Mathematics, Indian Institute of Space Science and Technology, Trivandrum, Kerala, India

B. Surendiran Department of CSE, National Institute of Technology Puducherry, Karaikal, India

T. Teenu Prinica Department of Computer Science and Engineering, Thiagarajar College of Engineering, Madurai, Tamil Nadu, India

Gogineni Viswa Sai Teja Department of Electronics and Communication Engineering, MLR Institute of Technology, Hyderabad, India

Prajwal Thakare Department of Computer Engineering, COEP Technological University, Pune, India

R. Thaneshwaran Velammal Institute of Technology Panchetti, Chennai, India

Dalton Meitei Thounaojam Department of Computer Science and Engineering, National Institute of Technology, Silchar, Silchar, India

R. Vanithamani Department of Biomedical Instrumentation Engineering, School of Engineering, Avinashilingam Institute for Home Science and Higher Education for Women, Coimbatore, Tamil Nadu, India

T. R. Abijeeth Vasra Department of Information Technology, Thiagarajar College of Engineering, Madurai, Tamil Nadu, India

E. Vasudha Department of Computer Science and Engineering, Anna University, Chennai, India

M. Vijayakumar Government College of Engineering, Dharmapuri, India

R. Vinoth Raj Velammal Institute of Technology Panchetti, Chennai, India; Electronics & Communication Engineering, Saveetha Engineering College, Chennai, India

Sleep Apnea Classification Using KNN

Surekha Kamath

Abstract Sleep is a state of reduced mental and physical activity, in which consciousness is altered. And all the sensory activity is stopped to certain extent. Maintenance of good sleep is very essential for each individual. Nowadays, diseases related to sleep are very common among the population due to individual living styles and stress related disorder. One such sleep disorder among them is obstructive sleep apnea (OSA), in which an individual experiences difficulty in breathing. Delayed or improper detection of this disorder can cause serious health issues. This work explains the classification of sleep apnea using K-nearest neighbor algorithm using various features. The proposed work is to extract the statistical features such as mean, median, standard deviation and mean square error of ECG signal to detect the abnormal pauses in breathing during sleep apnea period. Classification of sleep stages can be performed using K-nearest neighbor classifier. The KNN classifier was found to work with an accuracy of 95%.

Keywords Sleep · OSA · HRV · Polysomnography · Mixed sleep apnea

1 Introduction

To have mental and physical wellbeing of an individual, sleep is an essential physiological parameter. Sleep is associated with many parameters such as brain functions, hormones and the neurotransmitter pathways. Sleeping patterns are varied by various parameters such as change in the brain function as well as changes in physical and mental activity. Irregularity in sleep pattern may lead to many side effects such as cardiac diseases, depression and tiredness. Most important disorder related to sleep is sleep apnea and consists of irregular ceasing of breath during sleep [1, 2]. The disease is characterized by irregular breathing pattern during sleep. Common health

S. Kamath (✉)
Department of Instrumentation and Control Engineering, Manipal Institute of Technology, Manipal Academy of Higher Education, Manipal, India
e-mail: surekha.kamath@manipal.edu

S. Bandyopadhyay et al. (eds.), *Intelligent Computing Systems and Applications*, Lecture Notes in Networks and Systems 1010,
https://doi.org/10.1007/978-981-97-5412-0_1

issue related to sleep is obstructive sleep apnea (OSA). There will be disturbance in the upper airway during OSA, which leads to interruption of sleep during sleep. Sleep apnea is defined as interruption of sleep during night. This may occur many a times during night [3]. Common symptoms that are occurring during sleep apnea are fatigue, snoring, headache, dry mouth and restlessness.

Polysomnography (PSG) is laboratory test to diagnose diseases related to sleep [4]. In this method, various physiological parameters during sleep such as bio-signals from cardiac muscle, bio-signals from skeletal muscles and bio-signals from scalp are measured using many electrodes which are connected to different parts of the body. The results of this test are used to detect the apnea occurrence. To diagnose sleep-related disorders PSG is the gold standard. The test is conducted by a registered sleep Technologists in a certified sleep laboratory. This test may cause uneasiness during sleep as well as discomfort due to large number of sensors which are used to extract various physiological parameters. Because of involvement of many wires during testing this method suffers from many disadvantages.

This study explains a method which is very simple and easy tool to diagnose sleep disorders. The focus of the method lies in screening for sleep apnea using a single channel.

The paper is organized into different sections such as Sect. 2 explains about the methodology of the work that is carried out to achieve sleep detection from ECG data. Section 3 clearly explains various results that is obtained from the experiment, it lists the confusion matrix of onefold and twofold. Section 4 deals with the conclusion and future scope of the work.

2 Methodology

Block diagram of proposed methodology is shown in Fig. 1. The method explains clearly on the various steps that are carried to achieve sleep detection from ECG data. To detect sleep apnea, feature selection criteria can be used. The classifier utilized five different features extracted from ECG signal. The various steps that are followed in the proposed methodology are.

2.1 ECG Signal Recording

The ECG signals utilized for this study were sourced from the MIT-BIH website. In this work, a total of 20 datasets were considered which are identified as normal and apnea data by the annotation in the ECG database. The ECG data of both men and women from the database was included in the dataset. The age of subjects was ranging from 27 to 63 years, and the weights ranging from 53 to 135 kg's. These ECG signals were sampled at 200 Hz and later decimated to 100 Hz to obtain the consistency over the signal [20].

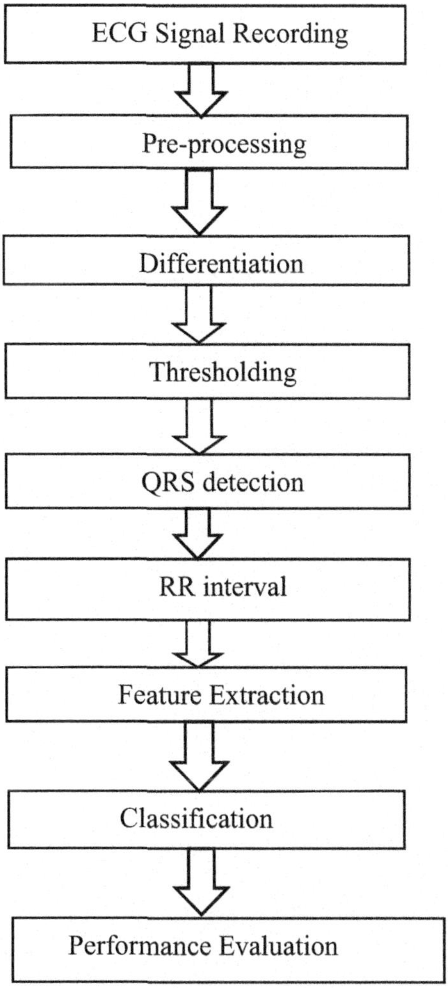

Fig. 1 Methodology of sleep apnea detection

2.2 Data Pre-processing

This step on ECG data is used to eliminate the unwanted noise in the signal. Various filters such as band pass filters are used to eliminate the high-frequency noise signals. Later differentiation, smoothing and thresholding are performed on ECG signal to eliminate other noises and to smoothen the signal. Incorporating sleep apnea annotations ensured that the patient intervals encompassed the apnea phases.

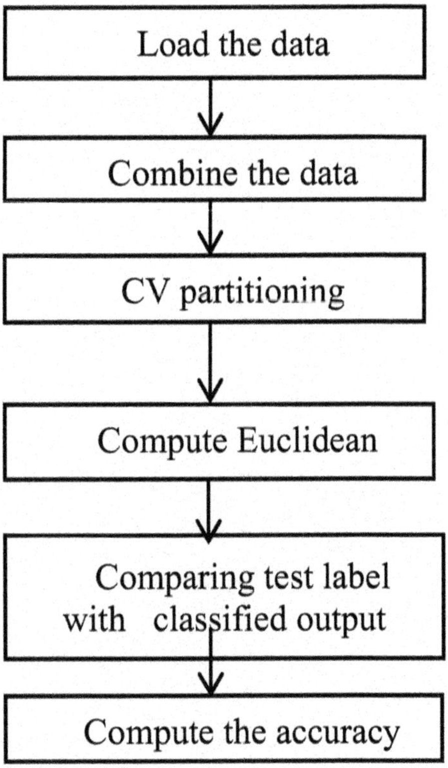

Fig. 2 Flowchart of the KNN algorithm

2.3 *Search Algorithm for Detecting R-Peaks*

This is the important stage in the algorithm. The important information is obtained is the QRS complex of the ECG signal. This is done using differentiation technique and averaging the signal. This is more helpful to extract the features of ECG signal (See Fig. 2).

2.4 *R-R Interval*

After applying thresholding, from the QRS complex, the R-peaks were extracted. The extracted R-peaks are stored in the program code in order to calculate the R-R interval of the ECG. The R peak proves highly useful in calculating the R-R interval, representing the duration between successive R-peaks. This is the common feature that has been extracted from ECG which helps in diagnosing most of the cardiac arrhythmias [2].

2.5 Feature Extraction

Statistical features are derived from the ECG signal's R-R intervals, demonstrating high efficacy in sleep apnea detection. This study employs a variety of feature types to construct the comprehensive feature set. Based on the feature set, the apnea detection and classification is carried out [2].

The important features extracted in this work which are considered to be highly related with sleep apnea are:

(a) Segmented mean of R-R interval.
(b) Segmented standard deviation of R-R interval (SD).
(c) Variance of R-R interval.
(d) Mean square error of R-R interval recording (MSE).
(e) Median of R-R interval recording.

In general, the mean, median, SD, variance and mean squared error has been explained below,

(a) *Segmented mean*: It is the ratio of sum of all datasets by the number of values present in the data.

The segmented mean is given by the Eq. (3.4),

$$\bar{x} = \frac{\sum_{i=1}^{n} X_i}{n} \tag{1}$$

where $\bar{x}$ the segmented mean, X_i is sum of all dataset and n is number of value present in the data.

(b) *Segmented standard deviation*: The amount of the variation in dataset is quantified by the measure known as standard deviation (SD). It is denoted as 'σ' and is given by the Eq. (3.5),

$$S = \sqrt{\frac{\sum_{i=1}^{n} (x - \bar{x})^2}{n - 1}} \tag{2}$$

where s is denoted as standard deviation, n is number of values in the samples and $\bar{x}$ is the mean.

(c) *Variance:* It is defined as the random variable from the expected value of squared deviation from the mean. This is denoted as σ2, and is given as in Eq. (3.6),

$$\sigma^2 = \frac{1}{n - 1} \sum_{i=1}^{n} (x - \bar{x})^2 \tag{3}$$

where σ^2 is denoted as variance, n is number of values in the samples and $\bar{x}$ is the mean of the sample.

(d) *Median*: It is defined as the mid value of the total ordered data in the dataset. This is given as,

Let us consider a set of data $P = P1, P3, P6, P2, P8, P4, P9$.

The ordered dataset is $P1, P2, P3, P4, P6, P8$, and $P9$.

The median value of the dataset P is $P4$.

(e) *Mean square error*: It is defined as average squared errors. The mean square error is expressed as,

$$\text{MSE} = \frac{1}{n}\sum_{i=1}^{n}(x_i - \tilde{x}_i)^2 \tag{4}$$

where n is the total number of data in a data set and $\sum_{i=1}^{n}(x_i - \tilde{x}_i)^2$ is the average squared error.

2.6 Classification

The data classification is divided into two main categories: parametric and nonparametric classifiers. Parametric classifiers categorize data by considering the statistical distribution of the classes, while nonparametric classifiers make use of conditions related to unknown density functions. Learning method is grouped into supervised and unsupervised learning. Supervised technique is defined as form of learning with the help of supervisor, whereas unsupervised learning is the learning method without supervisor. Kth nearest neighbor method of classification is a fundamental and simple classification method, and when there is small or no prior knowledge about the data, this method is the good choice to begin. The development of KNN classification arose from the necessity to conduct discriminant analysis in cases where accurate parametric estimations of probability densities are either unavailable or challenging to ascertain. KNN is a powerful supervised method of classification, which is considered as a reference classifier and also implemented in technological fields. This is one of the simplest methods of all the form of classification. In this method, the input is Kth nearest neighbor and class membership is the output where feature vector is assigned. KNN works very well with the test data and it depends on value of K. Choosing the appropriate value for K represents a crucial step in the development of any classifier. Opting for K values that are excessively small or large might result in either overfitting or underfitting the predictive model.

2.7 K-Nearest Neighbor (KNN) Classifier

Kth nearest neighbor method of classification is a fundamental and simple classification method, and when there is small or no prior knowledge about the data, this method is the good choice to begin. KNN is a powerful supervised method of classification, which is considered as a reference classifier and also implemented in technological fields. This is one of the simplest methods of all the form of classification. In this method, the input is *K*th nearest neighbor and class membership is the output where feature vector is assigned. KNN depends on the function based on distance called as Euclidean distance [23]. In the proposed work, Euclidean distance method is used because of its simplicity and accurate classification results.

Figure 2 shows the flowchart of KNN algorithm. The feature extracted from the normal and abnormal set of ECG is loaded into the MATLAB. These normal and apnea data are combined and stored in one location.

Picking the correct value for *K* stands as a pivotal stage in the development of any KNN classifier. Utilizing *K* values that are excessively small or large may potentially result in the prediction model suffering from overfitting or underfitting.

3 Results and Discussions

3.1 Comparison of Normal and Apnea ECG Data of Subject

3.1.1 ECG Signal Recording

The raw ECG signals, normal and abnormal (with apnea) obtained from the MIT-BIH database are as shown in Figs. 3a and b.

It is observed that over duration of 15secs there are 17 R-peaks in the normal ECG and 14 R-peaks in the abnormal (apnea) ECG signals. It is also observed that, there is a gradual increase in the duration of R-peaks, leading to bradycardia.

(a)

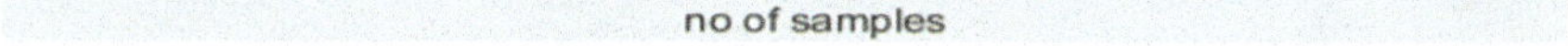

(b)

Fig. 3 **a** and **b** Raw input ECG signals of normal and sleep apnea

3.2 Extracted Features of Normal and Sleep Apnea ECG Data

The efficacy of the model on the apnea ECG database is evaluated by employing various records from the MIT-BIH database. The implementation of the model was conducted using MATLAB. The Tables 1 and 2 shows, the statistical features from ECG data (mean, median, variance, standard deviation and mean square error) extracted from 10 sleep apnea patients that include normal and sleep apnea ECG data.

Table 1 Feature extracted from normal ECG data

Mean	Median	Variance	SD	MSE
0.9044	0.89	0.0027	0.0516	0.8204
0.8625	0.875	0.0027	0.0523	0.7465
0.9771	0.96	0.0017	0.0414	0.9564
0.9607	0.99	0.0057	0.0753	0.9282
0.972	0.98	0.001	0.0321	0.9457
0.9227	0.94	0.0089	0.0944	0.859
0.8944	0.875	0.0042	0.0649	0.8039
0.885	0.9	0.001	0.032	0.7842
0.9479	0.95	0.0024	0.0489	0.9006
0.835	0.89	0.0025	0.045	0.8042

Table 2 Feature extracted from sleep apnea ECG data

Mean	Median	Variance	SD	MSE
1.0207	1.035	0.012	0.1097	1.053
1.03	1.015	0.0114	0.1068	1.0715
1.03	1.045	0.0014	0.0374	1.0622
1.06	1.04	0.0124	0.1114	1.1481
1.0493	1.04	0.0045	0.0674	1.1052
1.1675	1.175	0.0095	0.0977	1.3718
1.0562	1.03	0.0025	0.0501	1.1178
1.005	1.01	0.00072	0.0274	1.0107
1.05	1.035	0.0059	0.077	1.108
1.1234	1.034	0.0133	0.1154	1.1073

3.3 Performance Evaluation

The performance evaluation of the KNN-based classifier is carried out and the results are reported here. The confusion matrix of onefold is as shown in Fig. 4 which describe the error measures. The estimated sensitivity, specificity and accuracy values are 100%.

The confusion matrix of twofold is as shown in Fig. 5. Sensitivity and specificity is of 100% and 80% and accuracy is of 90%.

Finally, an average value of sensitivity, specificity and accuracy are calculated and their values are 100%, 90% and 95%, respectively.

It is found that the classifier classified the test data with an accuracy of 95%. The classifier performance is compared with the accuracy of classifiers implemented by various authors found in the literature is as shown in Table 3.

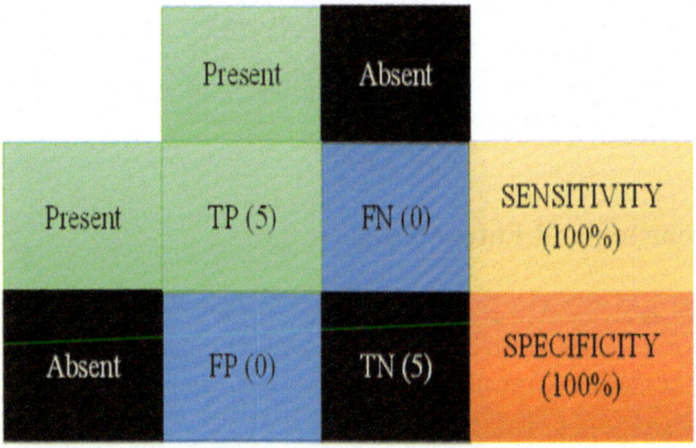

Fig. 4 Confusion matrix of onefold

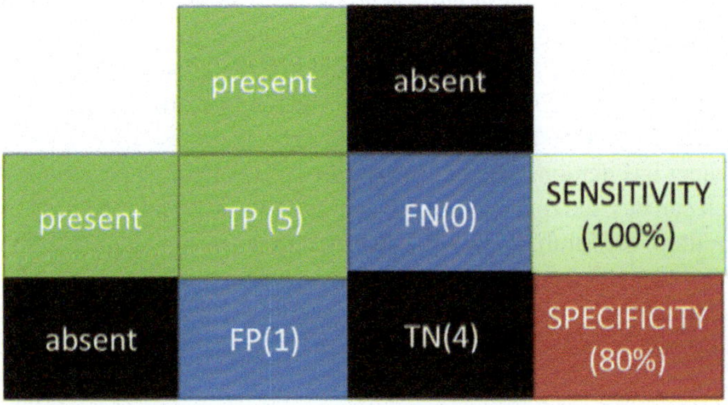

Fig. 5 Confusion matrix of twofold

Table 3 Comparison of results obtained from the proposed work with the results obtained from various methodologies by different authors

Authors	Classes	Methods	Classifier	Maximum accuracy (%)
Sezgina et al. [24]	Normal and sleep apnea	DWT, energy coefficients	ANN	86.84
Kaloniya and Sabherwal [2]	Normal and sleep apnea	Segmented mean of fixed interval, segmented SD of fixed interval, segmented peak, variance, mean square error of R-R interval, mean absolute error of R-R interval, median of R-R interval	SVM	93
Almazaydeh et al. [4]	Normal and sleep apnea	Mean epoch, SD, NN50 (variant 1 and variant 2), pNN50, SDSD, RMSSD	SVM	96.5
Thommandram et al. [25]	Normal and sleep apnea	Spectral variance and entropy of EDR signal	KNN	91.2
Proposed work	Normal and sleep apnea	Mean, median, standard deviation, variance, mean square error of R-R interval	KNN	95

4 Conclusion and Future Scope

Sleep apnea is a very serious sleep disorder that causes a pause while breathing for about 10s or more. The carbon dioxide increases and oxygen concentration decreases during sleep apnea, this chemical concentration of CO_2 and O_2 which will lead into sleep apnea. The sleep disorder may also lead to cardiac disease such as myocardial infraction and stroke, hence demanding a quick diagnosis and evaluation.

The proposed methodology is used as a noninvasive tool to detect apnea from ECG signal, and to classify ECG as normal or abnormal (apnea). The R-R interval of ECG signal is computed and some nonlinear ECG features are extracted such as mean, median, variance, standard deviation and mean square error from ECG signal. The KNN classifier has been implemented to perform the classification of ECG signal into apnea and normal signal. The classifier was found working with a sensitivity of 100%, specificity of 90% and accuracy of 95%.

The project has demonstrated the feasibility of detecting sleep apnea at an early stage which can be used to minimize the cardiac disorders and breathing problems to certain extent. In the future work, the other parameter along with ECG signal such as EMG, EEG and EOG signals can be taken for detecting the sleep disorder problem. The frequency domain features such power spectral density; HRV can be extracted from these signal.

References

1. Chang DW, Liu YD, Young CP, Chen JJ, Chen YH et al (2012) Design and implementation of a modularized polysomnography system. IEEE Trans Instrum Meas 27(7):1933–1944
2. Kaloniya R, Sabherwal P (2015) ECG signal recordings analysis for detection of sleep apnea. Int J Enhanced Res Sci Technol Eng 4(6):228–235
3. Varon C, Suykens J (2015) Sleep apnea classification using least-squares support vector machines on single lead ECG. IEEE Trans Biomed Eng 62(9):2269–2278
4. Almazaydeh L, Elleithy K, Faezipour M (2012) Detection of obstructive sleep apnea through ECG signal features. IEEE Int Conf Electro/Inf Technol (EIT) 25(4):967–983
5. https://commons.wikimedia.org/wiki/File:Airway_obstruction.png. Available on online. Accessed 23 Aug 2017
6. Lubeck VM, Lubecke OB (2009) Wireless technologies in sleep monitoring. IEEE Conf Radio Wireless Symp 10(3):135–138
7. Gopal S, Devi A (2014) A review on sleep apnea detection from ECG signal. IEEE J Biomed Health Inf 2(4):107–111
8. https://upload.wikimedia.org/wikipedia/comons/thumb/5/55/Qrs.svg/2000pxQrs.svg.png. Available online. Accessed 23 Aug 2017
9. Bacharova L, Traintafyllou E, Vazaios C, Tisko R (2014) QRS complex patterns in patients with obstructive sleep apnea. Int J Electro Cardiol Proc 41st Int Congr Electro Cardiol 23(5):195–198
10. Roebuck A, Monasterio V (2014) A review of signals used in sleep analysis. National Inst Health Public Access 35(1):1–71
11. Hofsøy DA, Clauss JF, Wolf B (2009) Monitoring and therapy of sleep-related breathing disorders. IEEE conference on wearable micro and nano technologies for personalized health, pp 1–4
12. Patel P, Warrier S, Bagal R (2012) ECG analysis and detection of arrhythmia using MATLAB. Int J Innovative Res Dev 1(11):59–68
13. Almazaydeh L, Faezipour M, Elleithy K (2012) A neural network system for detection of obstructive sleep apnea through SpO2 signal features. Int J Adv Comput Sci Appl 3(5):7–11
14. Mitiche I, Morison G, Stewart BG (2016) Electrocardiogram-based feature extraction for machine learning classification of obstructive sleep apnea. Int J Signal Process Syst 4(6):515–518
15. Jezzini A, Ayache M, Elkhansa L (2015) ECG classification for sleep apnea detection. Int Conf Advan Biomed Eng (ICABME) 29(9):301–304
16. Jaylalitha S, Susan D, Kumara S, Archana B (2014) KNN method for analyzing the ECG signal. Int J Appl Sci 14(14):1628–1632
17. Almazaydeh L, Elleithy K, Faezipour M, Abushakra A (2013) Apnea detection based on respiratory signal classification. Int Conf Curr Future Trends Inf Commun Technol Healthc 50(4):310–316
18. Ahamed MA, Ul-Ahad MA, Sohag MHA, Ahmad M (2015) Development of low cost wireless ECG data acquisition system. IEEE Conf Adv Electr Eng (ICAEE) 22(5):72–75
19. https://upload.wikimedia.org/wikipedia/commons/9/9d/UnoConnections.jpg/. Available online. Accessed 12 Jan 2017
20. https://www.physionet.org/physiobank/database/apnea-ecg/. Available online. Accessed 15 May 2017
21. https://upload.wikimedia.org/wikipedia/commons/thumb/6/6b/Bandwidth_2.svg/2000pBand width_2.svg.png/. Available online. Accessed 15 Sept 2017
22. Pan I, Tompkins WJ (1985) A real-time QRS detection algorithm. EEE Conf Biomed Eng 32(3):230–236

23. Timus O, Bolat E (2017) K-NN-based classification of sleep apnea types using ECG. Int Turk J Electr Eng Comput Sci 25(4):3008–3023
24. Sezgina N, Taglukb ME (2009) Energy based feature extraction for classification of sleep apnea syndrome. Comput Biol Med 39:1043–1050
25. Thommandram A, Mikael Eklund J, McGregor C (2013) Detection of apnoea from respiratory time series data using clinically recognizable features and kNN classification. Annual International Conference of the IEEE Engineering in Medicine and Biology Society (EMBC)

Disease Prediction Model Using Deep Transfer Learning

Anjana Jalnnavar, Suvarna G. Kanakareddi, Vidya S. Handur, Ashok Chikaraddi, and Shantala Giraddi

Abstract Pneumonia the word which is taken from the Greek word pneuma, breath is an infection that can be fatal as well as an inflammation of the lower respiratory tract such as the bronchioles and alveoli. It is typically brought on by inhaled bacteria as well as viruses such as Streptococcus pneumoniae, also known as pneumococcus, which has both of these characteristics. Machine learning technique called transfer learning is, where a model is applied to newly collected data instead of being learned from scratch using a convolutional neural network (CNN) model which is pre-trained that was trained on a dataset. The disease prediction model is built using transfer learning techniques. Image pre-processing methods such as image dilation, image erosion, Gaussian blur, grayscale and canny edge detection and they are carried out for extracting the characteristics of the CXR images containing 5826 image dataset which were taken from the Kaggle repository consisting of two subfolders normal and pneumonia. They were split into training, testing and validation dataset. The features are extracted from the CNN model which is built from the beginning after that transfer learning techniques are used for prediction of presence of pneumonia using the test set and in last and final step results were produced. The following three models such as CNN model got an accuracy of 78% and transfer learning

A. Jalnnavar (✉) · S. G. Kanakareddi · V. S. Handur · A. Chikaraddi · S. Giraddi
School of Computer Science and Engineering Department, KLE Technological University, Hubli, India
e-mail: 01fe21mcs007@kletech.ac.in

S. G. Kanakareddi
e-mail: suvarna_gk@kletech.ac.in

V. S. Handur
e-mail: vidya_handur@kletech.ac.in

A. Chikaraddi
e-mail: chikaraddi@kletech.ac.in

S. Giraddi
e-mail: shantala@kletech.ac.in

S. Bandyopadhyay et al. (eds.), *Intelligent Computing Systems and Applications*, Lecture Notes in Networks and Systems 1010,
https://doi.org/10.1007/978-981-97-5412-0_2

model VGG16 gave an accuracy of 88% followed by the ResNet50V2 model which gave an accuracy of 90%. In all three models, ResNet50V2 model gave the highest accuracy.

Keywords Convolution neural network (CNN) · Transfer learning · Visual geometry group16 (VGG16) · Residual networks 50V2 (ResNet50V2) · Pneumonia · Artificial intelligence

1 Introduction

Health care is the largest sections in the whole world, the health care sector has an impact on people's life quality. Health care refers to the identification, management, and decrease of disease, and other physical and mental diseases in the people. The tiny air sacs known as alveoli are the main target of pneumonia, a lung inflammation. Asthma, diabetes, heart failure, a history of smoking and a weakened immune system are all risk factors for pneumonia.

Figure 1 depicts the working of a machine learning technique called transfer learning. Transfer learning is where the model is applied to newly collected data instead of being learned from scratch using a pre-trained convolutions neural network (CNN) model that was trained on a separate dataset. In terms of the quantity of parameters to be learnt, training time, and hardware computation costs, transfer learning considerably lowers the cost of training. It also significantly removes the necessity for big data sets. Images from CXR are used in the detection of pneumonia via transfer learning. X-rays are electromagnetic radiation, just like visible light. But, X-rays can go deep into most objects, including the human body, as they have a higher energy than light.

A radiologist typically views and interprets the results. A number of researchers have recently suggested various artificial intelligence (AI)-based remedies for various medical issues. Even a skilled radiologist finds it challenging to examine chest X-rays. Therefore, in order to increase diagnosis precision, it is crucial to develop an automated system for spotting pneumonia. New method which seeks to remove unwanted background removing background noise from photographs allows deep learning models to focus on recognizing diseases from their specific properties. According to these findings, CT and X-ray scans cannot match the detection accuracy of ultrasound imaging [1].

To solve this, deep learning and deep transfer learning methods are used that can identify pneumonia. Many automated strategies utilizing machine learning methods have been suggested to aid in accelerating the diagnoses and enhancing the decision-making process [2]. Total of three models are used to detect the pneumonia where CNN model is trained from beginning and all the knowledge learned by the CNN model is transferred to other two models which are transfer learning models such as

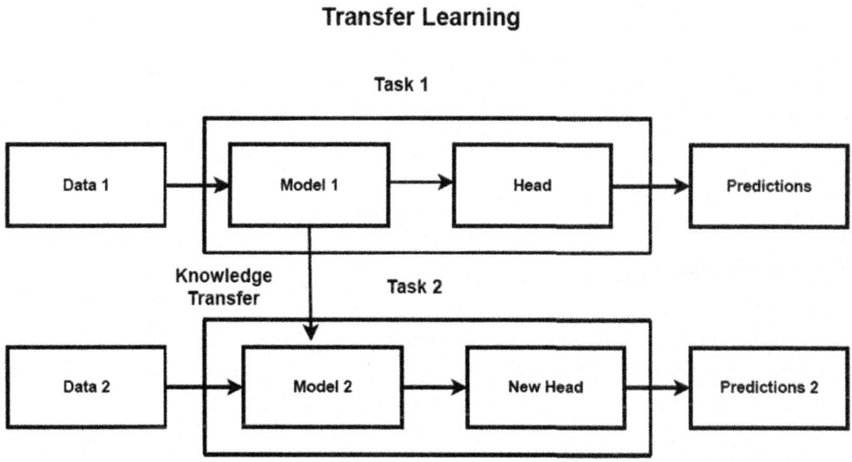

Fig. 1 Transfer learning

VGG16 and ResNet50V2 models to check if the accuracy is better from the CNN model or not. The models were trained using 5856 chest X-ray images in total (4273 scans of the pneumonia class and 1583 scans of the normal class). For evaluating a model, metrics including accuracy and precision, recall, and f1 score were given.

2 Related Work

Generative adversarial networks with deep convolutions were trained for the augmentation of synthetic images. The VGG16 foundation model was then used for convolutional neural networks to categorize images [3]. Neural network models were created to analyze chest X-ray pictures and identify pneumonia. Four models were created using transfer learning and CNN, including a simple (CNN), VGG16, VGG19, and InceptionV3 [4]. Mask R-CNN dataset is used on CXRs to identify and segment the region of the lungs that has pneumonia lung opacity [5]. New method which seeks to remove unwanted background noise from photos so deep learning models may concentrate on identifying the diseases with particular traits from them. According to these findings, CT and X-ray scans cannot match the detection accuracy of ultrasound imaging [1]. Many automated strategies utilizing machine learning methods have been suggested to aid in accelerating diagnoses and enhancing the decision-making process [2]. The study assesses the effectiveness of knowledge transfer achieved through a combination of deep learning models tailored to particular modalities in advancing the state of the art in TB diagnosis [6]. To learn feature representations specific to each modality, an unique convolutional neural network and a number utilizing publicly available CXR collections, of ImageNet pre-trained models are used for training and evaluated at the patient composure [7]. The research seeks to concentrate on migraine prediction. In this research, a machine learning method and model to identify the type of migraine is proposed or provided [8]. The authors of this work suggest developing and refining a lightweight CNN for categorizing diabetic retinopathy into one of five groups [9]. The authors suggest utilizing textural features and an SVM classifier to automate the diagnosis of glaucoma [10].

3 Proposed Work

This chapter depicts the system architecture of disease predictive model which arises from giving input of chest X-ray, dataset exploratory analysis, data splicing, and feature extraction and building a model and applying transfer learning techniques and results.

Figure 2 depicts detailed system design of disease prediction model.

Step 1: Chest X-ray dataset is given as input which contains jpeg images.
Step 2: In next step, the input is split into training and testing dataset and n next step feature extraction of the input takes place here in feature extraction image processing techniques which have been used and they are image erosion where the pixels from the images are removed using structuring element, image dilation where the pixels are added to the image using structuring element. Applying grayscale to images where only two colors are extracted from the image mainly gray and white which reduces the complexity of feature extraction from the images and then Gaussian blur is applied to the images so that the noise is removed from the images, and in next image processing technique, Canny edge detection is

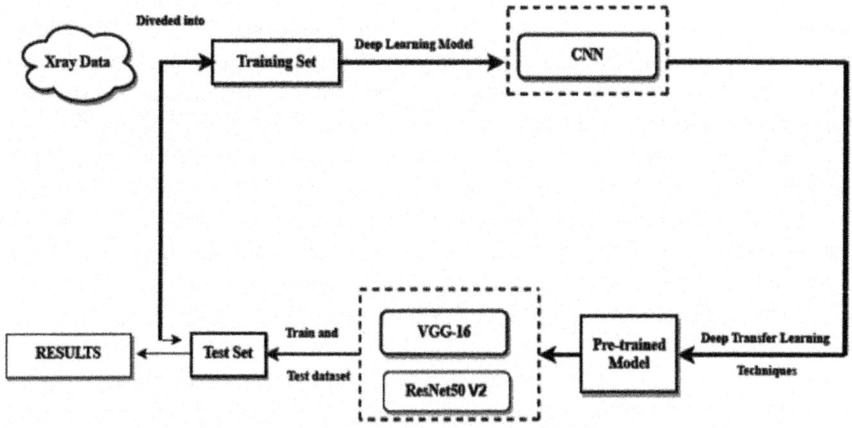

Fig. 2 Detailed system design of disease prediction model

applied to find the edges from the images of chest X-ray and removes all the unnecessary noise from the image. This last step helps the CNN model to extract features from edges where it is possible to detect the pneumonia in chest X-rays.

Step 3: Convolution neural network (CNN) model is built from the beginning and the training accuracy is identified after applying 20 epochs. After that dataset is tested on test data based on this confusion matrix and classification. A report is generated that includes accuracy metrics such as Accuracy and Precision, Recall, and F1-Score.

Step 4: Transfer learning techniques are used where two models such as VGG16 and ResNet50V2 model are used. For the VGG16 model, the training accuracy is identified after applying 20 epochs. After that dataset is tested on test data based on this confusion matrix and classification report consisting A report is generated that includes accuracy metrics consisting of Accuracy and Precision, Recall, and F1-Score. Next another model ResNet50V2 is trained after that training accuracy is identified after applying 20 epochs. Next the dataset is tested on test data based on this confusion matrix and classification report consisting of A report is generated that includes accuracy metrics which consists of Accuracy and Precision, Recall, and F1-Score.

Step 5: Image path of pneumonia infected lungs is given in the code which detects the 100 percent presence of pneumonia in the chest X-rays. Next correct predictions of images and incorrect prediction of images is done in the last step.

Figure 3 depicts detailed low-level design of disease predictive model. Following that, a pre-processing technique produces treated data. The classification and identification of the set of features happens in the following stage. The data are interpreted in the final step before the results and new insights are generated. In next step, splitting of training and testing dataset takes place and some image processing after features are extracted, techniques are utilized for feature extraction CNN model is built from the beginning after that transfer learning techniques are used for prediction of presence of pneumonia using the test set, and in the final step, result analysis will take place.

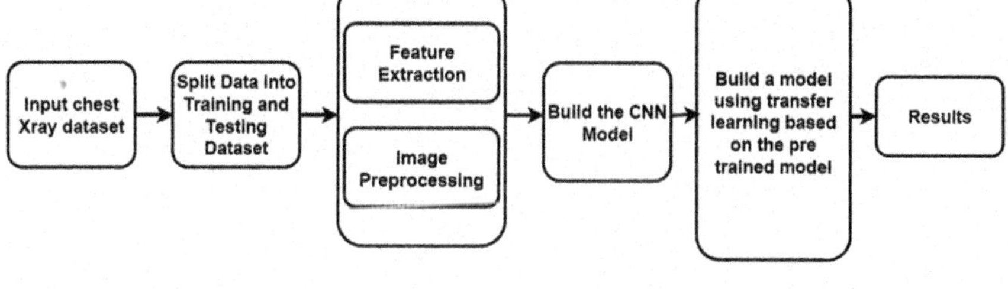

Fig. 3 Low-level design of disease predictive model

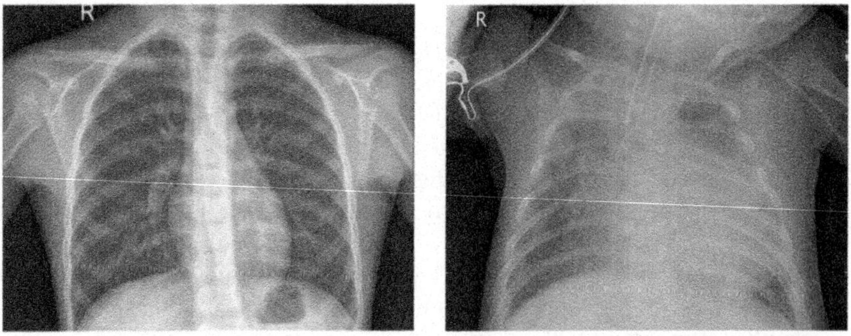

a. Normal CXR b. Pneumonia affected CXR

Fig. 4 **a** Normal CXR **b** Pneumonia affected CXR

4 Results Analysis

In this chapter, result gives an analysis of disease predictive model. At first dataset description of the chest X-ray images were done. In exploratory data analysis or image pre-processing techniques, data is correctly identified, accuracy graphs, and confusion matrix are produced. Along with the classification report are shown.

4.1 Dataset Description

CXR database is taken from the Kaggle repository. Dataset is divided into 2 subfolders for each image category and is organized into 3 folders (train, test, and validation). Create the training and testing set for the model based on the CXR images. The dataset is 2 GB in size 5863 X-ray pictures (JPEG) total. The two subfolders are Normal and Pneumonia.

Figure 4a depicts the Normal CXR Image which is in the format of (JPEG). This image repository is collected from the Kaggle repository.

Figure 4b depicts the pneumonia CXR image which is in the format of (JPEG). This dataset repository is collected from the Kaggle repository.

4.2 *Image Processing Techniques Used for Chest X-ray Feature Extraction.*

1. Image Erosion

Figure 5 depicts the CXR images after applying the image pre-processing techniques called image erosion. Corrosion removes tiny lines and floating pixels, leaving only solid objects. The remaining lines seem narrower, and the shapes seem more defined.

2. Image Dilation

Following the use of image dilation pre-processing procedures, Fig. 6 shows the chest X-ray pictures. In an image, dilation refers to adding pixels to object boundaries. The size and shape of the structuring element used to reuse the image determines how many pixels are added to the image boundary.

3. Grayscale and Gaussian Blur

After applying the picture pre-processing techniques known as image grayscale and Gaussian blur technique, Fig. 7 shows the chest X-ray images. Gaussian blurring is frequently employed when lowering the size of an image. It is typical to use a low-pass filter while down sampling an image.

4. Canny Edge Detection

Figure 8 shows the chest X-ray pictures, after applying the Canny edge detection algorithm for image pre-processing. A basic image processing method for identifying objects borders in images is edge detection. It functions by spotting irregularities in brightness. Canny is one of the edge detection techniques.

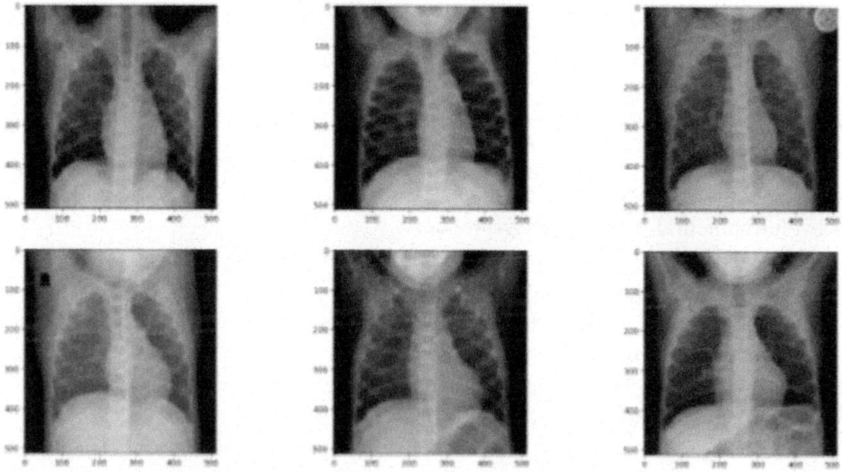

Fig. 5 Chest X-ray images after applying erosion technique

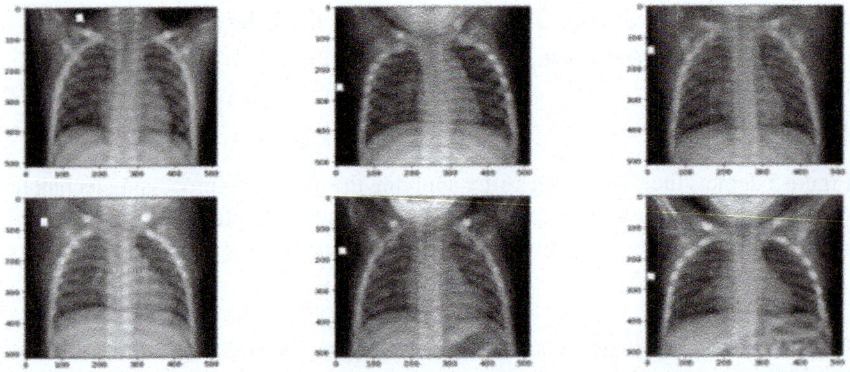

Fig. 6 Chest X-ray images after applying dilation technique

Fig. 7 Chest X-ray image after applying grayscale and Gaussian blur technique

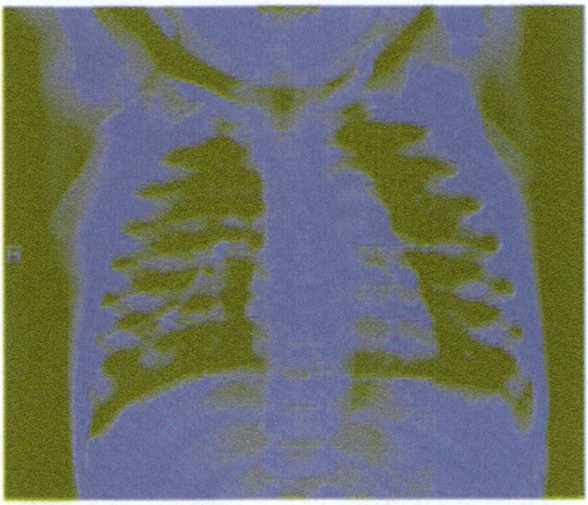

4.3 Training, Validation Accuracy and Loss of All Graphs

Figure 9a and b depicts the training as well as validation accuracy and loss of convolution neural network (CNN) model.

Figure 10a and b depicts the training as well as validation accuracy and loss of visual geometry Group16 (VGG16) model.

Figure 11a and b depicts the training as well as validation accuracy and loss of residual neural network (ResNet50V2) model.

Table 1 depicts the epochs of all three models such as CNN, VGG16 and ResNet50V2 models.

Table 2 depicts the performance and comparison of the pre-trained models such as VGG16 and ResNet50V2 models.

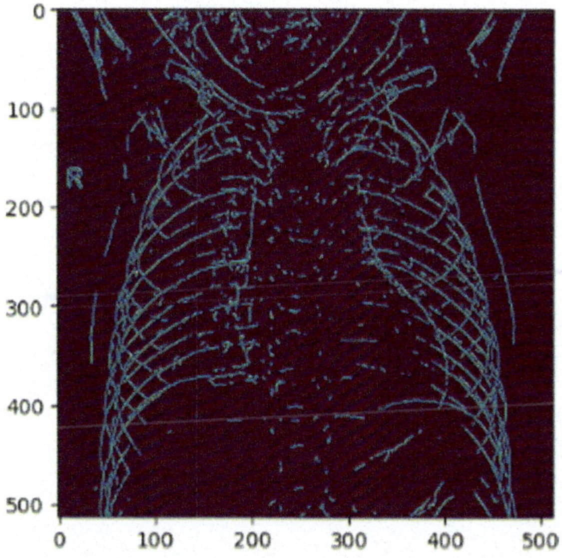

Fig. 8 Chest X-ray image after applying Canny edge detection technique

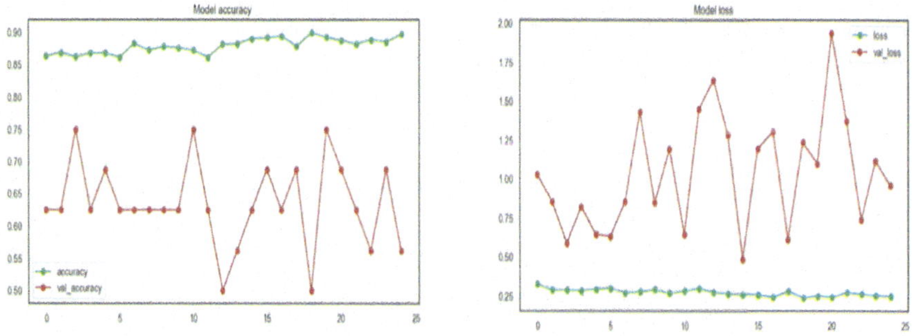

a b

Fig. 9 **a** Training, validation accuracy of CNN **b** Training, validation loss of CNN

4.4 Confusion Matrix of All Three Models

Figure 16 depicts the Confusion matrix of convolution neural network (CNN) model (Figs. 12, 13 and 14).

Figure 17 depicts the confusion matrix of visual geometry Group16 (VGG) model.

Figure 18 depicts the confusion matrix of residual network 50V2 (ResNet50V2) model.

Table 3 depicts the Precision and Recall, F1-score, and Support comparison of all three deep learning techniques.

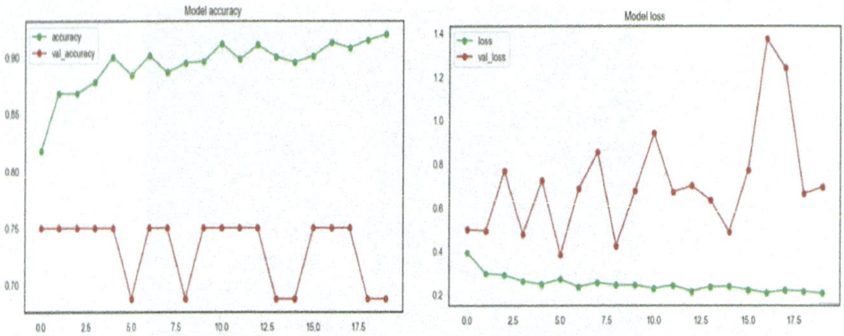

a b

Fig. 10 **a** Training, validation accuracy of VGG16 **b** Training, validation loss of VGG16

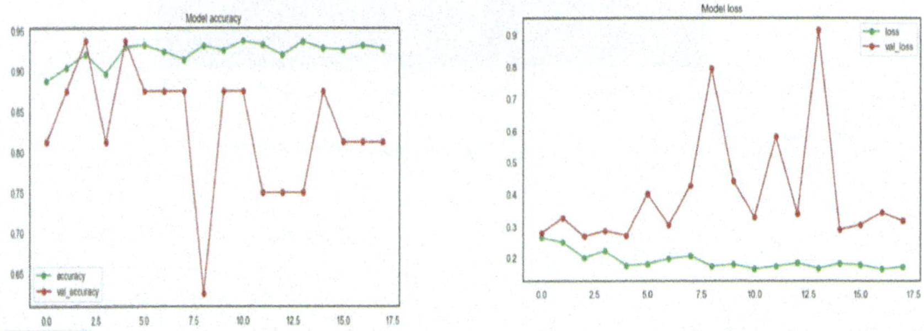

a b

Fig. 11 **a** Training, validation Accu of ResNet50V2 **b** Training, validation loss of ResNet50V2

Table 1 Epochs of all three models

Sl. No.	Models	Models training accuracy at the end of 20 epochs (%)
1	CNN	0.88
2	VGG16	0.90
3	ResNet50V2	0.92

Table 2 Performance of pre-trained models

Sl. No.	Pre-trained models metrics	VGG16	ResNet-50 V2
1	Training accuracy	0.90	0.92
2	Training loss	0.21	0.17
3	Validation accuracy	0.75	0.81
4	Validation loss	0.74	0.31
5	PARAMETERS	14,714,688	94,668,760

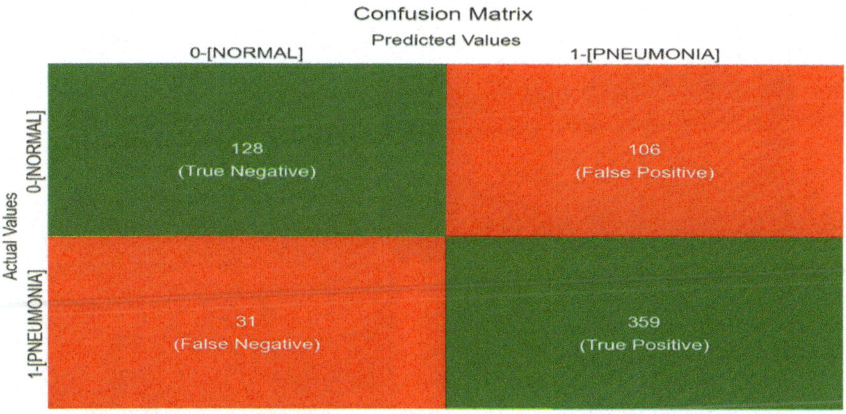

Fig. 12 Confusion matrix of CNN model

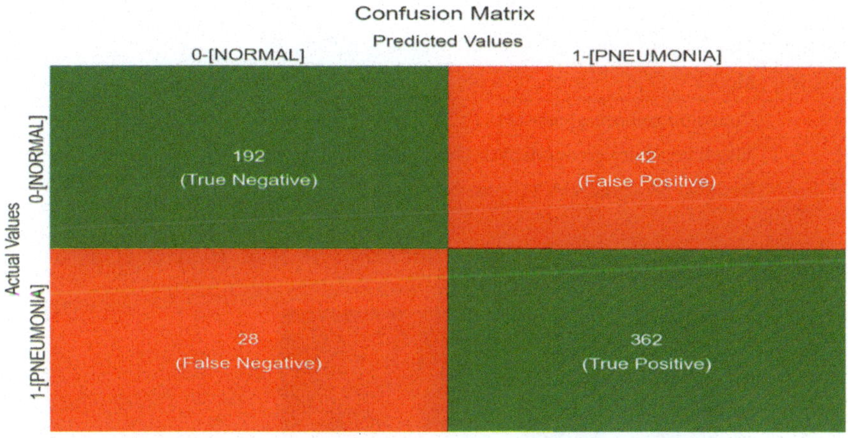

Fig. 13 Confusion matrix of VGG16 model

4.5 Prediction of Pneumonia

Figure 15 depicts the prediction of pneumonia in a chest X-ray image by residual network 50V2 (ResNet50V2) model.

4.6 Correct and Incorrect Predicted Classes

Figure 16 depicts the correct predictions of chest X-ray images the above image belongs to normal class 0 and prediction is also made a normal class.

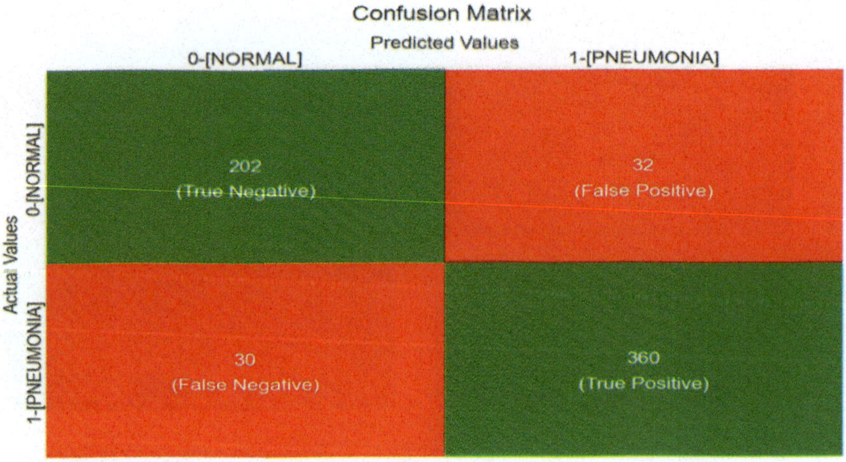

Fig. 14 Confusion matrix of ResNet50V2 model

Table 3 Classification report of all three deep learning techniques

SL. No.	Model	Precision	Recall	F1-score	Test accuracy
1	Convolutional neural networks (CNN)	0.77	78	0.92	0.83
2	Visual geometry group-16 (VGG16)	0.89	88	0.92	0.91
3	Residual networks 50 V2 (Resnet50v2)	0.91	90	0.92	0.92

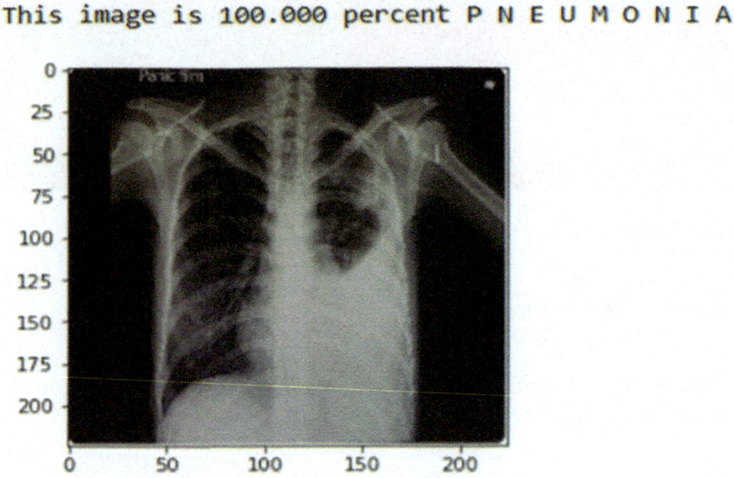

Fig. 15 Prediction of presence of pneumonia in an image by ResNet50V2 model

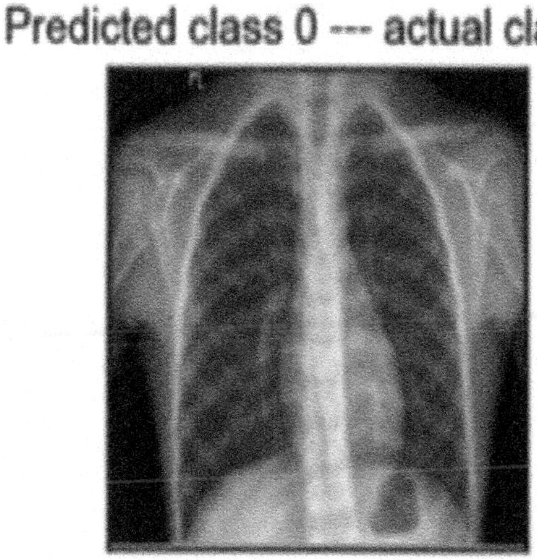

Fig. 16 Correct predictions

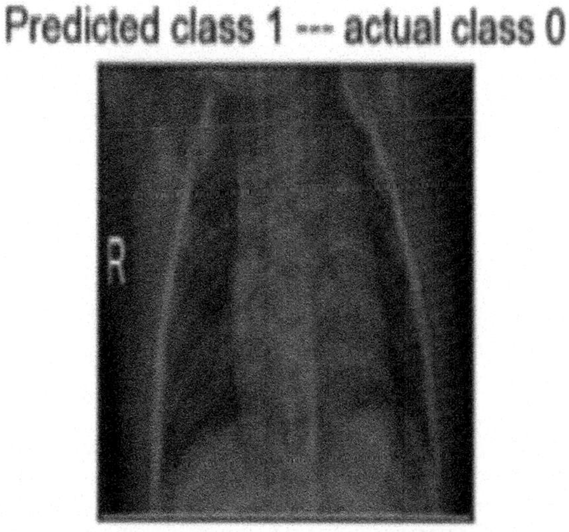

Fig. 17 Incorrect predictions

Figure 17 depicts the incorrect predictions of chest X-ray images the above image its predicted class is Pneumonia (1) but its actual class is Normal (0).

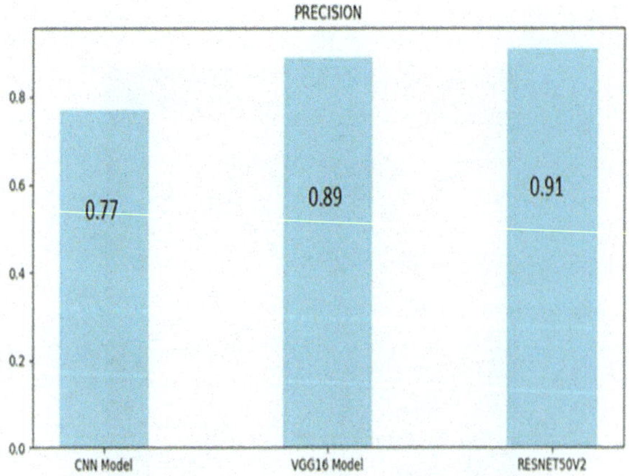

Fig. 18 Precision report of all three models

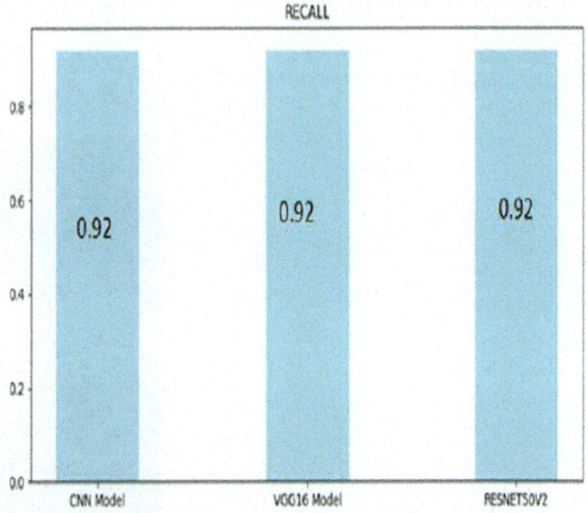

Fig. 19 Recall report of all three models

Figure 18 depicts the precision comparison of all three deep learning models such as CNN, VGG16 and ResNet50V2 models.

Figure 19 depicts the recall comparison of all three deep learning models such as CNN, VGG16 and ResNet50V2 models.

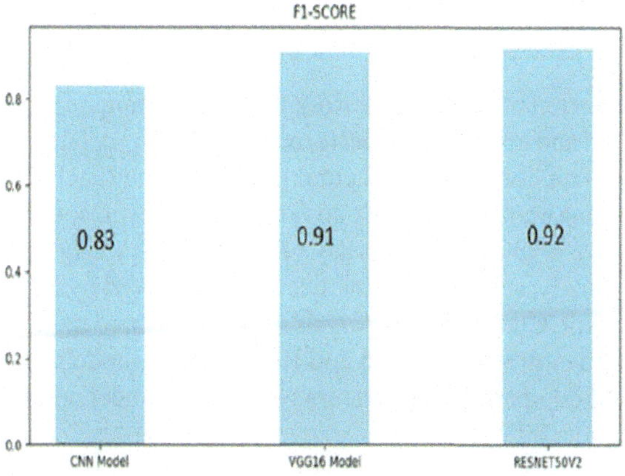

Fig. 20 F1-score report of all three models

Fig. 21 Test Accuracy report of all three models

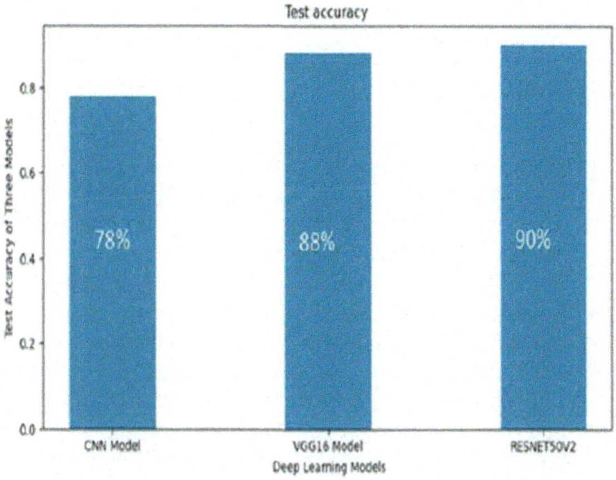

Figure 20 depicts the F1-score comparison of all three deep learning models such as CNN, VGG16 and ResNet50V2 models.

Figure 21 depicts the test accuracy report of all three deep learning models such as CNN, VGG16 and ResNet50V2 model.

5 Conclusion

The disease prediction model is built using transfer learning techniques. All 3 classifiers are compared and evaluated for performance based on accuracy, precision, recall, and F measure. With convolutional neural network (CNN) achieved an accuracy of 78%, after applying transfer learning to VGG16 model it obtained an accuracy of 88%, ResNet50V2 obtained an accuracy of 90%, respectively. ResNet50V2 Model gave a better accuracy, achieving a 90% accuracy, followed by VGG16 with an accuracy of 88% percent. Comparative analysis of all these three models shows that, after applying transfer learning to VGG16 and ResNet50V2 models' accuracy of the two models increased compared to the accuracy of CNN model and ResNet50V2 had a highest accuracy of (90%).

Future scope is to improve the accuracy of the transfer learning models and identifying the pneumonia disease using the symptoms of the patients. Pneumonia can be detected based on genders and it can also be detected age wise and severity of pneumonia can also be identified along with the prediction.

References

1. Horry MJ, Chakraborty S, Paul M, Ulhaq A, Pradhan B. COVID-19 detection through transfer learning using multimodal imaging data. Published in: 14 August 2020, IEEE
2. Alahmari SS, Altazi B, Hwang J, Hawkins S, Salem T (2022) A comprehensive review of deep learning-based methods for COVID-19 detection using chest X-ray images. IEEE Access 10:100763–100785. https://doi.org/10.1109/ACCESS.2022.3208138
3. Srivastav, Bajpai A, Srivastava P (2021) Improved classification for pneumonia detection using transfer learning with GAN based synthetic image augmentation. In: 2021 11th international conference on cloud computing, data science & engineering (Confluence), Noida, India, pp 433–437. https://doi.org/10.1109/Confluence51648.2021.9377062
4. Labhane G, Pansare R, Maheshwari S, Tiwari R, Shukla A (2020) Detection of pediatric pneumonia from chest X-ray images using CNN and transfer learning. In: 2020 3rd international conference on emerging technologies in computer engineering: machine learning and internet of things (ICETCE). Jaipur, India, pp 85–92. https://doi.org/10.1109/ICETCE48199.2020.9091755
5. Yadavendra, Chand S (2021) Pneumonia lung opacity detection and segmentation of chest x rays by using transfer learning of the mask R-CNN. In: 2021 7th international conference on advanced computing and communication systems (ICACCS). Coimbatore, India, pp 1035–1040. https://doi.org/10.1109/ICACCS51430.2021.9441864
6. Rajaraman S, Antani SK (2020) Modality-specific deep learning model ensembles toward improving TB detection in chest radiographs. IEEE Access 8:27318–27326. https://doi.org/10.1109/ACCESS.2020.2971257
7. Cortés E, Sánchez S (2021) Deep learning transfer with AlexNet for chest X-ray COVID-19 recognition. IEEE Lat Am Trans 19(6):944–951. https://doi.org/10.1109/TLA.2021.9451239
8. Jalannavar A, Kanakaraddi SG, Handur VS (2022) Migraine prediction using deep learning model. In: 2022 fourth international conference on emerging research in electronics, computer science and technology (ICERECT). Mandya, India, pp 1–8. https://doi.org/10.1109/ICERECT56837.2022.10059843

9. Giraddi S, Chickerur S, Patil MS, Kanakaraddi S, Giraddi V (2022) Lightweight convolution neural network for diabetic retinopathy grading. In: 2022 IEEE 2nd Mysore sub section international conference (MysuruCon). Mysuru, India, pp 1–5. https://doi.org/10.1109/MysuruCon55714.2022.9972448
10. Giraddi S, Mugdha S, Kanakaraddi S, Chickerur S (2022) Glaucoma diagnosis: handcrafted features versus deep learning. In: Shakya S, Ntalianis K, Kamel KA (eds) Mobile computing and sustainable informatics. Lecture notes on data engineering and communications technologies, vol 126. Springer, Singapore. https://doi.org/10.1007/978-981-19-2069-1_60

Optimized FOREX Rate Prediction Using Hybrid Machine Learning Algorithm

Challa Madhavi Latha, S. Bhuvaneswari, and K. L. S. Soujanya

Abstract This research examines the application of fuzzy time series (FTS) models and different machine learning techniques to anticipate changes in foreign exchange (FOREX) data. The statistics used in this investigation include the actual FOREX prices and the high and low values of four different currencies. These statistics begin from January 2012 to November 2022; the data is available on the Reserve Bank of India website. This article investigates the significant flaws present in the literature information on FTS. This article identifies fundamental flaws in all FTS-based algorithms presently available in the literature and shows how to fix them. The study then describes the unique structure of improvised FTS. The Hybrid Machine Learning Algorithm is the name given to this structure. The purpose is to solve the identified issues present in FTS-based models previously published in the literature. The result was compared with the before and after optimization processes. After implementing the proposed hybrid machine learning algorithm, the result shows the better accuracy.

Keywords Fuzzy time series · Foreign exchange · Machine learning · Fuzzy logic · Hybrid algorithm

C. M. Latha (✉) · S. Bhuvaneswari
Department of Computer Science, School of Engineering and Technology, Pondicherry University, Karaikal, Pondicherry, India
e-mail: saidatta2009@pondiuni.ac.in

K. L. S. Soujanya
Department of Information Technology, CMR College of Engineering and Technology, Kandlakoya, Medchal, Hyderabad, India
e-mail: klssoujanya@cmrcet.ac.in

S. Bandyopadhyay et al. (eds.), *Intelligent Computing Systems and Applications*, Lecture Notes in Networks and Systems 1010,
https://doi.org/10.1007/978-981-97-5412-0_3

1 Introduction

Accurate foreign exchange (FOREX) rate prediction remains critical in financial markets. The dynamic and intricate nature of currency markets presents challenges for investors and analysts seeking to make informed decisions. Traditional forecasting methods often need to catch up in capturing the complexities inherent in these markets. Integrating advanced methodologies becomes imperative to provide more reliable predictions and actionable insights. This study delves into the domain of FOREX rate prediction by introducing a novel approach that combines the principles of optimal portfolio strategy, fuzzy logic, and machine learning algorithms. The primary objective is to develop a comprehensive methodology that anticipates future currency exchange rates and provides investors with insights into risk tolerance and potential returns. By merging these methodologies, this research aims to address the limitations of existing forecasting models and contribute to the advancement of accurate and robust prediction techniques.

This research aims to enhance the prediction of changes in foreign exchange rates through a multi-faceted approach that draws from various disciplines. Specifically, this study aims to: Apply an optimal portfolio strategy to gather and structure data for analysis, leveraging the strategy's ability to account for risk preferences and return expectations. Employ machine learning classification algorithms and a fuzzification process to construct portfolios based on trend analyses and forecasted values. Categorize these portfolios to gain insights into different currency pairings' risk tolerance and return capabilities. Propose a unified framework synthesizing fuzzy logic and machine learning techniques to anticipate future currency exchange rates and refine predictions for optimal outcomes.

The significance of this research lies in its potential to revolutionize the field of FOREX rate prediction. By introducing a holistic methodology that incorporates optimal portfolio strategy, fuzzy logic, and machine learning, this study aims to overcome the limitations of conventional forecasting techniques. The resultant approach enhances prediction accuracy and provides investors valuable insights into risk management and potential gains. Moreover, integrating these methodologies contributes to the broader discussion on the synergy between financial analysis, artificial intelligence, and decision-making processes in currency markets.

2 Literature Survey

The quick technological change of the twenty-first century has changed our lives. It is quickly nearing a fully automated civilization that demands less labor but more innovation and inventiveness. Artificial Intelligence (AI), Machine Learning (ML), 3D printing, robotics, nanotechnology, and digital currencies have influenced the world. Research into phenomena in engineering, medicine, biology, physics, etc., which are typically challenging to forecast [1], has been stimulated by developing methodologies that can accurately anticipate regular behavior variables. Many researchers see employing hybrid approaches for time forecasting [2–4], such as fuzzy time series, as a viable option in this scenario due to the hybrid's effectiveness compared to the many other suggestions already developed in the literature. Fuzzy time series forecasting has spawned a variety of methodological ideas, including particle swarm optimization (PSO), genetic algorithms, and artificial neural networks. Many academics focus on time series forecasting approaches combining fuzzy sets theory [5]. Fuzzy sets theory may optimize stochastic modeling processes employed alone or with neural networks, evolutionary algorithms, PSO, etc. Solar activity [6], electricity load [7–9], oil demand [10], rice production [11–13], the Chinese stock index [14–16], the operation of nuclear power plant turbines [2, 17, 18], a thermal error compensation system, short-term load forecasting [19] the Taiwan Capitalization Weighted Stock Index (TAIEX) [8, 20, 21], ship traffic in the Strait of Istanbul [22], and daily wind velocity [4, 8] all use hybrid fuzzy forecasting models. The forecasting as mentioned earlier approaches converge on reducing forecasting error in the suggested algorithm.

3 Proposed Methodology

Applying the optimal portfolio strategy presented here is necessary to collect the data for the study within the current framework. This approach is applied to foreign exchange (FX) rates with domestic currency pairings and determines the trend analysis of that specific foreign currency. Using machine learning classification algorithms and a fuzzification process, the portfolios have been constructed based on the trend and performance of forecasted values. These categorized portfolios reveal the risk tolerance and return capabilities of various currencies. This approach will use a specific portfolio representing the investor's risk tolerance as input data. After categorizing the data, minimum and maximum values are determined from the FX price. To preserve the equal intervals, the values of datasets (DS_1 and DS_2) are specified to create the top and bottom boundaries of the data in similar sections. Then, split the FX rates into un_1, un_2, … un_n equal halves based on the difference in the middle values. Then create a new universe dataset, i.e., un'. The following procedure is to build new fuzzy sets for training data (i.e., NFS_{-n} to NFS_n). Next, establish a link between the months of April and May using the universal variable (i.e., $un_1 \rightarrow un_1$), and then select a link between the months of April and May using the NFSs (i.e., $NFS0 \rightarrow NFS0$). Then, collect all the recently introduced fuzzy relations into a single set, and project their values into the future. Future values are adjusted using the alpha and beta coefficients to lower the fitted squared errors. Finally, the optimum future anticipated values are found. The fitted squared errors are mistakes in the prediction between the data point and fitted line distance.

3.1 Hybrid Machine Learning Algorithm [23]

The Hybrid Time Series Portfolio (HTSP) algorithm is proposed in the present study. Step 1 of the algorithm divides the FX rates into multiple portfolios per each currency's trend. The trend was calculated based on the GARCH(1,1) process. Step 2 of the HTSP follows the Hybrid Fuzzy Time Series Forecasting model (HFTSF) [23], specifically tailored to enhance the precision of stock price predictions. The HFTSF model was applied to the S&P BSE IT index stock prices, exemplifying its potential to offer more accurate forecasting for data characterized by multiple trends. The S&P BSE IT index, a subset of the broader stock market, comprises technology-related stocks. Due to the volatility and complexity inherent in the technology sector, accurate predictions are challenging using conventional forecasting methods. The introduction of the HFTSF model seeks to provide a remedy that capitalizes on the unique features of stock index information, enhancing prediction accuracy and contributing to more informed investment decisions. The HFTSF model is a fusion of fuzzy time series forecasting techniques and optimization processes. This hybridization enables the model to effectively capture the intricate trends in stock price data.

Algorithm 1: Hybrid Time Series Portfolio (HTSP) Algorithm

```
Step1: Portfolio Construction
          Function Collect_portfolio(risk_capacity)
          {
            Step 1: Collect Foreign exchange rates
            Step 2: pre-process the data
            Step 3: Find the heteroskedasticity exists or not
            Step 4: If data is homoskedasticity, then
                   Apply Formula for Garch (1,1):
                   GARCH= (Alpha* SLR + Beta* LV + Gamma* LRV)
               (SLR: Squared Lagged Returns; LV: Lagged Variance; LRV: Long Run Variance),
Collect Residual errors, Gama, Alpha, and Beta values, and forecasted values.
            Step 5: logistic regression and fuzzy logic method)
            Step 6: generate portfolios1 – n based on the fuzzification process
            Step 7: Check If (risk_capacity < risk(portfolio))
                    Select portfolio, currencies as dataset
            Step 8: return dataset.
          }
Step 2: Prediction of the future FOREX rates using HFTSF Algorithm
```

By incorporating fuzzy logic, the model accommodates the inherent uncertainty and complexity of financial data, offering a more flexible and adaptive approach to prediction.

4 Implementation and Results

The proposed method, the hybrid machine learning algorithm, was analyzed using Python programming. The historical data is described in the following Table 1. The data has been collected from the Reserve Bank of India website; monthly data has been considered for the present study from January 2012 to November 2022; for training purposes, January 2012 to March 2022 and prediction and validation purposes, the data considered from April 2022 to November 2022.

Table 1 gives the descriptive statistics of collected data with the proxy of the Indian Rupee (INR) with Euro, Yen, Pound sterling, and US Dollar (USD). The highest mean was observed in a pound sterling; a high standard deviation means high volatility, which is very high in the USD/INR FOREX case. The difference between min and max is very high in the case of USD, i.e., 33.1537, which means the high fluctuations are identified in the USD/INR FOREX rate.

Step 1: Collect the portfolios: This step identifies the portfolios based on the Garch(1,1) process. For this need to check heteroscedasticity and estimate volatility using the GARCH(1,1) process (Fig. 1).

Table 1 Descriptive statistics of FOREX data

	Euro/INR	Yen/INR	Pound sterling/INR	USD/INR
Count	131	131	131	131
Mean	78.237193	61.963015	93.153749	66.713074
Std	6.148612	5.189087	6.65702	7.24204
Min	64.81515	51.615	77.5512	49.16175
25%	72.923775	58.23	87.697025	62.31835
50%	78.3895	61.72	93.71445	66.96735
75%	83.176525	65.65	99.122225	72.78265
Max	89.17385	71.015	103.1296	82.31545

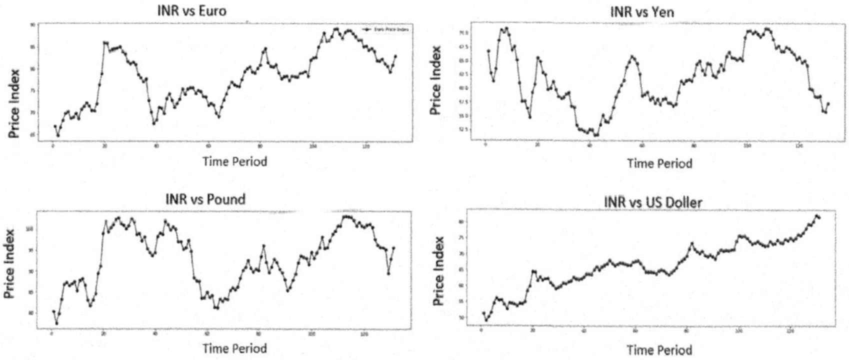

Fig. 1 Monthly movement for EURO, Yen, Pound sterling, and USD

4.1 Heteroskedasticity Identification

Model residuals that might indicate false assumptions would be explored in relation to the effects of heteroscedasticity. The main causes of heteroscedasticity would include flawed linear models, unreliable distributional assumptions, models with varying subgroup performance, and autocorrelation, which would lead to incorrect estimations of the standard error and p-value. Figure 2 shows that the data exhibits homoscedasticity; as a result, this data is suitable for reliable projections.

The FX rates do not exhibit heteroskedasticity, as seen in Fig. 2. Figure 2 shows the residuals for the duration of the considered period.

Table 2 displays the Garch(1,1) result for four omega, alpha, and beta currencies. The prediction value at the infinite prospect (alpha and beta) defines the omega (variance intercept), and variance targeting determines the asymptotic variance to avoid guessing the omega. Based on the value of the omega, omega must be more than 0 and alpha + beta must be less than 1. The first portfolio was created using the mean-max membership function of fuzzy logic. Table 3 gives the same thing. The mean-max value is 2.56.

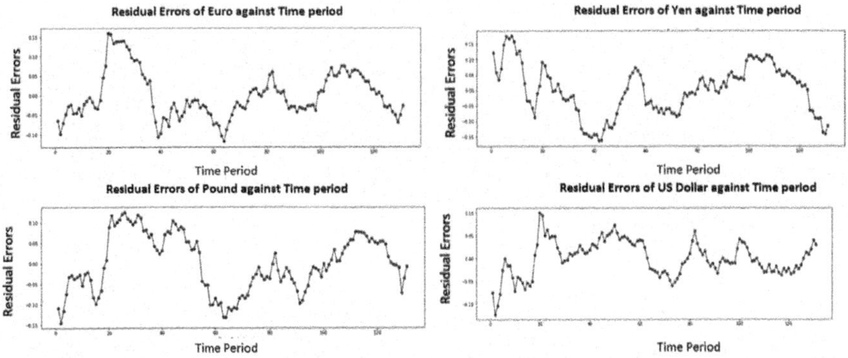

Fig. 2 Residual graph for 4 currencies

Table 2 GARCH(1,1) result for FOREX

Currency	Euro	Yen	Pound sterling	USD
Omega	1.365	3.8917	4.6119	0.3877
Alpha	0.8277	0.8297	0.8824	0.9998
Beta	0.1605	2.31E-16	2.74E-17	2.82E-17

Table 3 Portfolio construction

Portfolio1	Rank	Portfolio2	Rank
Euro	1	Yen	2
USD	2	Pound sterling	1

Table 4 Daily, monthly, annual volatility for FOREX

Currency	Daily volatility	Monthly volatility	Annual volatility
Euro	6.1486	28.176	97.606
Yen	5.1891	23.779	82.374
Pound sterling	6.657	30.506	105.68
USD	7.242	33.187	114.96

Table 3 gives a portfolio 1 list of currencies according to the mean-max membership function. The lesser values of 2.56 have been considered for this portfolio. The remaining currencies are greater values of mean-max value, so they are in portfolio 2.

Table 4 gives the daily, monthly, and yearly volatility for the 4 currencies. USD has high volatility, indicating daily, monthly and annual fluctuations as 7.24, 33.18, and 114.96, respectively. Less volatility was observed in Yen currency (5.18, 23.77, 82.37).

4.2 *Prediction of Future FOREX Rates*

Step 2: In this step, find each portfolio's minimum and maximum datasets and UN segment datasets. This follows the HFTSF algorithm which is already implemented in [23] (Table 5).

$$\begin{aligned} \text{DSmin} &= \min(\text{FX price}) \\ \text{DSmax} &= \max(\text{FX price}) \\ \text{UNT} &= [\text{DSmin} - \text{DST1}, \text{DSmax} + \text{DST2}] \end{aligned}$$

UNT might be divided into three segments for Euro, five segments for USD, three segments for Yen, and four segments for Pound sterling of equal length and 10 units each, namely UNT1, UNT2, UNT3, UNT4, and UNT5.

Step 3: UNT = [UNT1, UNT2, ...UNT5].

Step 4. define new universe dataset $\text{un}' = [-(\text{n}-1), (\text{n}-1)]$.

$\text{un}' = [-4, 4]$ and $\text{un}-4' = [-4.5-3.5]$, $\text{un}-3' = [-3.5, -2.5]$ ...., $\text{un3}' = [2.5, 3.5]$, $\text{un4}' = [3.5, 4.5]$.

Step 5. preparation of novel fuzzy sets for each training dataset noted as NFST.

$$\begin{aligned} \text{NFST}_{-(4)} &= 1/\text{un}'_{-(4)} + 0.5/\text{un}'_{-(3)} + \ldots + 0/\text{un}'_{(3)} + 0/\text{un}'_{(4)} \\ \text{NFST}_{-(5)} &= 0.5/\text{un}'_{-(4)} + 1/\text{un}'_{-(3)} + \ldots + 0/\text{un}'_{(3)} + 0/\text{un}'_{(4)} \end{aligned}$$

Table 5 Min and max values of FOREX rates against the INR

	Portfolio1		Portfolio2	
	Euro	USD	Yen	Pound sterling
Min value	64.81	49.16	51.61	77.55
Max value	89.17	82.31	71.01	103.12
DST1	4.81	9.16	1.61	7.55
DST2	0.83	7.69	8.99	6.88
UNT	60, 90	40, 90	50, 80	70, 110
UNT1	60–70	40–50	50–60	70–80
UNT2	71–80	51–60	61–70	81–90
UNT3	81–90	61–70	71–80	91–100
UNT4	–	71–80	–	101–110
UNT5	–	81–90	–	–

......

...

$$\text{NFST}_{(4)} = 1/\text{un}'_{-(4)} + 0.5/\text{un}'_{-(3)} + \ldots + 0.5/\text{un}'_{(3)} + 1/\text{un}'_{(4)}$$

Step 6. the new fuzzy sets and new fuzzy logic relations (NFLR) are shown in the following Table 6.

Table 6 NFS and NFLR for the period of January 2012 to November 2022

Month	Close	UNT	NFST	Date	FLR	NFLR
01–2012	66.94	UNT1	NA	01–12 → 02–12	NA	NA
02–2012	64.81	UNT1	NFST0	02–12 → 03–12	UN1-UN1	NFST0 → NFST0
03–2012	66.71	UNT1	NFST0	03 12 → 04–12	UN1-UN1	NFS0 → NFS0
04–2012	68.31	UNT1	NFST0	04–12 → 05–12	UN1-UN1	NFST0 → NFST0
05–2012	69.84	UNT1	NFST0	05–12 → 06–12	UN1-UN1	NFST0 → NFST0
06–2012	70.21	UNT2	NFST1	06–12 → 07–12	UN1-UN2	NFST0 → NFST1
...	...	...	...	...	...	...
...	...	...	...	...	...	...
07–2022	81.28	UNT3	NFST0	07–22 → 08–22	UN3-UN3	NFST0 → NFST0
08–2022	80.73	UNT3	NFST0	08–22 → 09–22	UN3-UN3	NFST0 → NFST0
09–2022	79.35	UNT2	NFST-1	09–22 → 10–22	UN3-UN2	NFST0 → NFST-1
10–2022	80.96	UNT3	NFST1	10–22 → 11–22	UN2-UN3	NFST-1 → NFST1
11–2022	83.01	UNT3	NFST0	11–22 → 12–22	UN3-UN3	NFST1 → NFST0

4.2.1 Grouping of Novel Fuzzy Logic Relations

NFST0 → 4NFST0, NFST1, 2NFST0, NFST1, 8NFST0, NFST1, 11NFST0, NFST-1, 4NFST0, NFST-1, NFST1, NFST-1, 17NFST0, NFST-1, NFST1, 8NFST0, 3NFST1, 6NFST0, NFST-1, 10NFST0, NFST1, 28NFST0, NFST-1.

NFST1 → NFST-1, 9NFST0.

NFST-1 → 3NFST0, NFST1, 2NFST0, NFST1.

Step 7: Calculate the forecasted output

$$
\begin{aligned}
\text{FV(Mar - 12)} &= \text{AV(Feb - 12)} + ((4(0) + 1 + 2(0) + 1 + 8(0) \\
&\quad + 1 + 11(0) - 1 + 4(0) - 1 + 1 - 1 + 17(0) - 1 \\
&\quad + 1 + 8(0) + 3(1) + 6(0) - 1 + 10(0) + 1 + 28(0) \\
&\quad -1/113) * 10 \\
&= 64.81 + 0.17 \\
&= 64.99
\end{aligned}
$$

Step 8: Dataset with the Optimization process.

This process is continued until the best value is obtained, and the best values (α and β) are used to forecast test results. This is decided to reduce the fitted squared errors in training data, which are employed internally in defuzzification. The fitted squared errors are straightforward in determining the distance between the data point and the fitted line. Figure 3 depicts the end outcome.

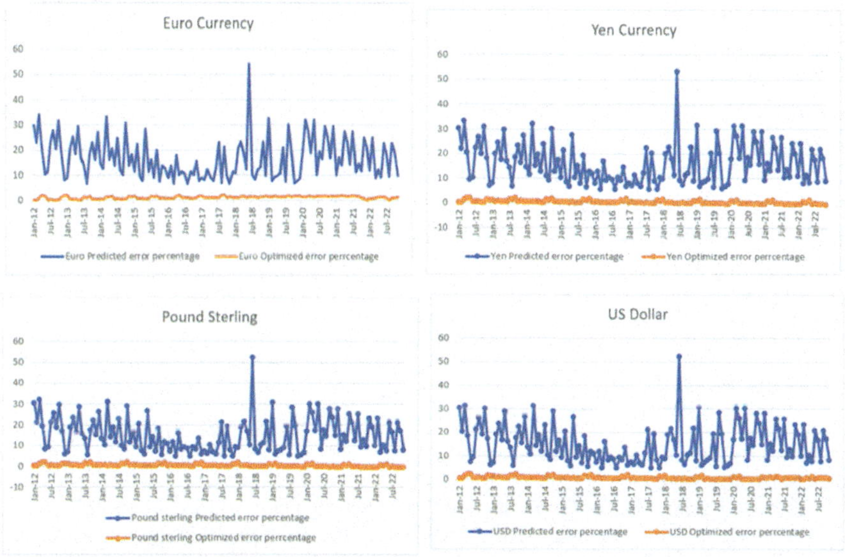

Fig. 3 Error percentage of predicted and optimized values

Figure 3 shows the graphical representation of each currency's predicted and optimal error percentages. As per Fig. 3, the error percentage of optimized forecasting has been reduced.

The advantages of the proposed method are as follows: the hybrid machine learning algorithm enables investors to choose the most advantageous international currency in which to place their money and earn returns that are changeable according to their preferences. Using fuzzy rule-based evidential reasoning, an enhanced automated trading system for the foreign exchange market may be developed as part of this study. The methodology that is used takes the use of fuzzy sets when applied to the setting of the evidence theory. This enables the concerns with operating rules that have been uncovered to be avoided and improves the method's overall performance. The optimization process, which consists of determining the most appropriate optimum solution based on the myriad of possible combinations of the projected data, has improved the accuracy of the predictions. A hybrid machine learning algorithm is used in the training of the model to optimize the purpose of optimizing the necessary parameters. Various applications employ the computational method, including an optimization strategy that does not need the function's gradient, similar to most Evolutionary Algorithms. In other words, the gradient is unnecessary for most of these Evolutionary Algorithms.

5 Conclusion

To estimate the internal trend rate, a hybrid machine learning algorithm was presented in the current research for use on the foreign exchange rates against other currencies. The suggested program used a unique optimization approach in tandem with hybrid NFST logic to boost the accuracy of its predictions. Results show enhanced data trend analysis and parameter optimization. The hybrid machine learning algorithm model converted the final predictions into high-performance, optimized forecasting values. Even though this study primarily emphasizes predicting FX rates, the findings may also be helpful in other contexts where forecasting is necessary. Stationarity in time series data may be improved by mapping conventional NFSTs to innovative NFLRs, as was shown in this work.

References

1. Kantz H, Schreiber T (2004) Nonlinear time series analysis, vol 7. Cambridge university press
2. Challa ML, Malepati V, Kolusu SNR (2020) S&P BSE Sensex and S&P BSE IT return forecasting using ARIMA. Finan Innov 6(1):47. https://doi.org/10.1186/s40854-020-00201-5
3. Challa ML, Malepati V, Kolusu SNR (2018) Forecasting risk using auto regressive integrated moving average approach: an evidence from S&P BSE Sensex. Finan Innov 4(1):24. https://doi.org/10.1186/s40854-018-0107-z

4. Venkataramanaiah M, Madhavi Latha C, Siva Nageswara Rao K (2018) Determinants of dividend policy in the Indian corporate sector: a study of companies listed on nifty 50, NSE. Indian J Finan 12(1):37. https://doi.org/10.17010/ijf/2018/v12i1/120740
5. Zadeh LA (1968) Probability measures of Fuzzy events. J Math Anal Appl 23(2):421–427. https://doi.org/10.1016/0022-247X(68)90078-4
6. Abdollahzade M, Miranian A, Hassani H, Iranmanesh H (2015) A new hybrid enhanced local linear neuro-fuzzy model based on the optimized singular spectrum analysis and its application for nonlinear and chaotic time series forecasting. Inf Sci (NY) 295:107–125
7. Efendi R, Ismail Z, Deris MM (2015) A new linguistic out-sample approach of fuzzy time series for daily forecasting of Malaysian electricity load demand. Appl Soft Comput 28:422–430
8. Wang L, Zeng Y, Chen T (2015) Back propagation neural network with adaptive differential evolution algorithm for time series forecasting. Expert Syst Appl 42(2):855–863
9. Lee W-J, Hong J (2015) A hybrid dynamic and fuzzy time series model for mid-term power load forecasting. Int J Electr Power Energy Syst 64:1057–1062
10. Rubinstein S, Goor A, Rotshtein A (2015) Time series forecasting of crude oil consumption using neuro-fuzzy inference. J Ind Intell Inf 3(2)
11. Askari S, Montazerin N (2015) A high-order multi-variable fuzzy time series forecasting algorithm based on fuzzy clustering. Expert Syst Appl 42(4):2121–2135
12. Madhavi L, Siva NRK (2016) Implementation of DRSMLA method for predicting profitable indices. I-manager's J Manage 11(1):20. https://doi.org/10.26634/jmgt.11.1.8071
13. Malepati V, Latha CM, Nageswararao KS (2021) Determinants of access to finance by MSMEs in Andhra Pradesh. SEDME (Small Enterprises Dev, Manage Extension J): Worldwide Window MSME Stud 48(2):203–219. https://doi.org/10.1177/09708464211066630
14. Cai Q, Zhang D, Zheng W, Leung SCH (2015) A new fuzzy time series forecasting model combined with ant colony optimization and auto-regression. Knowl Based Syst 74:61–68
15. Chen M-Y, Chen B-T (2015) A hybrid fuzzy time series model based on granular computing for stock price forecasting. Inf Sci (NY) 294:227–241
16. Sun B, Guo H, Karimi HR, Ge Y, Xiong S (2015) Prediction of stock index futures prices based on fuzzy sets and multivariate fuzzy time series. Neurocomputing 151:1528–1536
17. Baraldi P, di Maio F, Genini D, Zio E (2015) Reconstruction of missing data in multidimensional time series by fuzzy similarity. Appl Soft Comput 26:1–9
18. Malepati V, Challa ML, Kolusu SNR (2019) Stock market volatility modeling and forecasting with a special reference to BSE sensex, pp 105–118. https://doi.org/10.1108/S0196-382120190000035005
19. Sadaei HJ, Enayatifar R, Abdullah AH, Gani A (2014) Short-term load forecasting using a hybrid model with a refined exponentially weighted fuzzy time series and an improved harmony search. Int J Electr Power Energy Syst 62:118–129
20. Lu W, Chen X, Pedrycz W, Liu X, Yang J (2015) Using interval information granules to improve forecasting in fuzzy time series. Int J Approximate Reasoning 57:1–18
21. Wang N, Liu X, Yin J (2012) Improved Gath-Geva clustering for fuzzy segmentation of hydrometeorological time series. Stoch Env Res Risk Assess 26(1):139–155
22. Bulut E (2014) Modeling seasonality using the fuzzy integrated logical forecasting (FILF) approach. Expert Syst Appl 41(4):1806–1812
23. Latha CM, Bhuvaneswari S, Soujanya KLS (2022) Stock price prediction using HFTSF algorithm. In: 2022 sixth international conference on I-SMAC (IoT in social, mobile, analytics and cloud) (I-SMAC), IEEE, pp 1053–1059. https://doi.org/10.1109/I-SMAC55078.2022.9987378

Yarn Quality Assessment and Monitoring Using ML in Textile Industry

S. Selva Kumar and Rajesh Ramasamy

Abstract The monitoring and alert systems in most textile industries are manual or partially automated. Temperature and humidity are manually captured and recorded in textile units, but there is no use with this raw data without immediate analysis and necessary actions. Some of the issues in textile automation are lack of automated monitoring systems, poor data analysis after data collection, lack of real-time alerts and inability to integrate with existing systems. Live sensor-based data collection and data analysis are essential to take corrective or preventive measures in textile units. The proposed system collects the data from the sensors and stores it in a database. The raw material for each production run is determined by the technological value of cotton called the fibre quality index. During the daily production run, temperature, humidity and pollution level inside spinning units are measured and transferred by the sensor and monitoring unit. At the end of every production run, the imperfection level and count strength product are measured and updated by the quality analysts for every fibre quality index. Once the data was collected, it should be analysed to identify any patterns or correlations between the different variables. After analysing the data, a supervised machine learning (ML) algorithm was implemented to select high-quality yarn.

Keywords IoT · Sensors · MCU · Machine · Learning · SVM

1 Introduction

Textile industries traditionally relied on manual monitoring and recording of data, which often leads to delays in identifying issues and taking corrective action. Automated monitoring systems were needed to ensure that any anomalies are quickly

S. Selva Kumar (✉) · R. Ramasamy
Department of Electrical and Electronics Engineering, Amrita School of Engineering, Amrita Vishwa Vidyapeetham, Coimbatore, India
e-mail: s_selvakumar@cb.amrita.edu

R. Ramasamy
e-mail: cb.en.p2ebs21006@cb.students.amrita.edu

S. Bandyopadhyay et al. (eds.), *Intelligent Computing Systems and Applications*, Lecture Notes in Networks and Systems 1010,
https://doi.org/10.1007/978-981-97-5412-0_4

identified and addressed. Many textile factories lack the ability to properly analyse data and identify trends that could indicate potential problems. Without proper data analysis, it is difficult to predict and prevent problems before they become serious issues that affect the quality of fabric. Most textile industries do not have the capability to send real-time alerts to personnel when critical parameters are exceeded. Without real-time alerts, personnel may be unaware of issues until it is too late to take corrective action. Textile industries often have legacy systems that are not easily integrated with modern technologies such as IoT and machine learning. This makes it difficult to leverage the data collected to improve operations. Without proper visibility into the data collected, it is difficult to identify potential problems and take corrective action. This is especially true in facilities that use manual recording processes.

The quality of the textile fibres depends on various parameters, namely raw material, air quality, temperature and humidity. Cotton waste and count strength product (CSP) depend on all these parameters. Fully automated, periodical monitoring is needed to maintain these parameters within the threshold to have better yarn quality. This can be done by using sensors and other monitoring devices. These sensors can be used to measure temperature, humidity and air quality. Usually, these sensors are installed along with microcontroller units like Raspberry Pi, Node MCU and Arduino. It can be seen in many similar case studies using sensor-based monitoring. The MCUs are selected based on the requirement attributes like the distance between nodes and control unit, data type (analog or digital) and cost. This monitoring system helps to keep track of the parameters and to act on any deviation from the thresholds. It also helps to detect any potential problems in the early stages and helps to take corrective actions. This automation system is especially helpful for large-scale factories, where controlling the parameters manually can be a tedious task.

Live monitoring of the sensor data is accomplished using IBM Watson or Adafruit IO cloud platform. Users can find the fluctuations as a live trend report and take action. An alert mechanism like WhatsApp or SMS is used to send alerts to users when the sensor values deviate from the threshold. Few case studies are available with the integrated sensor data and monitoring mechanism. Sensor data capturing and best yarn selection using ML were implemented in this work. This end-end integrated system design includes sensor data capturing, monitoring, real-time alerts and data analysis for yarn selection. The system helps in reducing manual intervention, saving time and cost and increasing productivity. It also helps in providing a better understanding of the production process and improving the quality of the product.

2 Literature Review

Many related case studies were analysed on both monitoring systems and ML before proposing this solution. This system includes both a monitoring system and an ML-based analysis.

2.1 Monitoring Systems in the Industry

A case study [1] proposed a system with a turbidity sensor and air quality sensors, connected to Raspberry Pi, integrated water and air quality sensors and monitored using IBM Watson, and alerts are triggered based on the threshold. In another system to measure multiple power interrupts [2], cost impacts were measured with the help of the readings with ION 7650 power monitoring. In a case study, to predict dust that can get into the lungs in the work area [3], dust is measured with a cotton dust monitor called a vertical elutriator (VE). The results conclude knitting unit produces more cotton dust than any other unit. A water quality monitoring system was implemented with Arduino ATMEGA328 used as MCU which has in-build ADC and WiFi modules [4].

2.2 Applying ML for Predictions

A decision support system [5] combining the ANN model, AHP and Q-learning was proposed for supporting decision-making in textile units. 612 sets of experimental data (training and test data) were used to predict the total number of machine stoppages in the spinning unit.

To reduce the modelling effort in ML, several automated ML tools, namely H2O, AutoGluon and AutoKeras, are compared [6]. Two years of sensory data on machine downtime were taken for modelling to predict yarn breaks. Similarly, multiple fabric design features are used for the prediction [7] of yarn quality. AutoML was used to select the best among six state-of-the-art ML algorithms.

After data collection, some of the algorithms used for the hypothesis are K-nearest neighbours, adoptive boosting, random forest and support vector machine (SVM) [8]. Fault logs from the textile production units are collected and used for this case study. Thinkspeak an open-source platform is used for monitoring temperature and humidity for agriculture [9], and an email-based alert mechanism is implemented based on the threshold.

2.3 Applying Support Vector Machine

Image processing is used for data collection in defect detection with active contour and SVM [10]. The data were collected as images from which features are prepared and classified into linear and nonlinear boundaries. Also, image processing is used for classifying the features based on foreign fibre content (%) and these features are used for modelling using a decision tree—support vector machine [11]. In another case study, both artificial neural networks and SVM [12] are used to predict the characteristics of elastane yarn. SVM is effectively used for Yarn strength prediction

[13] in textiles. Water quality samples were taken from various geographical locations with sensors, and Raspberry Pi, applied logistic regression and SVM were used for water quality index (WQI) classification [14].

A system was proposed with Zigbee protocol for communication between water quality sensors and edge nodes, threshold-based data analysis and reporting implemented with Raspberry Pi in the edge node, and water quality index (WQI) is used for classification in SVM. Researchers used MLX90615 and TCRT5000 as temperature sensors, heart rate and SPO2 are measured using smartwatch [15]. Raspberry Pi is used for data transfer and analysis using ML algorithms like decision tree, random forest, SVM and Bayes. In a water management system [16], NodeMCU is used for uploading data in the Thinkspeak cloud platform, and ML algorithms KNN, decision tree, random forest and SVM are applied to control the motors for irrigation.

3 System Overview

The sensor unit (DHT22, MQ135) in this workflow is integrated with a NodeMCU microcontroller and WiFi modules, as shown in Fig. 1. The microcontroller is used to collect the sensor data (temperature, humidity and pollution) and upload them as feed to Adafruit IO cloud platform. The Adafruit IO also has a web user interface to generate live trend charts for all these three parameters. Adafruit trend charts can be viewed through web browser in laptop or mobile. Adafruit IO also has an API interface to publish the data, and this is called in a Python application to fetch the data and upload in SQLite3 database. A separate Python user interface module was developed for capturing thresh-holds for sensor parameters and initiating production with necessary parameters like CSP and FQI. Another user interface is used for processing the ML algorithm and generates yarn selection report before the quality check is completed by the quality engineers. Alert mechanism will send alerts to the user when any of the thresholds have been exceeded the limit.

User roles involved in this system are shown in Table 1.

3.1 Hardware

The NodeMCU V3 (Node Microcontroller unit) was used as a microcontroller for this work. The pin diagram is shown in Fig. 2. It is a System-on-a-Chip called the ESP8266 in which hardware and open-source software development environment is available with a CH340G USB-TTL chip.

DHT22 Temperature sensor is connected to NodeMCU, with a single-wire protocol as shown in Fig. 3. DHT22 uses a thermistor for temperature measurement. Resistance decreases with an increase in temperature. Its temperature measuring range is from −40 °C to +125 °C with ± 0.5 °C tolerance. DHT22 also accommodates a humidity sensing capacitor for humidity measurement which has two electrodes

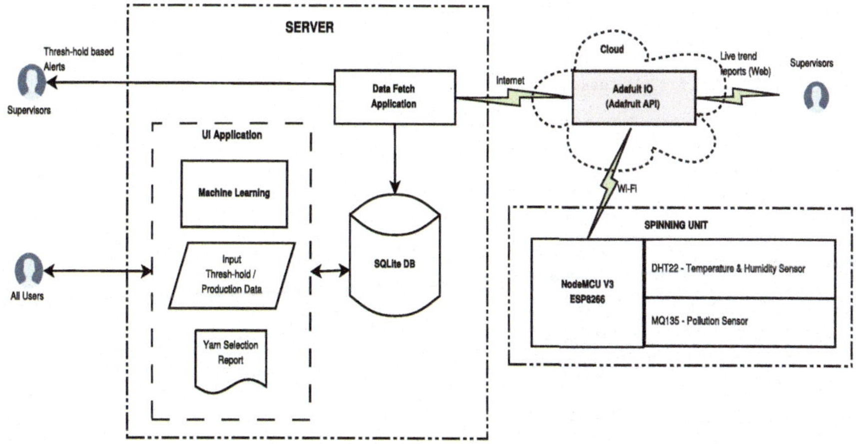

Fig. 1 System overview

Table 1 User roles

User role	Responsibilities in this system
Administrator	Create user, start sensor in units, initiate sensor input
Supervisor	Configure thresh-hold, configure production parameters, generate yarn prediction report, monitor trend for sensor values, receive thresh-hold-based alert and take preventive action
Management	Generate yarn prediction report values, monitor trend for sensor, recommend selected yarn for customer sample

with a moisture-holding substrate as a dielectric between them. Change in the capacitance value occurs with the change in humidity levels. Its humidity measuring range is from 0 to 100% with 2–5% accuracy. Just 3 jump wires are required to connect this sensor to ESP8266, namely Data, GND and Vin. The bias voltage required for this sensor is 3V DC.

The MQ135 air quality sensor module consists of a sensing element housed inside a steel exoskeleton as shown in Fig. 4. This sensing element is activated due to current through the connecting leads. It will measure the dust density. The first pin is the 3V DC for biasing, second pin is the ground, and the last pin is the data (A0—Analog) connected to NodeMCU.

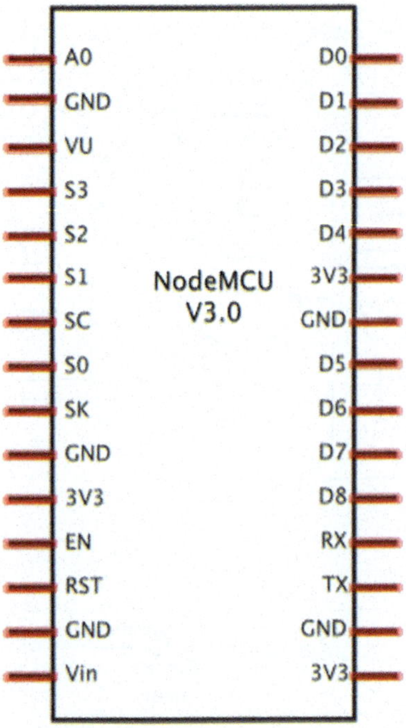

Fig. 2 NodeMCU V3 pin-out diagram

Fig. 3 DHT22 sensor pin-out diagram

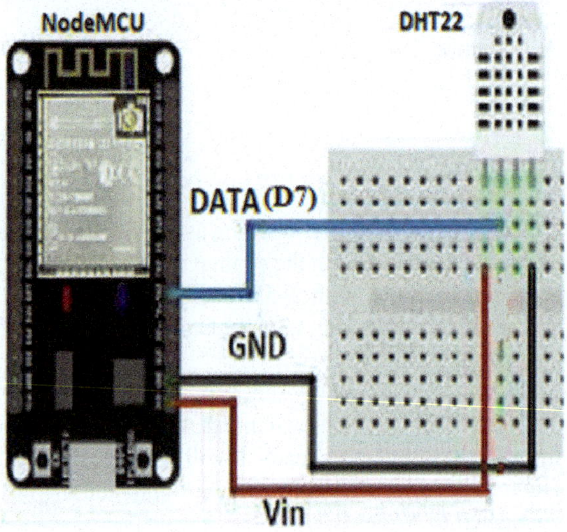

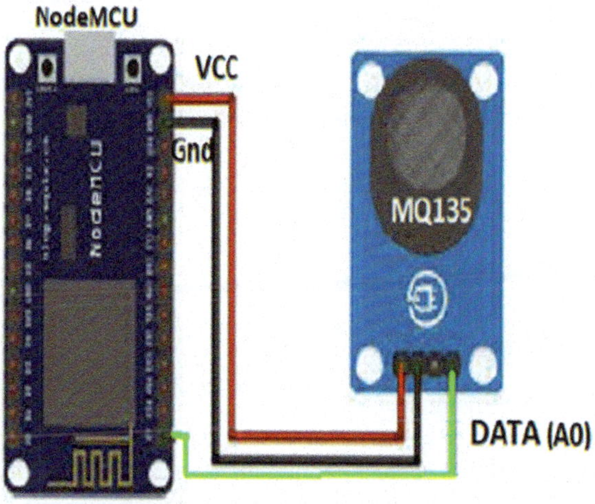

Fig. 4 MQ135 sensor pin-out diagram

3.2 *Software*

Platform IO in the visual studio is used to develop code (C Language) to fetch data from the sensors and send it through WiFi to the server. Unique Sensor feed identifications are configured (/Feed/Temperature/Feed/Humidity/Feed/PPM) in both MCU and Adafruit IO for mapping and data transfer.

Sensor data received from Adafruit IO API is captured in SQLite database tables using Python code. WhatsApp alerts are triggered using the PyWhat library in Python. The user interface application was developed using Python to apply the SVM machine learning algorithm for analysis and plot charts.

4 Machine Learning Implementation

4.1 *Algorithm Selection*

The supervised learning algorithm learns from the training dataset by making predictions iteratively on the data and adjusting till the correct result is met. To choose a suitable algorithm, prediction accuracy is analysed with sample data from the SKlearn machine learning library and checked with textile unit data for multiple supervised learning algorithms like KNN, linear regression and SVM.

KNN assumes the similarity between the new cases and available cases and put the new case into the category that is most similar to the available categories.

Table 2 Prediction scorecard

Data source	Data volume	Prediction scores		
		KNN	Linear regression	SVM
Sklearn/iris	Training = 120 Test =30	0.966	0.946	0.985
Sample spinning unit data	Training = 176 Test = 64	0.625	0.481	0.772

KNN Algorithm:

- Select the number K
- Calculate the Euclidean distance of all the neighbours from K
- Take the K nearest neighbours as per Euclidean distance
- Count the number of data points in each category out of the nearest neighbours
- Assign the new data points to the category which has the maximum neighbour.

The SVM algorithm is used to determine the best line (or) decision boundary called hyperplane between two categories of data. The vectors which are near the hyperplane are called support vectors. The SVM algorithm predicts the classes. One of the classes is identified as 1, and the other is -1.

Linear regression is the statistical regression method used for predictive analysis, showing the relationship between continuous variables. It shows the linear relationship between independent variables and dependent variables. Linear regression models the relationship between the dependent variables and independent variables as $y = mx + c$ where y is the dependent variable, x is the independent variable, m is the slope, and c is a constant.

The algorithms were simulated using Python, and based on the prediction scores as shown in Table 2, SVM was selected for yarn prediction.

4.2 *Algorithm Implementation*

Features used for SVM machine learning are fibre quality index, humidity, cotton dust level, count strength product, imperfection level and selected yarn (Label). Python application is developed and used for machine learning algorithm execution. The data processing workflow for machine learning is shown in Fig. 5. The count strength product and fibre quality index are captured through application user interface by supervisor, temperature, humidity and pollution are captured from sensors periodically. The yarn strength and imperfection level are captured by quality engineers.

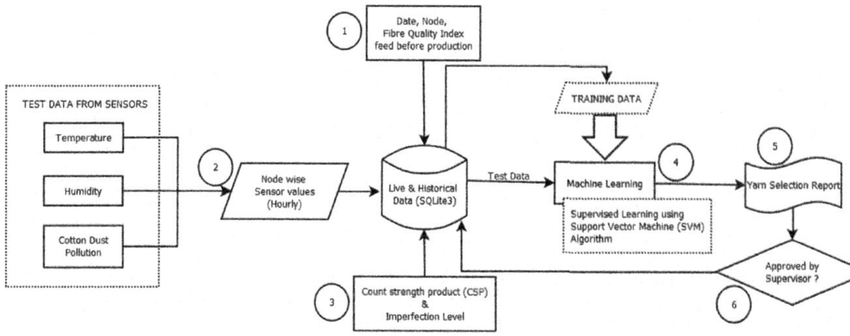

Fig. 5 Machine learning workflow

A user interface developed in Python for ML report criteria. Training data period, test data period and kernel are the key criteria. The kernel can be selected based on the data pattern. It is better to select the linear kernel if the data is more linear, and if it is nonlinear, we can select Poly or Sigmoid.

Two months of processed data (8 records/day from the sensor) were selected as training data, and one month of fresh data was selected as test data. Some of the possible combinations of observations from features are count strength product versus humidity, count strength product versus pollution, imperfection level versus humidity, imperfection level versus pollution and humidity versus pollution.

The linear SVM model was trained with the obtained dataset. During training, the linear SVM creates a hyperplane (decision boundary) given as $w^T x - b = 0$ and then optimizes it over parameters w, b so that it 'best' segregates the two classes in the training data. During prediction, the linear SVM classifies datapoint x_i as class '1' if $w^T x_i - b > 0$ and as the class '0' if $w^T x_i - b < 0$. The chart with the classification pollution (cotton waste) versus humidity is taken for analysis. The SVM prediction score for linear kernel is 0.7. A chart with overall training data and test data is plotted with a hyperplane as shown in Fig. 6 and observed the class pattern.

Since the classes are not linearly separable, the data (x_i) is mapped to a higher dimensional space through a nonlinear transformation $(\varphi(x_i))$ such that the data is linearly separable on the higher dimensional feature space. Then linear SVM is applied on the higher dimensional feature vectors $\varphi(x_i)$, which identifies the maximum-margin hyperplane $w^T \varphi(x) - b = 0$ in the transformed space. This hyperplane obtained in the transformed space is then used to predict the classes. We applied this nonlinear transformation based on the Sigmoid kernel and plotted the chart as shown in Fig. 7.

The SVM prediction score is 0.8 with the available data (highest = 1). The output (Predicted versus Actual) of the Kernel SVM using the Sigmoid kernel is shown in Table 3.

True Negative (TN): Predicted = No, Actual = No

False Positive (FP): Predicted = No, Actual = Yes

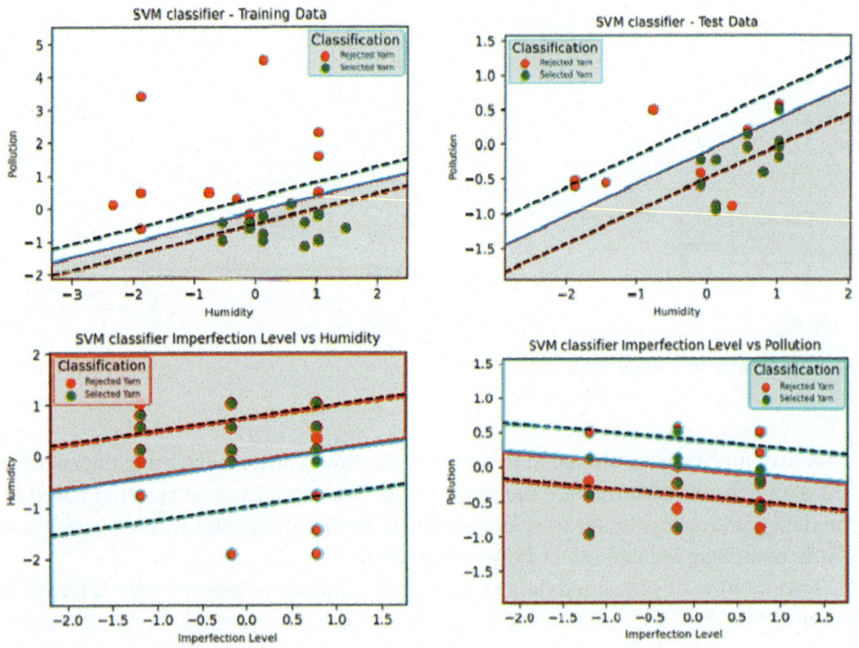

Fig. 6 SVM charts for linear kernel

True Positive (TP): Predicted = Yes, Actual = Yes
False Negative (FN): Predicted = Yes, Actual = No

The features of the applied algorithm can be obtained using the formulas stated below.

Classification Accuracy:

$$(\text{TP} + \text{TN})/(\text{TP} + \text{TN} + \text{FP} + \text{FN}) \times 100 = 80\%$$

Error Rate:

$$(\text{FP} + \text{FN})/(\text{TP} + \text{TN} + \text{FP} + \text{FN}) \times 100 = 19\%$$

Precision:

$$\text{TP}/(\text{TP} + \text{FP}) \times 100 = 81\%$$

Recall:

$$\text{TP}/(\text{TP} + \text{FN}) \times 100 = 86\%$$

This accuracy varies for every kernel. The fibre quality index (FQI) is used to determine the technological value of cotton. It is the product of fibre strength, length and uniformity index.

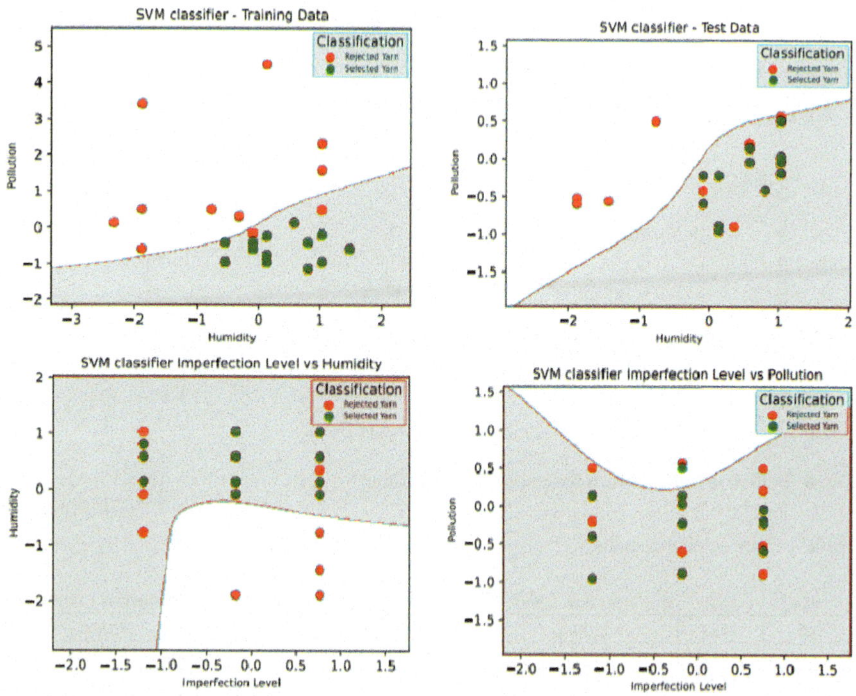

Fig. 7 SVM charts for sigmoid kernel

Table 3 Confusion matrix

		Actual		Total
		No	Yes	
Predicted	No	64	24	88
	Yes	16	104	120
	Total	80	128	

$$\text{FQI} = \text{Fibre strength} \times \text{Length} \times \text{Uniformity index},$$

where *Fibre Strength* is the breaking force of the yarn, *Length* is the mean length by the number of fibres in the largest half by the weight of fibres in a cotton sample, and *Uniformity Index* is the ratio of mean length with the upper half mean length

Selected yarn for the raw materials (FQI) chosen for test data is shown in Fig. 8. FQI used during this production run is 48, 55 and 63. Selected yarn count for FQI 48, 55 and 63 are 40, 42 and 72, respectively. This selection made by the algorithm matches the yarn selection based on expert judgement by the supervisors working in the textile unit.

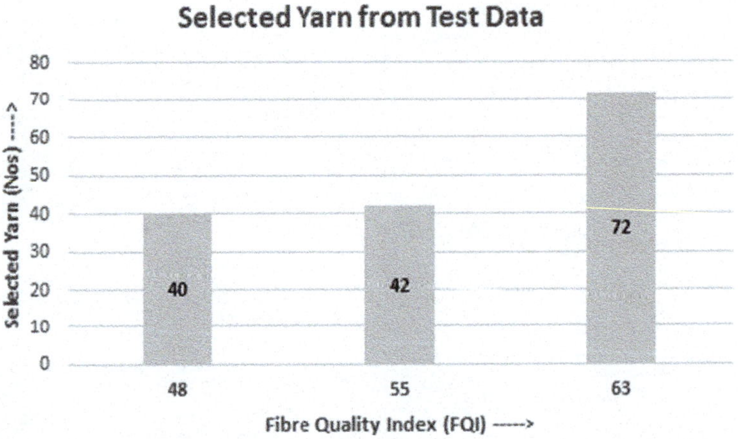

Fig. 8 FQI-wise selected yarn based on SVM algorithm

Table 4 Yarn selection status

FQI	CSP	Imperfection Level	Temperature	Humidity	Cotton Waste	Selected / Rejected
63	1782	1100	41	47	190	Rejected
63	1782	1100	40	45	170	Selected
48	2466	1370	38	60	190	Rejected
55	2285	1618	50	44	161	Rejected
55	2270	1618	45	55	175	Selected

Raw data with sample processed records from the ML algorithm has both the selected and rejected yarns as shown in Table 4. The features that crossed the threshold limits for rejected yarn are highlighted in amber.
Selected yarn samples are sent to the customers for approval and processing thereafter.

5 Live Reports and Alerts

Live trend reports of the different features are published from Adafruit IO, as shown in Fig. 9 from sensors. This data will be refreshed every three seconds.

Three different feeds are maintained in Adafruit to receive data from sensors, and they are:

\feeds\temperature
\feeds\humidity
\feeds\PPM

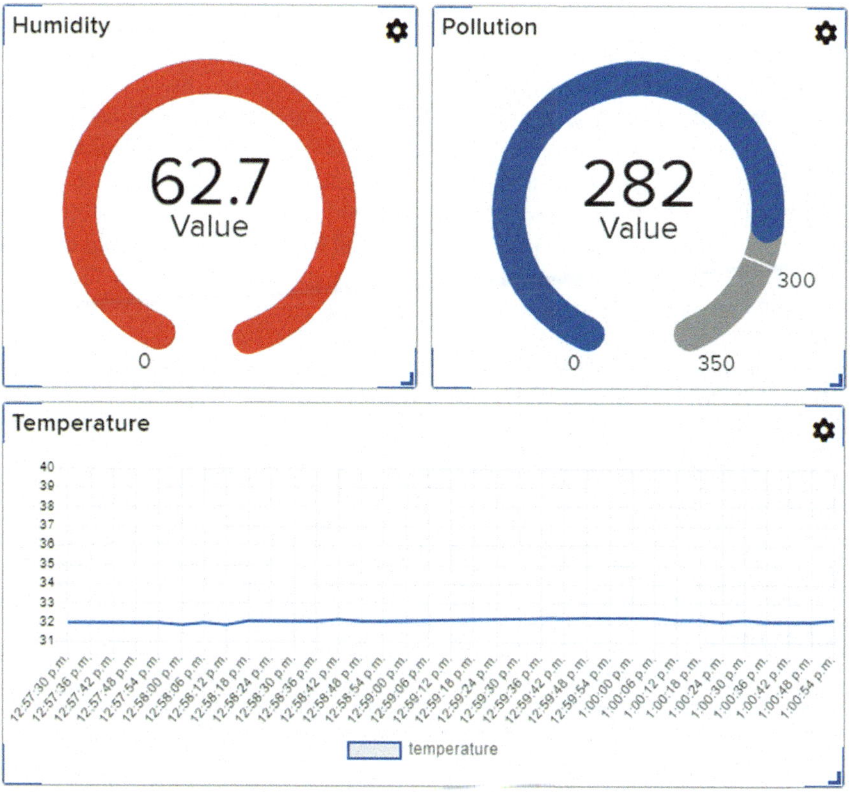

Fig. 9 Adafruit live report

The same feeds are used in Adafruit API to transfer data to the SQLite database. The thresh-holds are configured in Adafruit. The colour code will change automatically to RED when the data exceeds the threshold value.

Threshold-based alert

The threshold for independent sensor attributes is maintained in the database for every FQI. Threshold-based alerts are enabled using the WhatsApp Python library. Alert messages are triggered to the supervisor/support group.

6 Conclusion

In this work, the temperature, humidity and air pollution data were collected with the help of sensors integrated with Node MCU and captured in a database. Threshold-based alerts are triggered when there is a spike in these feature values. Also, live trend report is published using the Adafruit IO cloud platform. The captured data was categorized as training and test data, and the SVM machine learning algorithm was applied to select the best yarn (date-wise). On user approval, the selected records will be updated and used as training data in future predictions. With the currently available training data, the achieved prediction score for test data is 0.8. This accuracy will increase after multiple iterations with more data captured. The proposed system is expected to improve the quality of the yarn and reduce the manual effort of the production support group. The system is expected to be cost-effective and efficient in terms of time and resources. When we propose the same system for the dyeing unit, water quality sensor data will be taken for analysis.

Acknowledgements The authors thankfully acknowledge the Chancellor, Amrita Vishwa Vidyapeetham for motivating us to work on research that has an impact on small/medium-scale cotton textile industries. We are also grateful for getting this opportunity to make our humble efforts in applying the techniques learned to choose high-quality yarn for customers. We are highly indebted to the Raj Ranga Textile in Coimbatore to implement this system and gather data for analysis.

References

1. Saravanakumar G, Baalachandar C, Mohaideen Abdul Kadhar K (2019) Environmental condition monitoring system of textile industry for sustainable development goal. Int J Innov Technol Explor Eng (IJITEE) 9(2S2):986–991
2. Koçyiğit F, Yanıkoğlu E, Yılmaz A, Bayrak M (2009) Effects of power quality on manufacturing costs in textile industry. Sci Res Essays 4(10):1085–1099
3. Sangeetha B, Rajeswari M, Atharsha S, Saranyaa K, Ramya S (2013) Cotton dust level in textile industries and its impact on human. Int J Sci Res Publ 3(4):1–6
4. Lakshmikantha V, Hiriyannagowda A, Manjunath A, Patted A, Basavaiah J, Anthony AA (2021) IoT based smart water quality monitoring system. Global Trans Proc 2(2):181–186
5. Lu Z, He Z, Tran KP, Thomassey S, Zeng X, Hong M (2021) Decision support systems for textile manufacturing process with machine learning. In: Machine learning and probabilistic graphical models for decision support systems. CRC Press, pp 107–123
6. Azevedo J, Ribeiro R, Matos LM, Sousa R, Silva JP, Pilastri A, Cortez P (2022) Predicting yarn breaks in textile fabrics: a machine learning approach. Procedia Comput Sci 207:2301–2310
7. Ribeiro R, Pilastri A, Moura C, Rodrigues F, Rocha R, Morgado J, Cortez P (2020) Predicting physical properties of woven fabrics via automated machine learning and textile design and finishing features. In: Artificial intelligence applications and innovations: 16th IFIP WG 12.5 international conference, AIAI (2020) Neos Marmaras, Greece, June 5–7, 2020, proceedings, Part II 16. Springer, pp 244–255
8. Seçkin M, Seçkin AÇ, Coşkun A (2019) Production fault simulation and forecasting from time series data with machine learning in glove textile industry. J Eng Fibers Fabrics 14:1558925019883462
9. Bhanu K, Jasmine H, Mahadevaswamy H (2020) Machine learning implementation in IoT based intelligent system for agriculture. In: (2020) International conference for emerging technology (INCET). IEEE, pp 1–5
10. Bumrungkun P (2019) Defect detection in textile fabrics with snake active contour and support vector machines. J Phys: Conf Ser 1195(1):012006. IOP Publishing

11. Ji R, Li D, Chen L, Yang W (2010) Classification and identification of foreign fibers in cotton on the basis of a support vector machine. Math Comput Model 51(11–12):1433–1437
12. Doran EC, Sahin C (2020) The prediction of quality characteristics of cotton/elastane core yarn using artificial neural networks and support vector machines. Text Res J 90(13–14):1558–1580
13. Nurwaha D, Wang X (2011) Prediction of rotor spun yarn strength using support vector machines method. Fibers Polym 12:546–549
14. Adithya K, Ajith V, Nibi K, Ramesh MV (2022) Design of IoT and ml enabled framework for water quality monitoring. In: 2022 3rd international conference on electronics and sustainable communication systems (ICESC). IEEE, pp 452–460
15. Sreehari A, Nandakumar VG, Sankar VR, Venugopal A, Dhivvya J (2021) Machine learning based IoT framework for early detection of covid-19 like pandemics. In: 2021 12th international conference on computing communication and networking technologies (ICCCNT). IEEE, pp 1–6
16. Swetha TM, Yogitha T, Hitha MKS, Syamanthika P, Poorna S, Anuraj K (2021) IoT based water management system for crops using conventional machine learning techniques. In: 2021 12th international conference on computing communication and networking technologies (ICC-CNT). IEEE, pp 1–4

Tamper Proof Reconfigurable Speech Encryption Engine

R. Vinoth Raj, M. Ashwin, N. Dineshwar, R. Thaneshwaran, and M. Vijayakumar

Abstract In this paper, Rossler's attractor and the circular shifting method are used to create a hardware-based encryption scheme that can be changed to protect speech communication. In order to ratify the limitations of software-based speech encryption schemes like low throughput, being prone to attacks, poor quality in the deciphered signal, an inadequate key space, and high computational overhead. An enhanced reconfigurable hardware-based speech encryption architecture based on a three-dimensional Rossler's attractor that offers high randomness through highly sensitive initial conditions and control parameters. The keys for the diffusion process are generated by Rossler's attractor, and by using circular shifting, the scrambling has been achieved. The proposed speech encryption scheme is validated by conducting various analyses, such as histogram and spectrogram analyses, correlation analyses, signal-to-noise ratio analyses, and key space analyses. The obtained results show that the chosen approach is extremely safe, offers peer-to-peer encryption, and works well for real-time speech communication.

Keywords Circular shift · Field programmable gate array · Rossler's attractor and speech encryption

R. Vinoth Raj · M. Ashwin · N. Dineshwar · R. Thaneshwaran
Velammal Institute of Technology Panchetti, Chennai, India

M. Vijayakumar
Government College of Engineering, Dharmapuri, India
e-mail: mvk.gct@gcedpi.edu.in

R. Vinoth Raj (✉)
Electronics & Communication Engineering, Saveetha Engineering College, Chennai, India
e-mail: vinothraj@velammalitech.edu.in

S. Bandyopadhyay et al. (eds.), *Intelligent Computing Systems and Applications*,
Lecture Notes in Networks and Systems 1010,
https://doi.org/10.1007/978-981-97-5412-0_5

1 Introduction

Speech encryption is the process of turning speech data into a secure, encrypted format that can only be decoded by authorized recipients. To make sure speech encryption works, security analyses must be done to find and fix possible security risks. Here are some common security analyses for speech encryption: Strength of cryptography: The algorithm used to encrypt speech should be strong enough to resist attacks from hackers and other bad people. Experts in the field should look at the algorithm to make sure it is safe [1]. Key management: The encryption key used to encrypt the voice data should be kept in a secure location to avoid unauthorized access. The procedures of key generation, distribution, storage, and revocation should all be included in key management. Authentication: In order to avoid impersonation and man-in-the-middle attacks, the sender and recipient of the voice data should have their identities verified. Digital certificates and other authentication procedures can be used to achieve authentication. Integrity: It is important to verify that the speech data hasn't been altered or tampered with. Digital signatures or cryptographic hashes can be used to accomplish this. Confidentiality: In order to prevent unauthorized access, the confidentiality of the speech data should be guaranteed. Either symmetric or asymmetric encryption can be used to accomplish this. Replay attacks: To get unauthorized access, replay attacks entail recording encrypted voice data and playing it back afterward. Security studies should take replay threats into account and include countermeasures like time stamping and nonce values. Denial-of-service attacks: By overloading speech encryption services with requests or by bringing about system failures, denial-of-service attacks can be used to interfere with their operation. Security evaluations should take denial-of-service attack prevention methods like load balancing and rate limitation into consideration. These security studies can increase the security and efficacy of speech encryption in protecting private speech data [2].

The authors of [3] developed a secure method for encrypting speech. It makes use of both DNA sequences and chaotic maps. The eight alternative DNA sequence rules are used to a sampled speech signal in this suggested technique to perform inter- and intra-level substitutions. The keys are produced using a variety of chaotic maps, including the Henon, sine, logistic, and tent maps. The analysis of the histogram, spectrum, correlation, peak signal-to-noise ratio, NSCR, and UACI is used to assess this voice encryption method [3].

A dual multichoice DNA computation and self-adaptive bit scrambling audio encryption method was created in [4]. Three distinct keys are employed over the course of three phases in this suggested audio encryption system. Phase 1 employs the SHA512-H key and self-adaptive scrambling (cyclic shift). The second stage makes use of the dynamic DNA coding. DN calculations are performed during the last stage of encryption. The histogram and spectrogram between the plain and ciphered audio recordings were collected, along with key space and key sensitivity evaluations, to confirm the validity of the system. A new audio encryption method that uses chaotic systems and DNA coding to confuse and diffuse audio data was proposed in [5]. This technique uses DNA coding and chaotic systems to encrypt audio. The initial value of the chaotic system is determined by the hash value of the audio, hence rendering the subsequent trajectory of the system unpredictable. The strategies used are key and system initial value generation, obfuscation, and diffusion. A PSNR test, histogram analysis, spectrum analysis, and key space analysis were all completed.

In [6], a safe opto-audio cryptosystem was created using Hartley transforms and XORing masks. This work proposes a secure opto-audio encryption scheme based on the Hartley transform (HT) and XORing masks. Digital one-dimensional (1D) plain audio data is converted into a two-dimensional (2D PAM) plain audio map (PAM), which is then divided into two-dimensional (2D PAB) plain audio blocks (PABs). A mask is created by XORing a single image from a personal image database with each block of 2D PABs in the proposed opto-audio encryption scheme. This image can be viewed as an additional secret key in the opto-audio cryptosystem that is being suggested. The pixels of the 2D PABs are jumbled using a 2D chaotic map called a logistic adjusted sine map (LASM). A FPGA-based voice encryption equipment under the analog voice communication channel was suggested by the authors in [7]. In this article, the speech is divided into frames and the frames' contents are encrypted using FPGA. The frames are then reassembled in time. The test results demonstrate that the device can encrypt normally during various analog phone call types and that the voice quality, voice latency, and encryption effect are all assured. Time delay analysis, voice quality testing, and encryption strength analysis were the tests and analyses that were carried out. The system offers efficient remedies for mitigating speech channel noise, limited bandwidth, and restricted latency, so ensuring the reception of suitable, low-latency, and highly confidential voice signals at the receiving end.

In [8], an Efficient Speech Encryption Engine is created with a Dynamic and Secure Substitution Box (DS2B). Systems with limited resources and IoT devices can employ DS2B-based security. The DS2B has a simple design and reliable encryption. The DS2B was validated by an FPGA voice encryption device. Security was evaluated using the NIST statistical test set, NSCR, MSE, correlation, histogram, and spectrogram. A Xilinx Nexys 4 FPGA test board was utilized for implementation and verification.

The authors of [9] encrypted and decrypted speech and image data. Chaos systems are coordinated by TSK fuzzy brain emotional learning regulators. In chaos-based secure communication, the chaotic main system works as an emitter. It sends out a drive signal that is later encrypted to send the information message. Second, the receiver's recurrent Takagi–Sugeno–Kang (RTFBECAC) fuzzy brain emotional learning cerebellum model articulation controller instructs the slave system to obey the transmitter's master system. Third, the information payload is recovered by decoding the scrambled signal. For safe speech and image transmission, this research offers synchronizing the 4D hyper jerk memristive chaotic system. Online learning laws are established by the Lyapunov stability theorem, which also provides system stability and closure.

The FPGA's speech encryption method [10] makes use of a substitution box and a chaotic map. By assigning speech values to the keys that the Lorentz attractor creates, the S-box is constructed using the Lorentz chaotic attractor. The Xilinx

ISE 14.7 was used to implement the Verilog HDL model of the entire encryption technique. After that, a Virtex-5-xC51 × 50 T Xilinx FPGA is used to implement it. To confirm the proposed speech encryption approach, security evaluations were made using MATLAB. [11] describes the implementation of speech encryption using a Xilinx Virtex II Pro FPGA platform. The Viretx II Pro is used to initially train the ADC/DAC converters such that they sample the analog voice signal at 1 MHz. By selecting a random bit sequence from the sampled values, the speech encryption was carried out. Data is simultaneously transmitted in encrypted form from the DAC to the speakers. The 14-bit ADC output is converted into a 12-bit DAC input by rounding the LSB values.

There are several motivations for using FPGA for speech encryption: (1) High-speed encryption and decryption: Because FPGAs can process information in parallel, they can encrypt and decrypt speech signals quickly. FPGAs can use hardware to implement encryption algorithms, which is often much faster than using software. (2) Latency: Processing that encrypts and decrypts speech signals in real-time applications like voice-over-IP (VoIP) needs to be fast. FPGAs can handle processing quickly because encryption and decryption can be done in the hardware. (3) Customizable: FPGAs can be changed to use certain encryption algorithms and configurations, which can make speech signals more secure. (4) Hardware security: FPGA-based implementations can provide hardware security for speech signals by storing the encryption key and performing encryption operations within the FPGA. This can prevent attackers from accessing the key and the sensitive data. (5) Low power use: FPGAs can implement encryption algorithms in a way that uses less power, so portable speech encryption devices can use less power. (6) Scalability: FPGA-based implementations can be easily scaled up or down depending on the level of security needed and the processing needs of the application [12]. Overall, FPGA-based implementations of speech encryption algorithms can provide a high level of security, low-latency processing, and low-power consumption, making them suitable for a wide range of applications that require secure communication of speech signals. Because keys and encryption methods are available in the software framework itself, we chose FPGA as the implementation platform, whereas software-based security algorithms are insecure. Furthermore, all background applications share the CPU's frequency, which lowers throughput. The software-based security solutions are then executed in a sequential manner, which reduces execution time. The FPGA-based security architecture delivers higher throughput since it includes specialized modules and only provides full frequency to security solutions. The execution time is lowered because it executes concurrently. When utilized as independent hardware, FPGA is conceptualized as a tamper-resistant device.

2 Preliminaries

2.1 Rossler Attractor

A well-known example of a chaotic dynamical system is the Rössler attractor. A chaotic dynamical system is a deterministic system with complex, random behavior that is highly dependent on initial conditions. Nonlinear equations govern chaotic systems, which can show unpredictable, aperiodic behavior that appears random but is actually deterministic. They are modeled in Eqs. (1–3). Small differences in a chaotic system's initial conditions can result in drastically different trajectories over time, a trait known as sensitive dependence on initial conditions. The bifurcation diagram for the Rossler's attractor is shown in Fig. 1. As a result, chaotic systems are difficult to predict over long periods of time and can appear highly random and unpredictable [13, 14].

The Rössler attractor is a three-dimensional chaotic attractor (x, y, and z) that generates random number sequences. The FPGA realization of Rossler's attractor is depicted in Fig. 2. The resulting random numbers have very unpredictable float values. The equations for Rossler's attractor are given as.

The defining equations of the Rössler system are

$$dx'/dt = -y_n - z_n \tag{1}$$

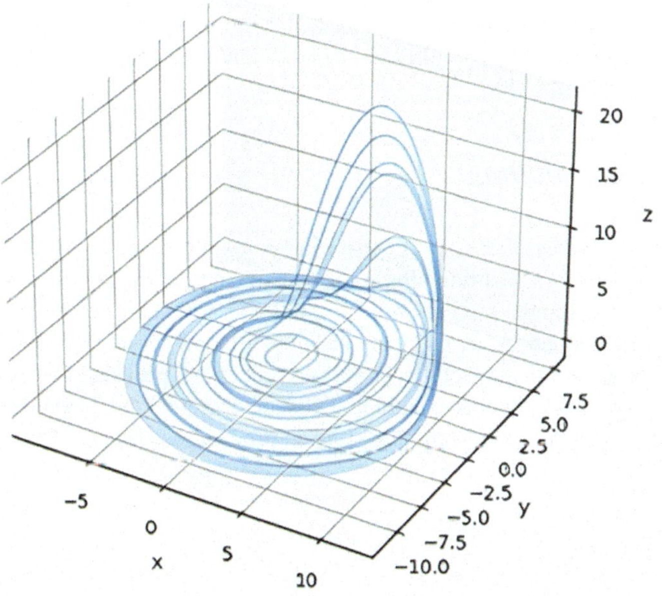

Fig. 1 Bifurcation diagram of Rossler's attractor

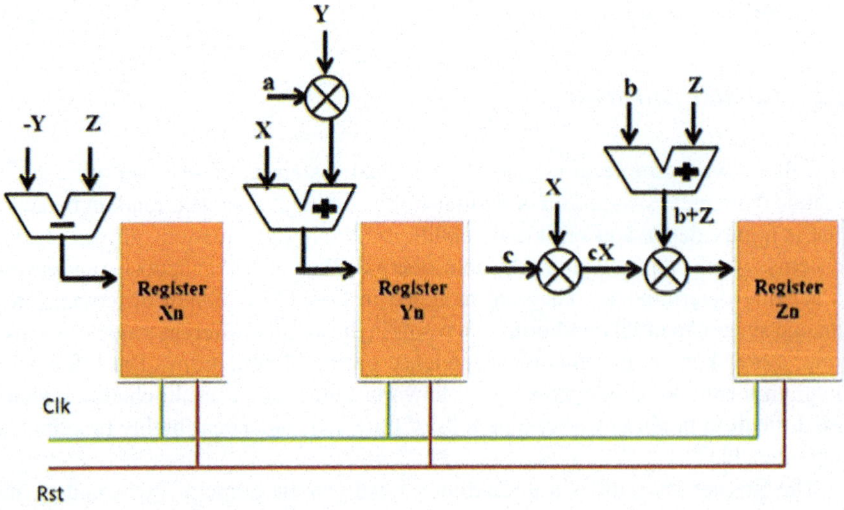

Fig. 2 Hardware realization of Rossler's attractor

$$\mathrm{d}y'/\mathrm{d}t = x_n + a_n * y_n \tag{2}$$

$$\mathrm{d}z'/\mathrm{d}t = b_n + x_n * (x_n - c_n) \tag{3}$$

Rössler evaluated the chaotic attractor with parameters $a_n = 0.2$, $b_n = 0.2$, and $c_n = 5.7$. Though $a_n = 0.1$, $b_n = 0.1$, and $c_n = 14$ have been more frequently used.

2.2 *Circular Shifting*

A circular shift is the process of moving the last entry into the first position and shifting all other entries to the position after it, or by carrying out the opposite operation. A circle shift is a particular kind of cyclic permutation, which is a permutation in and of itself [15]. Bit sequences are routinely permuted in encryption using circular shifts. Unfortunately, despite the fact that almost all processors provide bitwise operations, many programming languages, including C, lack operators, or standard methods for circular shifting. The sample circular shifting procedure is shown in Fig. 3.

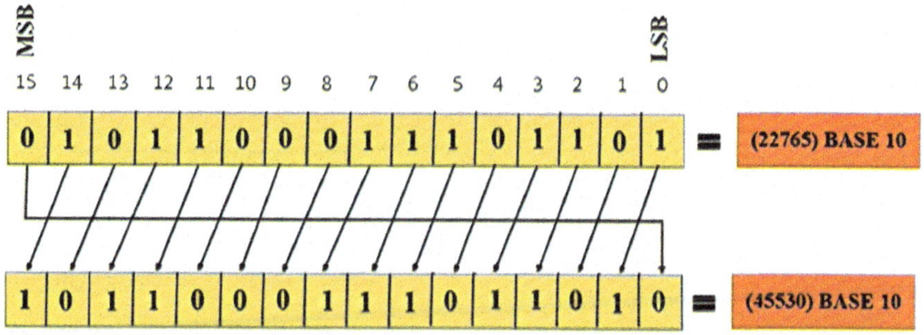

Fig. 3 Circular shifting

3 Proposed Methodology

The proposed speech encryption uses Rossler's attractor as a key generator for the diffusion process, and circular shifting is used to scramble the speech samples. Initially, the keys generated from the Rossler's attractor are digitized by using the IEEE 754 single-precision floating-point converter. The XOR operation is performed between the three planes of the Rossler's attractor to obtain a single 32-bit key. Then the digitized keys are XORed and XNORed with the speech samples. The above-mentioned diffusion methods are selected by using the 4-to-1 multiplexer. Based on the last two Least Significant Bit (LSB) of the speech samples, the XNOR and XOR operations are selected. For values 00 or 11, the XNOR operation is performed; otherwise, the XOR operation is performed. Later, the confusion is achieved on the fly from the circularly shifted values generated by the circular shifter. The address of the block RAM of the FPGA is shuffled. The encrypted speech samples are then obtained from the single-port RAM. ssss The implementation of the on-the-fly confusion technique results in a reduction in the amount of redundant memory required. Figure 4 displays the block diagram of the overall encryption architecture.

3.1 Encryption Procedure

1. Design the Rössler attractor-based diffusion and circular shifting-based confusion algorithms using a Verilog HDL.
2. Use an FPGA IP core, such as an Altera block RAM, to run the Rossler attractor-based diffusion algorithm. The core should be set up to make a random key and use that key to do the nonlinear function and diffusion operations.
3. Implement the circular shifting-based confusion algorithm using an FPGA shift register. The shift register should be configured to shift the bits of the speech signal by a certain number of positions determined by a counter that increments for each frame.

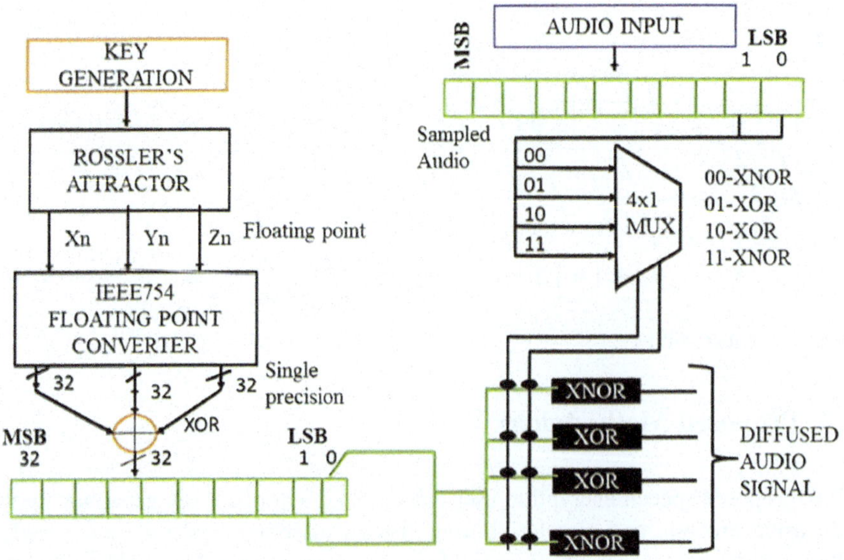

Fig. 4 Overall encryption architecture

4. Divide the speech signal into frames of equal size and store them in the FPGA's memory.
5. For each frame, increment the counter and use it to determine the number of positions to shift the bits of the speech signal using the circular shifting algorithm.

3.2 *Decryption Procedure*

1. Use the FPGA IP core and the key to apply the Rossler attractor-based diffusion to the speech signal that has been moved.
2. Store the encrypted frame back into the FPGA memory.
3. To decrypt the signal, read the encrypted frames from the FPGA memory.
4. For each frame, use the counter to figure out how many positions to move the bits of the encrypted speech signal based on circular shifting.
5. Use the FPGA IP core and the key to do the inverse operations of the Rossler attractor-based diffusion.
6. To acquire the final decrypted voice signal, it is necessary to concatenate the decrypted frame.

4 Results and Discussions

4.1 Security Analyses

A histogram of a speech signal can be used to analyze the statistical distribution of the signal. The histogram can give us information about the frequency of occurrence of different amplitude levels in the speech signal [16]. It can be useful for designing an encryption scheme based on the statistical properties of the signal. Correlation analysis can be used to study the relationship between different parts of a speech signal [17]. The obtained results of peak signal-to-noise ratio (PSNR), correlation, Number of Samples Changing Rate (NSCR), and Unified Average Changing Intensity (UACI) are reported in the Table 1 By analyzing the correlation between different segments of the signal, we can identify patterns that can be used to design an encryption scheme based on the statistical properties of the signal. The equations of SNR, PSNR, and correlation are given in Eqs. 4–6, respectively. A spectrogram can be used to visualize the frequency content of a speech signal over time. This can be useful for designing an encryption scheme that takes the frequency distribution of the signal into account. Figure 5 shows the original, encrypted, and decrypted speech signal.

$$\mathrm{SNR} = 10 * \log 10(P_s/P_n) \tag{4}$$

$$\mathrm{PSNR} = 10 * \log 10((A_m^2)/\mathrm{MSE} \tag{5}$$

P_s = power of the signal, P_n = power of the noise, A_m = maximum amplitude of the signal, MSE = mean squared error. The mean squared error (MSE) is calculated as the mean of the squared deviations between the original signal and the processed or encoded signal.

$$\mathrm{corr} = \left(\sum (x_n - x)(y_n - y)\right)/\mathrm{sqrt}\left(\sum (x_n - x)^2 * \sum (y_n - y)^2\right) \tag{6}$$

Table 1 Security analyses for the proposed speech encryption scheme

S.no	Samples	Correlation	PSNR	SNR	NSCR	UACI
1	Sample_1	0.0026	5.9112	– 11.7462	99.99816	33.3232
2	Sample_2	0.0022	5.9273	– 11.2546	99.99816	33.6764
3	Sample_3	0.0012	5.9548	– 13.3986	100	33.6868
4	Sample_4	0.0016	5.7228	– 12.4231	100	33.8322
5	Sample_5	0.0014	5.7883	– 11.4790	100	33.7672

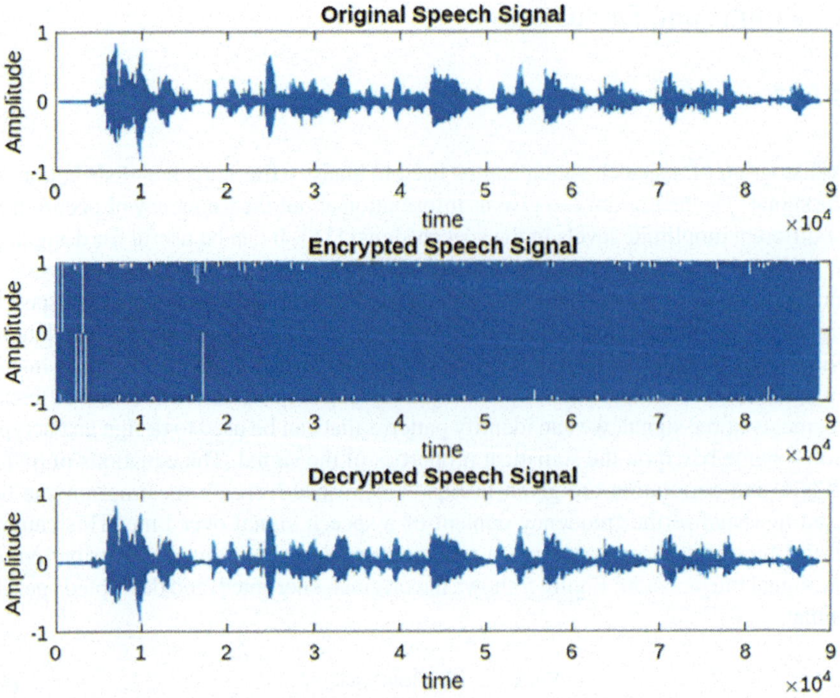

Fig. 5 Plain, ciphered, and reconstructed speech signal

corr = correlation coefficient; x_n = value of the n-th element of x signal; x = mean value of x signal; y_n = value of the n-th element of y signal; y = mean value of y signal. From -1 to 1, the correlation coefficient is available. A value of -1 means perfect negative correlation, a value of 1 means perfect positive correlation, and a value of 0 means there is no relationship.

The signal-to-noise ratio (SNR), which compares the strength of the signal to the strength of the noise that distorts it, is a metric for signal quality. SNR can be used in speech processing to assess how well a manipulated speech signal compares to the plain unprocessed signal. Peak signal-to-noise ratio (PSNR) is a metric for signal quality that compares a signal's maximum possible power to the power of the noise that distorts the signal [18]. When speech encryption is utilized, PSNR can be used to compare the encrypted signal's quality to the original, unencrypted transmission. The histogram of the plain, ciphered, and reconstructed speech signals is displayed in Fig. 6.

NSCR stands for Number of Samples change Rate. It is a measure of the compression performance of a data compression algorithm. In speech encryption, NSCR can be used to evaluate the effectiveness of the encryption algorithm in terms of the amount of compression achieved while maintaining a low level of distortion in the recovered speech signal. UACI stands for "Unified Average Changing Intensity". It

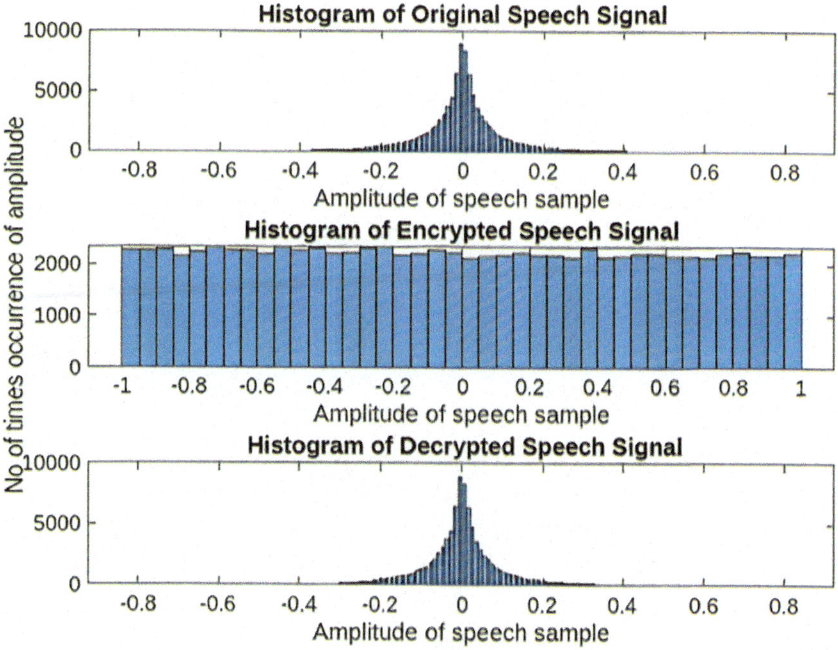

Fig. 6 Histogram of plain, ciphered, and reconstructed speech signal

is a measure of the effectiveness of a speech encryption algorithm in terms of the ability of a human listener to understand the recovered speech signal [19–25]. The equation for NSCR and UACI is depicted in Eqs. 7 and 8, respectively.

$$\text{NSCR} = 100 * ((C_n^2) - 1)/(1 - (C_n^2)) \tag{7}$$

where C_n = Correlation coefficient between the ratio of the plain signal to the processed or encoded signal

$$\text{UACI} = 100 * (1/(2 * n)) * \sum \left| F\left(x', y'\right) - G\left(x', y'\right) \right| /255 \tag{8}$$

n = number of pixels in the image F (x, y) = sample value of the plain speech signal at location x', y, $G\left(x', y'\right)$ = sample value of the processed or ciphered signal at location (x, y). The observed NSCR and UACI results are compared with similar types of encryption schemes. The proposed scheme outperforms the state-of-the-art methods as depicted in Figs. 7 and 8, respectively.

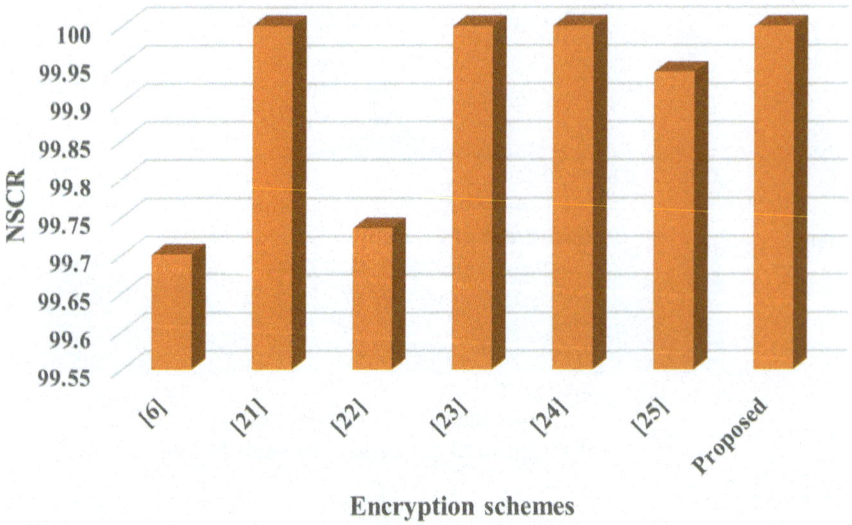

Fig.7 NSCR comparision

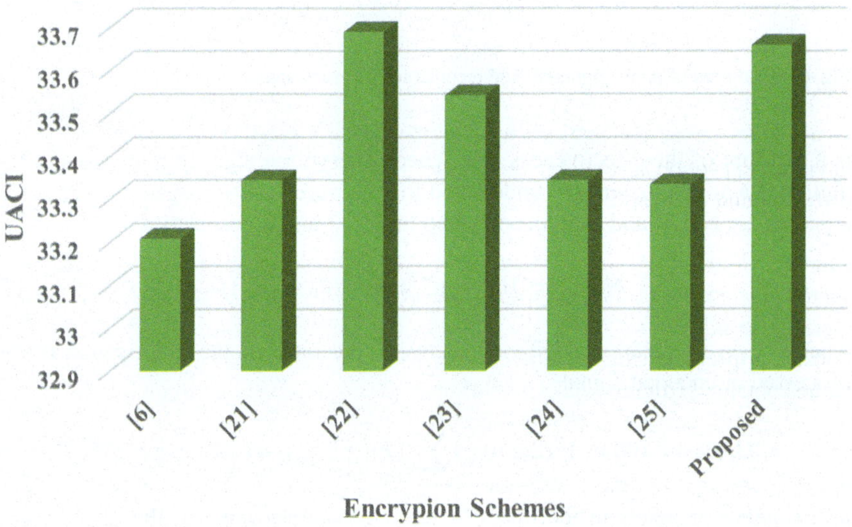

Fig.8 UACI comparison

4.2 *Hardware Analyses*

The proposed speech encryption scheme is designed using the Verilog HDL, simulated in ModelSim 6.3, synthesized in Quartus II 13.0, and implemented in an Intel Cyclone III FPGA. The Intel Quartus II software provides a comprehensive design

Table 2 Resource utilization details

Sample	Total logic elements	Total combinational function	Dedicated logic registers	Total memory (Bits)
Speaker_1	306	235	219	131,072

Table 3 Power dissipation details

Sample	Static power dissipation (mW)	Dynamic power dissipation (mW)	I/O Power dissipation (mW)	Total power dissipation (mW)
Speaker_1	51.75	1.37	18.95	72.07

```
000000  B1 E4 9E 84 32 41 0F 1D 3C C9 FC 72 27 E5 CD
000026  FB 99 DB 04 E1 67 03 E9 C2 65 01 0E BB 49 5F
00004c  8E 82 B8 82 72 81 69 4E 4B 7F CE 40 93 CA 9A
000072  C7 04 43 20 18 06 BA ED FE 07 97 4B 2F D9 E6
000098  42 ED EF E4 FE BF B5 DB D9 A4 AD 05 D5 1B B5
0000be  30 A7 F6 D9 94 24 8F 7A EA 7F 10 5B 03 84 5E
0000e4  40 4C 2C 2B B2 6C 35 BE 93 7C E7 8D 5C F3 64
```

Fig. 9 Encrypted audio samples from single-port RAM

environment for designing, implementing, and verifying the Rössler attractor-based diffusion and circular shifting-based confusion algorithms. By implementing these algorithms on the Altera Cyclone III FPGA, the encryption and decryption processes can be performed with low-latency and high throughput, providing a high level of security for speech data. The results obtained from resource utilization and power dissipation are given in Tables 2 and 3. The proposed speech encryption architecture consumes minimal resources and dissipates less power. The encrypted speech sample values obtained from the single-port RAM are shown in Fig. 9.

5 Conclusion

To address the limitations of software-based speech encryption schemes, a reconfigurable hardware-based speech encryption scheme was designed using the Verilog HDL, simulated using ModelSim, synthesized in Quartus IDE II 13.0, and implemented in the Cyclone III EP3C16F484C6 FPGA. The proposed speech encryption scheme utilizes fewer resources with minimal power dissipation. Rossler's attractor for diffusion, an IEEE 754 floating-point converter for key digitization, and on-the-fly confusion from circular shifting sequences that shuffle the Block RAM address of the Cyclone III FPGA provide high speech signal security. Military communications, business transactions, and healthcare applications can employ this strong encryption approach. The Rössler attractor-based diffusion technique on an FPGA allows real-time encryption and decryption with minimal latency and great throughput. Hardware acceleration secures sensitive data by storing the key and encryption procedures in the FPGA. For real-time speech encryption and decryption, the Altera Cyclone III FPGA is cost-effective and low-power.

References

1. Tolba MF, Sayed WS, Radwan AG, Abd-El-Hafiz SK (2018) FPGA realization of speech encryption based on modified chaotic logistic map. In: 2018 IEEE international conference on industrial technology (ICIT), Lyon, France, pp 1412–1417. https://doi.org/10.1109/ICIT.2018.8352387
2. Tolba MF, Sayed WS, Radwan AG, Abd-El-Hafiz SK, Soliman AM (2018) Permutation-only FPGA realization of real-time speech encryption. In: 2018 25th IEEE international conference on electronics, circuits and systems (ICECS), Bordeaux, France, pp 93–96. https://doi.org/10.1109/ICECS.2018.8618006
3. Nagakrishnan R, Revathi A (2020) A robust cryptosystem to enhance the security in speech-based person authentication. Multimed Tools Appl. https://doi.org/10.1007/s11042-020-08846-1
4. Abdelfatah RI (2020) Audio encryption scheme using self-adaptive bit scrambling and two multi chaotic-based dynamic DNA computations. IEEE Access 8:69894–69907. https://doi.org/10.1109/ACCESS.2020.2987197
5. Abd El-Maksoud AJ, Abd El-Kader AA, Hassan BG, Rihan NG, Tolba MF, Said LA, Radwan AG, Abu-Elyazeed MF (2019) FPGA implementation of sound encryption system based on fractional-order chaotic systems. Microelectron J 90:323–335. https://doi.org/10.1016/j.mejo.2019.05.005
6. Faragallah OS, El-Sayed HS (2021) Secure opto-audio cryptosystem using XORing mask and Hartley transform. IEEE Access 9:25437–25449. https://doi.org/10.1109/ACCESS.2021.3055738
7. Ge X, Sun G, Zheng B, Nan R (2021) FPGA-based voice encryption equipment under the analog voice communication channel. Information 12(11):456. https://doi.org/10.3390/info12110456
8. Tolba MF, Saleh H, Salami YA, Al-Qutayri M, Mohammad B (2021) DS2B: dynamic and secure substitution box for efficient speech encryption engine. IEEE Access 9:93902–93915. https://doi.org/10.1109/ACCESS.2021.3093247
9. Lin C-M, Pham D-H, Huynh T-T (2022) Encryption and decryption of audio signal and image secure communications using chaotic system synchronization control by TSK fuzzy brain emotional learning controllers. IEEE Trans Cybernet 52(12):13684–13698. https://doi.org/10.1109/TCYB.2021.3134245
10. Yassin HM, et al (2019) Speech Encryption on FPGA Using a Chaotic Generator and S-Box Table. In: 2019 fourth international conference on advances in computational tools for engineering applications (ACTEA), Beirut, Lebanon, pp 1–4. https://doi.org/10.1109/ACTEA.2019.8851086
11. Edwards C. FPGA-based speech encrypting and decrypting embeddedsystem. https://www.techdesignforums.com/practice/technique/fpga-based-speech-encrypting-and-decrypting-embedded-system
12. Raj V, Janakiraman S, Amirtharajan R (2021) Optimal concurrency on FPGA for lightweight medical image encryption. J Intell Fuzzy Syst 40(6):10385–10400

13. Raj V, Janakiraman S, Amirtharajan R (2023) Reconfigurable color medical image encryptor using hardware accelerated chao(s)-box triplets. J Real-Time Image Proc 20:27. https://doi.org/10.1007/s11554-023-01278-8
14. Ibrahim KM, Jamal RK, Ali FH. Chaotic behaviour of the Rossler model and its analysis by using bifurcations of limit cycles and chaotic attractors. https://doi.org/10.1088/1742-6596/1003/1/01209
15. Pradhan D, Som S, Rana A (2020) Cryptography encryption technique using circular bit rotation in binary field. In: 2020 8th international conference on reliability, infocom technologies and optimization (trends and future directions) (ICRITO), Noida, India, pp 815–818. https://doi.org/10.1109/ICRITO48877.2020.9197845
16. Wang X, Su Y (2020) An audio encryption algorithm based on DNA coding and chaotic system. IEEE Access 8:9260–9270. https://doi.org/10.1109/ACCESS.2019.2963329
17. Stoyanov B, Ivanova T (2021) Novel implementation of audio encryption using pseudorandom byte generator. Appl Sci 11:10190. https://doi.org/10.3390/app112110190
18. Li J, Fu T, Fu C, Han L (2022) A novel image encryption algorithm based on voice key and chaotic map. Appl Sci 12:5452. https://doi.org/10.3390/app12115452
19. Abdallah H, Meshoul S (2022) A multilayered audio signal encryption approach for secure voice communication. Electronics 12:2. https://doi.org/10.3390/electronics12010002
20. Mosa E, Messiha N, Zahran O, El-Samie F (2011) Chaotic encryption of speech signals. Int J Speech Technol 14:285–296. https://doi.org/10.1007/s10772-011-9103-7
21. Khalid I, Shah T, Almarhabi KA, Shah D, Asif M, Ashraf MU (2022) The SPN network for digital audio data based on elliptic curve over a finite field. IEEE Access 10:127939–127955. https://doi.org/10.1109/ACCESS.2022.3226322
22. Khoirom MS, Laiphrakpam DS, Tuithung T (2021) Audio encryption using ameliorated ElGamal public key encryption over finite field. Wirel Pers Commun 117(2):809–823
23. Naskar PK, Bhattacharyya S, Chaudhuri A (2021) An audio encryption based on distinct key blocks along with PWLCM and ECA. Nonlinear Dyn 103(2):2019–2042
24. Farsana FJ, Devi VR, Gopakumar K (2020) 'An audio encryption scheme based on fast Walsh Hadamard transform and mixed chaotic keystreams. Appl Comput Inform. https://doi.org/10.1016/j.aci.2019.10.001
25. Sathiyamurthi P, Ramakrishnan S (2020) Speech encryption algorithm using FFT and 3D-Lorenz–logistic chaotic map. Multimed Tools Appl 79:17817–17835. https://doi.org/10.1007/s11042-020-08729-5

Design of LLC Resonant Converter for EV Charging Using Neural Network-Based PI Controller

Royyuru Manasa Lakshmi and K. R. M. Vijaya Chandrakala

Abstract The proposed work presents the design and modeling of an LLC converter for electric vehicle battery charging applications, specifically using a 12 V lead-acid battery output. The LLC converter has opted for its soft switching characteristics, ability to produce smooth waveforms, high power density, low electromagnetic interference, and high efficiency, which are desirable for electric vehicle battery charging. The design process includes establishing the suitable transformer turns ratio and selecting resonant tank components such as the resonant inductor, resonant capacitor, and magnetizing inductor. These choices are critical for achieving optimal performance in the LLC converter. Furthermore, this work explores the implementation of closed loop control using a neural network controller. The advantages of neural networks, such as improved control performance and adaptability, are discussed. MATLAB/SIMULINK software is utilized to simulate the LLC converter, and the results demonstrate the benefits of employing a neural network-based controller.

Keywords Electric vehicle · LLC resonant converter · Neural network controller

1 Introduction

Recent advancements in power electronics and battery technology, however, have resulted in the resurrection of electric vehicles, owing to their environmentally beneficial and emission-free operation, cheaper maintenance costs, smooth and silent running, and simple charging alternatives [1, 2]. Developing charging infrastructure that matches the convenience and accessibility of typical gasoline stations remains a key technical problem for electric vehicles. To overcome this obstacle, a dense

R. M. Lakshmi · K. R. M. V. Chandrakala (✉)
Department of Electrical and Electronics Engineering, Amrita School of Engineering, Amrita Vishwa Vidyapeetham, Coimbatore, India
e-mail: krm_vijaya@cb.amrita.edu

R. M. Lakshmi
e-mail: cb.en.p2we21015@cb.students.amrita.edu

S. Bandyopadhyay et al. (eds.), *Intelligent Computing Systems and Applications*, Lecture Notes in Networks and Systems 1010,
https://doi.org/10.1007/978-981-97-5412-0_6

network of charging stations, advances in fast-charging technology, and the introduction of standardized charging protocols are required to support the broad adoption of electric vehicles [3, 4]. To address this issue, power electronics are utilized in charging applications for AC to DC or DC to DC conversion [5, 6]. Non-isolated and isolated converters transform and manage energy efficiently in power electronic automobiles [7, 8]. In general, the above converters tend to operate at low frequencies and have high switching losses, especially at high conversion ratios. Additionally, the converter volume increases with the power level. Overcoming these difficulties resonant soft switching converters such as LLC converters are gaining popularity due to their good performance and minimal EMI [9]. This makes it perfect for applications where the input voltage might vary greatly, such as server power supplies, LED lights, and renewable energy systems [10, 11]. This converter achieves zero voltage and current switching by using a resonant tank circuit made up of a capacitor, an inductor, and a transformer. As a result, the design process is made simpler, and fewer additional EMI filtering components are required. Compared to conventional hard-switching converters, this enables lower switching losses and greater efficiency [12, 13]. Its pros over traditional hard-switching converters such as the buck, boost, and flyback converters have made it a preferred choice in many high-performance applications [14, 15]. Control for power converters has become equally crucial for optimal efficiency as power converter design. While traditional methods for voltage or power control, such as PID and SM, are still employed, recent advances in machine learning have made data collecting and application easier. Because of their potential to learn and improve accuracy through training data, artificial neural networks (ANNs) have grown in popularity. ANNs are made up of interconnected nodes that process input data and modify weights and biases during training, making them useful in a variety of domains such as image and audio recognition, natural language processing, and predictive analytics. Artificial Neural Networks (ANNs) are favored for power converter voltage regulation due to their capacity for nonlinear handling, adaptability to varied conditions, parameter optimization, and effective generalization from training data. ANNs' flexible architecture and efficient memory usage offer advantages over specialized Convolutional Neural Network (CNN) which is used for solving computer-related vision problems and recurrent neural network (RNN) for speech recognition and natural language processing where they may involve complex memory requirements [16]. The main focus of the work is to look at the significance of ANNs in power electronic converters and contribute to research in this area. The configuration of LLC and its detailed design without and with NN controller is discussed in Sect. 2 and Sect. 3 discusses the results obtained.

2 LLC Resonant (Half-Bridge) Converter

The three-element resonant converter is an improved circuit of the series–parallel resonant converter that is suitable for front-end DC–DC applications. This configuration of inductor-inductor-capacitor (LLC) resonant converter shown in Fig. 1

consists of four parts. They are (a) Switching bridge. Converters DC to square AC and be half bridge for less than one-kilo watt and full bridge for more than one-kilo watt, (b) Resonant tank. Filters input square wave and eliminate the harmonics outputting the sine wave and eliminate the harmonics outputting the sine wave of fundamental frequency which goes into the transformer and consists series resonant capacitor (*Cr*), series resonant inductor (*Lr*), and a parallel resonant inductor (*Lm*), (c) Transformer. The voltage is scaled down to an appropriate level through scaling ratio n and isolating the input and output of the converter, (d) Rectifier. Converters reduced sine wave back to DC signal for the output [17, 18].

To comply with specifications, the converter's operating design calls for operation in the zero-voltage region. The total gain of the circuit is defined by transformer gain and resonant tank gain. As the transformer turns are fixed, only the resonant tank gain is considered. Three main parameters influence the resonant tank response. They are as follows: (a) the load, as expressed by the quality factor, (b) the normalized frequency, which is the ratio between the switching frequency of power switches and the tank's resonant frequency, and (c) the normalized inductance, which is the relationship between the resonant and magnetizing inductor. Resonant tank gain is given by:

$$M_g(Q, L_n, F_n) = V_o/V_{in} = \frac{(L_n - 1).F_n^2}{\sqrt{(F_n - 1)^2 + ((F_n - 1)^2.(L_n - 1)^2.Q^2)}} \quad (1)$$

where M_g is the gain of the resonant tank which is dependent on the quality factor Q, normalized inductance L_n, and normalized inductance F_n.

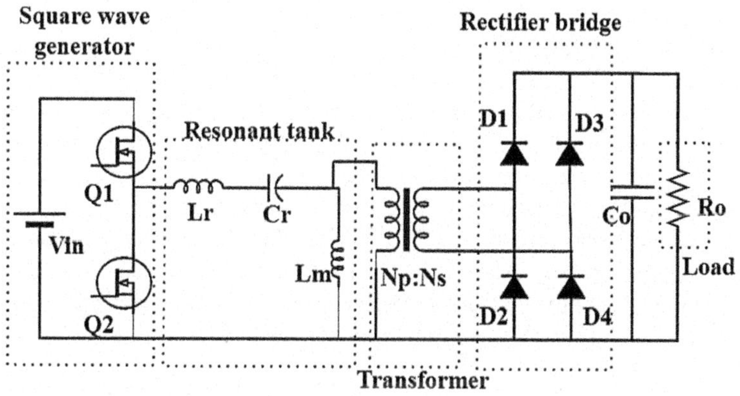

Fig. 1 Configuration of LLC resonant converter

2.1 Designing of LLC Converter

The steps listed below can be used to design the LLC converter [19, 20]. Table 1 displays the converter's specifications for input and output.

Step 1: Defining the Operating Boundaries of Converter

From the specified minimum and maximum input and output voltage to calculate the minimum and maximum gain and also calculate the transformer turns ratio (n).

$$n = V_o/V_{innom} = 16 \tag{2}$$

where V_o is the output voltage and V_{innom} is the nominal input voltage.

$$M_{gmin} = V_{inmax}/V_{innom} = 0.9364 \tag{3}$$

$$M_{gmax} = V_{inmin}/V_{innom} = 1.0746 \tag{4}$$

where M_{gmin} and M_{gmax} are minimum gain and maximum gain respectively.

Step 2: Selecting the Value of the Quality Factor (Q)and Normalized Inductance (L_n).

The Q and L_n must be chosen so that they satisfy the converter's boundary requirements. To find the precise values of Q and L_n, a graph is plotted between them. From the graph, the values of Q and L_n are selected in such a way that they meet minimum and maximum gain requirements. The following three figures show gain versus switching frequency plot for three different values of normalized inductance, respectively, $L_n = 3$, $L_n = 6$, $L_n = 15$.

From Fig. 2a–c, it is observed that the normalized inductance ratio $L_n = 3$ has narrower bandwidth which becomes difficult for the converter to operate at lighter loads and the normalized inductance ratio normalized inductance ratio $L_n = 15$ have larger bandwidth which becomes difficult for the converter to operate at heavy loads. The normalized inductance ratio $L_n = 6$ assures that it can produce higher gains at a wider range of loads it means that it helps the converter to have an appropriate frequency range and flexible regulation. Hence value of normalized inductance ratio is chosen as $L_n = 6$. High value quality factor refers to heavy load and low value

Table 1 Design specifications of LLC converter

Parameter	Designated	Value
Input voltage	$V_{inmin} - V_{inmax}$	360–410 V
Nominal input voltage	V_{innom}	380 V
Output voltage	V_o	12 V
Output current	I_o	7.2 Ah

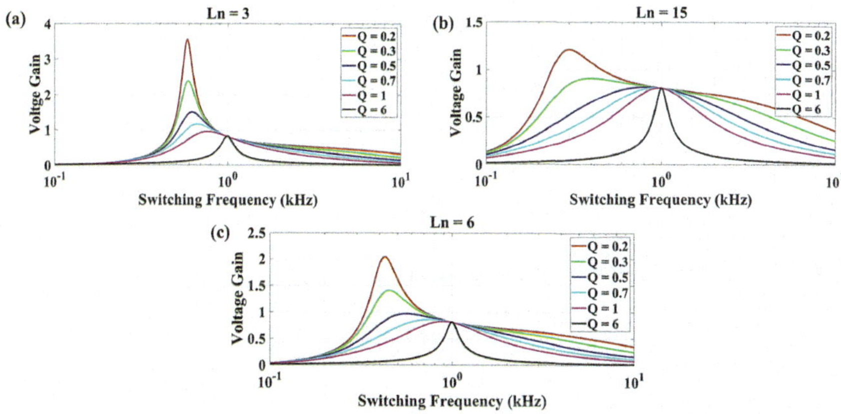

Fig. 2 Gain versus switching frequency of LLC converter under varying inductance ratio. **a** Gain versus switching frequency plots for normalized inductance ratios $L_n = 3$, **b** Gain versus switching frequency plots for normalized inductance ratios $L_n = 15$, **c** Gain versus switching frequency plots for normalized inductance ratios $L_n = 6$

refers to light load. By observing the plot of $L_n = 6$, the value of quality factor $Q = 0.38$ chosen is meeting the minimum and maximum gain requirements of the converter which helps the converter to under both light load and heavy loads.

Step 3: Obtaining the Frequency Range for the Converter.

Initially, the normalized frequency value is estimated by substituting the values of quality factor, magnetizing inductance ratio and maximum gain value in (1).

$$1.0796(0.38,6, F_n) = \frac{(L_n - 1).F_n^2}{\sqrt{(F_n - 1)^2 + ((F_n - 1)^2.(L_n - 1)^2.Q^2)}} = 1.314 \quad (5)$$

The value of F_n obtained is 1.314 which is higher than the maximum gain M_{gmax} = 1.0796 of the converter. Now the calculation of switching frequency is carried out by assuming the value of resonant frequency as 100kHz using (6).

$$F_n = f_s/f_r = 131\text{kHz} \quad (6)$$

where $F_n = 1.314$, f_r is resonant frequency 100kHz and f_s is switching frequency 131kHz.

Step 4: Designing the Resonant Tank Elements Values.

1. The values of ac load resistance is obtained by from (7).

$$R_{ac} = \frac{8}{\pi^2} \bullet \frac{N_1^2}{N_2^2} \cdot \frac{V_o^2}{P_o} = 354.84\Omega \quad (7)$$

where N_1 and N_2 are primary and secondary transformer turns and V_o is voltage output and P_o is the power output.

2. The values of L_r and C_r is calculated using (8).

$$Q = \sqrt{(L_r/C_r)/R_{ac}} \tag{8}$$

where L_r is resonant inductor 209.19 μH, C_r is resonant capacitor 12nF and R_{ac} equivalent ac resistance.

3. The value of L_m is calculated using (9).

$$L_n = L_m/L_r \tag{9}$$

where L_n is magnetizing inductance ratio 6, L_r is resonant inductance 209.19 μH and L_m is magnetizing inductance 1255 μH.

4. The second resonant frequency is obtained using (10).

$$f_m = 1/2\pi\sqrt{L_r + (L_m \cdot C_r)} \tag{10}$$

where f_m is the second resonant frequency, magnetizing inductor is L_m. By substituting the values of the components, the value of f_m obtained is 81kHz. The converter designed parameters are given in Table 2.

Table 2 Resonant tank parameters for LLC resonant converter

Parameter	Designated	Value
Minimum gain–maximum gain	$M_{gmin} - M_{gmax}$	0.9364–1.0746
Transformer turns ratio	n	16
Normalized inductance ratio	L_n	6
Quality factor	Q	0.38
AC load resistance	R_{ac}	354.84 Ω
Load resistance	R_o	1.66 Ω
Resonant inductor	L_r	209.19 μH
Resonant capacitor	C_r	12 nF
Magnetizing inductor	L_m	1255 μH
DC link capacitor	C_o	100 μF
Minimum and maximum switching frequency	$f_{smin} - f_{smax}$	81–131 kHz
Resonant frequency	f_r	100 kHz

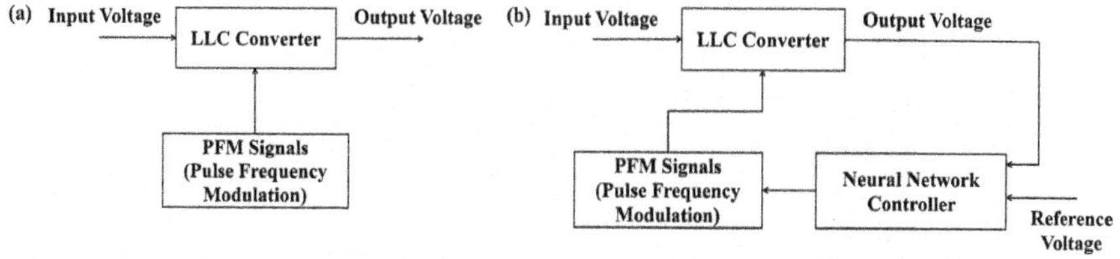

Fig. 3 LLC converter block diagram. **a** Open block diagram of LLC Converter; **b** Closed loop block diagram of LLC converter using neural network controller

2.2 Neural Network (NN)-Based Voltage Regulator for Fast EV Battery Charging

Figure 3a depicts block layout of with open loop having an LLC converter and pulse frequency modulation generator that generates PFM signals to the converter, whereas Fig. 3b depicts the block layout of closed loop system with neural network-based controller that is interconnected to PFM generator to produce signals.

The neural network predictive controller employs a nonlinear plant neural network model to forecast its performance and determine the optimized control input. The first step in model predictive control is to identify the neural network plant model using system identification. The controller then uses the plant model to anticipate future performance. The network can be trained using algorithm such as multi-layer shallow networks and back propagation.

A neural network with numerous layers is referred to as a multi-layer perception. A multi-layer perceptron contains one input layer with one neuron (or node) for each input, one output layer with one node for each output, and it can have any number of hidden levels with any number of nodes.

Backpropagation is a technique used to fine-tune neural network weights based on the error rate attained in the previous epoch (i.e., iteration). This method calculates the gradient of a loss function w.r.t to all of the network's weights. It computes one layer at a time and does it efficiently. It computes the gradient but makes no mention of how it will be used. It generalizes the delta rule computation. Using (11), calculate the error in the output. Returns to the hidden layer from output layer to alter the weights in order to lessen the error.

$$\text{Error} = \text{Actual output} - \text{Desired output} \tag{11}$$

Figure 4a gives the flow chart of overall procedure taking place and Fig. 4b gives the structure of NN network of the system. The system is designed with 10 hidden layers, which helps in capturing complex relationships between the input and output variables. The sampling interval for the system is set as 0.005, indicating that the input and output data is sampled at this interval. To train the system, a training process is carried out for 1000 epochs. During training, the ANN adjusts its weights and biases iteratively to reduce the difference between the predicted output and

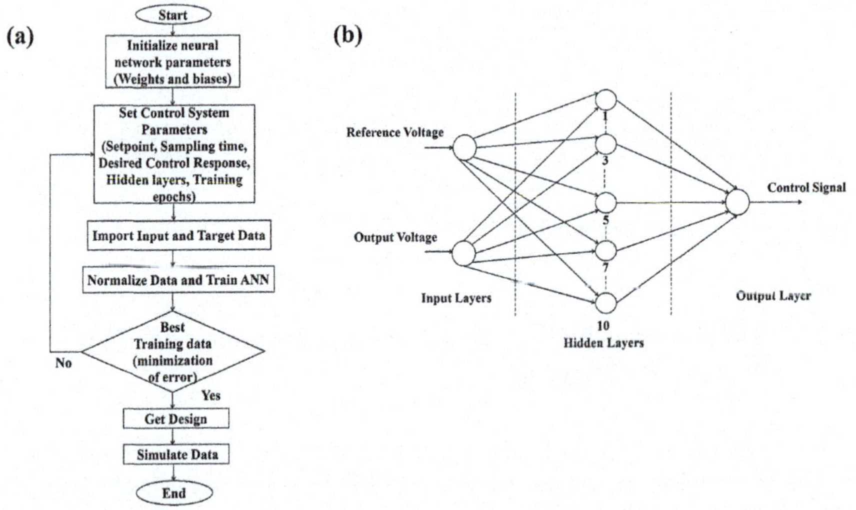

Fig. 4 Neural network configuration. **a** Flowchart of NN controller system, **b** NN system structure

the desired output. This process involves forward propagation of input through the network, calculation of errors and backward propagation of errors using techniques like backpropagation. The control signal generated by the system is responsible for generating switching pulses to the MOSFET, which controls the operation of the system. Once the system is trained, the best set of weights and biases is determined by the ANN itself based on the training data and the desired output. This configuration represents the optimized parameters for the network, allowing it to make accurate predictions. Finally, the overall model is simulated using the trained ANN.

The neural network training window indicates that the trained network required multiple iterations to attain accurate output and also provides information on the network's training state and performance.

Figure 5a depicts the neural network controller block's neural network training performance plot, which provides insight into the best validation performance attained. Figure 5b, on the other hand, shows information about the controller block's gradient performance and validation tests.

3 Results

The nominal input voltage is set to 380 V, and the acceptable input voltage range is specified as 360–410 V. The converter is required to operate within the frequency range of 81–131 kHz. For the closed loop performance of the converter, the neural controller block's inputs are the reference and plant outputs, and it has a control signal output. The controller reference input data is input voltage range and required

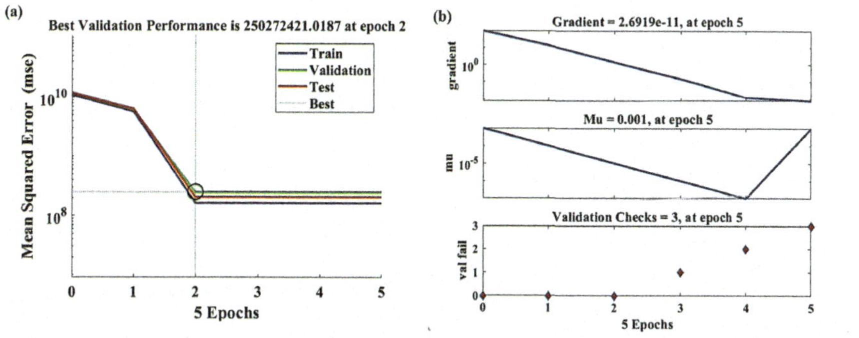

Fig. 5 Performance plot of NN. **a** Training network performance plot of NN controller block, **b** Testing data performance of neural network controller block

output voltage. Preferably the input voltage is considered as 380 V and the control signal is considered based on plant output data which is operating in frequency range as mentioned above. This control signal output is coupled to the simulink developed pulse frequency modulation subsystem (which provides signals to turn on the switches for the converter). The converter is linked to the battery, and its operational performance was evaluated by analyzing the switching responses of the converter components such as MOSFETs and didoes, the magnetizing inductor and resonant inductor current flow. The switching voltage and current waveforms of the MOSFET employed in the converter are shown in Fig. 6a, b. The voltage across the switch appears as a 380 V alternating current (AC) square wave, while the current through the MOSFET is around 2.5 A. The capacity of the converter to step down the input voltage results in the input voltage exceeding the voltage output and the input current being less than the output current. The LLC resonant converter's characteristic operation can be linked to its function as a voltage step-down converter. In the full bridge rectifier circuit, the diode plays a crucial role in converting alternating current (AC) voltage to direct current (DC) voltage. Figure 6c depicts the diode's switching voltage waveform, which appears as a sine wave with an amplitude of about 12 V. This value corresponds to the secondary side's desired output voltage requirement. Figure 6d also shows the current waveforms through the resonant and magnetizing inductors.

The circuit's magnetizing inductance stores energy during the ON-time of the switches and then subsequently releases it during the OFF-time. During each switching cycle, this energy transfer mechanism causes the current in the magnetizing inductance to rise and fall, resulting in a dynamic waveform. The approximate value of magnetizing inductance current is 0.4 A. The resonant inductance current, on the other hand, runs within the resonant tank circuit and allows energy oscillation between the capacitor and the resonant inductor. As a result, the current waveform in this situation seems sinusoidal with an amplitude of 3.2 A. Figure 7 shows an assessment of the battery output voltage with and without neural network control and Fig. 8 highlights the State of Charge (%Soc) of the battery. Table 3 gives the information

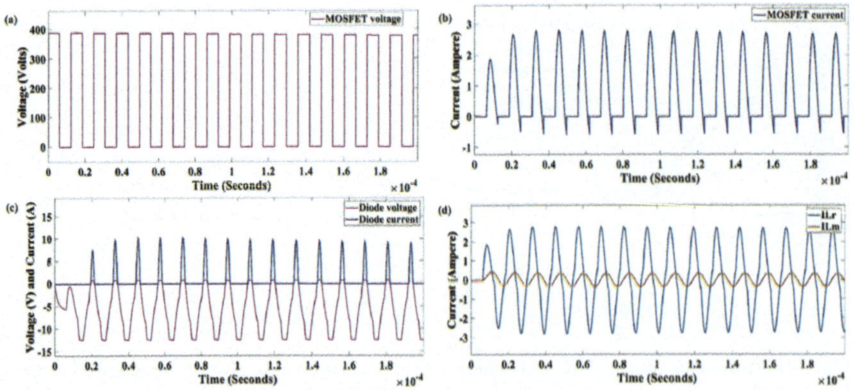

Fig. 6 Responses of proposed LLC converter. **a** Switching pulses of LLC resonant converter (Voltage), **b** Switching pulses of LLC resonant converter (Current), **c** Voltage and current across rectifier bridge of LLC resonant converter, **d** Current through resonant inductor and magnetizing inductor in LLC resonant converter

about peak overshoot and settling time of the battery voltage with and without neural network controller. Effective output voltage regulation is critical for reducing load damage in power electronics converters. Without a controller, the open loop voltage may exhibit unwanted ripples and peaks, rising as high as 13.2 V before settling at a steady-state value of around 11.2 V. Such voltage fluctuations may affect the system's stability and reliability. The switching frequency for the pulses can be calculated as $1/(T_{\text{on}} + T_{\text{off}})$ for MOSFET in the given circuit which by value obtained from graph is 81 kHz. The incorporation of a neural network (NN) controller, on the other hand, considerably improves system performance. With the NN controller integrated, the voltage immediately stabilizes at 11.6 V, eliminating any overshoot or peaks in the battery voltage waveform. As a result, a considerably smoother voltage waveform is obtained, illustrating the NN controller's potential to efficiently improve battery performance.

The starting point for this analysis is a battery with an initial state of charge (SoC) of 30% to avoid the excess stress on the battery. Deep discharges (near 0% SoC) can contribute to battery degradation, leading to increased wear and reduced cycle life. To avoid the extremes of both low and high states of charge, the battery experiences less wear and tear, ultimately improving its longevity. In practical applications, the SoC is typically not allowed to exceed 50% to ensure safe battery charging and discharging operations. As the converter charges the battery, the SoC percentage gradually increases, as the battery is acting as the load. Conversely, when the battery is functioning as a supply, the SoC percentage decreases.

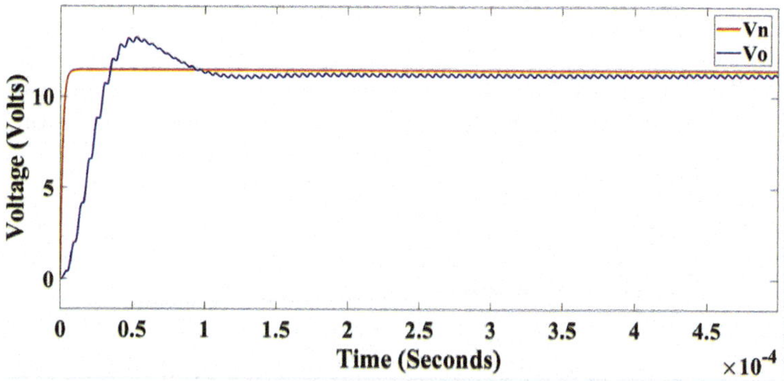

Fig. 7 Voltage of battery with NN controller and without controller

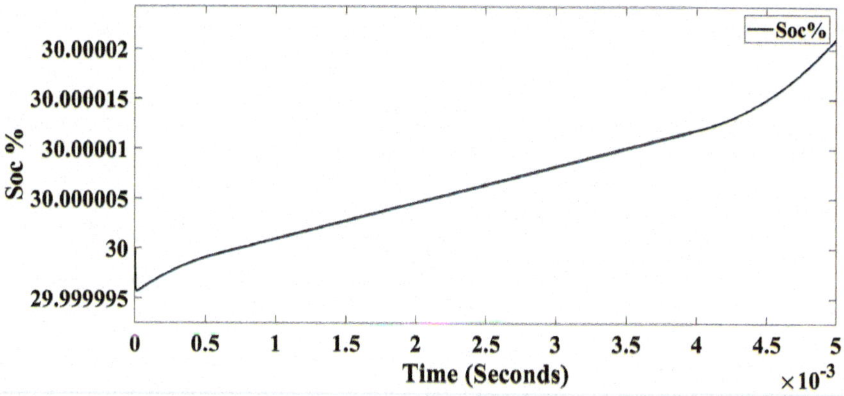

Fig. 8 State of charge of battery

Table 3 Comparison of battery voltage with and without neural network controller

Parameter	Peak overshoot	Settling time
Vo (Voltage without NN controller)	13.2 V	243.177 μs
Vn (Voltage with NN controller)	No overshoot	19.244 μs

4 Conclusion

The work focuses on the design of a LLC resonant (half-bridge) converter suitable for faster onboard battery charging. A 12 V, 7.2 Ah lead-acid battery is used for charge by the LLC resonant converter which is designed and validated the performance using simulation. A frequency-controlled switching technique is used to generate switching pulses at high frequencies. The utilization of a neural controller tuned LLC resonant controller has been found to be useful in regulating a battery's output voltage to a desired value of 12 V, even with input voltage variations ranging between 360 and 410 V. The neural controller successfully reduced output ripple and peak overshoot thus smoothening the response of the system, which ultimately aids in ensuring the battery is free from sudden peak voltage during charging. With a stable voltage, the charging current can be increased, resulting in a more rapid charging process where fast charging is necessary to reduce downtime and maintain system performance. The performance of resonant tank elements in the proposed LLC converter operation was analyzed on the bases of resonant frequency operation. Also, it was observed that the LLC converter gives the accurate value of output voltage when the converter's switching frequency is closer to its resonant frequency. The performance of the proposed half-bridge LLC converter makes it useful for battery charging applications in electric vehicles.

References

1. Hussain I (2003) Electric and hybrid vehicles design fundamentals, 3rd edn. CRC Press
2. Lipu MSH, Faisal M, Ansari S, Hannan MA, Karim TF, Ayob A, Hussain A, Miah MS, Saad MHM (2021) Review of electric vehicle converter configurations, control schemes and optimizations: challenges and suggestions. Electronics 10(4):477
3. Habib S, Khan MM, Abbas F, Tang HJ (2018) Assessment of electric vehicles concerning impacts, charging infrastructure with unidirectional and bidirectional chargers, and power flow comparisons. Int J Energy Res 42(11):3416–3441
4. Habib S, Khan MM, Abbas F, Ali A, Faiz MT, Ehsan F, Tang H (2020) Contemporary trends in power electronics converters for charging solutions of electric vehicles. CSEE J Power Energy Syst 6(4):911–929
5. Yilmaz M, Krein PT (2013) Review of battery charger topologies, charging power levels, and infrastructure for plug-in electric and hybrid vehicles. IEEE Trans Power Electron 28(5):2151–2169
6. Sabri MFM, Danapalasingam KA, Rahmat MF (2016) A review on hybrid electric vehicles architecture and energy management strategies. Renew Sustain Energy Rev 53:1433–1442
7. Peer BMA, Vijaya CKRM (2019) Dual input interleaved SEPIC converter based ripple current reduction in hybrid electric vehicle application. In: 2019 innovations in power and advanced computing technologies (i-PACT), pp 1–5. IEEE, Vellore
8. Deepa K, Sharika M, Mamatha I, Kumar MV (2012) Digital simulation of SISO-ZVS-push pull quasi resonant converter for different loads. Int J Eng Res Appl 2(4):896–901
9. Choi H (2007) Analysis and design of LLC resonant converter with integrated transformer. In: Twenty-second annual IEEE applied power electronics conference and exposition, pp 1630–1635. IEEE, Anaheim, CA, USA
10. Arazi M, Payman A, Camara MB, Dakyo B (2018) Study of different topologies of DC–DC resonant converters for renewable energy applications. In: 2018 thirteenth international conference on ecological vehicles and renewable energies (EVER), pp 1–6. IEEE, Monte Carlo, Monaco
11. Deng J, Li S, Hu S, Mi CC, Ma R (2014) Design methodology of LLC resonant converters for electric vehicle battery chargers. IEEE Trans Veh Technol 63(4):1581–1592
12. Yildiran N (2020) Design methodology and implementation of half-bridge LLC resonant converter. In: 2020 international conference on electrical, communication and computer engineering (ICECCE), pp1–6. IEEE, Istanbul, Turkey
13. Saji Babu D, Ilango Karuppasamy PS (2019) SoC estimation of lithium ion battery using combined coulomb counting and fuzzy logic method. In: 2019 4th international conference on

recent trends on electronics, information, communication technology (RTEICT), pp 948–952. IEEE, Bangalore
14. Sai Lakshmi KM, Sailaja V (2019) Resonant converter-based PV battery charger for EV applications. In: 2019 2nd international conference on power and embedded drive control (ICPEDC), pp 219–224, IEEE, Chennai
15. Daniel S, Chandrakala KRMV (2021) Design of an isolated onboard plug-in electric vehicle charger. In: 2021 7th international conference on electrical energy systems (ICEES), pp 147–149. IEEE, Chennai
16. Madhanakkumar N, Vasanth T, Sriramkumar V, Kanchana C (2020) Performance analysis of LLC resonant converter incorporating PI Fuzzy and neural controllers. Int J Electr Eng Technol 11(4):45–56
17. Yang B (2003) Topology investigation of front end DC/DC converter for distributed power system (Ch.4)
18. Maurya R, Saha R (2022) Design and simulation of a half-bridge LLC resonant converter for battery charger in EV. In: 2022 IEEE Delhi section conference (DELCON), pp 1–9. IEEE, New Delhi, India
19. Naraharisetti K, Channegowda J, Green PB (2021) Design of resonant DC–DC LLC converter. In: 2021 IEEE Kansas power and energy conference (KPEC), pp 1–6. IEEE, Manhattan, KS, USA
20. Krishna VV, Kumar PA, Chandrakala KRMV (2019) Development of hybrid energy storage system for DC motor powered electric vehicles. In: 2019 international conference on smart structures and systems (ICSSS), pp1–4. IEEE, Chennai, India

Optimized Ensemble Learning Framework for Prioritizing Biomolecular Markers in Yield Prediction Efficiency Enhancement

Aditi R. Durge, Deepti D. Shrimankar, and Sony K. Ahuja

Abstract Biomolecular markers play a pivotal role in detecting various outputs present in living organisms. Each of these marker's aids in gaging the existence of specific proteins, thereby enhancing output detection. Currently, these markers are employed to detect diverse outputs, such as those related to different cancer and disease types. The algorithms developed to discern these outputs operate by ranking these biomolecular markers based on their genomic significance. A review of existing research reveals that present models for marker ranking target a single output type, constraining their adaptability and efficacy for real-world clinical use. Additionally, these models suffer from extended training and validation periods, impeding their practical utility. To overcome these limitations, our research introduces an advanced ensemble learning approach to amplify output detection proficiency. We applied our approach to various outputs, including different disease types. Our findings underscore the model's superior accuracy and speed compared to current top-tier methods. This enhancement is primarily due to the integration of GoogLeNet and Inception Net architectures, optimizing feature differentiation, thus heightening ranking efficiency levels. The proposed model has better precision, higher accuracy, and better recall than existing methods.

Keywords Biomarker · Prioritization · Machine learning · Convolutional neural · GoogLeNet · Inception Net · Gene network

1 Introduction

An examination of a person's genetic code can provide information regarding that person's state of health. When these sequences are disrupted, the body reacts in an aberrant manner, which can have severe repercussions such as coronary disease,

A. R. Durge (✉) · D. D. Shrimankar · S. K. Ahuja
Department of Computer Science and Engineering, Visvesvaraya National Institute of Technology (VNIT), Nagpur, India
e-mail: durge.cse@gmail.com

S. Bandyopadhyay et al. (eds.), *Intelligent Computing Systems and Applications*, Lecture Notes in Networks and Systems 1010,
https://doi.org/10.1007/978-981-97-5412-0_7

liver dysfunction, lung failure, and other conditions. Researchers have devised a wide variety of health monitoring measures, some of which are invasive while others are non-intrusive, in order to combat the spread of these diseases. Electroencephalograms (EEG), electrocardiograms (ECG), and magnetic resonance imaging (MRI) are all examples of noninvasive methods that provide an indication of a person's state of health. Because the majority of these studies are based on estimates, they are most beneficial when used for preliminary screening rather than real-time evaluation. The examination and categorization of biomarkers is one example of such an approach; it entails the creation of stratification algorithms for organ-specific peptide studies. In order to discern between the genetic sequences of the various organs, these methods make use of biomarkers, also known as patterns of expression. It is envisaged that the calculation and ranking of biomarkers will be particularly valuable for the early detection of progress and the exact measurement of yield. The LVYE1 biomarker is utilized for the diagnosis and monitoring of over 273 distinct diseases and ailments, some of which are non-lymphoma Hodgkin's (NHL), melanoma, and a great number of other conditions. Estimating the activity of this receptor and making adjustments to its physiological make-up are both helpful in arriving at an appropriate assessment of various yields. In the example model for ranking biomarkers given in Fig. 1, the expression profile of the yield as well as a weighted gene list are employed as inputs for different scenarios.

A rating model will be created if this data is analyzed using machine learning (also known as deep learning). The model assigns a score to each of these signals according to the effects they have on the immune system. After the ranking markers have been input into a classification algorithm, the area under the curve (AUC) score

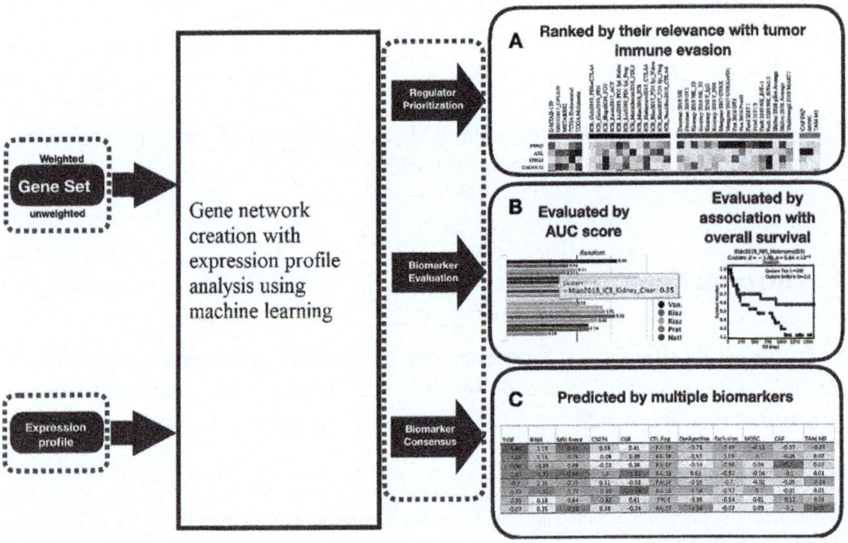

Fig. 1 Biomarker prioritization process

of each biomarker is calculated [1]. The AUC value is used to construct a consensus model, and a large number of biomarkers are chosen for effective categorization. Prioritization, evaluation, and consensus modeling are three areas in which numerous algorithms have been proposed to be used. Each of these algorithms has its own specific field of application, accuracy range, scalability, and so on. The performance of the suggested model was evaluated using a variety of measures across a large number of datasets. These metrics included accuracy, precision, recall, and end-to-end latency levels. It is possible to increase accuracy by prioritizing biomarkers [2], in which case each biomarker is prioritized according to the degree to which it influences a particular yield. It is possible that we will be able to improve the detection efficacy for the underlying yield if we investigate changes in biomarkers with lower ranks but higher priorities.

2 Literature Review

The engineering field of biomarker prioritization combines machine learning and biological research. It achieves high levels of accuracy by classifying yields using highly dynamic biomarkers. According to the research in [2], by tailoring the prioritization of cognitive traits, brain morphometric data may be used to determine Alzheimer's yield. The model is claimed to be highly accurate, but because of the repeated feature extraction steps, it is very slow and can only be used to one type of yield. The models discussed in [1, 3, 4] can be used to identify many cancers, including leukemia, CoVID-19, and hepatocellular carcinoma. Blood counts, metabolic studies, and protein interactions are used to build machine learning algorithms, which are subsequently used to categorize the various types of yield. Methods of biomarker prioritization with a medium level of scalability are investigated in [2, 5]. With only a moderate degree of accuracy, these biomarkers can be used to detect a variety of diseases, including Alzheimer's disease, pharmacogenomics, medications that have been repurposed, heart yield, lung yield, liver yield, and others. A technique for developing drugs and vaccines for liver-based outcomes is also suggested by the model in [5], which may speed up the body's natural healing process. The utilization of techniques suggested in [6–8] is to find lymph node metastases, breast cancer, prostate cancer, kidney cancer, and other diseases. The following techniques must be applied (AHP) to increase the accuracy of these models even more. These models are more adaptable and have shorter delays than the typical single yield detection approaches, according to a cross-parametric study.

Reference [9] suggests using a technique called "learning to rank" to prioritize amyloid imaging symptoms when making an Alzheimer's disease diagnosis. Due to its high accuracy and constrained scalability, this model cannot be applied for the great majority of the clinical applications discussed in [10]. Similar methods are discussed in [11], which employs a big dataset that has been painstakingly screened to uncover interesting biomarker candidates for the early detection of Alzheimer's disease. RDF, or resource description framework, is used to give

Table 1 Characteristics of the proposed model over the existing models

Aspect	Prior work	Solution presented in the paper
Output type	Specialized for a single type of yield or output [2, 5]	Designed for multiple yield types
Scalability and performance	Limited due to specificity to a single yield type, affecting real-time application [8–10]	Enhanced scalability and performance, suitable for real-time clinical applications
Training and validation	Long training and validation delays, reducing immediate applicability [7, 22, 23]	Efficient training likely due to the ensemble approach and integration of advanced architectures
Model architecture	Single network models [2, 4, 5, 24]	Augmented ensemble learning model with multiple neural network architectures
Efficiency and accuracy	Might vary, but generally lower due to the mentioned limitations [10, 11, 13, 25]	Improved efficiency and high average accuracy of 91% across various datasets, better precision and recall performance

semantic biomarker data, which improves the user experience for those performing data analysis. Following established protocols will probably produce a more consistent result for this experiment, according to [12]. The program can accurately categorize datasets [13, 14] thanks to the integration of 12 different biomarkers. Additional selection models are presented and assessed in [15–17] for the assessment of low-grade chondrosarcoma, enchondroma, pancreatic cancer, age, and age-related diseases. Enchondroma, pancreatic cancer, and low-grade chondrosarcoma may all be evaluated using these models [18–20]. Despite the fundamental design of these models' restricted scalability, they are effective in locating single or double yields. According to the research's findings, the majority of these models perform superbly when used with a single type of yield; but, as the number of yields increases, so does their accuracy [21]. The solution to this issue will be presented in the following section, which will present a novel method for prioritizing biomarkers. This model is appropriate for a variety of yield kinds since it can achieve high accuracy with a short latency for rapid clinical diagnosis. The key characteristics of the proposed model in comparison with existing methods can be observed from Table 1.

3 Proposed Model Design for Real-Time Scenarios

More research on the subject reveals that several biomarker prioritization algorithms are created to meet the needs of various fields. Differential analysis, Bayesian multi-task multi-kernel learning, kernelized rank learning, and customized cognitive feature prioritization are just a few of the machine learning techniques used in these models. All of these methods could be categorized as machine learning methods.

When employed with a range of yield types, these systems' adaptability and scalability are greatly limited. This section aims to provide a superior ensemble learning model that combines GoogLeNet and Inception Net convolutional neural networks (CNNs) in order to enhance the efficacy of biomarker prioritization strategies. These networks are extensively featured extracted from gene datasets, and the features are then ranked using a distinctive variance evaluation method. We may enhance the model's performance on the testing set by incremental learning, which also raises the testing set's ranking. The suggested approach for biomarker prioritization entails collecting genomic data that is yield independent from several sources and feeding it to an ensemble CNN layer, as shown in Fig. 2. By retrieving a wide range of properties in this way, the genome is better able to represent features. To assess population variance, these features are included into a state-of-the-art computational framework. This section calculates the inter-class variance of a set of biomarkers, and the biomarkers are then sorted in decreasing order of their inter-class variance. An ensemble classification technique is used to analyze the performance traits of the best biomarkers. Recall, precision, and accuracy are a few examples of these metrics. Looking more closely at these factors may help you increase the efficiency of your ensemble feature extraction model. Through iterative refinement, the method improves classification outcomes and the effectiveness of biomarker selection. The model is made up of these components, although they can also be used independently to improve existing methods for prioritizing biomarkers. It is necessary to finish this step before continuing. The genetic information must be transformed into a string of bytes, with each byte being stored in an array of 224 by 224, in order to finish this process. Many convolutional neural network (CNN) models based on Inception Net v3 and GoogLeNet are given access to this 2D array in order to extract features from the data. Despite the use of these models, the classification step is avoided by omitting the deep neural network layers that are typically present in such architectures. Figure 3 illustrates an ensemble model as an example. In this model, a feature reduction block is used to integrate the outputs of the two CNN models. The GoogLeNet model is used to extract coarse features from genomic data, whereas the Inception Net v3 model is used to extract a more extensive collection of genomic properties. Both concepts are blended in real-world circumstances. The shared foundation of both of these models, the standard CNN model with an inception layer, adds to their strength in feature extraction. By including a differential analysis layer, we may eliminate redundant features that might have developed as a result of training on the Google Net and Inception Net v3 models. This layer makes careful feature selection possible. Equation 1 manages feature selection by computing the optimal separation between Inception Net and GoogLeNet features.

$$d_{\mathrm{GI}_{i,j}} = \sqrt{\left(f_{1_{\mathrm{G}_i}} - f_{1_{\mathrm{I}_i}}\right)^2 + \left(f_{2_{\mathrm{G}_i}} - f_{2_{\mathrm{I}_i}}\right)^2 + \cdots + \left(f_{j_{\mathrm{G}_i}} - f_{j_{\mathrm{I}_i}}\right)^2} \tag{1}$$

where $d(\mathrm{GI}(i,j))$ stands for the Euclidean distance between features of the ith GoogLeNet and the jth Inception Net v3, and where the ith GoogLeNet and the jth Inception Net v3 are the two networks under consideration (f). Features with a

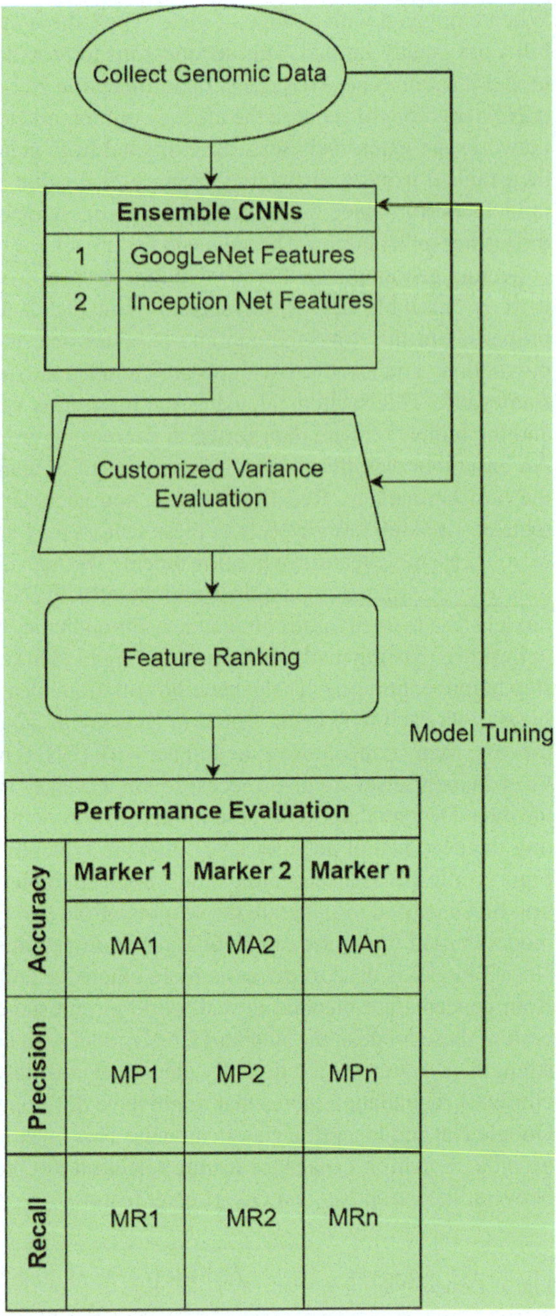

Fig. 2 Design of the proposed model for processing genome sequences

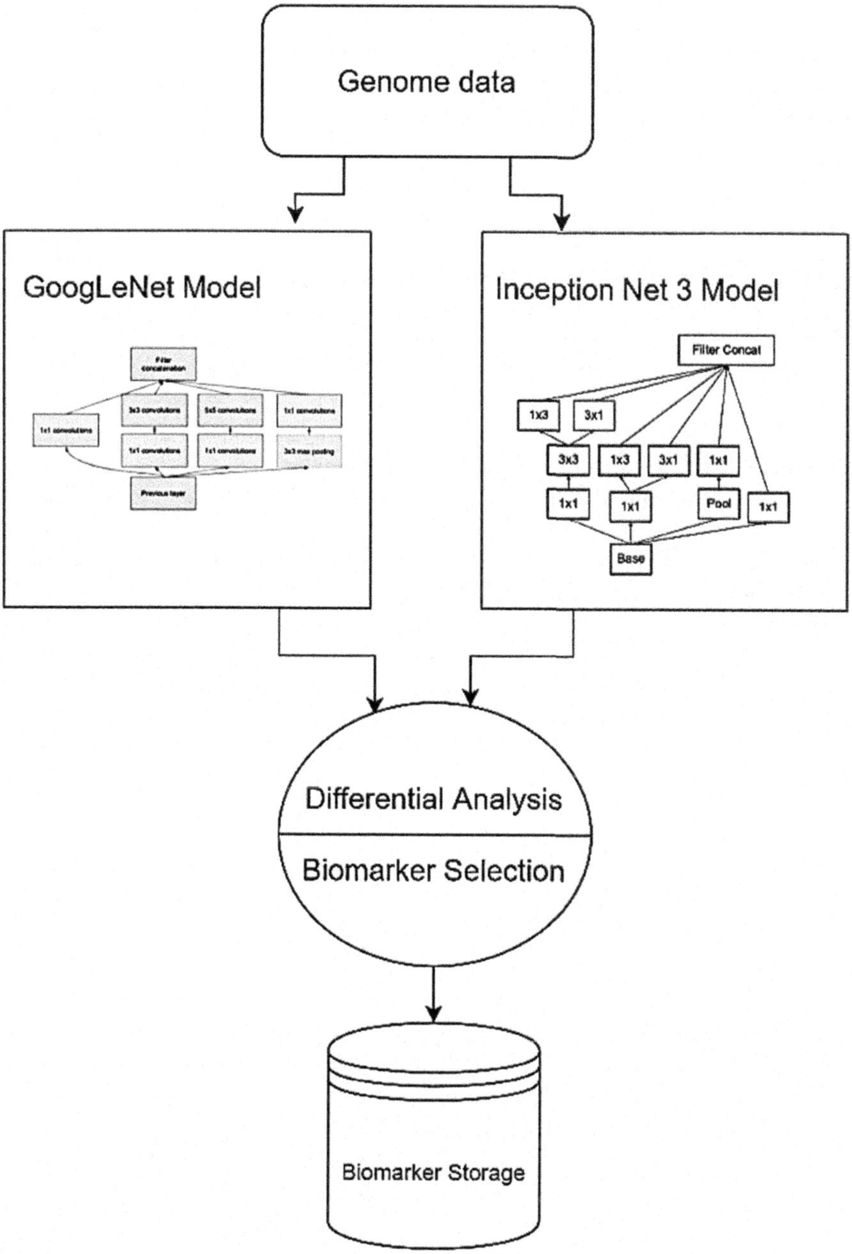

Fig. 3 Design of the proposed CNN for processing genomes

distance below a specified threshold are removed from the feature set of GoogLeNet model, but being included in the feature set of Inception Net model, Eq. 2 takes into consideration a tuning parameter and an average Euclidean distance to get a rough estimation of this threshold for different scenarios.

$$d_{\mathrm{TH}} = \frac{\sum_{i=1}^{N_{\mathrm{FG}}} \sum_{j=1}^{N_{\mathrm{FI}}} d_{\mathrm{GI}_{i,j}}}{N_{\mathrm{FG}} * N_{\mathrm{FI}}} * L_{\mathrm{tuning}} \tag{2}$$

For the total number of GoogLcNet features extracted, the total number of Inception Net v3 features extracted, and the tuning factor, respectively, the letters "N_{FG}," "N_{FI}," and "L_{tuning}" are used.

The tuning factor controls how many features (biomarkers) are acquired by this module by assessing how effectively the different biomarkers work. The variance evaluation model is given features from a subset (approximately 25% of all characteristics) in order to rank biomarkers since the default value of this tuning factor is set to 0.5. By doing this, the model is made as accurate as feasible for different scenarios.

In addition to the biomarkers that have been selected, it have access to a variance evaluation engine. This strategy starts by grouping each extracted biomarker into the appropriate yield category. In order to distinguish between the biomarkers of the various kinds of yield, these categories are included in the training set. Equation 3 may be used to calculate a custom variance rank ($\mathrm{CVR_d}$) for each yield class based on the distribution of biomarkers.

$$\mathrm{CVR_d} = \sqrt{\frac{\sum_{i=1}^{m_{\mathrm{d}}} \left(b_i - \frac{\sum_{j=1}^{m} \sqrt{\frac{\sum_{k=1}^{n} \left(b_k - \frac{\sum_{l=1}^{n} b_l}{n} \right)^2}{n-1}}}{m_{\mathrm{d}}} \right)^2}{m_{\mathrm{d}} - 1}} \tag{3}$$

where *b*, *m*, and *n*, respectively, denote the values for the extracted biomarker, the number of elements in the current yield class, and the number of elements in other yield classes, respectively. This CVR is evaluated for each class and compared to the CVR that was determined for each biomarker ($\mathrm{CVR_b}$), which was determined by following the steps outlined in Eq. 4

$$\mathrm{CVR_b} = \sum_{i=1}^{m_{\mathrm{a}}} \sqrt{\frac{\sum_{m=1}^{m_{\mathrm{a}}} \left(b_m - \frac{\sum_{l=1}^{m_{\mathrm{a}}} b_l}{m_{\mathrm{a}}} \right)^2}{m_{\mathrm{a}}}} \tag{4}$$

The CVR value of a biomarker is thrown out of consideration if it has a CVR that is lower than the cutoff value ($\mathrm{CVR_d}$). This causes the remaining $\mathrm{CVR_b}$ values to be arranged in decreasing order. In general, a higher CVR is illustrative of the fact that the biomarker may be used for yield detection with better precision. Increasing accuracy levels are evaluated via Eq. 5, while reducing levels are evaluated via Eq. 6 as follows,

$$\mathrm{LF}_{\mathrm{inc}} = \frac{A_{\mathrm{current}}}{A_{\mathrm{previous}}} \tag{5}$$

$$\mathrm{LF}_{\mathrm{dec}} = \frac{A_{\mathrm{previous}} * N_i}{\sum_{i=1}^{N_i} A_i} \tag{6}$$

where A_{current}, A_{previous}, A_i, and N_i are current, previous, instantaneous accuracies, and number of elements present in the genomes, where N_i is evaluated as per Eq. 7

$$N_i = \frac{L(T_{\mathrm{s}}) + L(V_{\mathrm{s}})}{N_{\mathrm{c}}} \tag{7}$$

The notations $L(T_{\mathrm{s}})$, $L(V_{\mathrm{s}})$, and N_{c}, respectively, stand for the lengths of the training set, validation set, and the total number of output categories that may be formed. The technique is deemed effective when the model's accuracy converges, as shown by the A_{current} and A_{previous}. This incremental learning method supports the model's capacity to rank biomarkers according to their high effectiveness and low average latency. The performance of this model across several datasets will be examined in the next section, along with comparisons to other state-of-the-art models.

4 Results and Comparative Analysis

The effectiveness of the suggested biomarker prioritization strategy was assessed using datasets for breast cancer, pancreatic cancer, brain cancer, and Alzheimer's disease. Both the Cancer Genome Atlas and the Alzheimer's Disease Study from the National Institute on Aging were included in these datasets. To access the Cancer Genome Atlas, click to https://www.cancer.gov/about-nci/organization/ccg/research/structural-genomics/tcga. The Alzheimer's Disease Study can be accessed at https://www.niagads.org/. These extracted samples were used in both the training and validation stages of the proposed model. We evaluated the efficacy of strategies similar to biomarker prioritization, such as those recommended by the models shown in [2, 14], using the identical datasets in each example. Table 2 illustrates how accurate the suggested model is at predicting the effects of Alzheimer's disease samples.

Table 2 demonstrates that the proposed model outperforms the referenced literature by 11% in terms of accuracy and by 8% in terms of Alzheimer's yield. Table 3 presents the results of a performance evaluation of pancreatic cancer disease types.

Table 3 presents that the proposed model outperforms [2, 14] by 5% in terms of accuracy when predicting pancreatic cancer. Evaluation of effectiveness for brain cancer is also done, and the results are given in Table 4.

Table 2 Average accuracy for Alzheimer's disease types

Number of samples	Acc. (%) [2]	Acc. (%) [14]	Acc. (%) proposed
100	76	71.25	81.8
200	77.9	74.01	84.39
400	78.57	74.39	84.97
600	78.95	75.15	85.6
800	79.23	75.24	85.81
1000	79.52	76.35	86.59
1200	80.15	76.74	87.16
1400	80.44	77.13	87.53
1600	80.72	77.51	87.91
1800	81.02	77.9	88.28
2000	81.3	78.29	88.65

Table 3 Average accuracy for pancreatic cancer types

Number of samples	Acc. (%) [2]	Acc. (%) [14]	Acc. (%) proposed
100	76.76	74.81	79.77
200	78.68	77.71	82.31
400	79.35	78.11	82.87
600	79.73	78.91	83.5
800	80.02	79	83.7
1000	80.31	80.17	84.46
1200	80.95	80.58	85.02
1400	81.24	80.98	85.38
1600	81.53	81.39	85.75
1800	81.82	81.8	86.12
2000	82.11	82.2	86.48

Table 4 presents that when it comes to brain cancer, the suggested model has a success rate that is 6% higher than [2] and 8% higher than [14]. Performance evaluations for breast cancer are quite similar to those for other cancers, as given in Table 5.

Table 5 presents that when it comes to breast cancer, the suggested model has a prediction accuracy that is 9% better than [2] and 6% better than [14]. Similarly, Table 6 presents the results of an examination of the average accuracy and recall for all yield kinds.

Table 6 presents that across a range of yields, the proposed model performs better than [2, 14] in terms of accuracy (by 8 percentage points) and recall (by 6 percentage points) which we can observe this from Fig. 4a, b.

Table 4 Average accuracy for brain cancer types

Number of samples	Acc. (%) [2]	Acc. (%) [14]	Acc. (%) proposed
100	77.54	74.91	82.02
200	79.48	77.8	84.62
400	80.16	78.2	85.2
600	80.55	79	85.83
800	80.84	79.1	86.05
1000	81.13	80.27	86.83
1200	81.78	80.67	87.4
1400	82.07	81.08	87.77
1600	82.37	81.49	88.15
1800	82.66	81.89	88.53
2000	82.95	82.3	88.9

Table 5 Average accuracy for breast cancer types

Number of samples	Acc. (%) [2]	Acc. (%) [14]	Acc. (%) proposed
100	75.51	72.45	79.87
200	77.4	75.25	82.4
400	78.06	75.64	82.96
600	78.43	76.41	83.58
800	78.72	76.5	83.79
1000	79	77.63	84.55
1200	79.63	78.03	85.1
1400	79.91	78.42	85.47
1600	80.2	78.81	85.84
1800	80.49	79.21	86.2
2000	80.78	79.6	86.57

Table 6 Evaluation of precision and recall levels for different disease types

Yield type	*P* (%) [2]	*P* (%) [14]	*P* (%) proposed
Alzheimer's	79.29	75.58	86.03
Pancreatic	80.08	79.36	83.91
Brain	80.89	79.46	86.27
Breast	78.77	76.86	84.01
Yield type	*R* (%) [2]	*R* (%) [14]	*R* (%) proposed
Alzheimer's	80.09	76.65	87.07
Pancreatic	80.88	80.47	84.92
Brain	81.71	80.57	87.31
Breast	79.56	77.93	85.02

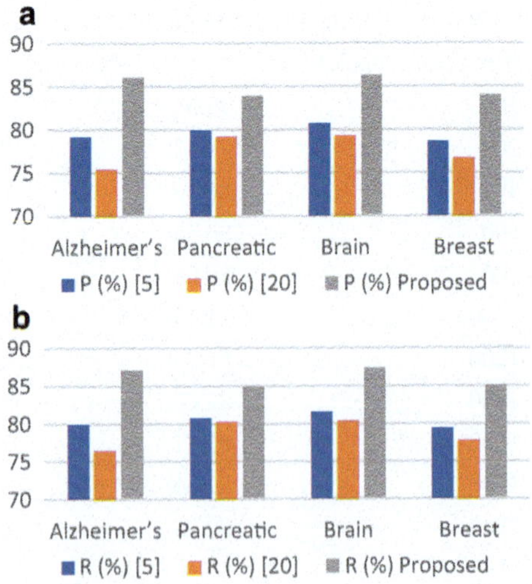

Fig. 4 **a** Average precision levels. **b** Average recall levels

This characteristic allows for little downtime and great classification accuracy while using the model in the clinical scenarios.

To quantify the superiority of our proposed model over model [2], a paired *t*-test was conducted based on the accuracy metrics for the Alzheimer's disease dataset. This test is utilized to determine if there's a statistically significant difference in the performance of the two models. Considering the first three sample sizes (100, 200, 400) as representative, we calculate the difference in accuracy:

- At 100 samples: $\Delta = 81.80\% - 76.00\% = 5.80\%$
- At 200 samples: $\Delta = 84.39\% - 77.90\% = 6.49\%$
- At 400 samples: $\Delta = 84.97\% - 78.57\% = 6.40\%$.

Resulting in an average difference $\Delta^{-} = 6.23\%$, the standard deviation of the differences was computed to be 0.353%, and from the above metrics, the *t*-statistic was found to be:

$$\Delta^{-}0.353/3 = 10.31$$

At a confidence level of 95% ($\alpha = 0.05$) and 2 degrees of freedom, the critical *t*-value (sourced from standard *t*-tables for a two-tailed test) is approximately 2.92 for different use cases. Given the computed *t*-statistic (10.31) considerably surpasses the critical *t*-value (2.92), we reject the null hypothesis. This indicates a statistically significant enhancement in accuracy for our proposed model over model [2] concerning the Alzheimer's disease dataset samples. It is pertinent to mention that similar statistical evaluations can be pursued against model [14], and for different metrics such as precision or recall. Furthermore, analogous tests can be executed for other datasets, encompassing pancreatic, brain, and breast cancer datasets and samples.

5 Conclusion

It is necessary to do in-depth study on a variety of genetic markers before using those indicators to create a category in order to prioritize biomarkers. Priority biomarkers should have a high degree of precision and be able to predict yield with accuracy. This is due to the fact that high-priority biomarkers exhibit larger feature variance than low-priority markers. Based on this discovery, the body of work that was presented offered a special augmented ensemble learning method for prioritizing biomarkers in order to improve yield detection effectiveness. The model used a very high number of distinct CNNs to accomplish its goals of feature extraction, variance-based selection, and performance optimization. It has been shown that the suggested model outperforms the models described in [2, 14] in terms of accuracy, precision, and recall since it incorporates incremental learning. Additionally, it has been shown that the iterative learning approach utilized for model tweaking minimizes latency in comparison with earlier initiatives. This was accomplished by using an iterative learning loop. This model that was developed for use in clinical settings and has the potential to be applied to a broad range of different sorts of diseases, such as Alzheimer's disease, brain cancer, pancreatic cancer, and breast cancer. It will be crucial to assess various feature selection methods in the upcoming feature extraction tests for CNN models. Principal component analysis (PCA) and intrinsic mode functions (IMFs) are two examples of these techniques. It is possible to compare the results of the suggested technique with those of the model modifications and transfer learning that were previously discussed. If the Q-incentive learning components are applied, there is a potential that less mistakes will be made and that the learning process will go more quickly. These two possibilities are both plausible for real-time scenarios.

6 Future Work

While the results presented in this study underscore the potential of our augmented ensemble learning model in biomarker prioritization, there remains an expanse of uncharted territory warranting exploration. Firstly, integration with newer deep learning architectures could further enhance the model's precision and reduce computational time. Next, the applicability of our model can be expanded to encompass rare and less-studied diseases, presenting a comprehensive understanding of their genetic underpinnings. Another promising avenue to pursue is the incorporation of longitudinal data, which would allow the model to track the progression and evolution of diseases over time, and hence anticipate shifts in biomarker relevance. Furthermore, exploring the interplay between genetic biomarkers and environmental or lifestyle factors could provide a more holistic view of disease prediction and progression. Lastly, collaborations with clinicians would be invaluable, translating our computational findings into actionable medical insights, and refining our model through continuous feedback from real-world applications.

References

1. Park H, Shiraishi Y, Imoto S, Miyano S (2017) A novel adaptive penalized logistic regression for uncovering biomarker associated with anti-cancer drug sensitivity. IEEE/ACM Trans Comput Biol Bioinform 14(4):771–782. https://doi.org/10.1109/TCBB.2016.2561937
2. Sheng J, Li F, Wong STC (2015) Optimal drug prediction from personal genomics profiles. IEEE J Biomed Health Inform 19(4):1264–1270. https://doi.org/10.1109/JBHI.2015.2412522
3. Metsis V, Makedon F, Shen D, Huang H (2014) DNA copy number selection using robust structured sparsity-inducing norms. IEEE/ACM Trans Comput Biol Bioinform 11(1):168–181. https://doi.org/10.1109/TCBB.2013.141
4. Paul A, Sil J (2019) Identification of differentially expressed genes to establish new biomarker for cancer prediction. IEEE/ACM Trans Comput Biol Bioinform 16(6):1970–1985. https://doi.org/10.1109/TCBB.2018.2837095
5. Rezaul Karim M, Cochez M, Zappa A, Sahay R, Rebholz-Schuhmann D, Beyan O, Decker S (2020) Convolutional embedded networks for population scale clustering and bio-ancestry inferencing. IEEE/ACM Trans Comput Biol Bioinform 19:369382. https://doi.org/10.1109/TCBB.2020.2994649
6. He Q, Qiu Z, Tong Y, Song K (2020) A new TTZ feature extracting algorithm to decipher tobacco related mutation signature genes for the personalized lung adenocarcinoma treatment. IEEE Access 8:89031–89040. https://doi.org/10.1109/ACCESS.2020.2993118
7. Sadighbayan D, Ghafar-Zadeh E (2021) Portable sensing devices for detection of COVID-19: a review. IEEE Sens J 21(9):10219–10230. https://doi.org/10.1109/JSEN.2021.3059970
8. Pouryahya M, Oh JH, Javanmard P, Mathews JC, Belkhatir Z, Deasy JO, Tannenbaum AR (2022) aWCluster: a novel integrative network-based clustering of multiomics for subtype analysis of cancer data. IEEE/ACM Trans Comput Biol Bioinform 19(3):1472–1483. https://doi.org/10.1109/TCBB.2020.3039511
9. Tharmakulasingam M, Gardner B, La Ragione R, Fernando A (2023) Rectified classifier chains for prediction of antibiotic resistance from multi-labelled data with missing labels. IEEE/ACM Trans Comput Biol Bioinform 20(1):625–636. https://doi.org/10.1109/TCBB.2022.3148577
10. Yang C-H, Chuang L-Y, Lin Y-D (2019) Detecting SNP–SNP interactions in imbalanced case-control study. IEEE Access 7:143036–143045. https://doi.org/10.1109/ACCESS.2019.2943614
11. Chen H, Zhang Z, Li G (2019) Relating disease-gene interaction network with disease-associated ncRNAs. IEEE Access 7:133521–133528. https://doi.org/10.1109/ACCESS.2019.2941955
12. Alexandrou G, Moser N, Mantikas K-T, Rodriguez-Manzano J, Ali S, Coombes RC, Shaw J, Georgiou P, Toumazou C, Kalofonou M (2021) Detection of multiple breast cancer ESR1 mutations on an ISFET based lab-on-chip platform. IEEE Trans Biomed Circuits Syst 15(3):380–389. https://doi.org/10.1109/TBCAS.2021.3094464
13. Unverricht-Yeboah M, Giesen U, Kriehuber R (2018) Comparative gene expression analysis after exposure to 123I-iododeoxyuridine, γ- and α-radiation—potential biomarkers for the discrimination of radiation qualities. J Radiat Res 59(4):411–429. https://doi.org/10.1093/jrr/rry038
14. Yuan X, Zhang J, Wang Y (2010) Probability theory-based SNP association study method for identifying susceptibility loci and genetic disease models in human case-control data. IEEE Trans Nanobiosci 9(4):232–241. https://doi.org/10.1109/TNB.2010.2070805
15. Khaire UM, Dhanalakshmi R (2019) Effects of random forest parameters in the selection of biomarkers. Comput J 64(12):1840–1847. https://doi.org/10.1093/comjnl/bxz161
16. Hossain MT, Reza MS, Peng Y, Feng S, Wei Y (2023) Identification of key genes as potential drug targets for gastric cancer. Tsinghua Sci Technol 28(4):649–664. https://doi.org/10.26599/TST.2022.9010035
17. Bhadra T, Mallik S, Sohel A, Zhao Z (2022) Unsupervised feature selection using an integrated strategy of hierarchical clustering with singular value decomposition: an integrative biomarker

discovery method with application to acute myeloid leukemia. IEEE/ACM Trans Comput Biol Bioinform 19(3):1354–1364. https://doi.org/10.1109/TCBB.2021.3110989
18. Chi C, Tian R, Liu H, Wang H, Wei J, Guo J, Guo F, Li S (2013) Follow-up study of abnormal biological indicators and gene expression in the peripheral blood of three accidentally exposed persons. J Radiat Res 54(5):840–851. https://doi.org/10.1093/jrr/rrt022
19. Zhang L, Liu H, Huang Y, Wang X, Chen Y, Meng J (2017) Cancer progression prediction using gene interaction regularized elastic net. IEEE/ACM Trans Comput Biol Bioinform 14(1):145–154. https://doi.org/10.1109/TCBB.2015.2511758
20. Subramanian A, Shackney S, Schwartz R (2013) Novel multisample scheme for inferring phylogenetic markers from whole genome tumor profiles. IEEE/ACM Trans Comput Biol Bioinform 10(6):1422–1431. https://doi.org/10.1109/TCBB.2013.33
21. Yeh S-J, Chen B-S (2022) Systems medicine design based on systems biology approaches and deep neural network for gastric cancer. IEEE/ACM Trans Comput Biol Bioinform 19(5):3019–3031. https://doi.org/10.1109/TCBB.2021.3095369
22. Madhumita, Paul S (2022) A feature weighting-assisted approach for cancer subtypes identification from paired expression profiles. IEEE/ACM Trans Comput Biol Bioinform 19(3):1403–1414. https://doi.org/10.1109/TCBB.2020.3041723
23. Zhu X, Suk H-I, Huang H, Shen D (2017) Low-rank graph-regularized structured sparse regression for identifying genetic biomarkers. IEEE Trans Big Data 3(4):405–414. https://doi.org/10.1109/TBDATA.2017.2735991
24. Wang W, Song J, Zhang W, Tang Y, Luo R, Ran L, Song F (2019) Identification of long non-coding RNA signatures for specific disease-free prognosis in clear cell renal carcinoma. IEEE Access 7:99290–99298. https://doi.org/10.1109/ACCESS.2019.2929588
25. Zhao L, Yan H (2020) MCNF: a novel method for cancer subtyping by integrating multi-omics and clinical data. IEEE/ACM Trans Comput Biol Bioinform 17(5):1682–1690. https://doi.org/10.1109/TCBB.2019.2910515

DEAN: Improving the Efficiency of Disease Network Using Selective Ensemble Network Aggregation Model

Sony K. Ahuja, Deepti D. Shrimankar, and Aditi R. Durge

Abstract Disease networks facilitate the investigation of common linkages between the genetic sequences of distinct diseases. Researchers can analyze the implications of inter-disease interactions by discovering common gene links. The construction of these networks requires extensive data collection, preprocessing, variance analysis, grouping of features, and connection. Variance analysis and feature clustering needs huge processing time and are responsible for successfully linking genetic sequences. Existing high AUC (area under the curve) performance designs allow both the detection and analysis of dual disease links, limiting their applicability to networks including several illnesses. This article proposes the development of DEAN, a framework for choosing ensemble networks for aggregation, to improve the scalability of existing disease networks. This approach analyzes and chooses aggregate Kolmogorov–Smirnov (KS), differential expression (DE), empirical mode decomposition (EMD), and DE variance (DE_{Var}) coefficients to assess the relationship between various disease types. These coefficients are processed using adaptable hierarchical clustering (AHC). The AHC approach optimizes cluster assessment by maximizing the sum of squared distances. The optimum clusters model is first enlarged from individual coefficient analysis to integrated coefficient analysis to improve AUC performance. On datasets pertaining to breast cancer, brain cancer, and post-traumatic stress disorder (PTSD), the proposed DEAN model evaluates, and the produced results were compared to those obtained using the EMD, DE, DE_{Var}, and KS coefficient analysis methods. On these datasets, the proposed DEAN model performed 20% better than DE, 18% better than DE_{Var}, 14% better than KS, and 5% better than EMD models. In contrast to individual models, which are only scalable to a single dataset, the proposed model is scalable to a variety of illness types due to its consistent performance across diverse datasets. In addition, this paper provides several recommendations for enhancing DEAN's AUC performance and incorporating more datasets for better scalability levels.

S. K. Ahuja (✉) · D. D. Shrimankar · A. R. Durge
Department of Computer Science and Engineering, Visvesvaraya National Institute of Technology (VNIT), Nagpur, India
e-mail: ahujasony.cse@gmail.com

S. Bandyopadhyay et al. (eds.), *Intelligent Computing Systems and Applications*, Lecture Notes in Networks and Systems 1010,
https://doi.org/10.1007/978-981-97-5412-0_8

Keywords Disease network · Gene analysis · Selective aggregation · Ensemble · Clustering

1 Introduction

To effectively combat disease, networks must include models from a wide variety of disciplines. Biotechnology, artificial intelligencc, and the visualization of genetic data are only a few examples. To construct a disease network, it is possible to make educated guesses about the common genetic variables among different types of diseases. Genomic information, such as phenomes and genomes, is processed using feature extraction, feature selection, clustering, and linking methods to estimate shared features across diseases. Cancers that have a phenotype with breast, prostate, ovarian, and lymphoma all share genetic risk factors associated with AR, ATM, BRCA1, CDH1, and MSH2. Disease gene networks (DGNs), shown in Fig. 1 as an abbreviation for "disease phenome connections linked to genome-phenome linkages," are the consequence of such a pairing for different scenarios. Multiple disease network implementations have been built, so that researchers may study these relationships.

This study has the following contributions:

- The paper proposes the development of DEAN (selective ensemble network aggregation model), a novel framework with the goal of enhancing the scalability of

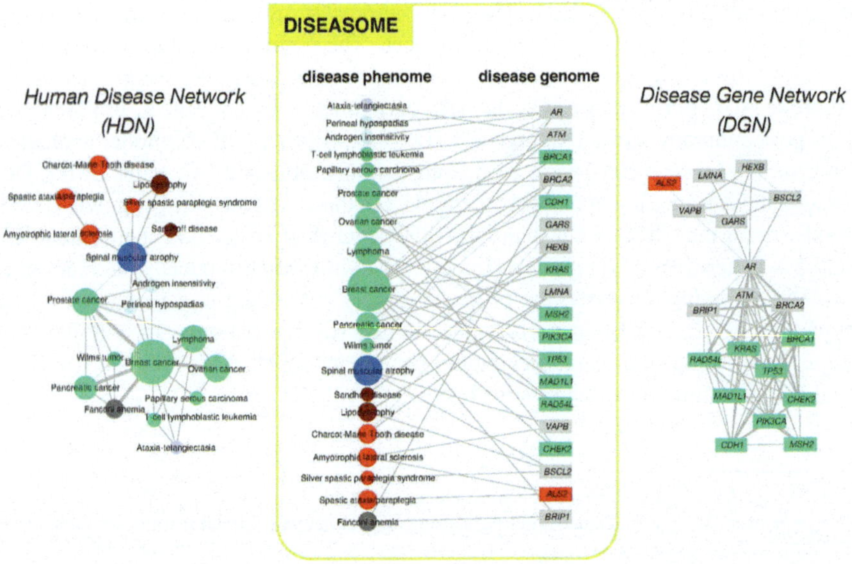

Fig. 1 Design of a DGN model

disease networks. DEAN addresses the limitations of existing designs with high AUC performance by enabling the detection and analysis of dual disease links in networks with multiple diseases.

- The paper introduces the use of aggregate Kolmogorov–Smirnov (KS), differential expression (DE), empirical mode decomposition (EMD), and DE variance coefficients for evaluating the relationship between various disease types. These coefficients provide valuable information regarding inter-disease interactions and common gene links.
- DEAN uses AHC, an adaptable hierarchical clustering method, to maximize the sum of squared distances in cluster evaluation. AHC facilitates the grouping and analysis of features, thereby improving the precision of disease network construction.
- On datasets pertaining to breast cancer, brain cancer, and post-traumatic stress disorder (PTSD), the proposed DEAN model is evaluated for its performance. The evaluation compares the performance of DEAN to that of individual models based on the DE, DE variance, KS, and EMD coefficient analysis techniques.
- In terms of AUC performance, the DEAN model outperforms the individual models, as demonstrated by the results. DEAN is 20% more efficient than DE, 18% more efficient than DE_{Var}, 14% more efficient than KS, and 5% more efficient than EMD models. This performance enhancement demonstrates the efficacy of the ensemble network aggregation strategy.
- In contrast to individual models, which are typically applicable to a single dataset or disease, DEAN demonstrates consistent performance across diverse datasets. This scalability and generalizability make DEAN applicable to a wide range of disease types, thereby increasing its potential impact in disease network analysis.
- The paper provides several suggestions for enhancing the AUC performance of DEAN and suggests incorporating additional datasets to further improve its scalability. These suggestions provide important insights for future research and development in the field.

The next section provides an overview of many approaches and discusses review of existing disease networks. This analysis helps choose approaches that are likely to provide the best realistic models of disease transmission networks. This study's results suggest that most of these treatments are used to treat just two separate illnesses, and that their utility is limited when applied to the treatment of a broader range of ailments.

In Sect. 3, we delve into the inner workings of the proposed selective ensemble network aggregation model (DEAN architecture). All these sections were included, so that we could address this issue. Researchers may use this section to identify gaps in the structures of their own models and decide whether to adopt the DEAN model fully or partially. The next step is to evaluate the suggested model using datasets for breast cancer, brain cancer, and post-traumatic stress disorder. Through comparison with other methods, this assessment helps assess the efficacy of the provided model and calculate an approximation of the real-time capabilities of the suggested remedy.

Section 4 discusses results of this work and compares them with existing methods. Finally, this paper is concluded with some contextual observations about the proposed model and recommends methods to improve its performance for real-time scenarios.

2 Existing Work

There is a significant amount of variation in the expected efficacy, adaptability, and scalability of datasets of disease network design approaches. Examples of illness networks that make use of human route theory, sickness module theory, and a bi-random walk to bacteria are provided in [1–3]. Because of their adaptability and scalability, these models may be effectively used to a wide variety of datasets, including those linked to breast cancer, brain cancer, post-traumatic stress disorder (PTSD), Alzheimer's disease (AD), and a great deal of other conditions and illnesses. Empirical mode decomposition (EMD), differential expression (DE), heterogeneous low-dimensional vector representation (HLVR), and multiple information generative adversarial network are some of the techniques that are used in the models that are proposed in [4–6]. These techniques are employed to increase the efficiency with which networks are constructed (MIGAN). Training and validating these networks take a significant amount of time and effort, even though they perform well in terms of recall and accuracy. Recurrent neural networks (RNNs) with extended long short-term memory (LSTM) are used, so that disease networks may be constructed, and the latency can be reduced [7]. In contrast to EMD and DE models, LSTM RNN models offer superior performance and lower latency levels, thanks to the incremental learning that they use. Other models such as bacterial foraging optimization-based radial basis function neural networks, random walk-through integrating network topological similarity, and the manifold regularization learning framework show promise for minimizing delay without sacrificing accuracy. These models are described in [8–10]. The radial basis function neural network, which was inspired by bacteria's foraging optimization, has the longest lag time because of its feedback learning and bio-inspired principles. This network was also the one that sparked the research that led to its development.

Networks based on ncRNAs, single-molecule magnetic resonance imaging, and a disease pathway adjacency matrix are proposed by the authors of [11–14]. To fine-tune the precision of network development, they make use of high-dimensional neural networks, convolutional neural networks, random walks with restart, and page rank models. The research described in [15–18] uses low-rank matrices and similarities to make predictions about the effectiveness of drug-miRNA networks in illness, the efficacy of artificial neural networks (ANNs), and the relationships between microorganisms and diseases [19, 20]. This research also predicts the efficacy of ANNs. These models predict the interconnectedness of disease networks by using highly analytical methodologies, which results in extraordinary accuracy but large processing delays [21–24]. There are a few different approaches to constructing sickness networks; however, the quirks of the underlying data restrict the output of these approaches. In the next part, we will discuss the process of developing a high-efficiency disease network architecture using an appropriate ensemble network aggregation model, as well as the methodology for determining the efficacy of that model for real-time scenarios.

3 Proposed Methodology

3.1 Design of the Proposed Selective Ensemble Network Aggregation Model for Designing Disease Networks

Different models, including empirical and statistical procedures like differential expression and empirical mode decomposition, are proposed as potential avenues of inquiry for the construction of illness networks within the scope of the research. These avenues of inquiry are practicable. Despite this, the disease-specific network architecture that is used by these analytical models makes it possible for them to attain an adequate level of accuracy. This section discusses the development of a selective ensemble network aggregation model (DEAN) that is capable of effectively developing networks for breast cancer, brain cancer, and post-traumatic stress disorder (PTSD). The purpose of this part is to make disease networks more effective. To optimize the use of illness networks, this was carried out. Figure 2 presents a diagrammatic representation of the DEAN model. It demonstrates how a variety of analytical methods, such as empirical mode decomposition (EMD), differential expression (DE), and DE variance, in addition to Kolmogorov–Smirnov (KS) coefficients, are applied to an input dataset to produce an ensemble model based on the comparative efficacy of these various methods (AUC). After that, a variety of datasets are used to explore various measures of the overall effectiveness of this methodology in the next step. This artificial clustering approach calculates the ideal number of groups to use for a given illness dataset by calculating the sum-of-squared-distances between each pair of groups. After that, these ideal clusters are used to establish a sickness network, and then, the performance of the sickness network is evaluated based on the AUC. After that, the values of the AUC are sent to a weight assessment engine, which is a component of the decision-making process about the ultimate design of the models that are selected. A human-created clustering technique is used to determine AUC. This is accomplished by passing the ensemble weights for each component of the disease network into the algorithm sets.

Many statistical models, including EMD, DE, DE_{Var}, and KS, employ user-submitted disease data as part of the grading system. By offering a new viewpoint on the relative importance of the illnesses that are connected, each of these models contributes something unique to the overall picture of sickness. The EMD model employs the values of the local minimum and maximum to fit Eqs. (1) and (2), therefore fitting the upper and lower envelopes, as explained in the following sentence. The model's goal is to fit the envelopes, and this aids in that goal for different scenarios.

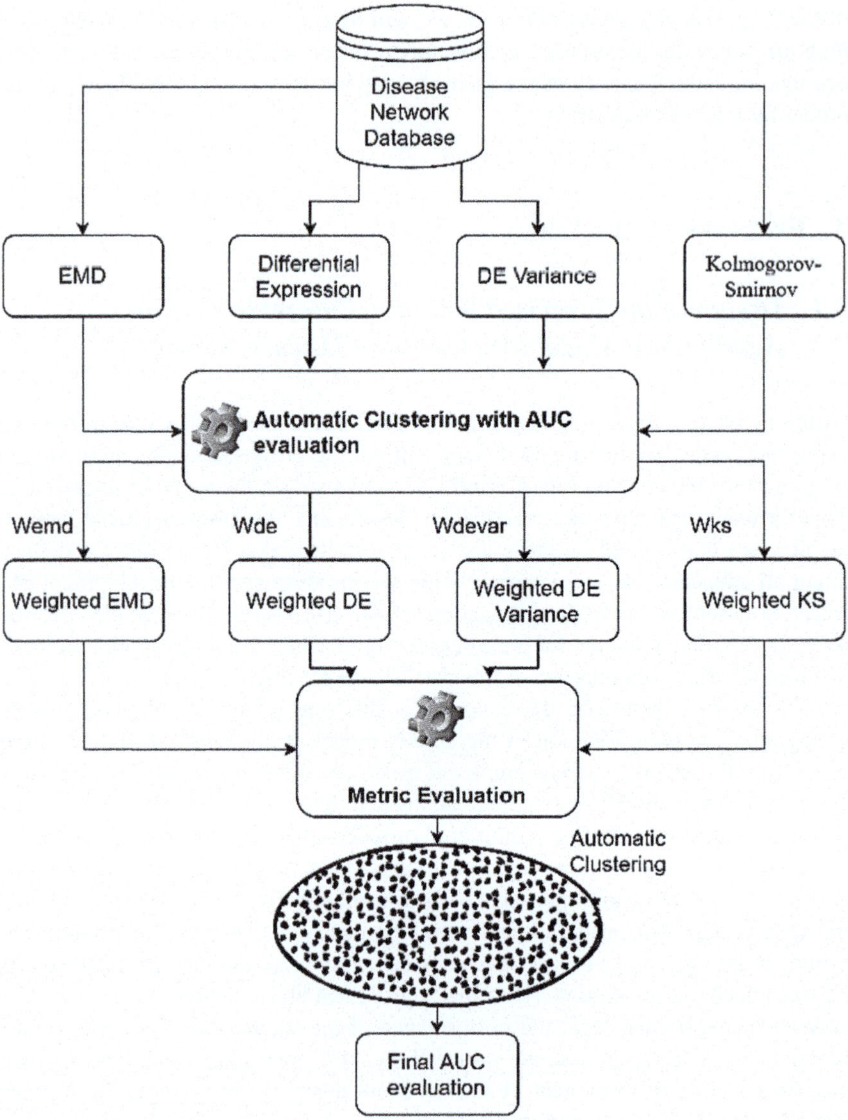

Fig. 2 Flow of the proposed DEAN Model for different scenarios

$$r_n = r_{n-1} - \text{Min}_n \tag{1}$$

$$h_{n,k}^{\text{EMD}} = h_{n,k-1} - \text{Max}_{n,k} \tag{2}$$

In this case, the symbols $\text{Min}_{(n)}$, $h_{(n,k)}$, and $\text{Max}_{(n,k)}$, respectively, are used to represent the input disease linkage intensity (r_n), its local minima, the output disease linkage intensity ($h_{(n,k)}$), and the local maxima ($\text{Max}_{(n,k)}$) between n and k gene sequences.

Equation (3) uses the disease linkage intensity as input to determine empirical mode as follows,

$$\text{EM}_{\text{val}} = \sum_{n=0}^{N} \frac{\left|h_{n,k-1} - h_{n,k}\right|^2}{h_{n,k-1}^2} \tag{3}$$

After a predetermined number of iterations, the evaluation of $h_{(n,k)}$ is terminated if and only if the value of EM_{val} is lower than the standard deviation of all the values of $h_{(n,k)}$. The following equation was used to get this estimate of the standard deviation sets,

$$\text{SD} = \frac{\sqrt{\sum h_{n,k} - \sum \frac{h_{n,k}}{N}}}{N} \tag{4}$$

where N is the whole network's connection count. In a similar way to how Eq. (5) is used to evaluate differential expressions,

$$h_{n,k}^{\text{DE}} = \mathrm{e}^{\sum_{n=1}^{N} r_{n,k} * \partial_{n,k}} \tag{5}$$

If we use the following equation, we may express the relative change in gene expression as the log 2 mean difference between n and k inputs,

$$\partial_{n,k} = \log 2\left(\delta\left(r_{n,k}\right)\right) \tag{6}$$

where the delta symbol indicates that there will be a new step in the input sequence. To determine the DE variance based on this sequence, Eq. (7) is used as the primary tool.

$$h_{n,k}^{\text{DE}_{\text{var}}} = \frac{\left(\sum h_{n,k}^{\text{DE}} - \sum \frac{h_{n,k}^{\text{DE}}}{N}\right)^2}{N-1} \tag{7}$$

Evaluation of this KS is done via Eq. (8),

$$h_{n,k}^{\text{KS}} = \sqrt{N} * \max\left|F_n - F_0\right| \tag{8}$$

These components are evaluated via Eq. (9),

$$F_n = \frac{1}{N} * \sum_{i=1}^{N} I_i, \text{ where } I_i = r_i, \text{ if } r_i > \frac{\sum_{i=1}^{N} r_i}{N} \tag{9}$$

To evaluate cluster weights by minimizing sum-of-squared-distances, all of these parameters are then given to an automated cluster model using the k-Means approach. This improves inter-cluster distance while reducing intra-cluster distance. Using the network intensities of the dataset and with k initially set to 2, it performs k-Means clustering. It then verifies for each cluster member and uses Eq. (10) to calculate the distance inside each of the clusters.

$$D_{\text{intra}} = \frac{\sum_{i-1}^{N} \sum_{j=1}^{N} \sqrt{\begin{matrix}(x_i - x_j)^2 + (y_i - y_j)^2 \\ + \cdots + (z_i - z_j)^2\end{matrix}}}{N^2} \quad (10)$$

Find the sum of the squared distances using Eq. (11). The distance between clusters may then be calculated using Eq. (10), which makes use of the cluster centroids.

$$d_s = \frac{\sum_{i=1}^{N} D_{\text{intra}} + \sum_{i=1}^{N} D_{\text{inter}}}{N} \quad (11)$$

Calculations are performed for all feasible cluster configurations, ranging from $k = 2$ to $k = 30$, using the aforementioned separations. A plot of d_s versus the number of clusters used in the clustering process is shown in Fig. 3 to provide a visual representation of these distances. The cluster configuration with the lowest d_s value is used to determine the final ratings. This clustering includes all of the EMD, DE, DE_{Var}, and KS data. Each method's AUC performance is evaluated after the cluster results have been chronologically arranged by cluster number. To aid selective assembly and ultimately provide more efficient disease network designs, these AUC values are presented in a weighted disease network architecture. To get the mean AUC, it is essential to first compute the AUC for each model and then use Eq. (12),

$$\text{AUC}_{\text{avg}} = \frac{\begin{pmatrix}\text{AUC}_{\text{EMD}} + \text{AUC}_{\text{DE}} \\ + \text{AUC}_{\text{DE}_{\text{var}}} + \text{AUC}_{\text{KS}}\end{pmatrix}}{4} \quad (12)$$

Methods having an AUC larger than the average $\text{AUC}_{(\text{avg})}$ are prioritized during network connection establishment. Using Eq. (13), the chosen network's connections are normalized and concatenated as follows,

$$\text{NC}_{\text{norm}_i} = \text{NC}_i * \text{AUC}_i \quad (13)$$

where i is the selected mode, which is used in Eq. (14) to estimate combined results as follows,

$$\text{NC}_{\text{combined}} = \frac{\sum_{i=1}^{N_{\text{Sel}}} \text{NC}_{\text{norm}_i}}{N_{\text{Sel}}} \quad (14)$$

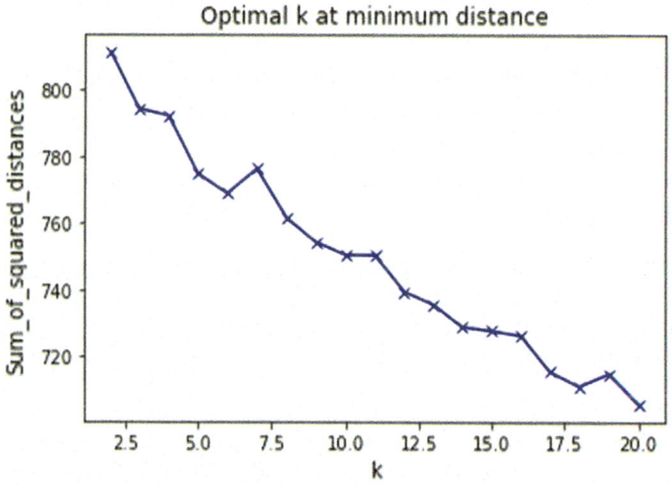

Fig. 3 Set of distances for different number of clusters

The total number of unique components used in the finished product is indicated by the number of selected models (N_{Sel}). The usefulness of the network connection weights given in Eq. (14) depends on how many clusters can be formed automatically for different use cases. The effectiveness of datasets for breast cancer, brain cancer, and post-traumatic stress disorder (PTSD) is assessed using the (AUC), precision, recall, and accuracy methods. The analysis is presented in the next section of this text.

3.2 System Requirements

Certain system requirements and software specifications must also be met in order to simulate the proposed DEAN (selective ensemble network aggregation model) for disease network analysis. Here is a concise summary of these requirements:

Hardware: A computer system with sufficient computational power and memory is required to meet the simulation processing requirements. Depending on the size and complexity of the datasets being analyzed, the necessary hardware specifications may vary.

Depending on the availability of the necessary software tools, the simulation can be conducted on a variety of operating systems, including Windows, macOS, and Linux.

Software Prerequisites:

Programming Language: The implementation and simulation of the DEAN model may require programming skills. For scientific computing and machine learning tasks, Python, R, and MATLAB are popular programming languages.

Data Processing and Analysis Libraries: Relevant libraries and frameworks must be installed to perform the required data preprocessing, variance analysis, and clustering tasks. The simulation involves statistical analysis, machine learning algorithms, and performance evaluation techniques.

Network Analysis Libraries: Since disease network analysis is the focus of the simulation, network analysis libraries can be advantageous. The Python libraries NetworkX and igraph, as well as the R package igraph, can aid in the construction and analysis of disease networks.

Optional Visualization Tools: Visualization tools may aid in the comprehension and presentation of simulation results. Matplotlib, Seaborn, and ggplot in Python and ggplot2 in R are libraries that enable the creation of informative plots and visualizations.

Depending on the needs of the DEAN model and its simulation, additional software or libraries may be required. If the DEAN model relies on deep learning techniques, for instance, frameworks such as TensorFlow or PyTorch are needed for simulations.

4 Results and Discussion

4.1 Findings and Comparative Analysis

Comparisons were made between the DE, DE_{Var}, EMD, and KS models and the proposed DEAN model in terms of accuracy, precision, recall, and AUC parameters. To evaluate the precision of this comparison, we employed Eq. (15) as follows,

$$\text{Acc} = \frac{T_{\text{ce}}}{T_{\text{e}}} \tag{15}$$

where T_{ce} is the total edge count upon successful detection and T_{e} is the total edge count of the disease network. The findings of the accuracy study of these samples are summarized in Table 1. Gene Expression Omnibus (https://www.ncbi.nlm.nih.gov/geo/) datasets on brain cancer, breast cancer, and post-traumatic stress disorder (PTSD) were utilized to analyze this accuracy. Each of the diseases had 2 k instances in the dataset that was utilized for the analysis.

Table 1 Average accuracy for different disease types

Size	*A* (%) DE and DE_{Var}	*A* (%) EM	*A* (%) KS	*A* (%) DEAN
200	88.73	90.51	88.41	95.34
400	89.15	90.82	88.73	95.76
600	89.57	90.13	89.04	96.08
800	90.83	90.97	89.78	96.92
1000	88.73	90.92	88.83	95.87
1200	91.56	92.54	90.3	97.44
1400	92.4	93.13	90.83	97.97
1600	93.24	93.65	91.35	98.6
1800	88.73	93.85	89.25	96.29
2000	94.08	94.28	91.98	99.23

From the findings in Table 1 and Fig. 4, it seems that the proposed model outperforms the competition on a wide range of datasets by an average of 14% compared to KS, 9% compared to DE, 5% compared to DE_{Var}, 15% compared to EMD, and so on.

Table 2 displayed the results of a similar analysis performed to establish the proper diagnosis of certain diseases and their types. When applied to a wide range of datasets, the suggested model demonstrates performance that is about 8% better than that of KS, 3% better than that of DE_{Var}, 12% better than that of EMD, and approximately 8% better than DE overall.

The information on the results of a similar examination for memories of various forms of sickness is included in Table 3, which shows that the proposed model consistently beats DE, DE_{Var}, EMD, and KS by 5%, 4%, 14%, and approximately 9%, respectively, on a wide range of datasets.

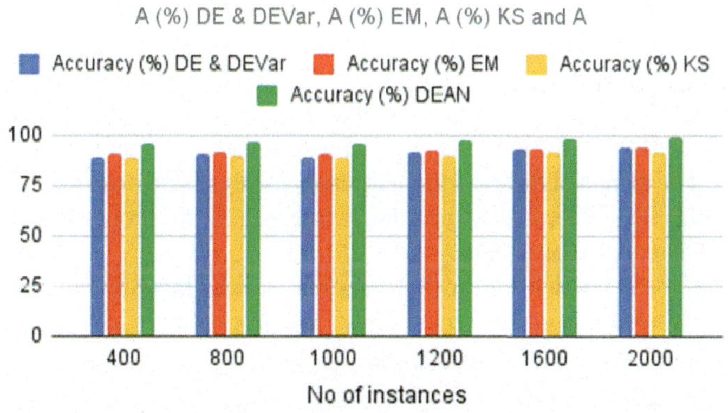

Fig. 4 Average accuracy for different disease types

Table 2 Average precision for different disease types

Size	P (%) DE_{Var}	P (%) EM	P (%) KS	P (%) DEAN
200	71.19	75.71	70.88	76.44
400	71.51	76.02	71.09	76.76
600	72.14	76.44	71.51	77.18
800	71.82	76.34	71.4	77.07
1000	72.14	76.55	71.61	77.28
1200	73.61	77.49	72.45	78.12
1400	74.24	77.91	72.87	78.65
1600	72.77	77.18	72.24	77.91
1800	73.08	77.49	72.45	78.23
2000	75.29	78.65	73.61	79.38

Table 3 Average recall for different disease types

Size	R (%) DE_{Var}	R (%) EM	R (%) KS	R (%) DEAN
200	79.91	85.16	79.59	85.89
400	80.33	85.37	79.91	86.21
600	80.85	85.79	80.33	86.63
800	81.38	86.21	80.64	86.94
1000	80.43	85.79	80.22	86.52
1200	82.53	86.94	81.38	87.78
1400	83.27	87.47	81.9	88.31
1600	82.95	87.47	81.8	88.2
1800	80.96	86.42	80.85	87.26
2000	84.63	88.52	82.74	89.36

The findings of a similar study for the AUC of brain cancer are summarized in Table 4, which shows that the suggested model achieves an average improvement of 8% over state-of-the-art methods across all datasets, i.e., higher than the percentages for Delaware (4%), Delaware County (6%), Eastern Marion County (9%), and Kansas State (8%). The results of this performance analysis suggest that the suggested model may be applicable in a clinical context that involves real-time processing of data from many disease networks and scenarios.

Similarly, the delay needed for these evaluations can be observed from Table 5. As per this evaluation, it can be observed that the proposed model is able to perform high-speed computations, thus can be used for delay-aware scenarios.

Table 4 AUC for different disease types

Size	AUC (%) DE_{Var}	AUC (%) EM	AUC (%) KS	AUC (%) DEAN
85.5	89.21	83.38	89.98	85.5
85.91	89.43	83.71	90.31	85.91
86.13	89.87	84.15	90.75	86.13
86.35	90.31	84.48	91.08	86.35
86.46	89.87	84.04	90.64	86.46
86.68	91.08	85.25	91.96	86.68
86.9	91.63	85.8	92.51	86.9
87.12	91.63	85.69	92.4	87.12
87.34	90.53	84.7	91.41	87.34

Table 5 Delay for different disease types

Size	$D(s)$ DE_{Var}	$D(s)$ EM	$D(s)$ KS	$D(s)$ DEAN
85.5	255.21	208.38	207.98	116.75
85.91	265.43	203.71	191.31	96.96
86.13	201.87	248.15	222.75	122.07
86.35	222.31	177.48	217.08	126.18
86.46	257.87	199.04	199.64	116.23
86.68	252.08	212.25	225.96	88.34
86.9	193.63	212.80	260.51	94.45
87.12	194.63	179.69	230.40	94.56
87.34	220.53	237.70	198.41	108.67
87.45	276.73	242.68	234.61	99.73

4.2 State-of-the-Art Comparisons

DEAN employs ensemble network aggregation, coefficient analysis, and adaptable hierarchical clustering, whereas N-GP [25] is primarily dependent on network topology and gene annotations. The selective ensemble approach of DEAN enables the detection and analysis of dual disease links, resulting in a more thorough comprehension of disease interactions. DEAN is more scalable than N-GP and achieves greater AUC performance across a variety of datasets.

MDA [26] focuses on identifying gene modules, whereas DEAN emphasizes aggregate coefficient analysis and hierarchical clustering. The DEAN framework enables a finer-grained analysis of the relationship between disease types, thereby shedding light on gene connections. While MDA provides information on co-expression patterns and functional relationships, DEAN enhances the accuracy of disease network construction by considering network aggregation and variance analysis.

GCNs [27, 28] employ deep learning techniques to analyze disease networks, employing graph neural networks for efficient gene prioritization. In contrast, DEAN integrates ensemble network aggregation, coefficient analysis, and hierarchical clustering. While GCNs offer the benefit of learning representations from graph structure, DEAN provides a systematic framework that employs multiple coefficient analysis methods and adaptable hierarchical clustering to enhance AUC performance. DEAN is distinguished from GCNs, which may be more specific to specific disease networks, by its scalability across diverse datasets. Combining ensemble network aggregation, coefficient analysis, and adaptable hierarchical clustering, the DEAN model presented in this paper offers a novel method for disease network analysis. This method differs from N-GP, MDA, and GCNs in that it allows for enhanced scalability, enhanced AUC performance, and a comprehensive understanding of inter-disease interactions.

5 Conclusion and Future Research Directions

The DE, DE_{Var}, EMD, and KS models are combined in the proposed DEAN model for efficient network design. This is augmented by automatic clustering, selected AUC based ensembling, and other criteria to choose the best set of models for the input dataset supplied. As a consequence of this choice, the proposed DEAN model achieves an accuracy of almost 90%, precision of 83%, recall of 80.5%, and AUC of 80.8% across many datasets. Due to its superior performance compared to most of the previously discussed models, DEAN is quite helpful for real-time deployments. The proposed model is proven to outperform state-of-the-art models on a number of metrics, including accuracy, precision, recall, and area under the curve (AUC), with improvements of 8% over DE, 3% over DE_{Var}, 12% over EMD, and about 8% over KS. Despite the fact that the paper proposes a novel framework for enhancing the scalability of disease networks, there are a number of limitations to consider.

The paper compares the performance of DEAN to that of existing methods, including DE, DE variance, KS, and EMD coefficient analysis. However, it does not provide a comparative analysis of the strengths, weaknesses, and applicability of these methods to various disease networks. A more thorough comparison would enhance comprehension of the proposed model's benefits and limitations. While the proposed DEAN model claims to be scalable to a variety of disease types, the paper does not provide detailed insights into how the framework can be adapted and extended to manage larger and more complex disease networks. The scalability of the model must be investigated further and validated on a variety of datasets. The paper does not validate the DEAN model using clinical or experimental data in the real world. The evaluation is based solely on computational analysis, which may not accurately reflect the complexity and variability of disease networks in real-world scenarios. Incorporating real-world data and validation would enhance the model's credibility and applicability. The paper does not delve deeply into the interpretability of the DEAN model's results. Even though it achieves superior performance compared to individual models, it is essential to comprehend the biological or genetic insights gleaned from the aggregated results. Increasing the model's interpretability would increase its utility for researchers and clinicians. The paper lacks an open-source implementation of the DEAN framework, making it difficult for other researchers to replicate and expand upon the proposed work.

By addressing these limitations, the proposed DEAN framework would be strengthened and its capabilities and potential impact on disease network analysis would be better understood. However, combining deep learning models like convolutional neural networks (CNN), bio-inspired models like genetic algorithms (GAs), and Q-learning models can help improve the performance of the proposed model in the future. Reinforcement learning algorithms and their analogs may also improve performance across datasets and their samples.

References

1. Ni P, Wang J, Zhong P, Li Y, Wu F-X, Pan Y (2020) Constructing disease similarity networks based on disease module theory. IEEE/ACM Trans Comput Biol Bioinform 17(3):906–915. https://doi.org/10.1109/TCBB.2018.2817624
2. Yan C, Duan G, Wu F-X, Pan Y, Wang J (2020) BRWMDA: predicting microbe-disease associations based on similarities and bi-random walk on disease and microbe networks. IEEE/ACM Trans Comput Biol Bioinform 17(5):1595–1604. https://doi.org/10.1109/TCBB.2019.2907626
3. Qin R, Duan L, Zheng H, Li-Ling J, Song K, Zhang Y (2021) An ontology-independent representation learning for similar disease detection based on multi-layer similarity network. IEEE/ACM Trans Comput Biol Bioinform 18(1):183–193. https://doi.org/10.1109/TCBB.2019.2941475
4. Xuan P, Zhan L, Cui H, Zhang T, Nakaguchi T, Zhang W (2022) Graph triple-attention network for disease-related LncRNA prediction. IEEE J Biomed Health Inform 26(6):2839–2849. https://doi.org/10.1109/JBHI.2021.3130110
5. Zhao Y, Ma B, Jiang P, Zeng D, Wang X, Li S (2021) Prediction of Alzheimer's disease progression with multi-information generative adversarial network. IEEE J Biomed Health Inform 25(3):711–719. https://doi.org/10.1109/JBHI.2020.3006925
6. Gao J, Tian L, Wang J, Chen Y, Song B, Hu X (2020) Similar disease prediction with heterogeneous disease information networks. IEEE Trans Nanobiosci 19(3):571–578. https://doi.org/10.1109/TNB.2020.2994983
7. Wang T, Tian Y, Qiu RG (2020) Long short-term memory recurrent neural networks for multiple diseases risk prediction by leveraging longitudinal medical records. IEEE J Biomed Health Inform 24(8):2337–2346. https://doi.org/10.1109/JBHI.2019.2962366
8. Luo J, Long Y (2020) NTSHMDA: prediction of human microbe-disease association based on random walk by integrating network topological similarity. IEEE/ACM Trans Comput Biol Bioinform 17(4):1341–1351. https://doi.org/10.1109/TCBB.2018.2883041
9. Sabo A, Gorodetsky C, Fasano A, Iaboni A, Taati B (2022) Concurrent validity of zeno instrumented walkway and video-based gait features in adults with Parkinson's disease. IEEE J Transl Eng Health Med 10:1–11. Art no. 2100511. https://doi.org/10.1109/JTEHM.2022.3180231
10. Gao X, Shi F, Shen D, Liu M (2022) Task-induced pyramid and attention GAN for multimodal brain image imputation and classification in Alzheimer's disease. IEEE J Biomed Health Inform 26(1):36–43. https://doi.org/10.1109/JBHI.2021.3097721

11. Basher A, Kim BC, Lee KH, Jung HY (2021) Volumetric feature-based Alzheimer's disease diagnosis from sMRI data using a convolutional neural network and a deep neural network. IEEE Access 9:29870–29882. https://doi.org/10.1109/ACCESS.2021.3059658
12. Zhao X, Yang Y, Yin M (2021) MHRWR: prediction of lncRNA-disease associations based on multiple heterogeneous networks. IEEE/ACM Trans Comput Biol Bioinform 18(6):2577–2585. https://doi.org/10.1109/TCBB.2020.2974732
13. Ashtari-Majlan M, Seifi A, Dehshibi MM (2022) A multi-stream convolutional neural network for classification of progressive MCI in Alzheimer's disease using structural MRI images. IEEE J Biomed Health Inform 26(8):3918–3926. https://doi.org/10.1109/JBHI.2022.3155705
14. Ghulam A, Lei X, Guo M, Bian C (2020) Disease-pathway association prediction based on random walks with restart and PageRank. IEEE Access 8:72021–72038. https://doi.org/10.1109/ACCESS.2020.2987071
15. Tanveer M, Rashid AH, Ganaie MA, Reza M, Razzak I, Hua K-L (2022) Classification of Alzheimer's disease using ensemble of deep neural networks trained through transfer learning. IEEE J Biomed Health Inform 26(4):1453–1463. https://doi.org/10.1109/JBHI.2021.3083274
16. Song Y, Cui H, Zhang T, Yang T, Li X, Xuan P (2022) Prediction of drug-related diseases through integrating pairwise attributes and neighbor topological structures. IEEE/ACM Trans Comput Biol Bioinform 19(5):2963–2974. https://doi.org/10.1109/TCBB.2021.3089692
17. Grani G, Madeddu L, Velardi P (2022) A network-based analysis of disease modules from a taxonomic perspective. IEEE J Biomed Health Inform 26(4):P1773–P1781. https://doi.org/10.1109/JBHI.2021.3106787
18. Zhang W, Li Z, Guo W, Yang W, Huang F (2021) A fast linear neighborhood similarity-based network link inference method to predict microRNA-disease associations. IEEE/ACM Trans Comput Biol Bioinform 18(2):405–415. https://doi.org/10.1109/TCBB.2019.2931546
19. Ding Y, Lei X, Liao B, Wu F-X (2022) Predicting miRNA-disease associations based on multi-view variational graph auto-encoder with matrix factorization. IEEE J Biomed Health Inform 26(1):446–457. https://doi.org/10.1109/JBHI.2021.3088342
20. Kamal U, Zunaed M, Nizam NB, Hasan T (2022) Anatomy-XNet: an anatomy aware convolutional neural network for thoracic disease classification in chest X-rays. IEEE J Biomed Health Inform 26(11):5518–5528. https://doi.org/10.1109/JBHI.2022.3199594
21. Zhang X, Han L, Zhu W, Sun L, Zhang D (2022) An explainable 3D residual self-attention deep neural network for joint atrophy localization and Alzheimer's disease diagnosis using structural MRI. IEEE J Biomed Health Inform 26(11):5289–5297. https://doi.org/10.1109/JBHI.2021.3066832
22. Ouyang J et al (2021) Longitudinal pooling & consistency regularization to model disease progression from MRIs. IEEE J Biomed Health Inform 25(6):2082–2092. https://doi.org/10.1109/JBHI.2020.3042447
23. Ma Y (2022) DeepMNE: deep multi-network embedding for lncRNA-disease association prediction. IEEE J Biomed Health Inform 26(7):3539–3549. https://doi.org/10.1109/JBHI.2022.3152619
24. Xuan P, Zhan L, Cui H, Zhang T, Nakaguchi T, Zhang W (2022) Graph triple-attention network for disease-related LncRNA prediction. IEEE J Biomed Health Inform 26(6):2839–2849. https://doi.org/10.1109/JBHI.2021.3130110
25. Jiang C, Tang M, Jin S, Huang W, Liu X (2023) KGNMDA: a knowledge graph neural network method for predicting microbe-disease associations. IEEE/ACM Trans Comput Biol Bioinform 20(2):1147–1155. https://doi.org/10.1109/TCBB.2022.3184362
26. Nandy S, Adhikari M, Hazra A, Mukherjee T, Menon VG (2023) Analysis of communicable disease symptoms using bag-of-neural network at edge networks. IEEE Sens J 23(2):914–921. https://doi.org/10.1109/JSEN.2022.3167416
27. Liu J, Kuang Z, Deng L (2023) GCNPCA: miRNA-disease associations prediction algorithm based on graph convolutional neural networks. IEEE/ACM Trans Comput Biol Bioinform 20(2):1041–1052. https://doi.org/10.1109/TCBB.2022.3203564
28. Verma S, Razzaque MA, Sangtongdee U, Arpnikanondt C, Tassaneetrithep B, Hossain A (2021) Digital diagnosis of hand, foot, and mouth disease using hybrid deep neural networks. IEEE Access 9:143481–143494. https://doi.org/10.1109/ACCESS.2021.3120199

An Ensemble Technique for Predicting the Human Heart Disease

Uttej Kumar Nannapaneni, Satish Kumar Satti, B. Himaja, K. Naga Poojitha, and K. Harshini

Abstract Heart disease remains a pressing global health issue, demanding the development of precise and efficient predictive models for early detection and prevention. In this investigation, we explore the potential of a voting classifier algorithm for forecasting heart disease in individuals. The dataset utilized in this research comprises comprehensive clinical and demographic data gathered from medical examinations of patients. A voting classifier harnesses a diverse set of base classifiers to make heart disease predictions. The assessment of the voting classifier model involves the use of various performance metrics. The test results validate the accuracy of the voting classifier in predicting heart disease. The model attains a notably high accuracy rate, surpassing individual classifiers and illustrating the benefits of ensemble methods in enhancing predictive accuracy.

Keywords Heart disease · Voting classifier · Ensemble learning

1 Introduction

Nowadays heart-related diseases are increasing vastly and they are leading to death. To reduce them we need to make a case study on heart-related diseases. Many people are dying without knowing the cause of the disease heart diseases encompass a wide range of conditions, including cardiovascular disease, high blood pressure, coronary artery disease, aortic disease, peripheral arterial disease, and valve disease [1]. Heart disease has received a lot of attention in medical research among other potentially fatal illnesses. Predicting a patient's heart condition through computer-generated predictions has emerged as a challenging yet crucial task, aiming to enhance the effectiveness of subsequent treatment interventions. Heart disease prediction has been researched for many years and new findings are under the development process. However, the new findings indicate that may not be wise to limit certain types of trans-fats in the diet, especially when considering certain health conditions. The

U. K. Nannapaneni · S. K. Satti (✉) · B. Himaja · K. Naga Poojitha · K. Harshini
Department of CSE, VFSTR, Deemed to be University, Guntur, Andhra Pradesh, India
e-mail: sskumar789@gmail.com

S. Bandyopadhyay et al. (eds.), *Intelligent Computing Systems and Applications*, Lecture Notes in Networks and Systems 1010,
https://doi.org/10.1007/978-981-97-5412-0_9

most important task in the healthcare profession is detecting the illness. If an illness is detected earlier, then we can save many people's lives from death. Nowadays, the prevalence of heart disease has witnessed a significant rise globally in recent years. It is widely acknowledged that heart disease is a condition that can be effectively managed and prevented. Heart strokes primarily occur as a result of arterial blockages and the inefficient pumping of blood by the heart throughout the body [2].

Various factors contribute to the development of heart disease, including inadequate nutrition, obesity, high cholesterol levels, and a lack of physical activity. Nowadays, human beings' lifespans are reduced because of this heart disease. Detecting and predicting heart disease was a huge difficulty, has historically been difficult. Although numerous devices are available for heart disease prediction, their high cost and limited effectiveness have posed significant drawbacks. Prediction of heart disease in the early stage means it plays a crucial role in giving treatment to the patient. So, if it was predicted before only it leads to stopping the deaths of the patient due to heart disease, and also, we can provide efficient treatment to the patients [3]. The symptoms of CVD varied for different genders. For example, the chest pain will be more for the male patient rather than the female patient. female patient will have other symptoms like shortness of breath and extreme fatigue. The main key point of those medical systems is obtaining accurate rates efficiently at low cost. Machine learning is an efficient technology for testing and training the data.

Based on the input information, the models are trained to process data and make predictions. Biological parameters such as blood pressure, age, cholesterol, and gender are used in the testing data in this study. By training the data with various algorithms, the accuracy values will be compared to assess their performance. Recently machine learning models have been working very efficiently in analyzing, evaluating, and predicting different medical conditions. Machine Learning could be an excellent alternative for forecasting heart disease in humans, which is a huge difficulty in medical science. Machine learning algorithms [4], including Random Forest (RF), Support Vector Machine Classifier (SVC) [5], Logistic Regression, and K-Nearest Neighbors (KNN) [6], can be employed for heart disease prediction. RF is one of the ML algorithms [7] that is used to predict heart disease. The random forest classification technique is used to evaluate of IHDSSO model. In random forest (RF), the prediction capability is enhanced through the utilization of bagging, which involves aggregating multiple decision trees (DT).

In this approach, each learner is independently trained through the process of bagging, where random samples of the data are generated with replacement. This technique improves the overall performance of the random forest model. The distinct model of data will be trained on each DT and, the unique datasets. During the creation of each tree, features are randomly selected. To enhance the accuracy of the random forest (RF), parameters such as the number of estimators, the number of features used to split nodes, and the number of characteristics can be optimized. We will be getting a few benefits and challenges while using this random forest algorithm. Here, are the few benefits of RF, it reduces the risk of overfitting, provides flexibility, and makes it easy to determine the importance of the feature. Here, are the few challenges we are facing while using this RF algorithm it is a time-consuming process, requires more resources, and is more complex. Now, SVC is one of the machine learning algorithms that can be used for heart disease prediction. SVC can be used to build a predictive model that can classify patients into different categories, such as those with or without heart disease which is based on the set of input features.

This SVC will be trained on a labeled dataset which consists of patient statues associated with the features. The training data was used to learn the relationship between input and output features. It is one of the most popular ML algorithms and it is also a supervised learning technique. Logistic regression was employed to predict the categorical dependent variable based on a given set of independent variables. This method enables the prediction of the output for the categorical dependent variable. In the logistic regression, we will have values of either yes or no, 0 or 1, true or false, etc.

2 Literature Survey

Dhai et al. [8] suggested a method to predict heart disease. It uses 3 data analytics techniques: neural networks, KNN, and SVM. These were applied to a different dataset of various sizes among these 3 they achieved a maximum accuracy of 93% by using a neural network. The dataset they used contains 1200 attributes. Guo et al. [9] an improved RF algorithm is employed and obtained 96.6% accuracy, 96.8% stability ratio, and 96.7% F-measure. The dataset they used was taken from the administrative dataset and named the cardiovascular disease dataset. Anbuselvan [10] together the authors used 3 algorithms they are random forest which gives the best accuracy of 86.89%, k-nearest neighbors which gives the worst accuracy of 57.83%, and XGBoost of accuracy of 78.69%. The dataset contains 303 attributes acquired from the UCI repository. The authors also describe the health conditions of people these days. Rababa et al. [11] used a dataset containing 299 patient data to predict heart disease.

Li et al. [12] the system employs a variety of classification techniques, including SVM, LR, ANN, KNN, NB, and DT, to reduce unnecessary and duplicate variables. Cenitta et al. [13] proposed utilizing a squirrel search algorithm with a random forest algorithm to detect cardiac disease. Williams et al. [14] have proposed Heart Disease Prediction Using Machine Learning Techniques, Artificial Neural Networks, Decision Trees, naive Bayes, Random Forests, Support Vector Machines, and XGBoost. This model is trained and tested on 14 attribute data sets. KNN classifier with 67.21% accuracy, NB classifier with 85.25% accuracy, Decision tree classifier with 81.97% accuracy, SVM with 81.97% accuracy, logistic Regression classifier with 85.25% accuracy, Random Forest classifier with 95.08% accuracy and XGBoost classifier with 80.33% accuracy.

Saw et al. [15] have suggested the prediction of heart disease using a LR model. Pan et al. [16] have suggested a CNN to predict heart disease. Karayılan [17] proposed a neural network backpropagation technique to detect heart disease.

Javeed et al. [18] authors have used a random forest algorithm and performed two types of experiments using online heart database failure which has 7 features, compared to other methods this method produces 3.3% higher accuracy and attained 93.33% accuracy. Wang et al. [19] the dataset is obtained from the hospital of 35,101 hospitalizations with HF diagnosis in the previous 18 years and employed a random forest with an accuracy of 93.60%, logistic regression with an accuracy of 92.333%, and deep neural network with an accuracy of 93.70%.

Ali et al. [20] the author mainly focused on overfitting the data and introduced DNN and evaluated the performance with ANN and DNN models and obtained an accuracy of 93.33% and predicted accurate results which can also be used by physicians.

Ordonez's [21] set of association rules is reduced to some set of rules so that we can reduce the time to predict the outputs. The author created an algorithm that successfully addresses limits by applying search constraints to minimize the number of rules generated while exploring association rules on real datasets. The procedure entails looking for association rules in a training set and then validating them on an independent test set. To assess the medical relevance of the identified rules, assessment measures such as support, confidence, and lift are used. The application of association rules is demonstrated using a genuine dataset consisting of medical records from patients diagnosed with heart disease.

Mohan et al. [22] the prevalence of cardiovascular disease as a major contributor to mortality underscores the criticality of accurate prediction and analysis of this condition in clinical data. Machine learning has emerged as a well-established and effective approach in healthcare decision-making, offering valuable support in this domain.

3 Proposed Methodology

Heart disease constitutes a significant contributor to global mortality, and its early detection plays a pivotal role in averting severe health complications. Leveraging machine learning techniques, we have harnessed the power of patient data analysis, encompassing factors such as medical history, demographics, and lifestyle choices, to predict the onset of heart disease. This intricate process entails the meticulous collection and preprocessing of patient information to render it compatible with machine learning algorithms. Subsequently, we identify the most pertinent features for heart disease prediction through a combination of statistical tests, domain expertise, and cutting-edge machine-learning methodologies. Armed with this knowledge, we meticulously select an appropriate machine learning algorithm and train the model using a dataset, optimizing its parameters and minimizing prediction errors. The model's efficacy is then thoroughly evaluated using a dedicated dataset, employing a suite of metrics including accuracy, precision, recall, and the $F1$ score to comprehensively gauge its performance. Finally, the fully-fledged model finds its place

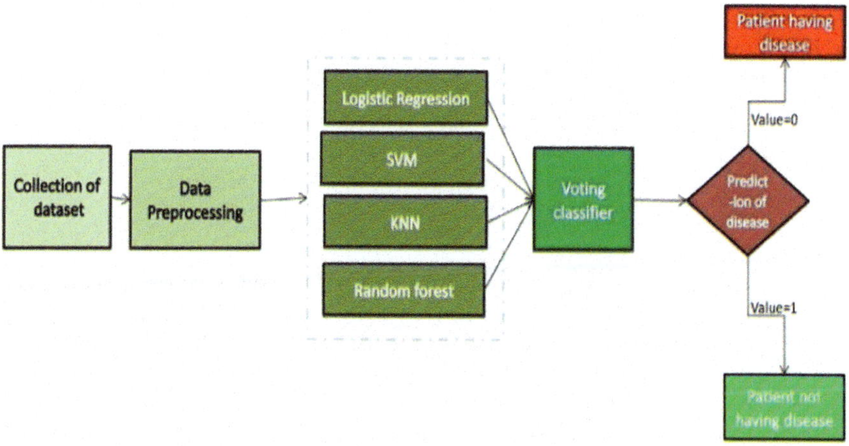

Fig. 1 Proposed pipeline

in clinical settings, aiding healthcare professionals in assessing individual patient's risk of developing heart disease. The proposed heart disease prediction pipeline is visually illustrated in Fig. 1.

3.1 Data-Set

The data set used in the proposed method contains 1191 patient records and has 13 features which are given in Table 1.

3.2 Data Preprocessing

To predict heart disease using machine learning, data processing is essential. It involves collecting patient data, cleaning, normalizing, selecting relevant features, and encoding categorical data. Initially, the data collected goes for a check of null values and duplicates in the dataset [23, 24]. This data preprocessing is done in such a way that the null values are replaced manually by filling in the required constraints and duplicates are deleted by the machine. Then, the data is split into training, validation, and testing sets. A suitable algorithm is selected, and the model is trained using the training data. The model's performance is evaluated using the testing data, and the final model is deployed in a clinical setting. By following these steps, machine learning can assist physicians in predicting heart disease risk for individual patients, leading to earlier detection and prevention of heart disease.

Table 1 Features details from data-set

S. No.	Features	Description
1	Age	Patient ages in years
2	Sex	Male or female; 1-male, 0-female
3	Cp	Chest pain type experienced by the patient contains 4 levels: 0, 1, 2, 3 0 means chest pain whereas 3 means high chest pain
4	Thesbps	Resting blood pressure
5	Chol	Cholesterol level
6	Restecg	Resting electrocardiographic results of the patient, measured in 3 levels: 0, 1, 2
7	Fbs	Fasting blood sugar level, if the level is greater than 120 mg/dl, then 1 is assigned, otherwise 0 is assigned
8	Thalach	Maximum heart rate
9	Exang	Exercise-induced angina. If angina was present during exercise, then 1 is assigned, otherwise, 0 is assigned
10	Oldpeak	Exercise causes ST depression compared to rest
11	Slope	The pea exercise ST segment's slope. It is measured in 3 levels: 0, 1, 2
12	Ca	There are no significant vessels (0–3) fluorescence coloring
13	Thal	Thalassemia. It is measured in three levels: 1, 2, 3

3.3 *Ensemble Classifier*

The following are machine learning algorithms that opted for developing an ensemble classifier [25] to predict heart disease.

Heart disease prediction using Logistic Regression LR is a statistical approach used to estimate the likelihood of a binary event, such as whether heart disease exists or not. It uses multiple predictor variables, including demographic, lifestyle, and clinical factors, to estimate the probability of heart disease. The first step is to gather a dataset with these variables and an outcome variable indicating heart disease. The data is separated into two sets to enable model training: a training set and a validation set. Following that, the model is fitted to the training data using maximum likelihood estimation, yielding coefficients that estimate the influence of each predictor variable on the log chances of heart disease. These coefficients can be exponentiated to produce odds ratios. The performance of the model is assessed by employing metrics such as accuracy. Upon successful validation, the model becomes capable of predicting the probability of heart disease for new individuals based on their unique personal characteristics. The model can also identify factors strongly associated with heart disease, helping clinicians and researchers understand and prevent the condition.

Heart disease prediction using Random Forest It is a ML algorithm that uses previous patient data to estimate the probability of heart disease, the model utilizes the patient's risk factors as input, enabling it to predict the likelihood of developing the condition. To use this method, data on various risk factors such as age, gender, family history, smoking status, cholesterol levels, and blood pressure are collected and prepared. The acquired data is utilized to train the random forest algorithm, which allows it to recognize patterns and relationships between risk variables and the prevalence of heart disease. Once trained, the algorithm can forecast the likelihood of heart disease in new patients by analyzing their risk variables. It is important to note that while random forests can be useful in predicting heart disease, they are not perfect and other factors must be considered in making a diagnosis or treatment decision. Therefore, clinical judgment and additional testing should also be taken into account. In summary, the random forest can be an effective tool for predicting heart disease but should be used in conjunction with other methods and medical expertise.

Heart disease prediction using Support Vector Machine SVMs are a type of machine learning algorithm that may predict heart disease in people by analyzing data from past patients, such as risk factors and whether or not they got heart disease [26]. To use SVMs for prediction, data must be gathered and prepared, including identifying associated risk factors. After collecting and cleaning the data, the SVM algorithm is trained on prior patient data to discover the patterns and correlations between risk variables and heart disease. The SVM model determines the best hyperplane that successfully divides the two classes throughout the training process. This separation is accomplished by using a kernel function to translate the input data into a higher-dimensional space. Once the SVM is trained, it can accurately predict the probability of heart disease in a new patient by evaluating their risk factors. This prediction is made by utilizing the decision boundary learned during the training phase. It's important to note that while SVMs can be useful for predicting heart disease, they are not perfect and may not accurately predict every case. Clinical judgment and other factors should also be considered in making a diagnosis or treatment decision.

Heart disease prediction using K-Nearest Neighbors The KNN algorithm predicts outcomes based on input data. In predicting heart disease, KNN identifies patients similar to those who have or have not developed heart disease in the past. The process starts by gathering and preparing data on the risk factors associated with heart disease [27]. The determination of the value K, representing the number of neighbors to consider during prediction, follows. The algorithm creates a model using training data by calculating the distance between each patient and others based on their risk factors. It then selects the K patients closest to the new patient and predicts whether they have heart disease based on their labels. If the majority of the K neighbors have heart disease, the algorithm predicts the new patient has heart disease as well, and vice versa. However, KNN may not accurately predict every case, so it's essential to use clinical judgment and consider other factors.

Heart disease prediction using Voting Classifier This study used four different machine learning models: KNN, SVM, RF, and LR. These models were trained independently on the same dataset, yielding unique predictions. The voting classifier technique was used to combine these individual model predictions into a final prediction, leveraging the diversity and knowledge of each model. A voting strategy, either hard or soft voting, is applied to determine the final prediction. Hard voting chooses the class label that receives the most votes, while soft voting relies on the average predicted probabilities of the individual models. By combining the insights of these models, the voting classifier produces a more informed prediction for heart disease classification. The voting strategy and the weights assigned to each model, if applicable, may impact the performance of the voting classifier. Experimenting with different strategies and weights may be necessary to achieve the optimal configuration for the heart disease prediction task.

4 Results and Discussions

The experimentation was carried out exclusively on Windows 10 Pro workstations, each equipped with an NVIDIA Quadro RTX 4000 graphics card, and the experiments were programmed using Python 3.0. After gathering the essential data set, data augmentation techniques are used to increase its size. To predict human cardiac disorders, an ensemble voting classifier is utilized. Figures 2, 3, 4 and 5 depict the confusion matrix of all the adopted ML models trained on our dataset.

Figure 2a provides a visual representation of the training confusion matrix for the logistic regression model. In the x-axis predicted rate and y-axis, true labels are taken. Where the label 0 implies that there is no heart illness and 1 indicates that there is heart disease. We evaluated 99 people who did not have an illness; 88 of them were accurately predicted, whereas 11 were predicted as having heart disease.

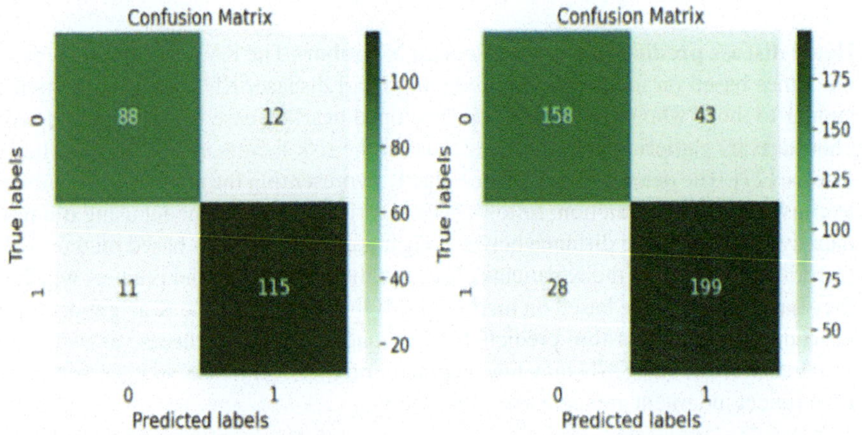

(a) Logistic regression-Training (b) Logistic regression-Testing

Fig. 2 Confusion matrix of logistic regression

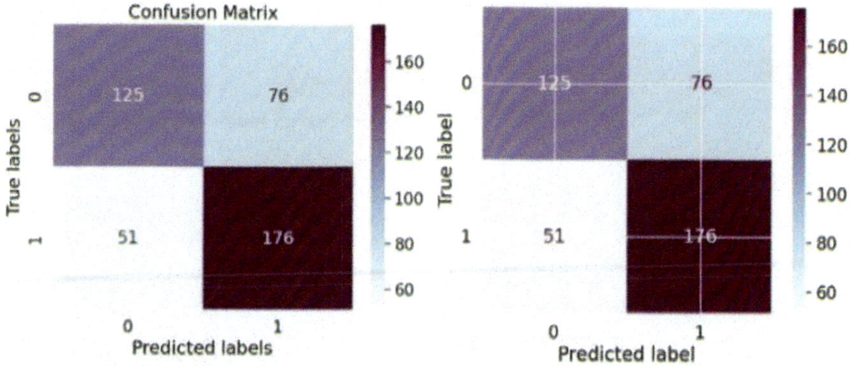

(a) SVM-Training (b) SVM-Testing

Fig. 3 Confusion matrix of SVM

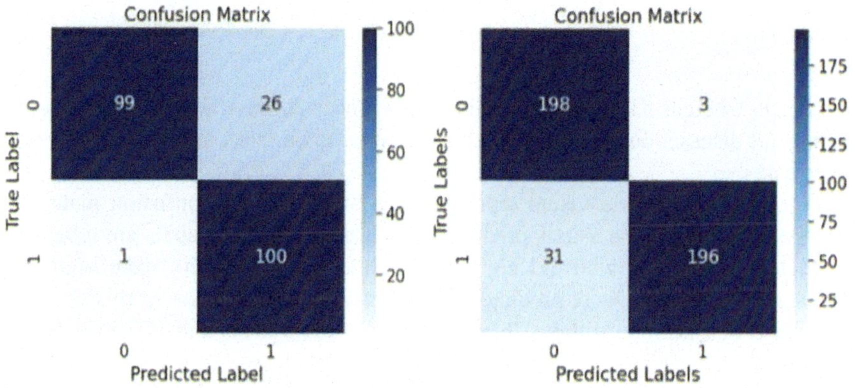

(a) KNN-Training (b) KNN-Testing

Fig. 4 Confusion matrix of KNN

There are 127 people with heart disease, out of them 12 anticipated to be false, and 115 expected to be true. Figure 2b provides a visual representation of the testing confusion matrix for the logistic regression model. In the x-axis predicted rate and y-axis, true labels are taken. Where the label 0 implies that there is no heart illness and 1 indicates that there is heart disease. We evaluated 186 people who did not have an illness; 158 of them were accurately predicted, whereas 28 were predicted as having heart disease. There are 242 people with heart disease, out of them 43 anticipated to be false, and 199 expected to be true.

Figure 3a provides a visual representation of the training confusion matrix for the SVM model. In the x-axis predicted rate and y-axis, true labels are taken. Where the label 0 implies that there is no heart illness and 1 indicates that there is heart disease. We evaluated 176 people who did not have an illness; 125 of them were accurately

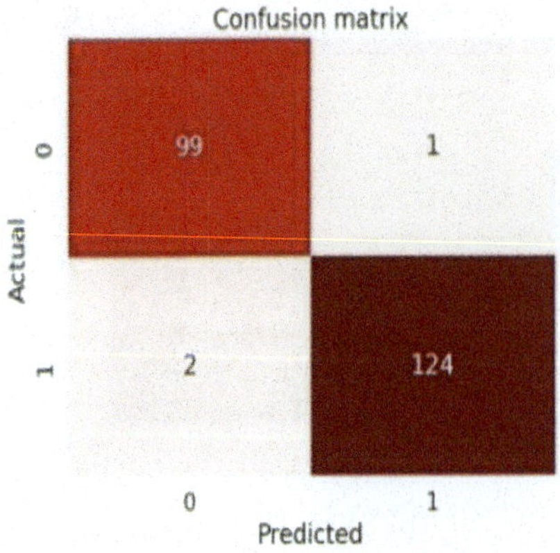

(a) RF-Training

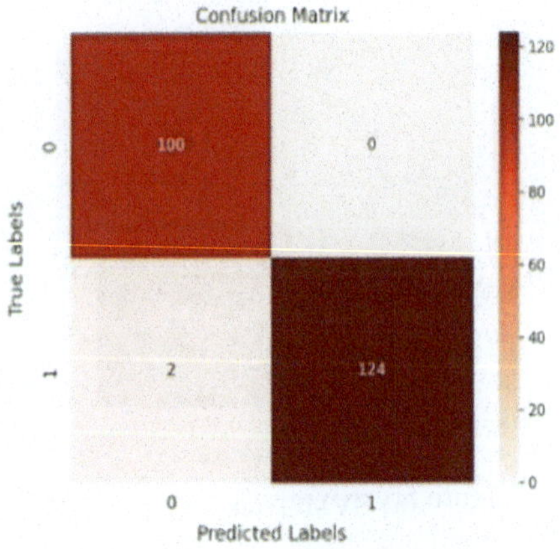

(b) RF-Testing

Fig. 5 Confusion matrix of RF

predicted, whereas 51 were predicted as having heart disease. There are 252 people with heart disease, out of them 76 are anticipated to be false, and 176 are expected to be true.

Figure 3b provides a visual representation of the testing confusion matrix for the SVM model. In the *x*-axis predicted rate and *y*-axis, true labels are taken. We evaluated 176 people who did not have an illness; 125 of them were accurately predicted, whereas 51 were predicted as having heart disease. There are 252 people with heart disease, out of them 76 are anticipated to be false, and 176 are expected to be true.

Figure 4a provides a visual representation of the training confusion matrix for the KNN model. In the *x*-axis predicted rate and *y*-axis, true labels are taken. We evaluated 100 people who did not have an illness; 99 of them were accurately predicted, whereas 1 was predicted as having heart disease. There are 126 people with heart disease, out of them 26 are anticipated to be false, and 100 are expected to be true.

Figure 4b provides a visual representation of the testing confusion matrix for the KNN model. In the *x*-axis predicted rate and *y*-axis, true labels are taken. We evaluated 229 people who did not have an illness; 198 of them were accurately predicted, whereas 31 were predicted as having heart disease. There are 199 people with heart disease, out of them 3 are anticipated to be false, and 196 are expected to be true. Figure 5a provides a visual representation of the training confusion matrix for the RF model. In the *x*-axis predicted rate and *y*-axis, true labels are taken. We evaluated 101 people who did not have an illness; 99 of them were accurately predicted, whereas 2 were predicted as having heart disease. There are 125 people with heart disease, out of them 1 is anticipated to be false and 124 are expected to be true.

Table 2 Comparative analysis of proposed model

Method	Precision		Recall		Accuracy	
	Train (%)	Test (%)	Train (%)	Test (%)	Train (%)	Test (%)
Logistic regression	90	84	90	83	90	83
SVM	72	70	72	70	73	70
KNN	92	91	92	89	92	88
Random forest	95	93	93	92	94	93
Ensemble classifier	97	95	94	93	96	95

Figure 5b provides a visual representation of the testing confusion matrix for the RF model. In the x-axis predicted rate and y-axis, true labels are taken. We evaluated 102 people who did not have an illness; 100 of them were accurately predicted, whereas 2 are predicted as having heart disease. There are 124 people with heart disease, out of them 0 anticipated to be false, and 124 expected to be true.

In logistic regression for precision, we obtained 90% for training and 84% for testing, for recall we got 90% for training and 83% for testing, and we got 90% and 83% of training and testing accuracy respectively. In SVM for precision, we have obtained 72% for training and 70% for testing, for recall, we got 72% for training and 70% for testing, and we got 73% and 70% of training and testing accuracy respectively. In KNN for precision, we obtained 92% for training and 91% for testing, for recall, we got 92% for training and 89% for testing, and we got 92% and 84% of training and testing accuracy respectively. In Random Forest for precision, we obtained 95% for training and 93% for testing, for recall we got 93% for training and 92% for testing, and we got an accuracy of 94% for training and 93% for testing. In the voting classifier for precision, we obtained 97% for training and 95% for testing, for recall we got 94% for training and 93% for testing, and we got 96% and 95% of training and testing accuracy respectively (Table 2).

5 Conclusions

Machine learning, through the analysis of extensive datasets gathered from diverse sources, has displayed significant potential in predicting heart disease. It unveils patterns and risk factors that conventional approaches may struggle to discern. Multiple techniques, including decision trees, logistic regression, random forests, and neural networks, have been employed to build predictive models for cardiac conditions. These models have consistently delivered high levels of accuracy and specificity in identifying individuals susceptible to heart disease. Additionally, machine learning algorithms hold promise in forecasting distinct cardiac ailments, such as coronary artery disease and congestive heart failure.

References

1. Singh A, Kumar R (2020) Heart disease prediction using machine learning algorithms. In: 2020 international conference on electrical and electronics engineering (ICE3), Feb 2020. IEEE, pp 452–457
2. Satti SK, Devi KS, Dhar P, Srinivasan P (2021) A machine learning approach for detecting and tracking road boundary lanes. ICT Express 7(1):99–103
3. Yadav A, Gediya L, Kazi A (2021) Heart disease prediction using machine learning. Int Res J Eng Technol (IRJET) 8(09)
4. Satti SK, Devi KS, Dhar P, Srinivasan P (2020) Enhancing and classifying traffic signs using computer vision and deep convolutional neural network. In: Machine learning, image processing, network security and data sciences: second international conference, MIND 2020, Silchar, India, 30–31 July 2020, proceedings, part I 2. Springer Singapore, pp 243–253
5. Boukhatem C, Youssef HY, Nassif AB (2022) Heart disease prediction using machine learning. In: 2022 advances in science and engineering technology international conferences (ASET), Feb 2022. IEEE, pp 1–6
6. Sun H, Pan J (2023) Heart disease prediction using machine learning algorithms with self-measurable physical condition indicators. J Data Anal Inf Process 11(1):1–10
7. Satti SK, Maddula P, Ravipati NV (2021) Unified approach for detecting traffic signs and potholes on Indian roads. J King Saud Univ Comput Inf Sci
8. Salhi DE, Tari A, Kechadi MT (2021) Using machine learning for heart disease prediction. In: Advances in computing systems and applications: proceedings of the 4th conference on computing systems and applications. Springer International Publishing, pp 70–81
9. Guo C, Zhang J, Liu Y, Xie Y, Han Z, Yu J (2020) Recursion enhanced random forest with an improved linear model (RERF-ILM) for heart disease detection on the internet of medical things platform. IEEE Access 8:59247–59256
10. Anbuselvan P (2020) Heart disease prediction using machine learning techniques. Int J Eng Res Technol 9:515–518
11. Rababa S, Yamin A, Lu S, Obaidat A (2023) Predicting heart disease and reducing survey time using machine learning algorithms. arXiv preprint arXiv:2306.00023
12. Li JP, Haq AU, Din SU, Khan J, Khan A, Saboor A (2020) Heart disease identification method using machine learning classification in e-healthcare. IEEE Access 8:107562–107582
13. Cenitta D, Arjunan RV, Prema KV (2022) Ischemic heart disease prediction using optimized squirrel search feature selection algorithm. IEEE Access 10:122995–123006
14. Williams R, Shongwe T, Hasan AN, Rameshar V (2021) Heart disease prediction using machine learning techniques. In: 2021 international conference on data analytics for business and industry (ICDABI), Oct 2021. IEEE, pp 118–123
15. Saw M, Saxena T, Kaithwas S, Yadav R, Lal N (2020) Estimation of prediction for getting heart disease using logistic regression model of machine learning. In: 2020 international conference on computer communication and informatics (ICCCI), Jan 2020. IEEE, pp 1–6
16. Pan Y, Fu M, Cheng B, Tao X, Guo J (2020) Enhanced deep learning-assisted convolutional neural network for heart disease prediction on the internet of medical things platform. IEEE Access 8:189503–189512
17. Karayılan T, Kılıç Ö (2017) Prediction of heart disease using neural network. In: 2017 international conference on computer science and engineering (UBMK), Oct 2017. IEEE, pp 719–723
18. Javeed A, Zhou S, Yongjian L, Qasim I, Noor A, Nour R (2019) An intelligent learning system based on a random search algorithm and optimized random forest model for improved heart disease detection. IEEE Access 7:180235–180243
19. Wang B, Bai Y, Yao Z, Li J, Dong W, Tu Y, He K (2019) A multi-task neural network architecture for renal dysfunction prediction in heart failure patients with electronic health records. IEEE Access 7:178392–178400
20. Ali L, Rahman A, Khan A, Zhou M, Javeed A, Khan JA (2019) An automated diagnostic system for heart disease prediction based on χ^2 statistical model and optimally configured deep neural network. IEEE Access 7:34938–34945

21. Ordonez C (2006) Association rule discovery with the train and test approach for heart disease prediction. IEEE Trans Inf Technol Biomed 10(2):334–343
22. Mohan S, Thirumalai C, Srivastava G (2019) Effective heart disease prediction using hybrid machine learning techniques. IEEE Access 7:81542–81554
23. Satti SK, Suganya Devi K, Dhar P, Srinivasan P (2019) An efficient noise separation technique for removal of Gaussian and mixed noises in monochrome and color images. Int J Innov Technol Explor Eng 8:588
24. Satti SK, Murthy RV, Srinivasan DP (2019) Efficient technique for removal of white and mixed noises in gray scale images. Int J Innov Eng Manage Res 8(09)
25. Satti SK, Suganya Devi K, Srinivasan P (2022) R-ICTS: recognize the Indian cautionary traffic signs in real-time using an optimized adaptive boosting cascade classifier and a convolutional neural network. Concurr Comput Pract Exp 34(10):e6796
26. Dhar P, Suganya Devi K, Satti SK, Srinivasan P (2021) Detail study of different algorithms for early detection of cancer. In: Health informatics: a computational perspective in healthcare, pp 207–232
27. Dhar P, Suganya Devi K, Satti SK, Srinivasan P (2022) Efficient detection and partitioning of overlapped red blood cells using image processing approach. Innov Syst Softw Eng 1–13

Enhancing Cybersecurity Metrics Evaluation Through the Application of Fuzzy Ahp Methodology

Seema Gupta Bhol, J. R. Mohanty, and P. K. Pattnaik

Abstract To successfully run an enterprise in this competitive era with new security threats emerging every day, a strong security system is the need of the hour. The objective of cybersecurity is to protect data and important information from unauthorized persons. They can be strangers or people working for the organization. Weakness in the existing system and potential threats from the attackers are important to build appropriate defense mechanism. The interaction between protection mechanism and potential risks leads to many situations. In the proposed paper, the hierarchy of security metrics is described. Security metrics evaluation depends upon various factors, and thus, multiple criteria decision-making methods can be applied. Various parameters defining cybersecurity need both qualitative and quantitative measurement and handling, and thus, concepts of fuzzy set theory can be incorporated to handle qualitative data successfully. Criteria priories are represented using fuzzy triangular scales. Fuzzy AHP techniques (Buckley's method) are used in the proposed paper for calculating metrics defining cybersecurity.

Keywords Security metrics · Multiple criteria decision-making · Analytic hierarchy process · Taxonomy · Fuzzy AHP · Buckley's method

S. G. Bhol (✉) · J. R. Mohanty · P. K. Pattnaik
Kalinga Institute of Industrial Technology, Bhubaneswar, India
e-mail: seemaguptabhol.fca@kiit.ac.in

J. R. Mohanty
e-mail: jmohantyfca@kiit.ac.in

P. K. Pattnaik
e-mail: patnaikprasantfcs@kiit.ac.in

S. Bandyopadhyay et al. (eds.), *Intelligent Computing Systems and Applications*, Lecture Notes in Networks and Systems 1010,
https://doi.org/10.1007/978-981-97-5412-0_10

1 Introduction

Cybersecurity can be defined as the mechanism of protecting systems, data, networks, programs, and other resources of the organization against malicious users and fraudulent transactions. It is achieved by combining people, technology, processes, and practices into a unified approach. Confidentiality, integrity, and availability of information are the objectives of security. According to NIST [1], metrics are tools that are designed to facilitate decision-making and improve performance and accountability through the collection, analysis, and reporting of relevant performance-related data. Metrics are the standard system of measuring quantifiable characteristics of some object, process, or activity. However, Savola [2] differentiated metrics from measurements. A measurement corresponds to a fixed view of particular factors, whereas metrics get computed by the comparison of several measurements during a period. Tashi and Solange [3] in his paper explained that counting yields measurements but analysis produces metrics. Cybersecurity cannot be measured directly, and there is no standard set of attributes available to identify derived measures [4]. The literature does not support any standard taxonomy to explain cybersecurity metrics. It is very important to have security metrics for cybersecurity processes and activities to guard organizations against data theft, hackers, and other fraudulent user. Assurance of uninterrupted business activities and minimizing the impact in case of any cybersecurity breach is crucial for all organizations [5]. Taxonomy was proposed, where security metrics were classified into four classes.

Vulnerabilities in the system, defense mechanism deployed, threats landscape, and the encounter outcomes [6].

Metrics for cybersecurity are defined by various criteria. These criteria include objective and subjective handling of data. The traditional methods available in decision theory are proven good for handling quantitative data. However, qualitative data needs special treatment. Fuzzy numbers are widely used to handle qualitative data. The proposed paper is making use of the FAHP method that embeds concepts of fuzzy set theory in the analytic hierarchy process (AHP) method. Subsequent paragraphs, however, are indented.

2 Related Work

First, Laarhoven in 1983 applied triangular fuzzy numbers (TFNs) in the formulation of a pair-wise comparison matrix for the computation of criteria weights [7]. After that, fuzzy AHP method is applied widely in a variety of problems in various fields. Alghassab [8] used fuzzy AHP and TOPSIS to evaluate cybersecurity evaluations for industrial control systems, in the energy sector. Abushark et al. [9] and Alharbi et al. [10] also implemented a fuzzy AHP-TOPSIS-based approach intrusion detection system. Xiao et al. [11] used FAHP in computing defense costs for the protection of high-voltage power substations from cyberattacks. Rokkas et al. [12] suggested the use of FAHP in a survey-based approach to know the factors influencing cyber-safety products in the energy and electrical systems. Information technology was the answer to many problems in the crisis world faced during COVID times. With the help of FAHP, Baz et al. [13] attempted to estimate the impact of COVID crisis in cybersecurity. Majid et al. [14] proposed FAHP and a machine learning-based method to evaluate the condition of cybersecurity in hospitals. Lan [15] made risk assessments for industrial systems using AHP and rank sum ration methods. Petrova [16] proposed a data-driven approach for risk assessment in information technology systems to optimize cybersecurity using fuzzy AHP methods.

Nasir et al. [17] proposed interval-valued complex intuitionistic fuzzy relations (IVCIFRs) to analyze the complex relationship between cybersecurity technology and potential cyberattacks. Cubukcu and Cantekin [18] used FAHP methods in ranking the firewalls as an attempt to evaluate cybersecurity. Fog computing brings faster results, but the security of devices is a matter of concern. Verma and Chandra [19] make use of FAHP in the assessment of Fog-IoT. Almotiri [20] suggested FAHP based method for the development of smart traffic detection by evaluating malicious traffic signals. Wang et al. [21] proposed an approach for risk evaluation in industrial systems by analyzing risk associated with cybersecurity in energy generation plants carried out using the FAHP method.

Most medical devices lack security features and are vulnerable when connected to cyberspace. The integrated fuzzy AHP-TOPSIS method is suggested for enhanced security of medical equipment [22]. Metrics of cybersecurity were evaluated using an integrated method of AHP and Electre-III [23]. Belgaum et al. [24], in their study, analyzed various aspects of smart cities that are sensitive to cyber threats and ranked them using the fuzzy AHP method. Cybersecurity risks associated with social media were evaluated using FAHP [25]. Yuen [26] used fuzzy AHP methods in analyzing the appropriate investment amount for cybersecurity. Information security matching of the organization is evaluated using the fuzzy AHP approach with focusing on security gaps [27].

3 The Fuzzy Set Theory

Zadeh and Dieter Klaua proposed fuzzy set theory for the first time in the year 1965 [28]. The fuzzy set theory derives concepts from traditional set theory to quantify the linguistic variables. Linguistic variables are not numeric but word/words written in natural language, e.g., tall, short, beautiful, not so beautiful, satisfaction, dissatisfaction, fast, moderate, slow, intelligent, dumb, smart, etc. The role of linguistic variables becomes very important in complicated and "not so well defined" problems. The uncertainty associated with human reasoning, especially in a situation where judgments are subjective, can be handled efficiently with the principles of fuzzy set theory. The fuzzy numbers help in making subjective judgments in vague and unclear situations.

Fig. 1 Representation of triangular fuzzy number

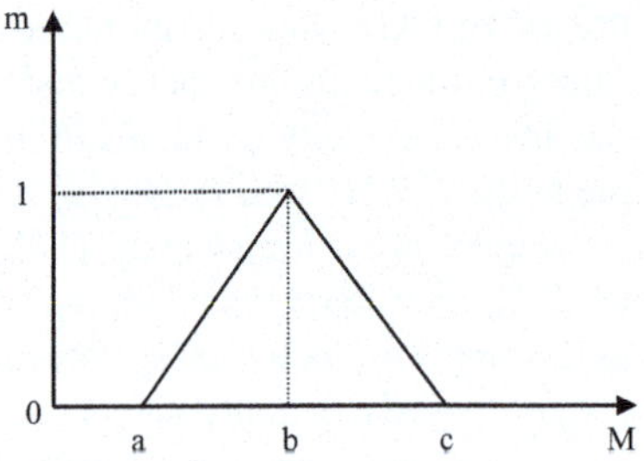

Fuzzy Number Definition—Let $M \in F(R)$ be called a fuzzy number, if the following two conditions are satisfied

(1) There exists $x_0 \in R$ such that $(x_0) = 1$.
(2) For any $0 \leq \alpha \leq 1$, $A\alpha = [x, \mu A\alpha\ (x) \geq \alpha]$ is a closed interval.

Where $F(R)$ represents a family of all fuzzy sets and R is the set of real numbers.

Triangular Fuzzy Number Definition—A fuzzy number M on R is said to be a triangular fuzzy number if its membership function (x): $R \rightarrow [0, 1]$ is defined as follows:

$$(x) = \{-\, l\, m - l,\ l \leq x \leq m - x + u\, u - m,\ m \leq x \leq u\, 0\},\ \text{otherwise} \quad (1)$$

where l and u stand for the lower and upper value of the support of M, respectively, and m represents the modal value. The triangular fuzzy numbers can be denoted by order triplet (l, m, u) of real numbers or regular numbers. The support of M is the set of elements $\{x \in R \backslash l < x < u\}$ (Fig. 1).

4 FAHP Method (Fuzzy Analytic Hierarchy Process)

MCDM provides various methods for making decisions when multiple criteria are involved. The criteria and alternatives are evaluated to select the best alternative among a set of alternatives. While taking decisions in the real world, one has to deal with both quantitative and qualitative data. Thus, an approach is required to integrate qualitative data with quantities data. Uncertainty and imprecise knowledge about real-world applications can be dealt effectively with the concepts mentioned in fuzzy set theory. The original analytic hierarchy process, proposed by Satty, cannot deal with this imprecise and uncertain information. The concepts of fuzzy set theory are beautifully imbibed in the AHP method and gave birth to the new method, fuzzy AHP. They are many techniques available for the implementation of fuzzy AHP. Buckley's method [2] is used in the present paper. Steps in Buckley's method:

Step 1 Development of the problem hierarchy

Developing a graphical representation of the problem in terms of the overall goal, the criteria, and the decision alternatives is done in the beginning. The AHP problem hierarchy consists of a goal (the decision), several alternatives for reaching that goal, and several criteria on which the alternatives can be judged that relate to the goal. Each of these criteria can be further divided into sub-criteria.

Step 2 Pair-wise comparison matrix for each criterion

Pair-wise comparisons are fundamental building blocks of the AHP. Element Ci,j of the matrix is the measure of preference of the item in row i when compared to the item in column j. AHP assigns a 1 to all elements on the diagonal of the pair-wise comparison matrix. AHP obtains the preference rating of Cj, i by computing the reciprocal (inverse) of Ci, j (the transpose position).

Step 3 Fuzzification

Pair-wise comparison between criteria is made, and data is represented with the help of triangular fuzzy numbers. The conversion table is used to get TFNs corresponding to Satty's scale of traditional AHP (Table 1).

In pair-wise comparison matrix, the inverse of TFN $\tilde{A}$ is written as $\tilde{A}^{-1}$.

$$\tilde{A}^{-1} = (l, m, u)^{-1} = \left(\frac{1}{u}, \frac{1}{m}, \frac{1}{l}\right) \tag{1}$$

Step 4 Determination of fuzzy geometric mean

Row-wise geometric mean of TFNs in the comparison matrix is calculated using the following formula

Table 1 Fuzzification scale based on pair-wise comparison for fuzzy numbers

Linguistic value	Satty scale	Triangular fuzzy scale
Absolutely strong (AS)	9	(8, 9, 9)
Very strong (VS)	7	(6, 7, 8)
Fairly strong (FS)	5	(4, 5, 6)
Slightly strong (SS)	3	(2, 3, 4)
Equal (E)	1	(1, 1, 1)
Slightly weak (SW)	1/3	(1/4, 1/3, 1/2)
Fairly weak (FW)	1/5	(1/6, 1/5, 1/4)
Very weak (VW)	1/7	(1/8, 1/7, 1/6)
Absolutely weak (AW)	1/9	(1/9, 1/9, 1/8)

$$\widetilde{r}_i = \left(\prod_{j=1}^{n} \widetilde{d}_{ij}\right)^{1/n} \quad (2)$$

For the calculation of the geometric mean, the fuzzy numbers are multiplied using the following formula

$$\widetilde{A}_1 \otimes \widetilde{A}_2 = (l_1, m_1, u_1) \otimes (l_2, m_2, u_2) = (l_1 * l_2, m_1 * m_2, u_1 * u_2) \quad (3)$$

Step 5 Fuzzy weights computation

Corresponding to each criterion, the geometric means are multiplied with the sum of reverse fuzzy geometric means in increasing order.

$$\widetilde{w}_i = \widetilde{r}_i \otimes \left(\widetilde{r}_1 \oplus \widetilde{r}_2 \cdots \oplus \widetilde{r}_j\right)^{-1} \quad (4)$$

Step 6 Defuzzification

Now, the fuzzy weights $\widetilde{w}_i$ corresponding to each criterion will be replaced by equivalent non-fuzzy weights w_i. The average of fuzzy weights is computed to get non-fuzzy weights. These fuzzy numbers are defuzzified to get crisp numerical values.

$$w_i = \left(\frac{l + m + u}{3}\right) \quad (5)$$

Step 7 Normalization of weights

The total of weights is usually not 1. So weights should be normalized to get total of 1. To achieve this, each of relative fuzzy weight is divided with the total weight to get normalized weight.

5 Application of Fuzzy AHP Method

Criteria priorities describing importance of one criterion over other are represented using fuzzy triangular scales. To made pair-wise comparisons between four criteria, we prepared a questionnaire and forwarded to some of our friends working in the field of cybersecurity; however, only three people responded to it. We took average of the data we collected from our decision-makers. We represented collected data using triangular fuzzy numbers. Table 2 describes fuzzy comparison matrix.

The inverse of TFN $\tilde{A}$ is written as $\tilde{A}^{-1}$, using Eq. 1, e.g., for cell ($c1$, $c2$), the TFN is (2, 3, 4), and hence, for cell ($c2$, $c1$), the data value will be $(2, 3, 4)^{-1}$.

Table 2 Pair wise comparison matrix

Criteria	S	PM	RM	EO
S	(1, 1, 1)	(2, 3, 4)	(4, 5, 6)	(6, 7, 8)
PM	(1/4, 1/3, 1/2)	(1, 1, 1)	(2, 3, 4)	(4, 5, 6)
RM	(1/6, 1/5, 1/4)	(1/4, 1/3, 1/2)	(1, 1, 1)	(2, 3, 4)
EO	(1/8, 1/7, 1/6)	(1/6, 1/5, 1/4)	(1/4, 1/3, 1/2)	(1, 1, 1)

$$(2, 3, 4)^{-1} = \left(\frac{1}{4}, \frac{1}{3}, \frac{1}{2}\right)$$

Now, the geometric means of TFN values in comparison matrix is calculated using Eq. 2, as follows:

$$\tilde{r}_i = (l_1 \times l_2 \times l_3 \times l_4)^{1/n} \cdot (m_1 \times m_2 \times m_3 \times m_4)^{1/n} \cdot (u_1 \times u_2 \times u_3 \times u_4)^{1/n}$$

where n is order of the matrix, i.e., $n = 4$.

Thus, multiply l value, m values, and u values of each of the four no. and taking their fourth roots, the calculations for first row are:

$$\begin{aligned}\tilde{r}_1 &= (1 \times 2 \times 4 \times 6)^{1/4} \cdot (1 \times 3 \times 5 \times 7)^{1/4} \cdot (1 \times 4 \times 6 \times 8)^{1/4} \\ &= (48)^{1/4} \cdot (105)^{1/4} \cdot (192)^{1/4} \\ &= (2.632) \cdot (3.201) \cdot (3.722)\end{aligned}$$

Similarly, $\tilde{r}_i$ values are calculated for other rows (Table 3).

Next, relative fuzzy weights are calculated corresponding to each criterion using geometric means calculated in the previous step. The geometric means are multiplied with the sum of reverse fuzzy geometric means in increasing order, using Eq. 4 (Tables 4 and 5).

$$\tilde{w}_i = \tilde{r}_i \otimes (\tilde{r}_1 \oplus \tilde{r}_2 \oplus \tilde{r}_3 \oplus \tilde{r}_4)^{-1}$$

$$\begin{aligned}\tilde{r}_1 \oplus \tilde{r}_2 \oplus \tilde{r}_3 \oplus \tilde{r}_4 = 2.632 + 1.189 + 0.537 + 0.269, \\ 3.201 + 1.495 + 0.669 + 0.312,\end{aligned}$$

Table 3 Geometric means of values in comparison matrix

Criteria	$\tilde{r}_i$		
S	2.632	3.201	3.722
PM	1.189	1.495	1.861
RM	0.537	0.669	0.841
EO	0.269	0.312	0.380

$$3.722 + 1.861 + 0.841 + 0.380$$
$$= (4.627, 5.677, 6.804)$$

$$(\widetilde{r_1} \oplus \widetilde{r_2} \oplus \widetilde{r_3} \oplus \widetilde{r_4})^{-1} = \left(\frac{1}{6.804}, \frac{1}{5.677}, \frac{1}{6.804}\right)$$
$$= (0.147, 0.176, 0.216)$$

Now, the fuzzy weights $\widetilde{w}_i$ corresponding to each criterion are converted into relative non-fuzzy weights w_i by taking the average of fuzzy numbers. These fuzzy numbers are defuzzified to get crisp numerical values, following Eq. 5 (Table 6).

$$\text{Total weight} = 0.585 + 0.280 + 0.126 + 0.059$$
$$= 1.050$$

We can see the total of weights is 1.050. So weights should be normalized to get total of 1. Each of relative fuzzy weight is divided with the total weight to get normalized weight (Table 7).

Comparison matrixes for four criteria are given in Tables 8, 9, 10 and 11. In the next step, the priority of alternatives will be computed for individual criteria, and then, the priority vector for the entire criterion will be calculated.

Table 4 Relative fuzzy weight calculations

Criteria	Fuzzy geometric mean $\widetilde{r_i}$	Fuzzy weight $\widetilde{w_i}$
S	(2.632, 3.201, 3.722)	(2.632, 3.201, 3.722) $\otimes\left(\frac{1}{6.804}, \frac{1}{5.677}, \frac{1}{6.804}\right)$
PM	(1.189, 1.495, 1.861)	(1.189, 1.495, 1.861) $\otimes\left(\frac{1}{6.804}, \frac{1}{5.677}, \frac{1}{6.804}\right)$
RM	(0.537, 0.669, 0.841)	(0.537, 0.669, 0.841) $\otimes\left(\frac{1}{6.804}, \frac{1}{5.677}, \frac{1}{6.804}\right)$
EO	(0.269, 0.312, 0.380)	(0.269, 0.312, 0.380) $\otimes\left(\frac{1}{6.804}, \frac{1}{5.677}, \frac{1}{6.804}\right)$

Table 5 Criterion-wise relative fuzzy weight

Criteria	$\widetilde{w_i}$
S	(0.387, 0.563, 0.804)
PM	(0.175, 0.263, 0.402)
RM	(0.079, 0.118, 0.182)
EO	(0.039, 0.055, 0.082)

Table 6 Non-fuzzy weights

Criteria	w_i
S	0.585
PM	0.280
RM	0.126

Table 7 Normalized relative weight of criteria

Criteria	Weights (w_i)	Normalized weights
S	0.585	0.585/1.050 = 0.557
PM	0.280	0.280/1.050 = 0.267
RM	0.126	0.126/1.050 = 0.120
EO	0.059	0.059/1.050 = 0.056

Table 8 Preference matrix for criteria 1 (susceptibility)

Criteria S	First Tech	Smart Techno	Digital Leaders
First Tech	(1, 1, 1)	(3, 4, 5)	(1/3, 1/2, 1)
Smart Techno	(1/5, 1/4, 1/3)	(1, 1, 1)	(1/7, 1/6, 1/5)
Digital Leaders	(1, 2, 3)	(5, 6, 7)	(1, 1, 1)

Table 9 Preference matrix for criteria 2 (protection mechanism)

Criteria PM	First Tech	Smart Techno	Digital Leaders
First Tech	(1, 1, 1)	(1, 1, 1)	(1/3, 1/2, 1)
Smart Techno	(1, 1, 1)	(1, 1, 1)	(1/3, 1/2, 1)
Digital Leaders	(1, 2, 3)	(1, 2, 3)	(1, 1, 1)

Table 10 Preference matrix for criteria 3 (risk measurement)

Criteria RM	First Tech	Smart Techno	Digital Leaders
First Tech	(1, 1, 1)	(1, 1, 1)	(1, 2, 3)
Smart Techno	(1, 1, 1)	(1, 1, 1)	(1, 2, 3)
Digital Leaders	(1/3, 1/2, 1)	(1/3, 1/2, 1)	(1, 1, 1)

To compute the normalized pair-wise comparison matrix, column values of the comparison matrix are divided by the sum of respective columns. Next, row averages are computed. These row averages of the normalized matrix are referred to as priority vectors. Priority vectors for all the criteria are given in Table 12.

Corresponding to each alternative, overall priorities are calculated. The weight vector is multiplied by the priority vector (Table 13).

Table 11 Preference matrix for criteria 4 (encounter outcomes)

Criteria EO	First Tech	Smart Techno	Digital Leaders
First Tech	(1, 1, 1)	(1/3, 1/2, 1)	(1/5, 1/4, 1/3)
Smart Techno	(1, 2, 3)	(1, 1, 1)	(1/3, 1/2, 1)
Digital Leaders	(3, 4, 5)	(1, 2, 3)	(1, 1, 1)

Table 12 Priority vectors for all criteria

	S	PM	RM	EO
First Tech	0.338	0.256	0.387	0.150
Smart Techno	0.089	0.256	0.387	0.297
Digital Leaders	0.573	0.488	0.225	0.553

Table 13 Priority calculations

	S	PM	RM	EO		S	0.557
FT	0.338	0.256	0.387	0.150		PM	0.267
ST	0.089	0.256	0.387	0.297	×	RM	0.120
DL	0.573	0.488	0.225	0.553		EO	0.056

Table 14 FAHP ranking of alternatives

Digital Leaders	Priority value
First Tech	0.323
Smart Techno	0.181
Digital Leaders	0.507

The products calculated above are added to get the overall priority. After calculations, resulted priority values are arranged from highest to lowest. FAHP rankings for three companies are given in Table 14.

6 Result and Analysis

The cybersecurity metrics are analyzed, and a hierarchical structure is developed. The problem of evaluation of cybersecurity metrics is considered as multi-criteria decision-making problem, and the first level of hierarchy is analyzed using fuzzy AHP techniques. To make pair-wise comparisons between four criteria, we prepared a questionnaire and forwarded it to some of our friends implementing cybersecurity policies in their respective organizations. The data collected from our decision-makers is averaged and then represented using concepts of triangular fuzzy numbers.

In this paper, we considered three companies First Tech, Smart Techno, and Digital Leaders. Their rankings are computed and presented in the form of a priority vector. The highest priority value in Table 14 is 0.507 followed by 0.323, and 0.181 is the least priority value. From this, it is evident that the ranking of criteria at different levels of hierarchy is shown in Fig. 2.

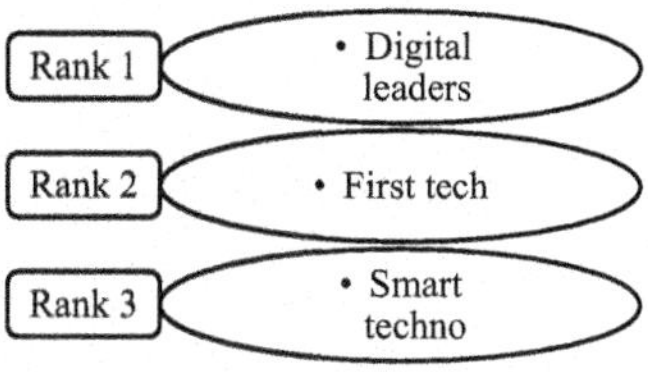

Fig. 2 Final ranking of companies

Thus, the company Digital Leaders has a strong cybersecurity system followed by First Tech. Smart Techno is having weakest cybersecurity system among these three companies.

7 Conclusion and Future Work

MCDM is the most popular tool among researchers and decision-makers to handle problems consisting of various contradictory criteria. FAHP is a powerful method to evaluate alternatives involving all or some parameters, expressed with natural language phrases having linguistic vagueness. Problem scenario in which subjective judgments are made by decision-makers, fuzzy numbers, and concepts from fuzzy set theory becomes very important. FAHP involves lengthy calculations for which simple software or procedure using Excel spreadsheets can be programmed. One of the advantages of using fuzzy AHP is the elimination of unnecessary criteria, in a situation where "absolutely not important" values are marked while comparing with other important criteria.

However, fuzziness in data does not permit the calculation of the consistency ratio, one of the major strengths of the traditional AHP method. Another disadvantage of FAHP is that data inputs from various decision-makers are articulated in the form of linguistic expressions that are based on one's views, experience and understanding of the problem domain, hence involving subjectivity. These linguistic data may vary from individual to individual. In such situations, fuzziness may result in risky decisions.

This paper proposes an efficient and effective decision-making tool for evaluating cybersecurity metrics with the help of the FAHP method. In this paper, the study has been conducted based on pilot data. In the future, we will be conducting the same study on the data collected by users and people working in this field using a questionnaire-based survey. To get better participation, we are planning to post a questionnaire on Website Analysis and Measurement Inventory (WAMMI).

References

1. Swanson M, Bartol N, Sabato J, Hash I, Graffo I (2003) Security metrics guide for information technology systems. National Institute of Standards and Technology Special Publication
2. Savola R (2008) A novel security metrics taxonomy for R&D organizations. ISSA 2008:1–12
3. Tashi I, Solange G (2011) Security metrics to improve information security management, a holistic approach. In: Management of technology series. EPFL Press
4. Vaughn R, Henning R, Siraj A (2003) Information assurance measures and metrics: state of practice and proposed taxonomy. In: Proceedings of 36th Hawaii international conference on system sciences
5. Seddigh N, Pieda P, Matrawy A, Nandy B, Lambadaris I, Hatfield A (2004) Current trends and advances in information assurance metrics. In: Proceeding of the second annual conference on privacy, security and trust, pp 197–205
6. Bhol SG, Mohanty JR, Pattnaik PK (2021) Taxonomy of cyber security metrics to measure strength of cyber security. Mater Today Proc
7. Laarhoven V, Pedrycz W (1983) A fuzzy extension of Saaty's priority theory. Fuzzy Sets Syst 11:229–241
8. Alghassab M (2022) Quantitative assessment of sustainable renewable energy through soft computing: fuzzy AHP-TOPSIS method. Energy Rep 8:12139–12152
9. Abushark YB, Khan AI, Alsolami F, Almalawi A, Alam MM, Agrawal A, Khan RA (2022) Cyber security analysis and evaluation for intrusion detection systems. Comput Mater Contin 72:1765–1783
10. Alharbi A, Seh AH, Alosaimi W, Alyami H, Agrawal A, Kumar R, Khan RA (2021) Analyzing the impact of cyber security related attributes for intrusion detection systems. Sustainability 13(22):12337
11. Xiao Y, Yang L, Li J, Xu J, Liu K (2021) Valuing the cyber-attacks budget in high voltage power substations to increase cyber-security; providing a method based on fuzzy analytical hierarchy process. Energy Rep 7:8322–8331
12. Rokkas T, Neokosmidis I (2020) Factors affecting the market adoption of cyber-security products in energy and electrical systems: the case of SPEAR. In: Proceedings of the 15th international conference on availability, reliability and security, pp 1–8
13. Baz M, Alhakami H, Agrawal A, Baz A, Khan RA (2021) Impact of COVID-19 pandemic: a cybersecurity perspective. Intell Autom Soft Comput 27(3):641–652
14. Majid N, Hatem S, Ehab A, Şaban O, Kemal P (2020) A healthcare evaluation system based on automated weighted indicators with cross-indicators based learning approach in terms of energy management and cybersecurity. Int J Med Inform 144
15. Lan J (2020) Research on cybersecurity risk assessment in SCADA networks based on AHP-RSR. In: International conference on communications, information system and computer engineering (CISCE), Kuala Lumpur, Malaysia, pp 361–364
16. Petrova V (2021) A cybersecurity risk assessment. Industry 4.0 6(1):37–40
17. Nasir A, Jan N, Gumaei A, Khan SU, Albogamy FR (2021) Cybersecurity against the loopholes in industrial control systems using interval-valued complex intuitionistic fuzzy relations. Appl Sci 11(16):7668
18. Cubukcu C, Cantekin C (2022) Using a combined fuzzy-AHP and TOPSIS decision model for selecting the best firewall alternative. J Fuzzy Extens Appl 3(3):192–200
19. Verma R, Chandra S (2021) Interval-valued intuitionistic fuzzy-analytic hierarchy process for evaluating the impact of security attributes in fog based internet of things paradigm. Comput Commun 175:35–46
20. Almotiri SH (2021) Integrated fuzzy based computational mechanism for the selection of effective malicious traffic detection approach. IEEE Access 9:10751–10764
21. Wang S, Ding L, Sui H, Gu Z (2021) Cybersecurity risk assessment method of ICS based on attack-defense tree model. J Intell Fuzzy Syst 40:1–14
22. Alzahrani FA, Ahmad M, Ansari M (2022) Towards design and development of security assessment framework for internet of medical things. Appl Sci 12(16)

23. Bhol SG, Mohanty JR, Pattnaik PK (2020) Cyber security metrics evaluation using multi-criteria decision-making approach. Smart Intell Comput Appl 160:665–675
24. Belgaum MR, Alansari Z, Jain R, Alshaer J (2018) A framework for evaluation of cyber security challenges in smart cities. In: Smart cities symposium 2018. IET, pp 1–6
25. Zhang KW (2013) Risk assessment of smart grid cyber security based on multi-level fuzzy comprehensive evaluation method. Adv Mater Res 605:2311–2317
26. Yuen KKF (2019) Towards a cybersecurity investment assessment method using primitive cognitive network process. In: 2019 international conference on artificial intelligence in information and communication (ICAIIC). IEEE, pp 068–071
27. Nasser AA, Al-Khulaidi AA, Aljober MN (2018) Measuring the information security maturity of enterprises under uncertainty using fuzzy AHP. Int J Inf Technol Comput Sci 10(4):10–25
28. Buckley JJ (1985) Fuzzy hierarchical analysis. Fuzzy Sets Syst 17(3):233–247

Personalized Recommender System Using Topic Modelling Approach

K. S. Gayathri, R. Mohana, and M. Sagana

Abstract A personalized recommender system provides recommendations to users based on their preferences, behaviors, and past interactions with a platform. Data sparsity, cold start, bias, privacy concerns, lack of transparency and overfitting are the challenges in this system. There are several types of recommendation systems in the market. Some of the most commonly used systems are content based, collaborative based filtering, hybrid, knowledge and demographic based systems. Many recommendation systems rely on popularity-based recommendations, where popular items are recommended more often than less popular ones. This can lead to over-representation of popular items and limit the diversity of recommendations and at the same time may lead to overfitting as it's tailored too closely to user's behavior. Personality aids recommendation systems in comprehending user preferences and creating personalized suggestions that complement each user's specific traits, increasing relevance and user contentment. It can bring novelty, fill in information gaps, and boost user engagement, creating a more satisfying and individualized experience. By employing effective techniques to manage a large amount of data and carry out computations fast, a personalized recommender system may be made scalable. To manage several users simultaneously, the task might be divided among several computers or servers. To save effort and enhance speed, it may also store and reuse previously computed suggestions. The system is hence capable of managing more users and data without slowing down or crashing. The proposed system used Topic modelling approach and MBTI (Myers-Briggs Type Indicator) personality classes to effectively suggest the user a restaurant that is suited for their personality and this is done by analyzing eWoM (Electronic Word-of-Mouth) of users already posted on the social platform. Extreme gradient boosting and similarity measures like Cosine similarity are employed here to match the user posts with various topics and ultimately to a personality class from the 16 available labels. This proposed system focuses on restaurant recommendation based on the grounds that each restau-

Authors R. Mohana and M. Sagana contributed equally to this work.

K. S. Gayathri (✉) · R. Mohana · M. Sagana
Department of Information Technology, Sri Sivasubramaniya Nadar College of Engineering, Kalavakkam, Chennai, Tamil Nadu 603110, India
e-mail: gayathriks@ssn.edu.in

S. Bandyopadhyay et al. (eds.), *Intelligent Computing Systems and Applications*, Lecture Notes in Networks and Systems 1010,
https://doi.org/10.1007/978-981-97-5412-0_11

rant has a personality associated with it and people with similar personalities have similar tastes. It exhibits a precision of 81.8% hence vouching for the efficacy of the system and giving insight for further innovations and improvements down the lane.

Keywords Recommendation system · Context based filtering · Personality classification · Clustering · Topic modelling

1 Introduction

"It is up to each person to recognize his or her true preferences."—Isabel Briggs Myers. It is often seen that recommendations are generic and based on mass reviews, which forces the user to jump on a bandwagon with others and end up not having a personalized experience. Different individuals have different needs and choices, sometimes they may be flabbergasted by the notion of common good rather than worry about personal good [1]. The goal is to provide each person a choice they would enjoy rather than let users be deceived by fabricated reviews because with today's influence of social media and high cost of living, it is necessary that one gets the value for money in accordance to preferences without external influence and this recommendation system serves just that cause by incorporating user personality to provide valuable recommendations to them.

The majority of early recommendation systems used collaborative filtering methods. Item-based collaborative filtering found related things based on user ratings or interactions, whereas user-based collaborative filtering analysed the preferences of similar users. Data sparsity and the cold-start issue were problems with these methods. The goal of content-based filtering was to generate suggestions using item properties. Based on their tastes, it developed user profiles and paired them with related things. Although this method addressed the "cold-start" issue, it had trouble recording a range of user preferences. To get beyond their respective shortcomings, hybrid recommendation systems integrated collaborative filtering strategies with content-based filtering. Ultimately, Personalized recommender systems replaced basic filtering methods with more sophisticated strategies [2].

Incorporating Personality traits can help in providing novel and diverse recommendations. Capturing user preferences, dealing with data sparsity, and making precise, pertinent suggestions based on unique traits and behaviours are the main points of focus of such systems [1]. Lately, there has been an increase in interest in food experience, with organisations in restaurants and hotels industry seeking insights into client eating behaviours and preferences in order to enhance decision making in areas such as marketing and recommendation. In this paper, we use implicit and explicit information from customers' Electronic Word of Mouth (eWoM) to enhance restaurant suggestion. Implicit information in reviews relates to textual statements that may be used to predict consumers' personalities and preferences, whilst explicit characteristics correspond to ratings of restaurants and type of cuisine served. The base of this work is to use Topic modelling and Myers-Briggs Type Indicator (MBTI) personality classifier.

Topic modelling is an unsupervised machine learning technique that's capable of scanning a set of documents, detecting word and phrase patterns within them, and automatically clustering word groups and similar expressions that best characterize a set of documents. Besides typical techniques of user preference extraction such as explicit inquiries, user ratings and historical records, there is yet another approach that uses Natural Language Processing (NLP) called user opinion analysis. eWoM represents customer opinions on items and services, and it has been widely employed in determining consumer preferences [3]. This is called Automatic Personality Recognition (APR).

APR is considerably is effective, since they can be applied to the user's current data without requiring the user to fill out lengthy questionnaires. In text-based APR analysis the frequency of word selection in social network comments or interactions is used to determine the user's personality attributes. Hence from user comments in forums that are publicly available, using the methodologies stated, a efficient recommender system with a personal touch can be implemented.

The paper is organized as follows: Sect. 2 provides an state of art approaches in modelling recommender system. Section 3 presents the proposed approach for modelling personalized recommender system using topic modelling approach. Experimental study is presented in Sect. 4. Finally, Sect. 5 concludes the proposed work and provides future recommendations in the field of study.

2 Literature Survey

This section provides an overview of several studies carried out by researchers in the field of Recommender Systems. The section also provides a brief summary of the problems history and current state, which is outlined by a review of earlier research into closely similar issues. The recommendation system does a thorough analysis of the data, classifies and arranges it in a sensible manner, applies some algorithms, and hence helps in quick and accurate decisions. As a consequence, it will save a lot of time and energy wasted on choosing a certain course of action [4]. The current User-User techniques analyze the ratings in a user-item rating matrix to determine how similar users are. Content-based and collaborative filtering are the two most used filtering techniques [5].

Systems that employ content-based recommendations examine a collection of documents and/or descriptions of goods that users have previously evaluated, and then create a model or user interest profile based on the characteristics of the goods that users have previously assessed. For a content-based recommender system to truly comprehend user preferences and provide precise recommendations, it has to have accumulated enough ratings. As a result, the algorithm will not be able to offer trustworthy suggestions when there are few ratings available, such as for a new user [6].

Collaborative Filtering (CF) technique employs a user who is similar to another to suggest content. However, CF approaches have the cold start problem, which arises when there is little to no information about a user and makes it impossible to find customers who are similar to them [7]. By combining characteristics of both of Content-based filtering (CB) and Collaborative filtering (CF) into a single prediction and hence generating final recommendations based on the weighted average, Hybrid Recommender System incorporates the best aspects of both CB and CF [8].

One of the systems used TF/IDF computation on a content-based, relevant restaurant and food information is processed with regard to users. A similarity calculation was also performed to examine the connections between customers, restaurants, and foods. The system will provide recommendations for the users based on all of this processing. Matrix factorization was presented as a solution to the issues due to the high-dimensional data in social networks and the difficulty of collecting common features from these data. In order to reduce the amount of storage space and processing time needed, Matrix factorization is defined as the decomposition of the initial matrix into two or more smaller measures. The most popular MF methods are SVD, nonnegative matrix factorization (NMD) and probabilistic matrix factorization [9].

One of the approaches to extract user preferences is to perform sentiment analysis by analyzing their opinions through comments or reviews. Compared to techniques like TF, the accuracy of collecting user preferences from user's comments was found to be much higher [10]. Lately, there has been a lot of interest in using consumers' personalities to better understand and respond to their requirements, since "Personality" refers to the perceptions, feelings and preferences of individuals [1].

A person's personality is made up of a variety of traits and behaviors. It has been demonstrated over time that a person's personality traits affect many aspects of his or her life, including motives, preferences, as well as consumer preferences and behavior. Based on the personality theory, which claims that words can convey some psychological states and the personality of the author of the text.

Personality prediction from text is a common automated technique. The two primary ways for personality prediction in personality-aware recommender systems are through surveys and automated mechanisms. In general, surveys are more accurate in determining personality, but the procedure is time-consuming. By using the user's pre-existing data which can include text, photographs, videos, likes and dislikes, etc. an automated method can be carried out more quickly and easily [3].

Thus the proposed work presents a personalized recommender system that uses the social media post and topic modelling approach in the design. It introduces several novel aspects that distinguish it from existing papers and conventional recommendation systems as shown in Table 1.

3 Proposed Methodology

The aim is to develop a system using topic modelling approach along with the MBTI personality classification to map a person's post to the respective personality

Table 1 Differences in recommendation approaches: prior work versus proposed methodology

Characteristic	Prior work	Proposed methodology
Recommendation approach	Mainly collaborative filtering and content-based methods	Integrates topic modelling and personality classification
Data source	Often based on user-item interaction data	Utilizes customer eWoM reviews from a social interaction platform
Preprocessing	Standard text preprocessing	Includes additional steps for personality label removal and lemmatization
Modelling technique	Traditional recommendation algorithms	Utilizes Linear Discriminant Analysis (LDA) for topic modelling and XGBoost for personality classification
Engagement and satisfaction	May not optimize user engagement	Aims to enhance user engagement and satisfaction with tailored recommendations
Limitations	May suffer from limited personalization and user engagement	Potential challenges in handling noisy text data and ensuring accurate personality classification

and then these classifications are used for personalized recommendations, to a user in relevance to their similar personality counterparts who are believed to have the same tastes in places. This way, a personalized solution can be provided to the insipid recommendation systems. The overall working of the system is presented in Figs. 1 and 2.

While some earlier studies may have overlooked the importance of user engagement, the suggested method tries to increase user satisfaction by providing individualized suggestions that are in line with user preferences and personality attributes. A better user experience is a result of this emphasis on user involvement.

The suggested solution offers the idea of cosine similarity-based clustering on three similarity matrices produced from several sources (explicit data, topic modelling preferences, and personality classifications). By locating and combining related things of interest, the resultant improved matrix improves the recommendation process and increases the relevancy of tailored recommendations.

Topic modelling and personality classification are two individual components that have been used in various domains separately; however, their integration into a personalized recommendation system for place recommendations based on user preferences and personalities is relatively new and broadens the application of these techniques.

The methodology employed in Figs. 1 and 2 is as follows descriptively. User feedback is gathered through electronic word of mouth and then subjected to preprocessing to create a document-term matrix. This matrix finds application in the

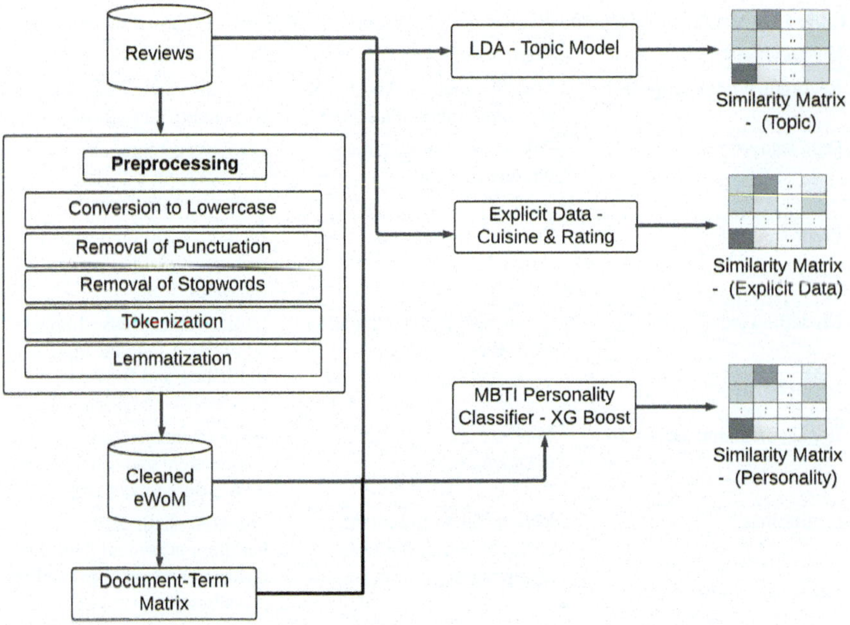

Fig. 1 Generation of similarity matrices

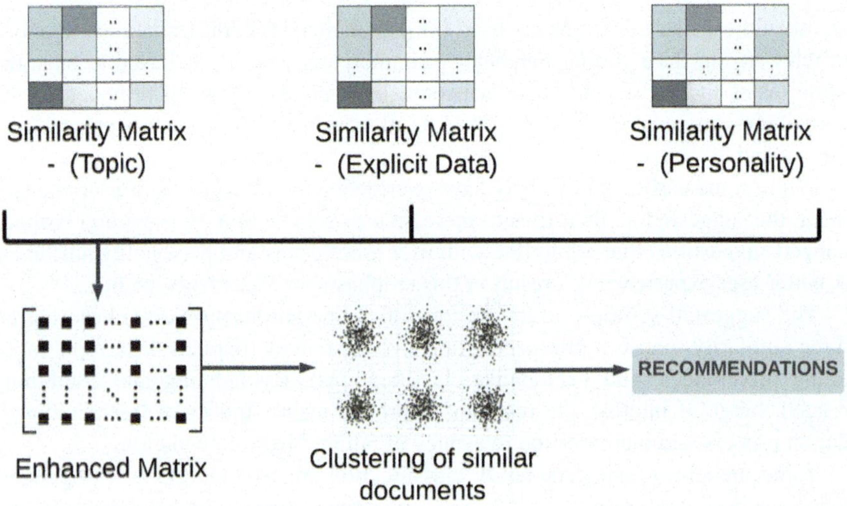

Fig. 2 Combining three factors

LDA topic model to calculate a similarity matrix grounded in the identified topics. Following this step, the electronic word of mouth is fed into the MBTI personality classifier, employing the XGBoost methodology. This generates a similarity matrix based on the resultant personality type obtained from the 16 distinct personality types identified by the MBTI testing model. In addition, explicit data like Cuisine and Rating is extracted from user reviews to build another similarity matrix.

The three generated similarity matrices is then used to produce an enhanced matrix, which is subsequently utilized to cluster akin documents. This clustering process ultimately yields tailored recommendations tailored to the user's preferences. The algorithm can be given as below:

Algorithm 1 Personalized recommendation system

Require: User eWoM
Ensure: Final Personalized Recommendation
Preprocess eWoM to clean and tokenize the text.
Generate a document term matrix (DTM) from the preprocessed reviews.

$TopicSimilarity$ ← Apply Latent Dirichlet Allocation (LDA) topic modeling on the DTM and compute similarity matrix based on similar topics

$PersonalitySimilarity$ ← Deploy the XGBoost methodology to classify the personality types based on the MBTI testing model and generate a similarity matrix based on personality types

$ExplicitDataSimilarity$ ← Extract explicit data such as Cuisine and Rating from the user reviews to generate a similarity matrix

$EnhancedMatrix$ ← (TopicSimilarity + PersonalitySimilarity + ExplicitDataSimilarity)
Perform clustering on the enhanced matrix to group similar documents.
Based on the clustering results, obtain the final personalized recommendations for users.

3.1 Topic Modelling Using Linear Discriminant Analysis

As a first step, gathering customer eWoM (reviews) through the social interaction platform. With the objective of making the text data ready for further analyses through Topic modelling and Personality classification. Preprocessing of eWoM is performed. This process is carried out using techniques like elimination of punctuation and URLs, dropping common stopwords, tokenization and lemmatization.

The development of a topic model utilising the eWoM text as a corpus to detect customer opinions and how they are related with each document: Two metrics are extracted from eWoM, the preferences of each user and the topics that characterize each venue. This is generated by choosing the Topic with the maximum probability. This shows the average customer perception for each venue.

To design topic model Linear Discriminant Analysis (LDA) approach is preferred and the various steps used in the design is presented below:

1. ***Dictionary Extraction***: After preprocessing the text data, a dictionary is extracted from the lemmatized reviews. This dictionary comprises all of the unique terms (tokens) in the reviews and provides each word a unique ID.
2. ***Document-Term Matrix***: After that, a document-term matrix is built. This is a table that shows how frequently each term appears in each review. Each row represents a review, while each column is a dictionary word. The table values show how many times each term appears in each review.
3. ***Training LDA Model***: Finally, utilising the document-term matrix the LDA model is trained. It estimates its parameters which include the topic proportions for each document and the word distribution for each topic using an iterative method.

3.2 Personality Identification Using XGBoost Machine Learning Approach

The preprocessing steps such as Stop words removal, Lemmatization and URL removal from the posts are done. In addition to this, the 16 personality labels explicitly specified in the posts are expunged in the MBTI dataset. This is used to train the classifier Models like Logistic regression, KNN (K-Nearest Neighbours) and XGBoost. XGBoost is chosen as the optimal model taking accuracy into account. The trained model is now used applied to the Preprocessed social reviews, where each review is mapped to a corresponding Personality Type. The steps followed for building classification mode are:

1. ***Feature Selection***: Choose the features that are crucial to the categorization process.
2. ***Data Splitting***: Partition the data into training and testing sets.
3. ***Training***: Run the corresponding model on the training set.
4. ***Evaluation***: Assess the model by applying it on the test set using performance measures like accuracy, recall and $F1$ score.

Three similarity matrices are generated as presented in Fig. 2, one from explicit data (Cuisine and Rating), the other from LDA Topic modelling approach where in user preference is obtained, the third similarity matrix generated is of the user personality that is obtained as result of the Personality classification where XGBoost is deployed. Cosine similarity measure is used for the clustering process. The three similarity matrices thus obtained are used to generate an enhanced matrix and a clustering of the similar item of interests are identified and recommended to ultimately provide relevant personalized recommendations.

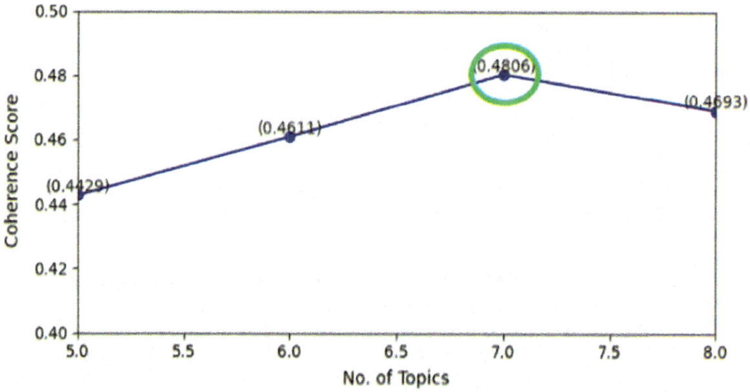

Fig. 3 Coherence score plot

4 Experimental Results and Discussion

The experimental study of the proposed approach is done for personalized restaurant recommendations. Topics generated from customer eWoM in TripAdvisor along with MBTI Personality classes are used to provide enhanced restaurant recommendations to the user. For restaurant data obtained from Trip advisor, the dataset used contains about 38,000 reviews written in English by people who visited restaurants in 8 cities and shared their experiences. To train the personality classifier, the Myers-Briggs Personality Types dataset was utilized, which contains data from the Personality Cafe forum, having around 8600 posts and their MBTI personality type.

The LDA topic model was modeled to ensure a strong coherence in the identification of customer topics on eWoM using the estimated desirable number of topics i.e. $K = 7$. The graph in Fig. 3 shows that coherence score increases with the number of topics, with a decline at $K = 8$. The Topic Coherence Metric measures how effectively a topic is supported by a text collection. Topic distribution for all the documents (reviews) is obtained as output of the LDA model and the Topic ID with the highest probability is assigned to the document. The topics in customer eWoM were carefully tagged based on domain expertise and the most popular terms associated with each topic.

The result of the LDA topic modelling on Trip Advisor dataset is presented in Table 2. This table displays the top frequent words identified by the LDA model for each of the topics.

With Reference to the Personality classifier in which the user posts are classified into 16 MBTI labels, techniques such as Logistic Regression, KNN and XGBoost were employed in the study. The results of the study are presented in Tables 3, 4 and 5. From the results it is found that XGBoost proved to be the best in comparison to the other two. After the topic modelling and personality modelling is done, the proposed system is evaluated with various test cases. Each test case specified contained a list of 10 restaurants arbitrarily given. The system is evaluated using

Table 2 Labelling of topics

Topic No.	Top frequent words	Label
0	Food, good, restaurant, service, place, great, nice, friendly menu, excellent	Ambience
1	Pizza, italian, pasta, topping, dough, crust, cheese, base, thin, ingredient	Pizza
2	Good, staff, great, food, place, friendly, service, nice, drink, burger	Welcoming staff
3	Review, time, experience, service, great, thank, lovely, team, food, feedback	Customer service
4	Food, good, chicken, place, time, service, order, restaurant, tasty, portion	Quality
5	Coffee, breakfast, good, place, cafe, tea, cake, sandwich, friendly, great	Breakfast and cafe
6	Raman, mexican, burrito, taco, guacamole, yen, cutlet, studio, broth, reais	Global cuisine

Table 3 Performance of the logistic regression model

Personality label	Train accuracy	Test accuracy
IE	0.78	0.78
NS	0.86	0.86
FT	0.77	0.72
JP	0.70	0.65

Table 4 Performance of the XGBoost model

Personality label	Train accuracy	Test accuracy
IE	0.99	0.76
NS	1.00	0.85
FT	0.99	0.68
JP	0.99	0.63

precision evaluation metric, where in the context of recommendation system, precision at k is the proportion of recommended items in the top-k set that are relevant as presented in Eq. 1.

$$\text{Precision} = \frac{\text{number of relevant item recommendations}}{\text{number of items recommended}} \tag{1}$$

The proposed approach gave an Mean Average Precision (MAP) of 0.818. In clustering, weightage is used to assign different levels of importance to the features or variables being used for clustering. Weightage helps in determining the relative

Table 5 Performance of KNN

K value	Train accuracy	Test accuracy
1	0.996	0.606
2	0.995	0.402

Table 6 Evaluation of personality classifier models

	Factor	Weightage
1	Topic	0.3
2	Personality	0.4
3	Cuisine	0.3

Table 7 Precision scores of the model without and with weightage applied to factors

Test case No.	Without weightage	With weightage
	Precision	Precision
1	0.630	0.795
2	0.780	0.817
3	0.670	0.793
4	0.821	0.819
5	0.799	0.855
6	0.820	0.815
7	0.685	0.787
8	0.671	0.837
9	0.663	0.857
10	0.758	0.806
Average precision score	0.730	0.818

influence of each feature on the clustering algorithm. The factors taken into consideration and their respective weightages have the following split up as specified in Table 6. The split up shows that personality has been given the highest weightage as this model's main focus is upon personality while topic and cuisine have been given equal weightages in pertinence to relevant importance. There is an evident distinction that can be conduced between the precision scores before and after the application of weights as can be evidently seen in Table 7. The showcased model here performs well and is able to accurately predict users' preferences effectively.

5 Conclusion and Future Work

The personalized recommender system using Topic modelling approach employing the 16 MBTI personalities has been successfully able to predict the user's personality based on their posts obtained from electronic word of mouth on social platforms and map effectively suggest the user tailored preferences by mapping it to other users' posts with similar personality configuration. Being different from the quotidian recommender systems it stands apart through accurate suggestions with an Precision score of 0.818 which is on the higher leagues in comparison to existing models.

Usage of long text adaptive methodologies like BERT can be adopted with the conditional necessity of long time and high processing capacity of systems used kept into consideration. By doing so the performance can be further enhanced a few notches. To improve user engagement features like "Not Interested" or "Irrelevant", this can be predominantly seen in streaming platforms like Youtube and Netflix and this will help in not recommending any more items of the same cluster or similar such items to the user. Hence if a user receives any irrelevant recommendation the model can be improved by Fine tuning the weights. This can be done by collecting user responses, asking if they are satisfactory with the recommendation provided. Once a sufficient number of user responses have been labelled, it can be used to train the ML model. This involves feeding the labelled data into the model and adjusting its weights and parameters until it can accurately predict the correct class for new, unseen responses.

References

1. Su W-Z, Lin P-H (2016) A study of relationship between personality and product identity. In: Rau P-LP (ed) Cross-cultural design. Springer, Cham, pp 266–274
2. Chavan P, Thoms B, Isaacs J (2021) A recommender system for healthy food choices: building a hybrid model for recipe recommendations using big data sets. https://doi.org/10.24251/HICSS.2021.458
3. Christodoulou E, Gregoriades A, Pampaka M, Herodotou H (2022) Personality-informed restaurant recommendation, pp 13–21. https://doi.org/10.1007/978-3-031-04826-5_2
4. Melese A (2021) Food and restaurant recommendation system using hybrid filtering mechanism. North Am Acad Res (NAAR) J 4:268–281. https://doi.org/10.5281/zenodo.4712849
5. Rajendran DPD, Sundarraj RP (2021) Using topic models with browsing history in hybrid collaborative filtering recommender system: experiments with user ratings. Int J Inf Manag Data Insights 1:2667–0968. https://doi.org/10.1016/j.jjimei.2021.100027
6. Gaudani H (2014) A review paper on machine learning based recommendation system. Int J Eng Dev Res 2:3955–3961
7. Fakhri A, Baizal A, Setiawan E (2019) Restaurant recommender system using user-based collaborative filtering approach: a case study at Bandung Raya region. J Phys Conf Ser 1192:012023. https://doi.org/10.1088/1742-6596/1192/1/012023
8. Khanal SS, Prasad PWC, Alsadoon A, Maag A (2020) A systematic review: machine learning based recommendation systems for e-learning. Educ Inf Technol 25:1–30. https://doi.org/10.1007/s10639-019-10063-9

9. Al-Nafjan A, Alrashoudi N, Alrasheed H (2022) Recommendation system algorithms on location-based social networks: comparative study. Information 13(4). https://doi.org/10.3390/info13040188
10. Asani E, Vahdat-Nejad H, Sadri J (2021) Restaurant recommender system based on sentiment analysis. Mach Learn Appl 6:2666–8270. https://doi.org/10.1016/j.mlwa.2021.100114

A Secure Fingerprint Transmission Protocol for Insecure Communication Channels

Laiphrakpam Dolendro Singh, Md. Arshad Ahmed, Teresa Louis, Azad Ali, Dalton Meitei Thounaojam, Ripon Patgiri, and Khoirom Motilal Singh

Abstract A fingerprint is a recognizable pattern of ridges and valleys on a person's finger surface. A person's fingerprints are distinct and unchangeable throughout a person's life unless distorted through external injury. The fingerprint is one of the most commonly utilized biometric identification techniques. As fingerprints are used in realistic authentication, securing fingerprints is a paramount concern. Moreover, enciphering the whole fingerprint image collected through a fingerprint scanner would require extensive computation. Research has shown that minutiae points are used to determine the uniqueness of a fingerprint. A method is proposed to store and transmit fingerprint data securely. The proposed technique enciphers only the internal angles subtended by the relevant minutiae points and the information about the type of minutiae points. The proposed techniques reduce the computation involved as only the internal angles information and type of minutiae points are enciphered rather than the whole scanned fingerprint image. Statistical and security analyses conducted on the cipher data suggest the proposed encryption is secure. A digital signature is incorporated while transmitting the fingerprint information, thereby maintaining the data integrity.

Keywords Security · Fingerprint · Minutiae points · Encryption · Digital signature

L. D. Singh (✉) · Md. Arshad Ahmed (✉) · T. Louis · A. Ali · D. M. Thounaojam · R. Patgiri
Department of Computer Science and Engineering, National Institute of Technology, Silchar, Silchar, India
e-mail: ldsingh@cse.nits.ac.in

A. Ali
e-mail: azad_ug@cse.nits.ac.in

D. M. Thounaojam
e-mail: dalton@cse.nits.ac.in

R. Patgiri
e-mail: ripon@cse.nits.ac.in

K. M. Singh
Department of CSE, Indian Institute of Information Technology, Senapati, Manipur, India

S. Bandyopadhyay et al. (eds.), *Intelligent Computing Systems and Applications*, Lecture Notes in Networks and Systems 1010,
https://doi.org/10.1007/978-981-97-5412-0_12

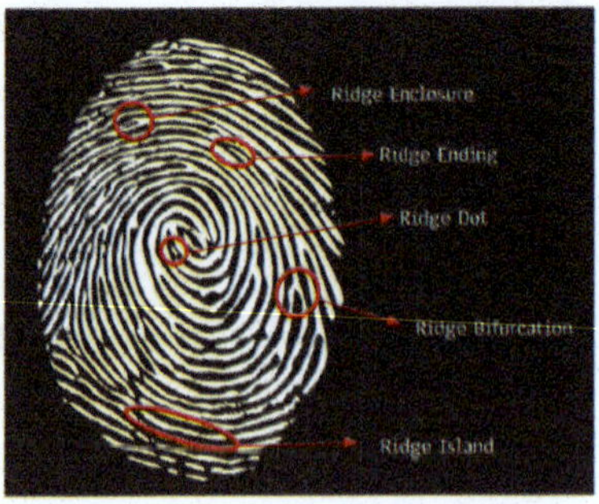

Fig. 1 Some common minutiae points

1 Introduction

Biometric identification is a practical and reliable method that utilizes unique physiological and behavioral characteristics like face, fingerprints, iris, DNA, signature, voice, and keystrokes. Among these, fingerprint recognition is widely used, as it relies on the distinct patterns of ridges and valleys on a person's finger, including minutiae points, to determine uniqueness and establish identification. Some common minutiae types are shown in Fig. 1. The white lines indicate the ridges, and the black space between the ridges are the valleys. The termini or forks of the ridge lines are known as minutiae.

Fingerprint templates are vulnerable to unauthorized access, posing a significant risk as individuals have limited biometrics. To enhance security, converting the templates into non-invertible and cancellable forms is recommended. Non-invertibility ensures the difficulty of recovering the original biometric from the encrypted template, while cancellable fingerprints make it computationally nearly impossible to get back the original biometric data from the transformed version. Trivedi et al. [1] proposed a non-invertible fingerprint template based on relative geometric information, while Jin et al. [2] introduced a technique for producing revocable templates using feature extraction and binarization. The proposed technique in Trivedi et al. [3] removes location and orientation information from minutiae points in a fingerprint template, ensuring non-invertibility and immunity to reconstruction. This enhanced security and reduced the risk of unauthorized access or tampering of the biometric data. In their research, Hong et al. [4] proposed a fast fingerprint enhancement algorithm that improved fingerprint clarity with a fast enhancement algorithm based on local ridge orientation and frequency.

Bansal and Garg [5] proposed a cancellable biometric template protection technique that utilizes format-preserving encryption and Bloom filters to secure the input biometric template and achieve high-performing biometric identification.

Abou Elazm et al. [6] introduces a novel biometric recognition method combining optical encryption, the 3D jigsaw transform, and template revocation for enhanced privacy and security. Cappelli et al. [7] evaluated fingerprint recognition algorithms using benchmark datasets from the Fingerprint Verification Competition (FVC) 2006, showcasing performance metrics like EER, FAR, and FRR. Top-performing algorithms achieved EERs as low as 0.4% and FARs as low as 0.05%, highlighting high accuracy and reliability. The paper also discusses the impact of factors like image quality, sensor technology, and fingerprint pattern variability, exploring alternative approaches beyond minutiae-based methods.

The contribution of our proposed method is to provide security for fingerprint template database transmission. Fingerprint as a whole is a large chunk of data. Every pixel in a fingerprint is not used for entity authentication, only the minutia information from a fingerprint can be used to ascertain entity authentication. This method secures only the needful data.

2 Dataset

The template is evaluated using the FVC2000 standard database [8] and FVC2006 [9] standard database. In the case of FVC2000, DB1, DB2, DB3, and DB4 are the four groups that make up the database. Each group is then split into two smaller sets, set A and set B, each of which contains the fingerprint photos of 100 and 10 distinct fingers, respectively, with 8 samples per finger in depth, i.e., 880 fingerprint images.

Meanwhile, for FVC2006, DB1, DB2, DB3, and DB4 are the four groups that make up the database, 150 fingers wide and 12 samples per finger in depth, i.e., 1800 fingerprint images.

3 Proposed Method

The outlined approach for the transmission protocol consists of the subsequent procedural steps:

1. Generating fingerprint template database.
2. Enciphering the fingerprint template database.
3. Generating Digital Signature for the enciphered fingerprint template database.
4. Digital Signature verification.
5. Deciphering ciphered fingerprint template.
6. Fingerprint template matching.

3.1 Generating Fingerprint Template Database

In order to convert a raw scanned fingerprint image database to a fingerprint template database, the following steps are performed:

1. Import a scanned fingerprint from a fingerprint database.
2. The imported scanned fingerprint image is prepossessed with the following operations: sharpening, binarization, color negation, thinning and deleting small components.

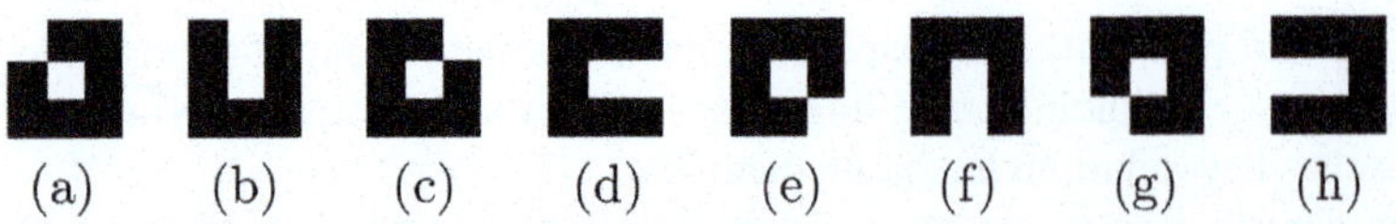

Fig. 2 **a–h** Possible 3 × 3 blocks for termination minutia

3. The prepossessed image is partitioned in blocks (b_i) of 3 × 3 with an offset of one.
   - Some of these b_i are of interest as it can denote termination minutia (m_{t}) or bifurcation minutia (m_{b}). Figures 2 and 3 show the possible 3 × 3 minutia termination blocks and minutia bifurcation blocks respectively.
4. Coordinate position of each 3 × 3 minutiae termination blocks (m_{t}) are stored in $\mathrm{PM_T}$. Similarly, coordinate position of each 3 × 3 minutiae bifurcation blocks (m_{b}) are stored in $\mathrm{PM_B}$. The combined minutiae points $\mathrm{PM_T}$ and $\mathrm{PM_B}$ that are of interest are stored in a list CMP.
5. Using the coordinate stored in combined minutiae points list CMP, a DelaunayMesh (DM) is generated.
6. From the generated DelaunayMesh (DM), triplet ($T_i : (p_1, p_2, p_3)$) consisting of points that make up the triangles are obtained.
7. From each $T_i : (p_1, p_2, p_3)$, the corresponding coordinates $p_1 : (x_1, y_1)$, $p_2 : (x_2, y_2)$ and $p_3 : (x_3, y_3)$ along with the type of minutiae ($\mathrm{type_m}$), minutiae termination (m_{t}) or minutiae bifurcation (m_{b}) are generated.
8. Using Euclidean distance, the distance between the three coordinate $p_1 : (x_1, y_1)$, $p_2 : (x_2, y_2)$ and $p_3 : (x_3, y_3)$ are computed where a is the distance between p_2 to p_3, b is the distance between p_1 to p_3 and c is the distance between p_1 to p_2.
9. If $a \neq 0$, $b \neq 0$ and $c \neq 0$, the angles in a triangle with sides as a, b and c are computed as:

$$\begin{aligned} \angle\mathcal{A} &= \mathrm{ArcCos}\left[(b^2 + c^2 - a^2)/(2 \times b \times c)\right] \\ \angle\mathcal{B} &= \mathrm{ArcCos}\left[(a^2 + c^2 - b^2)/(2 \times c \times a)\right] \\ \angle\mathcal{C} &= \mathrm{ArcCos}\left[(a^2 + b^2 - c^2)/(2 \times a \times b)\right] \end{aligned} \tag{1}$$

10. Angles $\angle\mathcal{A}$, $\angle\mathcal{B}$ and $\angle\mathcal{C}$ are converted to degrees. If $\angle\mathcal{A}$, $\angle\mathcal{B}$ and $\angle\mathcal{C}$ are all greater than 16°, the template is stored as:
    $\mathcal{T_I} = \{\angle\mathcal{A}, \angle\mathcal{B}, \angle\mathcal{C}, \mathrm{type_m}, \mathrm{type_m}, \mathrm{type_m}\}$, where $\mathrm{type_m}$ can be m_{t} or m_{b} corresponding to the respective angles $\angle\mathcal{A}$, $\angle\mathcal{B}$ or $\angle\mathcal{C}$.

Every scanned fingerprint for a particular person from the dataset is processed with the above Steps 1–10. After obtaining all the $\mathcal{T_I}$ of a particular person, the fingerprint template $\mathcal{T_I}$ is updated by removing all the duplicates entries in $\mathcal{T_I}$ to $\mathrm{UT}_{\mathcal{I}}$.

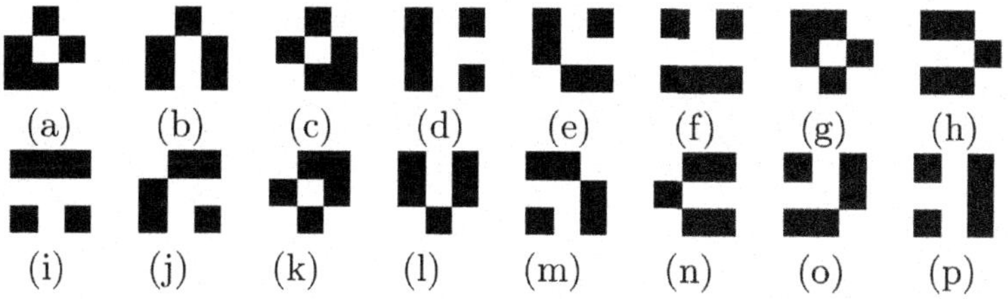

Fig. 3 **a–p** Possible 3 × 3 blocks for bifurcation minutiae

Table 1 512 bits Brainpool ECP

Parameter	Values
p	894896220765023255165660281515915342216260964409835451134459718720005701041355243991793430419195694276544653038642734593796389430992392853607053460781694 7
a	629486055797306322766642130647637932407471577062274622713691044545030191428127609802799096840798396269115185367856387783422183402743971823806572584426413 8
b	324578900832896705927484958434207791653190900963750191832832366873617917658326349646352512848882826115598007735069737717977648114988349952343415308622866 27
G	6792059140424575174435640431269195087843153390102521881468023012732047482579853077545647446272866794936371522410774532686582484617946013928874296844351522, 6592244555240112873324748381429610341312712940326266331327445066687010545415256461097707483288650216992613090185042957716318301180159234788504307628509330
n	894896220765023255165660281515915342216260964409835451134459718720005701041341852837898173064352495985745139837002928058309421561388204397335439211554416 9

3.2 Secure the Fingerprint Template Database

To secure the updated fingerprint template $\text{UT}_{\mathcal{I}}$, elliptic curve cryptography is deployed using 512 bit Brainpool elliptic curve parameter given in Table 1.

Each angles $\angle\mathcal{A}$, $\angle\mathcal{B}$ and $\angle\mathcal{C}$ can have their value in the range of [17–148] as each angle are taken into consideration only if all the angles are greater than 16. Instead of using a unique single integer value for m_t and m_b, each m_t and m_b entries in the updated fingerprint template $\text{UT}_{\mathcal{I}}$ are masked with random values ranging from [1–74] and [75–148] respectively. The following steps are performed to encipher the updated fingerprint template $\text{UT}_{\mathcal{I}}$.

1. Determine $\mathcal{LEN}$, $\mathcal{L}$, $\mathcal{Q}$ and $\mathcal{R}$.

$$\begin{aligned} \mathcal{LEN} &= \text{Length}[\text{IntegerDigits}[p, \text{base}]] - 1 \\ \mathcal{L} &= \text{Length}[\text{UT}_\mathcal{I}] \\ \mathcal{Q} &= \text{Quotient}[\mathcal{L}, \mathcal{LEN}] \\ \mathcal{R} &= \mathcal{L} \bmod \mathcal{LEN} \end{aligned} \tag{2}$$

where
$\mathcal{LEN}$: Length of digits generated when elliptic curve parameter prime modulo p is base converted with base = 149 minus 1.
$\mathcal{L}$: Length of flatten $\text{UT}_\mathcal{I}$.
$\mathcal{Q}$: Quotient when $\mathcal{L}$ is divided by $\mathcal{LEN}$.
$\mathcal{R}$: Remainder when $\mathcal{L}$ is divided by $\mathcal{LEN}$.

2. $\text{UT}_\mathcal{I}$ is left padded with 0 in two cases:
Case (i) If $\mathcal{Q}$ is even and $\mathcal{R}! = 0$, $\text{UT}_\mathcal{I}$ is left padded with 0, for a length of $\mathcal{L} + 2 \times \mathcal{LEN} - \mathcal{R}$.
Case (ii) If $\mathcal{Q}$ is odd and $\mathcal{R} == 0$, $\text{UT}_\mathcal{I}$ is left padded with 0, for a length of $\mathcal{L} + 2 \times \mathcal{LEN}$.
3. The fingerprint template $\text{UT}_\mathcal{I}$ are partitioned into blocks ($\mathcal{B}_\mathcal{I}$) upto length ($\mathcal{LEN}$).
4. Blocks (B_I) are paired up and converted to coordinate points $P_I : (x_I, y_I)$ using base = 149.
5. $P_I : (x_I, y_I)$ is encrypted using elliptic curve cryptographic operation as:

$$\begin{aligned} \mathcal{K} &= \text{SHA}(\text{UT}_\mathcal{I}) \\ \mathcal{C}_{\mathcal{KG}} &= \mathcal{K} \cdot \mathcal{G} \\ \mathcal{C}_\mathcal{I} &= (\mathcal{P}_\mathcal{I} + \mathcal{K} \cdot \text{PUB}_\text{R}) \end{aligned} \tag{3}$$

where
$\mathcal{K}$: 512 bit SHA hash value of $\text{UT}_\mathcal{I}$;
$\cdot$: represent point multiplication operation;
$\mathcal{G}$: Generator point;
PUB_R: receiver's public key computed as $\text{PRI}_\text{R} \cdot G$ where PRI_R is the receiver's private key.

6. Each $\mathcal{C}_\mathcal{I}$ is converted to cipher template CT using base = 149, maintaining length = $\mathcal{LEN} + 1$ with zero left padding if necessary.

Steps 1–6 are performed for all the individuals (1 to n) in the database generating the complete cipher template.

$$\text{CT}_\mathcal{I} = \{\text{CT}_1, \text{CT}_2, \text{CT}_3, \ldots, \text{CT}_n\} \tag{4}$$

3.3 Generating Digital Signature for the Enciphered Fingerprint Template Database

The complete cipher template $CT_{\mathcal{I}}$ for all the individuals (1 to n) in the database is stored as $ACT_{\mathcal{I}} = \{(\mathcal{C}_{\mathcal{KG}_1}, CT_1), (\mathcal{C}_{\mathcal{KG}_2}, CT_2), (\mathcal{C}_{\mathcal{KG}_3}, CT_3), \ldots, (\mathcal{C}_{\mathcal{KG}_n}, CT_n)\}$. A digital signature pair $(\mathcal{R}, \mathcal{S})$ is generated using elliptic curve cryptography on $ACT_{\mathcal{I}}$. $(\mathcal{R}, \mathcal{S})$ is computed as follows:

$$\begin{aligned} \mathcal{H} &= \text{SHA}(ACT_{\mathcal{I}}) \\ \mathcal{R} &= (\mathcal{K} \cdot \mathcal{G}) \bmod n[[1]] \\ \mathcal{S} &= (K^{-1} \bmod n \times \mathcal{H} + \mathcal{R} \times \text{PRI}_\text{S}) \bmod n \end{aligned} \tag{5}$$

where
$\mathcal{H}$: 512 bit SHA hash value of $ACT_{\mathcal{I}}$;
$(\mathcal{K} \cdot \mathcal{G}) \bmod n$ [[1]]: x-coordinate obtained after taking modulo n on point multiplication of $\mathcal{K}$ and $\mathcal{G}$;
PRI_S: Sender's private key.

3.4 Digital Signature Verification

To confirm the integrity of the complete cipher template $ACT_{\mathcal{I}}$, verification is performed using the digital signature pair $(\mathcal{R}, \mathcal{S})$ as follows:

$$\begin{aligned} \mathcal{H}' &= \text{SHA}(ACT_{\mathcal{I}}) \\ \mathcal{U} &= (\mathcal{H}' \times \mathcal{S}^{-1}) \bmod n \\ \mathcal{V} &= (\mathcal{R} \times \mathcal{S}^{-1}) \bmod n \\ (\mathcal{X}, \mathcal{Y}) &= \mathcal{U} \cdot \mathcal{G} + \mathcal{V} \cdot \text{PUB}_\text{R} \end{aligned} \tag{6}$$

If $\mathcal{X} == \mathcal{R}$, integrity of $ACT_{\mathcal{I}}$ is maintained else tempering performed on $ACT_{\mathcal{I}}$.

3.5 Deciphering Ciphered Fingerprint Template

The cipher fingerprint template from $ACT_{\mathcal{I}}$ is deciphered using the receiver private key PRI_R using the following steps:

1. Each $CT_{\mathcal{I}}$ is flatten and partition into blocks with length = len and base converted using base = 149 and paired up $\mathcal{B}_{\mathcal{I}}$: (x_I, y_I).
2. The deciphered fingerprint templates $\mathcal{T}_{\mathcal{I}}$ are obtained using the private key of the receiver PRI_R as:

$$\mathrm{UT}_{\mathcal{I}} = \mathcal{B}_{\mathcal{I}} - \mathrm{PRI}_{\mathrm{R}} \cdot \mathcal{C}_{\mathcal{KG}} \tag{7}$$

where $\mathrm{PRI}_{\mathrm{R}} \cdot \mathcal{C}_{\mathcal{KG}}$: elliptic curve point multiplication of $\mathrm{PRI}_{\mathrm{R}}$ and $\mathcal{C}_{\mathcal{KG}}$.

3. Base convert $\mathrm{UT}_{\mathcal{I}}$ using base = 149 and drop the left padded zero if exist.
4. The deciphered fingerprint template is generated by partitioning $\mathrm{UT}_{\mathcal{I}}$ into six tuples, $\{\angle\mathcal{A}, \angle\mathcal{B}, \angle\mathcal{C}, \mathrm{type_m}, \mathrm{type_m}, \mathrm{type_m}\}$.
5. Each $\mathrm{type_m}$ in deciphered data obtained in Step 4 is converted to m_{t} or m_{b} as follows:

$$\begin{aligned} &\mathrm{IF}(\mathrm{type_m} < 75) \\ &\qquad \mathrm{type_m} = m_{\mathrm{t}} \\ &\qquad \mathrm{ELSE} \\ &\qquad \mathrm{type_m} = m_{\mathrm{b}} \end{aligned} \tag{8}$$

3.6 Fingerprint Template Matching

The fingerprint template generated using the technique given in Sect. 3.1 or deciphered in Sect. 3.5 is stored in the fingerprint template database. In order to match a query fingerprint for a person, the query finger is processed with the process given in Sect. 3.1 to generate the fingerprint template. With the generated fingerprint template, an intersection operation is performed from the database. The query fingerprint is authenticated if the match is more than 75% with the database fingerprint template.

4 Simulation

The simulation of the proposed method is executed on a workstation with Intel(R) Xeon(R) processor 3.60 GHz with 32.0 GB using Wolfram Mathematica. A sample scanned fingerprint, the prepossessed image, and the corresponding Delaunay mesh are shown in Fig. 4.

Sample query scanned fingerprints are shown in Table 2.

5 Statistical and Security Analysis

5.1 PSNR

The peak-signal-to-noise ratio (PSNR) can be used to ascertain the accuracy of the data after the decryption. Similarly, PSNR can be used to see how deviated the cipher

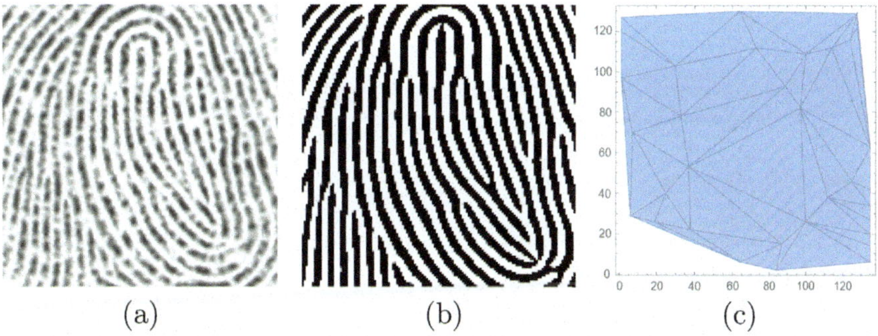

Fig. 4 **a** Scanned fingerprint. **b** Prepossessed scanned fingerprint. **c** Delaunay mesh of prepossessed scanned fingerprint

Table 2 Query fingerprint

Query fingerprint	P1	P2	P3	P4	P5	P6	P7	P8	P9	P10
Match % from Auth DB	78.79	97.87	89.16	92.23	77.24	77.90	86.42	87.81	96.93	77.78

Table 3 PSNR of encrypted and decrypted fingerprints for $\text{FVC2006}_{\text{DB1A}}$ dataset

Image ID	Encrypted PSNR	Decrypted PSNR
Person 1	8.25382	∞
Person 20	8.25154	∞
Person 40	8.16814	∞
Person 60	8.22809	∞

data is from the original data. The computed values are shown in Table 3. The PSNR is calculated using the below formula:

$$\text{PSNR} = 20 \times \log_{10} \frac{\text{Max}^{\text{m}}_{p_{\text{value}}}}{\text{Mean}_{s_{\text{error}}}} \tag{9}$$

$$\text{Mean}_{s_{\text{error}}} = \frac{1}{n} \sum_{i=0}^{n-1} [O_{\text{data}}(i) - C_{\text{data}}(i)]^2 \tag{10}$$

where $\text{Max}^{\text{m}}_{p_{\text{value}}}$: maximum supported value. $\text{Mean}_{s_{\text{error}}}$: mean squared error. n: length of the data. $O_{\text{data}}(i)$: original data at i. $C_{\text{data}}(i)$: cipher data or deciphered data at (i).

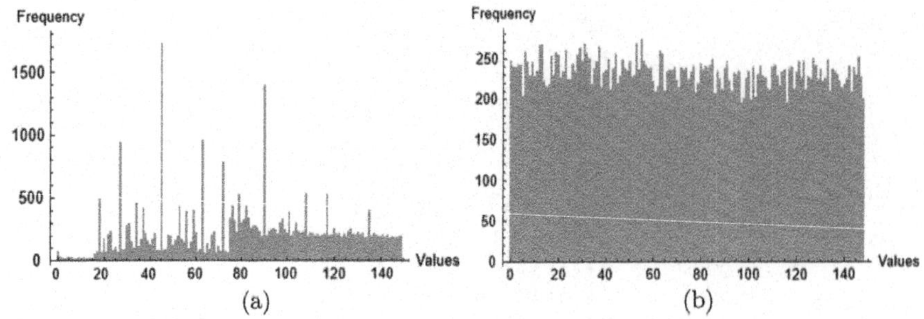

Fig. 5 **a** Original fingerprint data histogram plot. **b** Cipher fingerprint data histogram plot

5.2 Histogram Analysis

A histogram plots the frequency of each data value over the entire range of possible values. A cipher data with a uniform frequency distribution has good confusion and diffusion property. A good cipher data should have a consistent histogram. In order to protect against any possible attack based on the distribution of the cipher data, the cipher data must be uniformly distributed in the histogram plot. The histogram plot of the original fingerprint data and the cipher data are shown in Fig. 5a, b respectively.

5.3 Entropy Analysis

To protect against potential attacks, a cipher data values should be distributed randomly and have no relation to the plain data. The degree of randomness can be measured using entropy derived from the logarithm of a probability distribution. The higher the entropy, the higher the randomness of the pixels. Table 4 presents the entropy values for the original fingerprint data and their respective cipher data. Entropy can be calculated as follows:

$$\text{Entropy, } E(X) = -\sum_{i=1}^{n} P(x_i)\log_2 P(x_i) \tag{11}$$

5.4 Key Space

The concept of a key space refers to the complete set of potential keys available within an encryption system. For an encryption technique to be considered secure against brute force attacks, it is essential that the key space is extensive enough. In

Table 4 Entropy of original fingerprint data and cipher fingerprint data for FVC2006$_{DB1A}$ dataset

Fingerprint	Entropy	
	Plain	Cipher
Person 1	6.92098	7.20058
Person 20	6.81324	7.21556
Person 40	6.96714	7.20660
Person 60	6.98969	7.19788

the case of the proposed encryption scheme, the number of possible keys is on the order of 2^{512}. Such a vast keyspace ensures formidable resistance against any form of brute force attack.

5.5 *Brute Force Attack*

The brute force attack tries various methods for decrypting a fingerprint template by attempting to recreate every possible fingerprint template to match the original ones. To put it another way, it's a trial-and-error technique used by attackers to methodically attempt every conceivable combination until they discover the correct encryption key. In the proposed method, SHA512 along with 512 bits Brainpool ECP are used for the ECC encryption. Hence, the proposed method can withstand a brute-force attack.

5.6 *False Acceptance Analysis*

False accept analysis involves calculating the system's false acceptance rate (FAR), which is the percentage of impostor fingerprints that are incorrectly accepted as a match.

A low FAR indicates that the system is highly accurate and reliable, as it is able to correctly reject impostor fingerprints and avoid false positives. In contrast, a high FAR indicates that the system is less accurate and may be prone to false positives, which can lead to security breaches and privacy violations.

In our project, we did a false accept analysis on FVC2006 DB2-B, the corresponding result obtained is shown in Table 5.

As seen in Table 5, the false acceptance values for Person-4 and Person-7 are zero for almost all degrees of tolerance. This means that the fingerprint templates obtained for them are accurate and do not falsely identify with other fingerprint templates. However, as seen for Person-8 or Person-10, some templates falsely identify with

Table 5 False accept analysis for FVC2006 DB1B

i/p	0°	2°	4°	6°	8°
Person-1	0.2222	0.5556	0.4444	0.4444	0.2222
Person-2	0	0.2222	0.3333	0.2222	0.3333
Person-3	0.1111	0.3333	0.4444	0.5556	0.7778
Person-4	0	0	0	0	0.1111
Person-5	0.2222	0.8889	0.8889	0.8889	0.8889
Person-6	0.1111	0.7778	0.6667	0.6667	0.6667
Person-7	0	0	0	0	0
Person-8	0.4444	0.8889	0.8889	0.8889	1
Person-9	0.1111	0.6667	0.5556	0.3333	0.4444
Person-10	0.4444	0.8889	0.8889	0.7778	0.7778

the fingerprint templates of other persons. These variations arise due to the quality of the scanned fingerprint image, which often due to dust, cuts, sweat, etc., create false minutia points and hence hamper the evaluation process.

5.7 *Cipher-Text-Only Attack*

A cipher-text-only attack is a type of attack in which an attacker attempts to decrypt a message without any knowledge of the encryption algorithm or the key used to encrypt it. In a cipher-text-only attack, the attacker has access to the encrypted message but does not have access to the corresponding plain text message or the key used for encryption. The proposed method adopted ECC as the first layer of security where the security of the cryptographic system is based on the difficulty of solving the elliptic curve discrete logarithm problem (ECDLP). SHA512 along with 512 bits Brainpool ECP are also used for the encryption operation. Therefore, the proposed method can withstand cipher-text only attacks.

5.8 *Transmission Size Analysis*

For FVC2006 with 7200 fingerprints in total, the raw fingerprint images occupying a size of 950 MB, our method reduced the transmission size of the raw fingerprints from 950 MB to feature vectors of a total size of 129 MB.

5.9 *Irreversibility Analysis*

An irreversible fingerprint template means original fingerprint features cannot be recomputed from the fingerprint template. The proposed method has got two layers of security. The first layer of security is provided by the cipher data obtained using ECC that required solving a computationally hard problem called elliptic curve discrete logarithmic problem. Only if the private key is compromised, the cipher data can be deciphered to the fingerprint template. The fingerprint template of the proposed method consists of multiple six tuples of the form $\{\angle\mathcal{A}, \angle\mathcal{B}, \angle\mathcal{C}, \text{type}_{\text{m}}, \text{type}_{\text{m}}, \text{type}_{\text{m}}\}$, generated from the Delaunay Mesh of the pre-processed scanned fingerprint. From the tuples, it is computationally infeasible to get back to the original fingerprint features, providing a second layer of security.

6 Conclusion and Future Scope

The paper proposes creating and encrypting a secure non-invertible fingerprint template for transmission over insecure networks. Delaunay's triangulation method is used to generate the non-invertible template from minutiae points, consisting of angles and termination type. Encryption with Elliptic Curve Cryptography and digital signing ensures data integrity and tampering prevention during network transmission. The proposed method upon testing with FVC2006 gave an average matching score of 69.85 for 4° tolerance and 83.23 for 6° tolerance. The method is further put into security and false matching analysis techniques. Future prospects of our project include testing the effectiveness of our algorithm with respect to algorithms based on ridge patterns and working with real-time data using fingerprint scanners.

References

1. Trivedi AK, Thounaojam DM, Pal S (2018) A robust and non-invertible fingerprint template for fingerprint matching system. Forens Sci Int 288:256–265
2. Jin Z, Teoh ABJ, Ong TS et al (2010) A revocable fingerprint template for security and privacy preserving. KSII Trans Internet Inf Syst 4(6):1327–1342
3. Trivedi AK, Thounaojam DM, Pal S (2021) A novel minutiae triangulation technique for non-invertible fingerprint template generation. Expert Syst Appl 186:115832
4. Hong L, Wan Y, Jain A. Fingerprint image enhancement: algorithm and performance evaluation. Pattern Recognition and Image Processing Laboratory, Department of Computer Science, Michigan State University, East Lansing, MI
5. Bansal V, Garg S (2022) A cancelable biometric identification scheme based on bloom filter and format-preserving encryption. J King Saud Univ Comput Inf Sci 34(8, Part B):5810–5821. ISSN 1319-1578. https://doi.org/10.1016/j.jksuci.2022.01.014. https://www.sciencedirect.com/science/article/pii/S1319157822000271

6. Abou Elazm LA, Ibrahim S, Egila MG, Shawky H, Elsaid MK, El-Shafai W, Abd El-Samie FE (2020) Cancelable face and fingerprint recognition based on the 3D jigsaw transform and optical encryption. Multimed Tools Appl 79:14053–14078
7. Cappelli R, Maio D, Maltoni D, Wayman JL, Jain AK (2006) Performance evaluation of fingerprint verification systems. IEEE Trans Pattern Anal Mach Intell 28(1):3–18. https://doi.org/10.1109/TPAMI.2006.20
8. FVC2000. http://bias.csr.unibo.it/fvc2000/databases.asp
9. FVC2006. http://bias.csr.unibo.it/fvc2006/

Multi-frame Depth Estimation for Autonomous Driving Using Motion Masking and Self-supervised Learning

Om Singh, Anupam Biswas, and Rajdeep Paul

Abstract Fisheye cameras are often employed in automatic driving for 360o near-field sensing around the car. This paper suggests a multi-task visual perception network using rectified fisheye photographs and motion masking to assist the vehicle in detecting its surroundings. However, strong nonlinear distortions constitute a trade-off and necessitate more complex algorithms. Dense-depth supervision in practice has shown promise and may be able to address the problem, even though it can be difficult to offer accurate methods for self-supervised learning. The supervision signal from the neighbouring frames is used to train self-supervised depth estimation networks to forecast scene depth. Additionally, the information between the consecutive frames consists of moving objects which has relative motion between them with respect to the motion of the camera which will produce bias and must be tackled with motion masking annotations before feeding to the network. Also, for further ablation study, the proposed system is trained with the unstructured environments of roads that most developing countries have. Thus, we used AdasInd dataset along with WoodScape and Kitti dataset. Further, to measure the accuracy of the output from the depth pixel, a novel measure relative pixel depth alignment (RPDA) is proposed. Most monocular networks do not make use of this additional signal, ignoring crucial information that can enhance expected depth. Therefore, we propose multi-frame depth estimation from fisheye images using self-supervised learning, a dense adaptive depth estimation method that makes use of the sequential data accessible during test time.

Keywords Depth estimation · Fisheye cameras · Motion masking · Computer vision · Autonomous driving · Multiple frame

---

O. Singh (✉)
National Institute of Technology, Department of Mathematics, Silchar, India
e-mail: om_pg_22@math.nits.ac.in

A. Biswas · R. Paul
National Institute of Technology, Department of Computer Science and Engineering, Silchar, India
e-mail: anupam@cse.nits.ac.in

R. Paul
e-mail: rajdeep_pg_21@cse.nits.ac.in

S. Bandyopadhyay et al. (eds.), *Intelligent Computing Systems and Applications*, Lecture Notes in Networks and Systems 1010,
https://doi.org/10.1007/978-981-97-5412-0_13

1 Introduction

In autonomous driving industry, there has been a remarkable rise in the usage of fisheye cameras due to their large field of view. The fisheye camera gathers more information from the surrounding when compared with other cameras such as pinhole or digital cameras. Although, humans can remarkably adept at understanding the 3D scene of the environment from structure and motion of the surrounding over a short span of time, for a vehicle to navigate in the same scene needs the knowledge or perception of the distances of the objects, which is gathered from current and the nearby scenarios and is commonly known to be the depth perception for autonomous driving. Years of work in geometric computer vision research have not been able to replicate comparable modelling capabilities which human has with that for a vehicle.

In most of the depth estimation tasks, prediction from a single frame does not produce satisfactory results and thus using multiple frames came into picture. There are various uses for depth estimation in autonomous driving such as obstacle avoidance, simultaneous localization and mapping and some examples of 3D vision or robotics applications which can use depth prediction from a single image or multiple images. So, a predictive model is used which combines predictions from two or more frames generally known as sequential frames. Sequential frames are under the topic of supervised learning models because they may be taught and used to make predictions. One prerequisite must be met for the model to perform better than any other else and the model must be reliable and diverse, both when there is a surplus of data and when there is not enough, convolution-based systems can be used. Other computer vision issues have also experienced significant success using deep learning techniques. Single picture depth prediction was initially formulated as a supervised learning issue in early research. The lack of ground truth depth data makes the supervised technique challenging. So here we propose a model for depth estimation from multiple frames of fisheye images using self-supervised learning to improve distance prediction using on surrounding frames of fisheye images.

Our work eliminates the motion objects after doing preprocessing on the images using motion annotations corresponding to each frame to make the dataset more perfect for training. By creating a large number of hypotheses for a frame, a sequential frame improves its performance from the surrounding information and produces the best possible hypothesis for the job at hand. Alternatively, the nearby target frame is re-projected back relative to the source frame with the help of a collection of camera parameters. Thus, this work uses the sequential frames when they are available and combine the benefits of depth estimation using a single and multiple frame.

2 Literature Review

Depth estimation from monocular cases aims to quantify depth pixel wise in a single image. According to [6], dense supervision from depth sensors is used in supervised techniques, or human annotations to provide sparse supervision. The restriction of depth supervision requires ground truth and is removed by self-supervised algorithms, with losses like image reconstruction to be trained with utilizing stereo pairs [8] or video sequences which are monocular.

Existing self-supervised monocular models are being expanded in a rising number of works so that predicted depth as a quality can be enhanced by using during test time for temporal information. It's important to keep in mind that in addition to traditional SLAM-based methods [11] and methods based on SLAM that integrate a monocular depth estimation network, there are a number of techniques which are not based on deep learning and also strive to create depth estimates from consistent sequential. But in this case, we concentrate on cutting-edge depth estimation using neural networks.

In order to use information at test time sequentially, test-time refinement approaches adopt monocular methods. The disadvantage of this is that it calls for additional calculation of numerous forward and backward model runs for a set of test frames, which could take several seconds to complete per set, however one additional backward pass can be efficient and quick. A second large class of methods process sequences of frames using classic monocular networks and recurrent layers. A related method shares characteristics between the depth and posture [12] or computes depth using flow by using pairs of simultaneous frames during test time. These methods are significantly more effective than test-time refinement, but because they must sequentially extract features from numerous frames during training, they can be computationally taxing. Our tests demonstrate that while maintaining the efficiency of recurrent methods during inference, our approach frequently exceeds in terms of accuracy with regard to test-time refinement.

Multi-view depth estimation is a challenging task that involves predicting depth from multiple frames. Recent advancements in regularization methods and architectures have improved the efficiency of this process. While supervised variants outperform unsupervised stereo networks, some methods combine monocular depth cues with traditional stereo estimation, eliminating the need for stereo pairs. Multi-view stereo (MVS) techniques deal with image collections from unordered sets, using cost volumes for training with available ground truth depth and camera positions. However, certain limitations exist in current methods, such as the need for multiple input images during testing, assumptions of camera motion, and dependence on camera poses for training and testing. Our approach leverages sequence information during testing, combining the strengths of multi-view and monocular methods while addressing these limitations.

Our method improves multi-view depth estimation by utilizing sequence information, providing a balance between multi-view and monocular approaches, and being robust to scene motion.

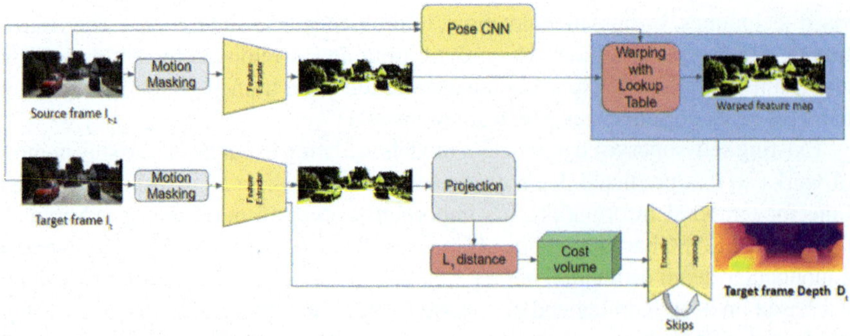

Fig. 1 Block diagram for the proposed system

3 Proposed Methodology

In the proposed system, the main methodology is using a multi-view cost volume and self-supervised re-projection-based training. Next, we present three crucial advancements that make cost volume matching possible when using self-supervision learning from video. Enhancement for static cameras, adaptive cost volumes, a method to prevent "cost volume overfitting" and single frame inputs are the three components of this pattern, and the corresponding blocks with their respective schema are shown in Fig. 1.

3.1 Preprocessing

In this preprocessing step, the distortion of the fisheye image is removed which includes polynomial mapping coefficients of projection function. To map from one coordinate system to another, the camera module is helpful, i.e. Raw to WorldPoint (where Raw is a point on a fisheye image). Fisheye image is converted into a clear image using a camera module. The camera's intrinsic and extrinsic parameters can be used for this. Also, the masking of the moving objects produces bias in multiple frame depth estimation due to the relative motion amongst them.

Motion Masking: For each corresponding images, we have the motion masking annotation for the moving objects which is used to neglect the moving objects during the training of the model so that it does not produces bias from the relative motion of the objects with respect to the relative motion of the camera. The pixel values of the original images is multiplied with the pixel values of the motion annotation so that the moving objects gets neglected.

Pose CNN: After masking of the moving objects, the features of the frames are extracted using ResNet18 and then those features are projected and re-projected from image coordinates to world coordinates and then back to image coordinates, respectively, using the intrinsic and the extrinsic parameters. The extrinsic parameters (rotation and the translation) matrix required for such conversion are considered from a separate pose network.

Warping: While considering fisheye images, preference must be given to take care of the distortions so that the re-projected image looks alike the original source image. For such cases, the root of the fourth-order polynomial distortion equation for each frame is recalculated and stored in a tabular format as a Python pickle file which is named as a lookup table so that the lookup features can be warped back to the current features.

Projection: The projection for a source image in 2D space to the 3D space takes place through conversion of the coordinates from one space to another. At first, the image coordinates are projected to the camera coordinates using the camera intrinsic parameters and then to the world coordinates using the extrinsic parameters. The re-projection of the world coordinates takes place using the pre-calculated values from the lookup table and the inverse intrinsic.

L_1 **Distance**: Following the projection of the source to target image, the difference of the distance between them is calculated and then stored in 3D structure named cost volume, which represents the difference of the depth values between the current and the nearby frames. At last, the minimum value from the cost volume is considered that represents the better estimation of depth. The features of the target frame is then extracted from the depth encoder and decoder so that the final target frame depth at that instance can be predicted.

Encoder–Decoder: The Encoder–Decoder architecture here uses ResNet18 for the feature extraction as the encoder until the first residual layer which results in a tensor of size quarter of the input resolution with 64 channels. These features are then aggregated into a cost volume for each input frame, using 96 depth bins. The output of the cost volume is then concatenated with the output of the feature extractor for the image. The decoder here takes the minimum value from the cost volume and skips the other values so that the target frame as a depth produces better results.

Lookup Table: Most computer vision algorithms require 100s–1000s of pixel mapping (forward or inverse) every frame. So, to optimize the application of the distortion model and to speed up the overall mapping, the lookup table is used to cover the entire field of view range for which the intrinsic parameters have been calibrated. Here, before training the model we generate the lookup table mapping using the calibration file corresponding to each input image file which contains the intrinsic and the extrinsic parameters. The mapped values are finally dumped to a Python pickle file, so that it can be used later by fetching the mapped values during training. For each pixel value of the image, angle from the lookup table is mapped to theta in the lookup table where angle of the lookup table denotes the coordinates for angle of incidence ray and theta of the lookup table denotes the coordinates for angle in the image plane. We obtain the angle of incidence theta by numerically calculating the fourth-order polynomial roots of Eq. 1 given below

$$\Theta = f_1\Theta + f_2\Theta^2 + f_3\Theta^3 + f_4\Theta^4, \tag{1}$$

using the distortion coefficients f_1, f_2, f_3, and f_4.

Cost Volume: In order to take advantage of multiple input frames, a cost volume is built that evaluates the compatibility between the pixels from source frames which are nearby and current frames at different depth values. Thus, we require understanding of pose between the frames which we calculate using a separate network. We establish a collection of ordered planes P with depths equally spaced between $d_{\min}$ and $d_{\max}$, each perpendicular to the optical axis at I_t, i.e. at instance t. We then encode a deep feature map from each frame based on the provided camera intrinsics and estimated pose and distorted to the perspective of current frame.

As a result, the feature map becomes twisted and the difference between the features from current frame and the warped features is used to create the final cost volume. For pixel point (i, j), what is the probability that the accurate depth is d for each depth belongs to P, the cost volume effectively asks. The input to a convolutional decoder is provided a cost volume combined with the features F_t and then used as the regressed depth D_t.

4 Experimental Set-Up

4.1 Base-Line Model

Earlier depth estimation methodologies from multiple frames was only carried out using images captured from pinhole cameras and was barely been explores for fisheye camera. In addition to that, the relative motion between the moving objects was not taken into consideration while training the model. In the proposed methodology, the images were preprocessed with the motion masking annotation to reduce the bias while training the model. Also, to train the model with fisheye camera requires highly computational roots of fourth-order polynomial equation for projecting the pixel values from image coordinate to camera coordinate and vice versa. These roots of the equation are pre-calculated for each pixel before training and stored in a Python pickle file which is named as the lookup table. The overall warping procedure through lookup table need not be carried out for pinhole images or the earlier depth estimation methods as the projection takes place linearly.

4.2 Evaluation Measures

There are two metrics, the Structural Similarity Index and the relative pixel depth alignment (RPDA), that are used to determine the accuracy of the depth.

Structural Similarity Index The Structural Similarity Index is a widely used image quality assessment metric that aims to quantify the perceived similarity between two images. It takes into account three factors: luminance, contrast, and structure. Luminance represents the brightness of an image, contrast measures the difference between the light and dark areas of an image, and structure refers to the patterns and textures present in the image.

The Structural Similarity Index metric compares two images by calculating the similarity between their local image patches, rather than their overall pixel values. This is because the human visual system is more sensitive to changes in local image features rather than overall pixel values. The Structural Similarity Index is calculated as the product of three terms: the luminance comparison term, the contrast comparison term, and the structural comparison term. The result is a value between 0 and 1, where a value of 1 indicates perfect similarity between the two images. The Structural Similarity Index metric has several advantages over other image quality assessment methods. It has been shown to be more consistent with human perception and can better detect distortions that are not easily detected by other metrics. Furthermore, it is more robust to changes in image size and scaling, making it a more reliable measure of image quality. As such, the Structural Similarity Index metric is commonly used in image and video compression, image processing, and computer vision applications. The overall equation for Structural Similarity between two image I and K is given in Eq. 2.

$$\text{Structural Similarity Index}(I, K) = \left[l(I, K)^{\alpha}\right]\left[c(I, K)^{\beta}\right]\left[s(I, K)^{\gamma}\right], \quad (2)$$

where

$$\left[l(I, K)^{\alpha}\right] = \left[\frac{(2\mu_I\mu_K + C_1)(2\sigma_{IK} + C_2)}{(\mu_I^2 + \mu_K^2 + C_1)(\sigma_I^2 + \sigma_K^2 + C_2)}\right]^{\alpha}$$

$$\left[c(I, K)^{\beta}\right] = \left[\frac{(2\sigma_I\sigma_K + C_2)}{(\sigma_I^2 + \sigma_K^2 + C_2)}\right]^{\beta}$$

$$\left[s(I, K)^{\gamma}\right] = \left[\frac{(\sigma_{IK} + C_3)}{(\sigma_I\sigma_K + C_3)}\right]^{\gamma},$$

and μ_I, μ_K, σ_I, σ_K, and σ_{IK} are the mean and standard deviation of the luminance values of images I and K, and the covariance between the luminance values of images I and K, respectively. C_1, C_2, and C_3 are small constants to avoid division by zero, and α, β, and γ are constants to control the relative importance of each term.

Relative Pixel Depth Alignment Relative pixel depth alignment or RPDA measure is a metric used in depth estimation to evaluate the accuracy of predicted depth maps. It measures the relative difference between the value of predicted depth map pixel and above and below pixel values in terms of the disparity or inverse depth of each pixel, and the value lies between 0 and 1. The basic idea behind this measure is that corresponding pixels in the predicted pixels should have similar disparities or inverse depths, as they represent the same point in 3D space. By measuring the difference in disparities or inverse depths between corresponding pixels, the relative pixel depth alignment measure provides a quantitative measure of the alignment. One commonly used method for computing the relative pixel depth alignment measure is the mean absolute relative error (RPDA), which is defined as the average of the difference among the pixel values. A higher value of the measure signifies better depth quality and lower value signifies improper depth quality. The equation for the same is given in Eq. 3.

$$\text{RPDA} = \frac{1}{N}\sum_{i=1}^{N}\left|\frac{d_a - d_{i-a}}{T}\right|, \tag{3}$$

where d_a, d_{i-a}, and T are the aligned pixel, non-aligned pixel, and the total number of pixel in the depth map, respectively. Overall, relative pixel depth alignment measures provide a useful tool for evaluating the accuracy of depth estimation algorithms and can be used to compare different methods and parameter settings.

4.3 *Parameter Settings and System Configuration*

In our training process, we initially used the Adam optimizer with a learning rate of 0.0001 for the first 20 epochs. After the initial 20 epochs, we applied a step-wise decay in the learning rate by a factor of 10 for the remaining 5 epochs. CPU: Intel(R) Xeon(R) W-2155 CPU @ 3.30GHz, CPU (RAM): 128 GB, GPU: Quadro RTX 5000, GPU (RAM) : 16 GB, Operating System: Ubuntu 20.04.4 LTS.

5 Results and Analysis

We have taken 1766 benchmark images for testing the perceptual robustness of our proposed system. The qualitative analysis of the predicted depth map from the disparities of the outputs is done using the colour mapping function on the Tensorboard. A detailed explanation of such is given below.

First, we create a SummaryWriter object and specify the directory where you want to store the TensorBoard logs and then to generate your predicted image tensors using your PyTorch model. Followed by converting the predicted image tensors to a format that can be displayed on TensorBoard. In this case, you can use functions like torch-vision utils to make a grid to convert the tensor to a grid of images or torchvision.transforms.To PIL Image to convert tensors to PIL images. And then, the converted image tensors to TensorBoard using the add images function of the SummaryWriter object. This function takes in a name for the image grid, as well as the tensor grid itself and an optional step number to associate with the image grid. Once all the image grids have been added to TensorBoard, we can visualize them

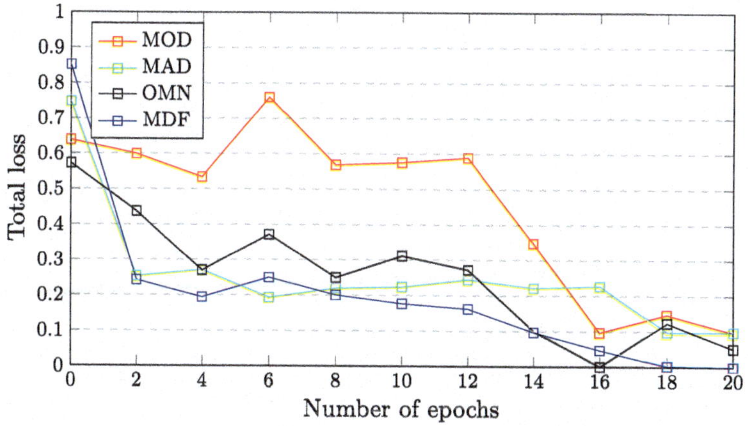

Fig. 2 Comparative analysis of MOD, MAD, OMN, and MDF based on total loss during training

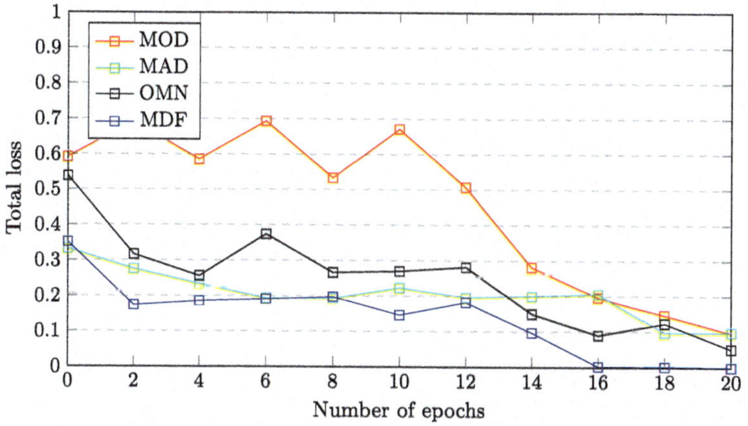

Fig. 3 Comparative analysis of MOD, MAD, OMN, and MDF based on total loss during validation

in the Image tab of the TensorBoard UI. You can interactively explore the images by zooming in, panning and adjusting the brightness and contrast, and comparing multiple image grids side by side. Now, for qualitative analysis of our self-supervised learning with other state-of-the-art models, we provide below a comparative study of total loss of our proposed system [MDF] with Zhou et al. [2] [MOD], Watson et al. [16] [MAD], and Godard et al. [1] [OMN] which is shown in Fig. 2 during training and Fig. 3 during validation below.

Structural Similarity Index: To elaborate the Structural Similarity Index measure in our work Manydepth fisheye [MDF], the proposed system is compared with Zhou et al. [2] [MOD], Watson et al. [16] [MAD], and Godard et al. [1] [OMN]. The input and

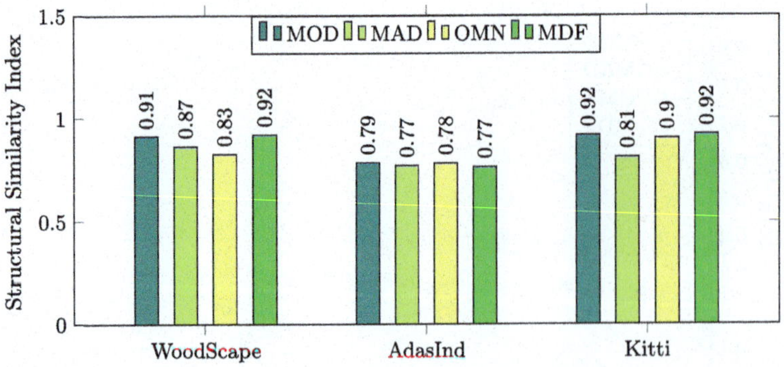

Fig. 4 Comparative analysis of MOD, MAD, OMN, and MDF based on structural similarity index on WoodScape, AdaInd, and Kitti dataset

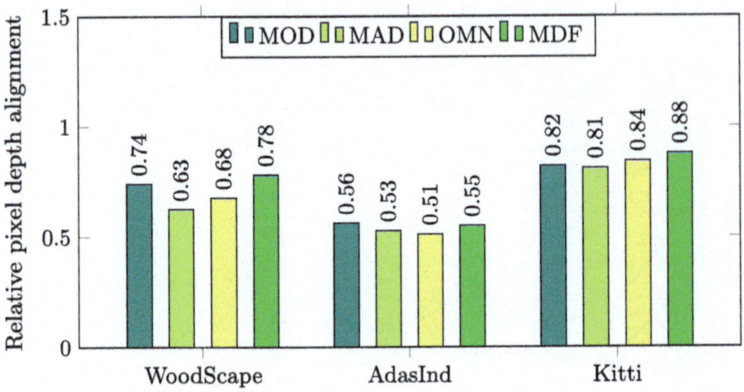

Fig. 5 Comparative analysis of MOD, MAD, OMN, and MDF based on relative pixel depth alignment on WoodScape, AdasInd, and Kitti dataset

the output from the model is fed to Structural Similarity function to get the similarity score. This measure is calculated by taking images from WoodScape, AdasInd, and Kitti dataset, and a graph plotted for comparative study is shown in Fig. 4.

Relative Pixel Depth Alignment: A comparative analysis of MOD, MAD, OMN, and MDF based on relative pixel depth alignment on WoodScape, AdasInd, and Kitti dataset is shown in Fig. 5.

Our proposed model is trained with two dataset, i.e. on WoodScape and AdasInd and their corresponding predicted depth measured from the disparity maps using color mapping function as shown in Fig. 6.

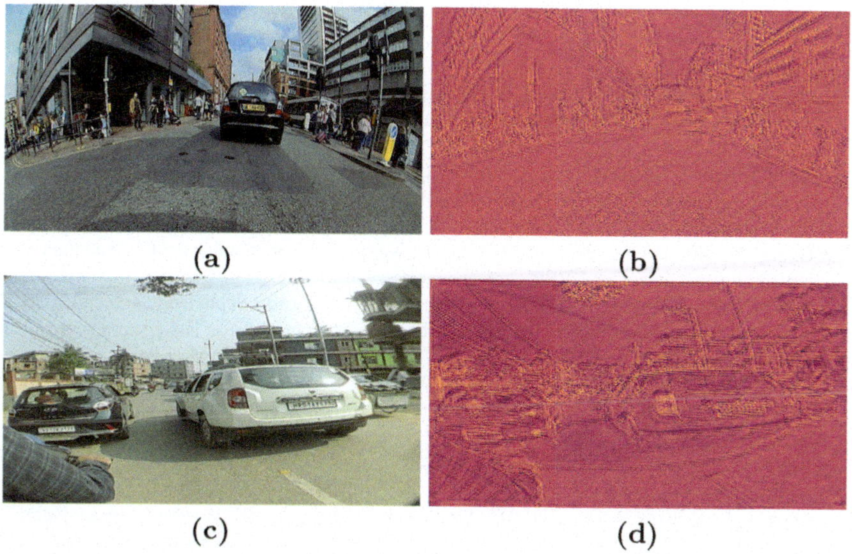

Fig. 6 Sample input and output image for our model trained with WoodScape and AdasInd dataset: **a** WoodScape input, **b** WoodScape output, **c** AdasInd input, **d** AdasInd output

6 Conclusion and Future Work

In this report, we introduced a fully self-supervised system that, given a single image or, in some cases, numerous photos, can accurately estimate superior depths. Comparing our approach to other states for implementing cost volume, the proposed system is dynamically handling moving objects and stationary cameras while achieving the advantages of both monocular and stereo techniques. On the Kitti and WoodScape databases, we provided the most recent findings. Although test-time refining techniques are closely matched by our methodologies to predict accuracy of depth, we have demonstrated that the approach substantially has better quality when making inferences. Through recent complementary developments in monocular depth estimation, such as discretized output depths or feature-based losses, we anticipate that our results will be substantially enhanced.

One potential area for future work in depth estimation is the handling of occluded regions. While current depth estimation methods can provide accurate depth maps for visible areas, they often struggle with regions that are occluded or hidden from view as shown below. This can lead to incomplete or inaccurate depth maps, which can limit the usefulness of the technique in applications such as autonomous driving or robotics. Possible approaches to address this challenge include the use of additional sensors or data sources, such as lidar or radar, to provide more complete information about the environment or the development of more sophisticated algorithms to infer depth in occluded regions based on contextual information.

References

1. Godard C, Aodha OM, Firman M, Brostow GJ (2019) Digging into self-supervised monocular depth estimation. In: Proceedings of the IEEE International conference on computer vision, pp 3828–3838
2. Zhou T, Brown M, Snavely N, Lowe DG (2017) Unsupervised learning of depth and ego-motion from video. In: Proceedings of the IEEE Conference on computer vision and pattern recognition, pp 1851–1858
3. Godard C, Aodha OM, Brostow GJ (2017) Unsupervised monocular depth estimation with left-right consistency. In: CVPR
4. Wang C, Miguel Buenaposada J, Zhu R, Lucey S (2018) Learning depth from monocular videos using direct methods. In: The IEEE Conference on computer vision and pattern recognition (CVPR), June 2018
5. Kumar VR, Hiremath SA, Milz S, Witt C, Pinard C, Yogamani SK, Mäder P (2019) Fisheye distance net: self-supervised scale-aware distance estimation using monocular Fisheye camera for autonomous driving. In: CoRR, vol. abs/1910.04076
6. Eigen D, Puhrsch C, Fergus R (2014) Depth map prediction from a single image using a multi-scale deep network. In: NeurIPS
7. Chen W, Fu Z, Yang D, Deng J (2016) Single-image depth perception in the wild. In: NeurIPS
8. Xie J, Girshick R, Farhadi A (2016) Deep3D: fully automatic 2D-to-3D video conversion with deep convolutional neural networks. In: ECCV
9. Gordon A, Li H, Jonschkowski R, Angelova A (2019) Depth from videos in the wild: unsupervised monocular depth learning from unknown cameras. In: ICCV
10. Aleotti F, Tosi F, Zhang L, Poggi M, Mattoccia S (2020) Reversing the cycle: self-supervised deep stereo through enhanced monocular distillation. In: ECCV
11. Newcombe RA, Lovegrove SJ, Davison AJ (2011) DTAM: dense tracking and mapping in real-time. In: ICCV
12. Wang J, Zhang G, Wu Z, Li X, Liu L (2020) Self-supervised joint learning framework of depth estimation via implicit cues. arXiv:2006.09876
13. Ji M, Gall J, Zheng H, Liu Y, Fang L (2017) SurfaceNet: an end-to-end 3D neural network for multiview stereopsis. In: ICCV
14. Wu Z, Wu X, Zhang X, Wang S, Ju L (2019) Spatial correspondence with generative adversarial network: learning depth from monocular videos. In: ICCV
15. Godard C, Aodha OM, Firman M, Brostow GJ (2019) Digging into self-supervised monocular depth estimation. In: ICCV
16. Watson J, Aodha OM, Prisacariu V, Brostow GJ, Firman M (2021) The temporal opportunist: self-supervised multi-frame monocular depth. In: CoRR, vol. abs/2104.14540

Detection of Brain Tumour Using Hybrid Features and Random Forest

Pranali Chanduka, Vibhav Prakash Singh, and Sumeet Sagar

Abstract Brain tumour is a peculiar entity of mass. Initially, it is a group of few cells but then expands at an accelerating rate. It may originate in the brain or in its surroundings. On a high level, the types of tumour can be benign or malignant. Benign types of tumour are in the initial stages and are curable, whereas the development of malignant types of tumour happens from the glial cells. With the help of various feature extraction techniques, machine learning can be incorporated to detect brain tumour. This paper introduces to application of some efficient feature extraction techniques for brain tumour detection using machine learning. For better results, hybrid features can also be extracted, which is received on combination of techniques. In the proposed model, the best results have been received by using Gabor Filters along with Local Binary Patterns. These techniques have been applied on four machine learning classifiers, i.e. k-Nearest Neighbour, Support Vector Machine, Logistic Regression and Random Forest. The performance of Random Forest is quite encouraging compared to the others.

Keywords Brain tumour · Grey level co-occurrence matrix · Local Binary Patterns · Gabor Filters · Machine learning · Random Forest

1 Introduction

Brain tumour is a distinctive bundle of mass. It is a group of few cells but then expands at an accelerating rate [13]. It may originate in the brain or in its surroundings. It

P. Chanduka (✉) · V. Prakash Singh
Computer Science and Engineering Department, Motilal Nehru National Institute of Technology, Allahabad, Prayagraj, India
e-mail: pchanduka09@gmail.com

V. Prakash Singh
e-mail: vibhav@mnnit.ac.in

S. Sagar
Thapar Institute of Engineering and Technology, Patiala, India
e-mail: ssagar012@gmail.com

S. Bandyopadhyay et al. (eds.), *Intelligent Computing Systems and Applications*, Lecture Notes in Networks and Systems 1010,
https://doi.org/10.1007/978-981-97-5412-0_14

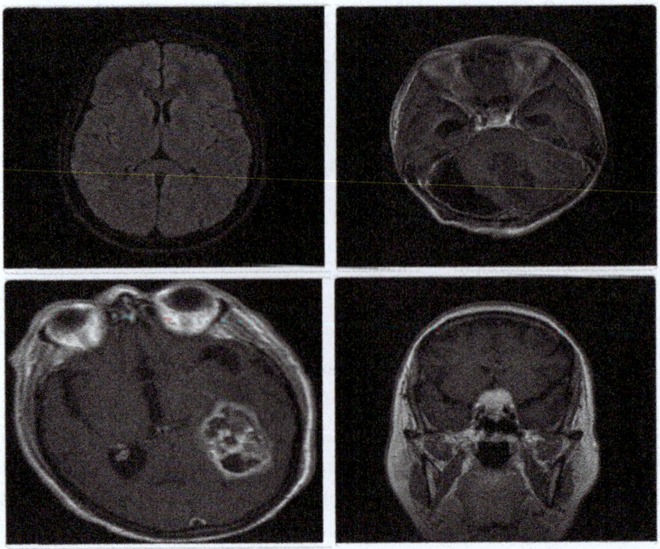

Fig. 1 Top left is MRI without tumour, top right represents Glioma, bottom left represents Meningioma, bottom right represents Pituitary [15]

is a fatal disease with a high mortality rate. As estimated by the Cancer Society of America, there will be 24,810 brain tumour cases which would result in 18,990 deaths [24]. Brain tumours are of two types, primary and metastatic. Primary brain tumours are further classified into benign and malignant, whereas metastatic tumours originate at different places and migrate to the brain. Figure 1 represents tumourous and non-tumourous MRIs. The top left is MRI without tumour, top right represents Glioma, bottom left represents Meningioma, bottom right represents Pituitary, respectively [15].

The World Health Organization (WHO) has developed a system for classification of brain tumour based on its degree of severeness [19]. There are four grades, namely Grade I, Grade II, Grade III and Grade IV. Grade I is the least malignant, and it is possible to cure it. Grade II is slow growing, and even after surgery, it may recur. Grade III is malignant and infiltrative, whereas Grade IV is the most fatal one, as it can recur and increase in size very rapidly [13]. The main cause of brain tumour has been predicted to be a malfunction genes in the chromosome of a cell. The blood and oxygen which is supplied to the brain for regular tissues, when reach the tumourous cells cause them to grow.

The commonly used imaging techniques used are Positron Emission Tomography (PET), Computed Tomography (CT scan) and Magnetic Resource Imaging (MRI). Each technique focuses on different level of details. PET is used to examine chemical properties of brain tumours like the blood flow, glucose production and absorption, lipid secretion, oxygen utilization and amino acid consumption. It is still thought to be one of the most efficient metabolic processes. A CT scan provides a clearer picture and aids in identifying the soft tissues, blood vessels and bones of several human body parts [2].

An MRI scan is performed to rigorously analyse the different regions of the body, but the main objective is to identify presence of tumours or any disorders in the brain, if any. Consequently, tumour segmentation is a difficult task due to the complexity of brain structures. MRI is used in clinical practice to image brain tumours and provides data on their size, granularity, microstructure, performance and functionality [12].

A lot of work has already been done on brain tumour detection like [1, 3–5, 10, 14, 22, 25]. Various feature extraction and feature reduction techniques have been applied along with different machine learning and deep learning models, each giving varying results. The fields of machine learning and deep learning are in growing phase and with time new algorithms and new techniques are developed, so there are still many combinations which need to be explored. Also, the more we explore brain tumour, we would find something new to study. So there is scope of evolution.

Some promising results were obtained by [1] where they suggested Local Binary Patterns to be applied in three orthogonal planes along with Histogram of Oriented Gradients and get a feature vector, this feature vector was used to train a Random Forest Classifier. Another methodology proposed in [3] creates a model for detection based on long short-term memory networks and achieves 98% Accuracy. Another piece of work by [4] included Gabor Wavelet Features, Histogram of Oriented Gradients, Local Binary Patterns and Segmentation-Based Fractal Texture Analysis for feature extraction and then used Random Forest Classifier. A model implemented by [5] uses GLCM for feature extraction followed by SVM for detection and segmentation.

A model suggested by [10] includes GLCM, Linear Discriminant Analysis and Principal Component Analysis along with supervised learning techniques like k-Nearest Neighbour and Artificial Neural Network with an Accuracy of 92.86%. Another model proposed in [14] is Histogram-Based Gravitational Optimization algorithm. This algorithm combines concepts from Brain Histogram Analysis and the Gravitational Optimization algorithm to address tumour and stroke detection and segmentation. In [22], Anisotropic Filtering is used for noise removal, followed by GLCM and wavelet features for feature extraction. A model proposed in [25] uses thresholding and advanced morphological techniques. Brain tumour detection is extremely essential in the field of Medical Sciences. It is important for the neurologists to identify the presence of tumour and detect its type for the respective treatment.

The models which have been discussed above have worked upon tumour detection, but their results have not been very promising. Therefore to have better Accuracy, the proposed model works upon hybridization of feature extraction techniques. The goal of this study is to increase the effectiveness of brain tumour detection. This paper intends to extract maximum features of the MRIs by changing their orientation, focusing on textures and edge features. The sections that follow go into great depth about how the suggested model is put into practice. In Sect. 2, i.e. Proposed Methodology, we have discussed the flow in detail. In the next section, i.e. Sect. 3, the results are discussed. The ultimate section, i.e. Sect. 4, deduces the paper with some insights.

2 Proposed Methodology

The flow of the proposed model is shown in Fig. 2, where firstly the MRI images will undergo preprocessing, followed by feature extraction with Gabor and LBP. Finally with the help of feature vector, we train and test the model. The ensuing subsections go into great depth about the steps.

2.1 Preprocessing

To achieve the best results when the techniques are applied on the data set, preprocessing of the data set is an essential step. With the help of preprocessing, we remove the basic dirt or unrequited elements from the image and change the shape, size, resolution, etc., of the images as per requirement [23]. Once all the images satisfy a convention, they are ready to be fed to the model. In the implementation here, we have resized the images of the data set to (64, 64) and converted them to greyscale.

2.2 Feature Extraction

This step aims at extracting all the information from the data set. Although every pixel is a piece of information, but the intensity of a pixel alone is not enough information; therefore, various techniques are available which draw the correct knowledge from the pixels. The multiple techniques used here for feature extraction are GLCM, LBP and Gabor.

Grey Level Co-occurrence Matrix Grey Level Co-ocurrence Matrix(GLCM) is a feature extraction technique which extracts texture features [26]. With the help of a probability matrix, Haralick brings forward fourteen important second order statistical features [8]. The joint probability distribution of pixel intensity pairings is represented by a two-dimensional matrix called a co-occurrence matrix, which is constructed by GLCM. The grey levels of the pixels in the image are represented by the rows and columns of the matrix. There are four standard orientations, 0°, 45°, 90° and 135°. Another parameter which is kept fixed is the distance between the pixels, by default we are keeping it unit here [17, 21]. Figure 3 gives a diagrammatic representation of the same:

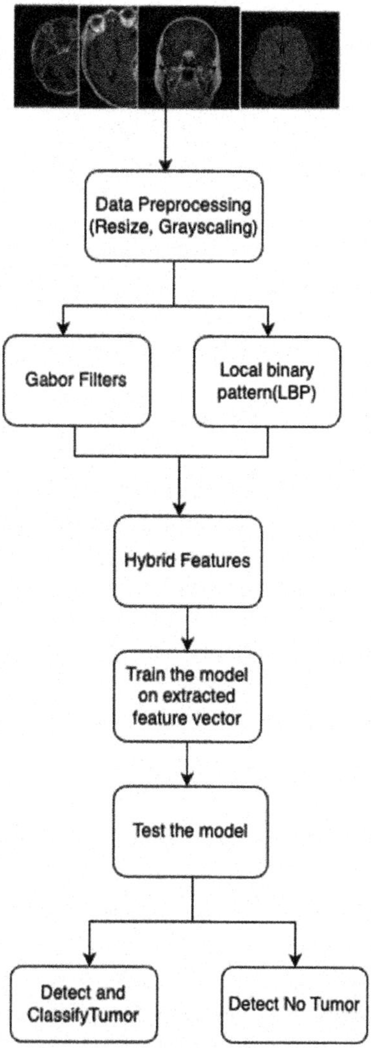

Fig. 2 Flow of the proposed system

These are some of the features which we would be extracting in this implementation:

$$\text{Contrast} = \sum_{i,j=0}^{N-1} (i-j)^2 * P(i,j), \tag{1}$$

$$\text{Dissimilarity} = \sum_{i,j=0}^{N-1} |i-j| * P(i,j), \tag{2}$$

$$\text{IDM} = \sum_{i,i=0}^{N-1} \frac{P(i,j)}{(1+(i-j)^2)}, \tag{3}$$

$$\text{Energy} = \sum_{i,j=0}^{N-1} P(i,j)^2, \tag{4}$$

$$\text{Correlation} = \sum_{i,i=0}^{N-1} (i-\mu_i)(j-\mu_j)\frac{P(i,j)}{(\sigma_i\sigma_j)}, \tag{5}$$

$$\text{Angular Second Moment} = \sum_{i=1}^{N-1}\sum_{j=1}^{N-1} P_{i,j}^2. \tag{6}$$

Local Binary Patterns Local Binary Patterns (LBP) is also a texture-based feature extraction technique. It also helps in reducing the local grey-level structure [18]. The k neighbours of a central pixel are given a threshold value, the central pixel. Based on whether the pixel is less than or greater than the central pixel, they are assigned '0' or '1' in clockwise or anti-clockwise manner. In the same manner, a k-bit binary number is formed, and its corresponding decimal value is given as the value of the central pixel in the resulting image [11, 20].

Gabor Filters Gabor Filter is a bandpass, linear filter. It is used for edge detection, texture extraction and feature extraction [9]. It works on the formula given below:

$$G(x, y, \lambda, \theta, \psi, \sigma, \gamma) = \exp(-\frac{(x'^2 + \gamma^2 y'^2)}{(2\sigma^2)}) \exp[i(2\pi\frac{x'}{\lambda}) + \psi], \tag{7}$$

where x, y represent kernel size, λ represents wavelength, θ represents orientation, ψ represents plane wave, σ represents standard deviation and γ represents aspect ratio that defines the ellipticity of the function support. When these values are put together in the above formula, it results in a convolution kernel. Depending on the different permutations of the values of the variables, different convolution kernels

0	1	1	0	0	2
0	0	2	1	0	0
2	0	2	2	1	0
1	1	2	2	0	1

	0	1	2
0	2	2	3
1	3	2	1
2	2	2	2

Fig. 3 First matrix represents an image with pixel intensities. The second matrix represents an output square matrix with N number of rows and columns, where N is one plus the highest intensity in the image

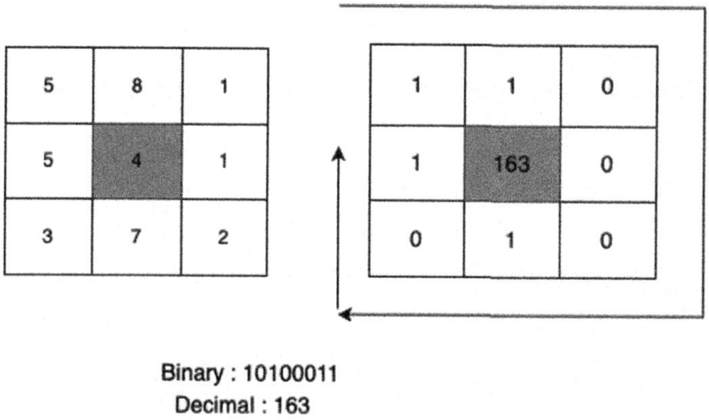

Fig. 4 First matrix represents an image with pixel intensities. The second matrix represents application of LBP operator

are formed, when an image passes through this convolution kernel, the low-level features of the image are extracted [16].

2.3 *Random Forest*

Random Decision Forest employs and ensemble learning approach to construct multiple decision trees [6]. This strategy is utilized for Classification, Regression and other tasks. Below is a detailed discussion of the Random Forest algorithm:

i Training
  (a) Input
    1 Training data set with labeled instances.
    2 Count of the trees in the forest (n_trees).
    3 The number of randomly selected characteristics for each tree (n_features).
  (b) For each tree in the forest (from 1 to n_trees):
    1 Create a bootstrap sample by randomly selecting instances from the training data set with replacements. The bootstrap sample size should be the same as the original data set.
    2 Create a random subset of features by randomly selecting features from the total available n_features.
    3 Using the bootstrap sample and the randomly chosen features, construct a decision tree.
    4 Choose the optimum feature and split point at each node based on an impurity metric (such as Gini impurity or information gain, for example).

5 Up until a stopping requirement (such as the maximum depth achieved or the minimum number of instances in a node) is satisfied, the data is split recursively depending on the chosen feature and split point.

ii Testing

(a) The Random Forest, which consists of a collection of decision trees. To classify a new instance using the trained Random Forest, take the following inputs
   1. Trained Random Forest.
   2. The new instance to be classified
(b) For each tree in the Random Forest:
   1. Traverse the decision tree by following the appropriate branches based on the feature values of the new instance.
   2. Assign the reached leaf node's class label as the projected class label.

iii Output

(a) The class label that is expected for the new instance (which can be chosen by majority vote or by averaging the class labels across all trees).

3 Results

In the proposed framework for brain tumour detection, we have worked upon a publicly available, multiclass data set. It comprises 7023 files [15], divided into training and testing folders, which are further divided into Glioma, Meningioma, No tumour and Pituitary categories, respectively. The training folder contains 5712 MRI's in total further divided into 1321, 1339, 1595 and 1457, respectively. The testing folder contains 1311 MRI's which contains 300, 306, 405 and 300 files of the respective types of tumours or non-tumour.

70% of the total data are used to train the model, while the remaining data are used to test it. For all the feature extraction techniques, the same split was performed. The data set went through preprocessing followed by feature extraction. Feature extraction has been done by GLCM, GLCM along with LBP and Gabor along with LBP. The feature vector obtained from GLCM comprised of 24 features, the one obtained from GLCM + LBP comprised of 83 features, whereas for Gabor + LBP, a feature vector of 33 features was obtained. Figure 5 shows a graph highlighting the feature vector size.

The metrics based on which the results have been compared are Confusion Matrix, Accuracy, Precision, Recall and F1-score. Confusion Matrix is a performance measurement tool for an ML algorithm [7]. Below are the explanations for the remaining metrics:

Fig. 5 Feature vector size for different feature extraction techniques

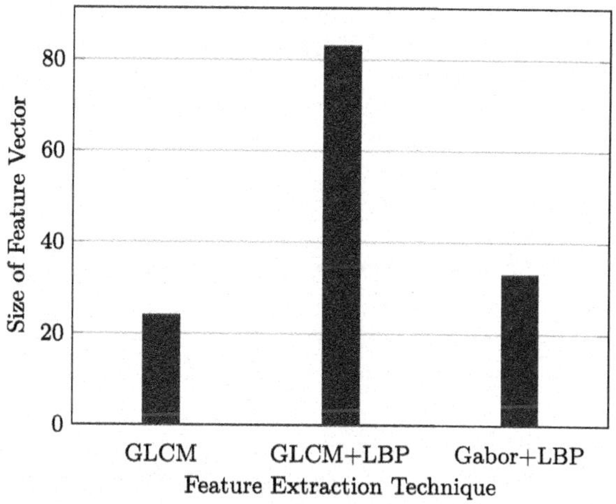

- Accuracy is used the measure the number of correct predictions.

$$\text{Accuracy} = \frac{\text{TP} + \text{TN}}{\text{TP} + \text{FP} + \text{TN} + \text{FN}}. \tag{8}$$

- Recall is popularly known as True Positive Rate. It is the actual positives identified correctly out of all the positives.

$$\text{Recall} = \frac{\text{TP}}{\text{TP} + \text{FN}}. \tag{9}$$

- Precision tells about the Positive predictions which are correct.

$$\text{Precision} = \frac{\text{TP}}{\text{TP} + \text{FP}}. \tag{10}$$

- Precision and Recall are used to determine the $F1$-score. It is the harmonic mean of the two metrics mentioned above.

$$F1\ \text{Score} = \frac{2 * \text{Precision} * \text{Recall}}{\text{Precision} + \text{Recall}}. \tag{11}$$

Figure 6 shows the confusion matrix of the best result combination, Gabor with LBP on Random Forest.

Tables 1, 2, 3 and 4 represent the Precision, Recall, Accuracy and $F1$-score values for the different feature extraction techniques for the ML algorithms applied.

Across all the feature extraction techniques and machine learning algorithms that were tried and tested for this implementation, Random Forest gave the best results

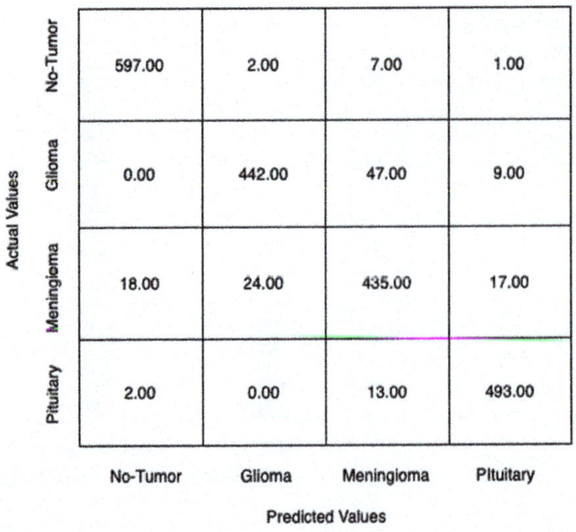

Fig. 6 Gabor and LBP with RF

Table 1 Results on applying KNN

Method	Accuracy (< 1)	Precision(<1)	Recall (< 1)	$F1$-score(< 1)
GLCM	0.7704	0.7508	0.7507	0.7507
GLCM + LBP	0.8637	0.8557	0.8897	0.8551
Gabor + LBP	0.9098	0.9065	0.9074	0.9069

Table 2 Results on SVM

Method	Accuracy (< 1)	Precision (< 1)	Recall (< 1)	$F1$-score(< 1)
GLCM	0.5148	0.5278	0.5123	0.5199
GLCM + LBP	0.7199	0.7239	0.7081	0.7054
Gabor + LBP	0.9321	0.9292	0.9295	0.9293

Table 3 Results on applying logistic regression

Method	Accuracy (< 1)	Precision (< 1)	Recall (<1)	$F1$-score (< 1)
GLCM	0.6445	0.6275	0.6358	0.6295
GLCM + LBP	0.7992	0.7872	0.7902	0.7879
Gabor + LBP	0.9297	0.9266	0.9268	0.9267

in all scenarios. The bar graphs in Figs. 7, 8 and 9 show the comparison of Accuracy across all the ML algorithms for GLCM, GLCM + LBP and Gabor+LBP, respectively.

Figure 10 shows a graph showing the effectiveness of the suggested model in comparison to the existing models. This clearly shows that the results are quite encouraging.

Table 4 Results on applying RF

Method	Accuracy (< 1)	Precision (< 1)	Recall (< 1)	$F1$-score (<1)
GLCM	0.8840	0.8809	0.8749	0.8778
GLCM + LBP	0.8951	0.8870	0.8910	0.8880
Gabor + LBP	0.9335	0.93320	0.9310	0.9310

Fig. 7 Accuracy of GLCM across ML algorithms

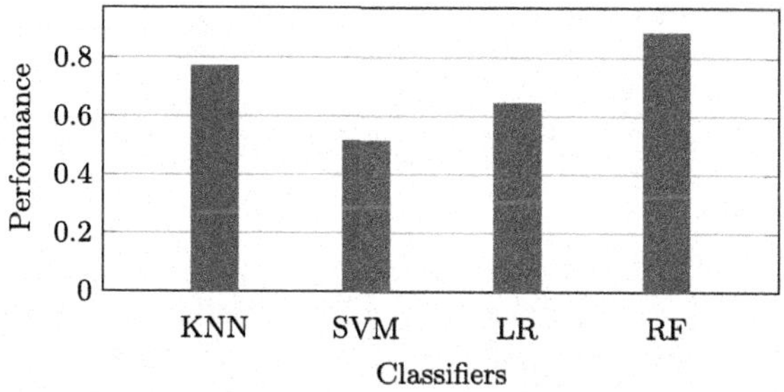

Fig. 8 Accuracy of GLCM + LBP across ML algorithms

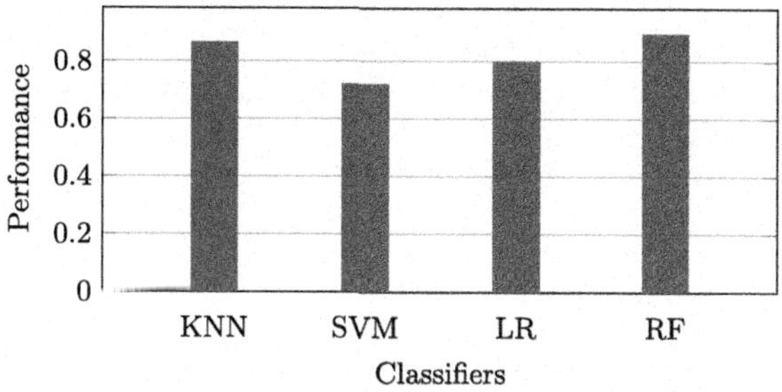

Fig. 9 Accuracy of Gabor + LBP across ML algorithms

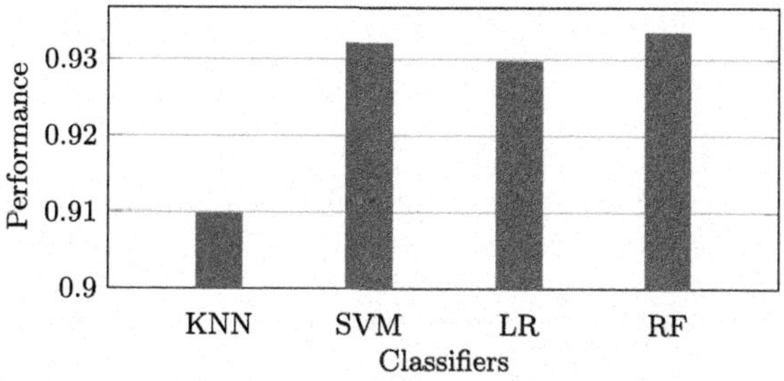

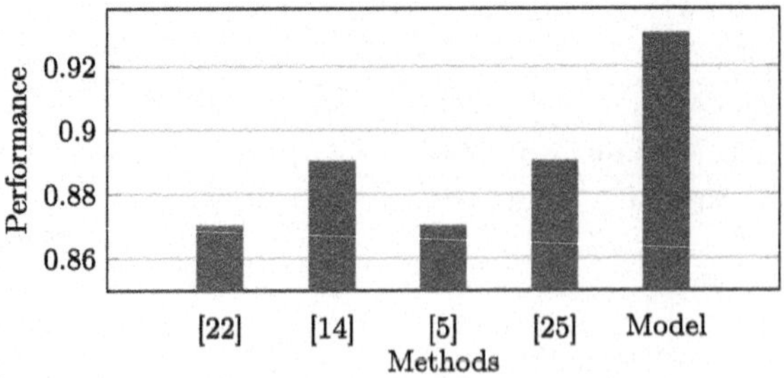

Fig. 10 Comparison of the existing state of the art and proposed model

4 Conclusion

This paper developed an algorithm for tumour detection. We discussed three feature extraction techniques in detail and applied their results on machine learning algorithms. The techniques for feature extraction used were Gabor Filters and LBP. The best results were obtained when Gabor Filters and LBP were applied together and Random Forest algorithm was used. The Accuracy of this model was 93.35%. Detection of tumour with the help of Gabor Filters can be truly insightful. Gabor can offer many kernels at different orientations and therefore increase the probability of locating the tumour.

References

1. Abbasi S, Tajeripour F (2017) Detection of brain tumor in 3D MRI images using local binary patterns and histogram orientation gradient. Neurocomputing 219:526–535. https://doi.org/10.1016/j.neucom.2016.09.051
2. Amin J, Sharif M, Haldorai A, Yasmin M, Nayak RS (2021) Brain tumor detection and classification using machine learning: a comprehensive survey. Complex Intell Syst 1–23
3. Amin J, Sharif M, Raza M, Saba T, Sial R, Shad SA (2020) Brain tumor detection: a long short-term memory (LSTM)-based learning model. Neural Comput Appl 32:15965–15973
4. Amin J, Sharif M, Raza M, Yasmin M (2018) Detection of brain tumor based on features fusion and machine learning. J Ambient Intell Human Comput 1–17
5. Amin J, Sharif M, Yasmin M, Fernandes SL (2020) A distinctive approach in brain tumor detection and classification using MRI. Pattern Recogn Lett 139:118–127. https://doi.org/10.1016/j.patrec.2017.10.036
6. Breiman L (2001) Random forests. Mach Learn 45:5–32
7. Erickson BJ, Kitamura F (2021) Magician's corner: 9. performance metrics for machine learning models
8. Haralick RM, Shanmugam K, Dinstein I (1973) Textural features for image classification. IEEE Trans Syst Man Cybern SMC-3(6):610–621. https://doi.org/10.1109/TSMC.1973.4309314
9. Ilonen J, Kamarainen JK, Paalanen P, Hamouz M, Kittler J, Kalviainen H (2008) Image feature localization by multiple hypothesis testing of Gabor features. IEEE Trans Image Process 17(3):311–325. https://doi.org/10.1109/TIP.2007.916052
10. Jafarpour S, Sedghi Z, Amirani MC (2012) A robust brain MRI classification with glcm features. Int J Comput Appl 37(12):1–5

11. Kaur N, Nazir N, Manik (2021) A review of local binary pattern based texture feature extraction. In: 2021 9th International conference on reliability, Infocom technologies and optimization (Trends and future directions) (ICRITO), pp 1–4. https://doi.org/10.1109/ICRITO51393.2021.9596485
12. Keerthana A, Kumar BK, Akshaya K, Kamalraj S (2021) Brain tumour detection using machine learning algorithm. J Phys Conf Seri 1937:012008
13. Li S, Wang C, Chen J, Lan Y, Zhang W, Kang Z, Zheng Y, Zhang R, Yu J, Li W (2023) Signaling pathways in brain tumors and therapeutic interventions. Signal Transduct Targeted Therapy 8(1):8
14. Nabizadeh N, John N, Wright C (2014) Histogram-based gravitational optimization algorithm on single MR modality for automatic brain lesion detection and segmentation. Expert Syst Appl 41(17):7820–7836. https://doi.org/10.1016/j.eswa.2014.06.043
15. Nickparvar M (2021) Brain tumor MRE dataset (2021).https://doi.org/10.34740/KAGGLE/DSV/2645886, https://www.kaggle.com/dsv/2645886
16. Nurhadiyatna A, Latifah AL, Fryantoni D (2015) Gabor filtering for feature extraction in real time vehicle classification system. In: 2015 9th International symposium on image and signal processing and analysis (ISPA), pp 19–24 (2015).https://doi.org/10.1109/ISPA.2015.7306026
17. O'Byrne M, Ghosh B, Pakrashi V, Schoefs F (2012) Texture analysis based detection and classification of surface features on ageing infrastructure elements. In: BCRI2012 Bridge and concrete research in Ireland
18. Ojala T, Pietikainen M, Maenpaa T (2002) Multiresolution gray-scale and rotation invariant texture classification with local binary patterns. IEEE Trans Pattern Anal Mach Intell 24(7):971–987. https://doi.org/10.1109/TPAMI.2002.1017623
19. Perkins A, Liu G (2016) Primary brain tumors in adults: diagnosis and treatment. Am Family Phys 93(3):211–217
20. Prakasa E (2016) Texture feature extraction by using local binary pattern. Jurnal INKOM 9:45. https://doi.org/10.14203/j.inkom.420
21. Shijin Kumar PS, Dharun VS (2016) Extraction of texture features using GLCM and shape features using connected regions. Int J Eng Technol 8:2926–2930. https://doi.org/10.21817/ijet/2016/v8i6/160806254
22. Shanthakumar P, Ganeshkumar P (2015) Performance analysis of classifier for brain tumor detection and diagnosis. Comput Electr Eng 45:302–311. https://doi.org/10.1016/j.compeleceng.2015.05.011
23. Sharma M, Sharma P, Mittal R, Gupta K (2021) Brain tumour detection using machine learning, vol 3. https://doi.org/10.36548/jei.2021.4.005
24. Siegel RL, Miller KD, Wagle NS, Jemal A (2023) Cancer statistics. CA Cancer J Clin 73(1):17–48
25. Sudharani K, Sarma T, Prasad KS (2015) Advanced morphological technique for automatic brain tumor detection and evaluation of statistical parameters. Procedia Technol 24:1374–1387 (2016).https://doi.org/10.1016/j.protcy.2016.05.153
26. Zulpe N, Pawar V (2012) GLCM textural features for brain tumor classification. Int J Comput Sci Issues (IJCSI) 9(3):354

SPECDUINO—A Secondary Sensory Vision

A. M. Rajeswari, V. Sreejith, N. Jyothi Prakash, J. V. Hari Harra Sudhan, and A. S. Bharath Kumar

Abstract Prior to the existence of Google Maps, or other similar online map providers, most people had a very difficult time navigating from one place to another. People with visual impairments, on the other hand, face a lot of problems when they try to navigate the outdoor. Although cutting-edge technologies have provided more applications for people with visual impairments to manage their daily lives, those applications take a long time. The authors are therefore proposing a Web application named 'SPECDUINO'. The proposed Web application will detect real-time objects like vehicles, animals, and the road divider via the mobile camera and give a voice warning message. We used deep learning to detect objects, and API directions to assist in the transit on both small and large scale. The technology used in SPECDUINO enhances the overall functionality of the system and helps visual impairments to reach their destination more accurately.

Keywords Visually challenged · Deep learning · Tensor flow · Object detection · Directions API · REST API

1 Introduction

Blindness is a disability that affects millions of people around the world, making it difficult for them to operate in public spaces independently and safely. To address this challenge, researchers and engineers have developed navigational guidance systems for visually impaired persons. The purpose of a navigational guidance system is to provide the visual impairments with a reliable and effective means of navigating their environment in a safe and independent manner. This is accomplished through the use of a variety of technologies, such as sensors, GPS, and audio signals, which work

A. M. Rajeswari (✉) · V. Sreejith · N. Jyothi Prakash · J. V. Hari Harra Sudhan · A. S. Bharath Kumar
Velammal College of Engineering and Technology, Madurai, India
e-mail: amrtce@gmail.com

S. Bandyopadhyay et al. (eds.), *Intelligent Computing Systems and Applications*, Lecture Notes in Networks and Systems 1010,
https://doi.org/10.1007/978-981-97-5412-0_15

together. These combos are used to detect obstacles, provide guidance, and transmit information on the environment to the user.

The specific objectives of a navigation guidance system may vary with the design and functionality of the system, but some common goals include:

- Provide accurate, real-time navigational information for persons who are blind, both inside and outside.
- Provide audible or tactile feedback to allow persons who are blind to navigate safely and with confidence.
- Identify potential environmental barriers and hazards and warn users of their presence.
- Provide a user-friendly interface that can be used by people who are blind on their own or with minimal assistance.
- Be cost-effective and scalable to ensure access to a wide range of users.
- Integrate state-of-the-art technologies and innovations to improve system accuracy and functionality over time.

By meeting these goals, the navigation-based guidance system can help people with visual disabilities overcome the challenges they face in daily navigation by improving their quality of life and independence.

To address the aforementioned issue, a Web application, "SPECDUINO" is proposed. It uses a mobile phone camera to capture live video in the visual impaired environment. The Directions API is used to provide detailed instructions from the user's starting point to where they are going. Next, a customized object detection model detects objects in the video and provides tags with the accuracy score. Text-to-voice technology converts objects into audio output. These can be used on any device regardless of the platform. This system improves the experience for users with visual disabilities while navigating.

The proposed SPECDUINO can be used in platforms like Android, iOS, Windows, Linux, Mac, and so on. The characteristics of the proposed system can be understood from its comparative relation with the existing system described in Table 1. Table 1 indicates that even though similar applications like BeMyEyes, SeeingAI, and TapTapSee are free, they are incapable of providing directions. Although the BlindSquare app is capable of delivering directions, it is economically unaffordable. The applications BeMyEyes, SeeingAI, TapTapSee, and BlindSquare are specifically tailored for a specific platform. With this in mind, the proposed SPECDUINO is designed to address the aforementioned concerns. The proposed SPECDUINO is designed to suit conventional platforms, guiding people with visual impairments to reach the destination by giving directions. Voice messages are provided instead of text messages for the detected objects (obstacles), which will be more helpful for the visually impaired.

The remainder of the document is organized in the following manner. Section 2 describes literature from similar studies. Section 3 deals with the framework and principles of operation of the proposed web application—SPECDUINO. In Sect. 4, the experimental assessment of SPECDUINO was mostly addressed. Finally, concluding remarks and future work are presented in Sect. 5.

Table 1 Comparisons of SPECDUINO with state-of-the-art applications

Applications	Cost	Direction provision	Live object direction	Tactile feedback	Platform supported
1. BeMyEyes	Free	No	Yes (manual done by volunteers)	Yes (volunteers provide descriptive comments for navigation)	Android, iOS
2. SeeingAI	Free	No	Yes	Provide the labels of the object detected live and read them out	iOS
3. TapTapSee	Free	No	Yes (detect static object upon double tapping to take photograph)	Provide the labels of the object detected from the static photograph, which is then read aloud	Android, iOS
4. BlindSquare	$39.99	Yes (speaks them out aloud)	No	Gives descriptive sentence for navigation	iOS
5. Proposed—SPECDUINO	Free	Yes (speaks them out aloud)	Yes	Reads aloud the labels of the specific objects	Web app, Android, iOS, Windows, Linux, Mac

2 Related Works

The ultimate aim of a navigation guidance system for persons who are blind is to improve their quality of life and increase their mobility. Also, to ensure visually impaired, more independence and confidence in navigation, the authors [1] investigate existing portable assistive devices. They gave an overview of possible research axes in the short/medium term that can improve existing devices. In addition, they assessed the devices through qualitative and quantitative measurements. The authors [2] came up with the idea of presenting a comparative study of portable assistive devices for visually impaired people. They discussed the developments in assistive technology for this group of individuals. Apart from that Grant, the authors [3] assessed the functional performance of the BrainPort V100, an FDA-approved sensory substitution system, for persons who are profoundly blind. The main purpose of this work [3] is to develop electronic travel assistance to help blinds identify obstacles in their navigation. This navigational aid consists of a single card processing system, a headset fitted with a vision sensor, and a pair of stereo headphones. The vision sensor captures the frame environment in front of the sensor. An alternative system was proposed by the authors [4] that collects the image and is processed by a new real-time image processing scheme using fuzzy clustering algorithms. The processed image is matched to specially structure stereo sound models and transferred to the system's stereo headphones. Persons who are blind have received training on the NAVI system and have undergone barrier identification tests. Here, the authors [5] came up with a technology that converts visual images into sound signals. Users also point to a detailed visual phenomenology that has developed over several months of immersive use and has continued to evolve over the years.

Hersh [6] suggested some features of a comprehensive new assistive technology model for the foundations of vision, mobility, communications, and access to information in daily life. Loomis [7] suggested a technology called digital map using Mapbox, and a navigational system is also used to help visually impaired people. The authors [8] present findings from human factors studies using a remote infrared sound signaling system. In this work, speaking signs were the basic tasks. The authors [9] proposed a technology called "Drishti" which uses a precise position measurement system, a wireless connection, a laptop, and a voice communication interface to guide blind users. In addition, the proposed system has assisted persons with visual impairments to travel in familiar and unfamiliar environments autonomously and safely. The authors [10] came up with a technology called "white cane" is a typical support tool for people with visual impairments. The author uses a white walking stick to detect obstructions. Their work also describes a new information transmission system based on light sensors for indoor positioning and navigation with special advantages for mobile and portable computers.

The authors [11] propose a system called SmartVision, whose overall aim is to give blind users the opportunity to travel in unfamiliar environments, whether indoors or outdoors. This article focuses mainly on the development of the computer vision module of the SmartVision system. A technology called Blavigator [12] has been put in place as a navigation system designed to assist visually impaired users. In an established location, the system used object recognition algorithms. The authors [13] proposed the ultra-wideband (UWB) technology which proved to be effective for interior positioning, offering a high level of precision and low installation complexity. The authors [14] introduced NAVI, the euro electron vision system that functions by extracting a depth map of the images from the camera. The aircraft assisted in the navigation of persons with point-to-point visual impairment by giving voice instructions. The authors [15] used the on-board RGB-D camera to develop effective obstacle detection and avoidance approach based on a time-stamped Kalman (TSM-KF) filter algorithm. A multimodal human–machine interface (HMI) is designed with audio-vocal interaction and robust hepatic interaction using an electronic SmartCane.

The authors [16] used the estimated posture and floor plan to locate the user of the device within a building and guided the user via a voice interface. The experimental results validate the effectiveness of the exposure estimation method and show that the system can greatly facilitate a domestic navigation task. The system senses dynamic obstacles and adjusts real-time route planning to improve navigation safety. The authors [17] put forward the Google Project Tango Tablet Development Kit. This kit is equipped with a powerful graphics processor and a number of sensors that allow it to follow its movement and orientation in 3D space in real time. The authors [18] proposed Microsoft Kinect sensor as a wearable device, which performs face detection. This device uses temporal coherence along with a simple biometric procedure to generate a sound associated with the identified person at his/her estimated 3D location. The authors [19] verified that the spatial images of the targets are formed when detected through extended touch and how accurate they are. The authors [20] have proposed a computer image recognition system, and a smart phone application have been integrated to form a single assisted guidance system. Two modes of online and offline operation may be chosen on network availability.

3 Proposed Methodology

The proposed system SPEDUINO use a device for capturing video live from the camera of the mobile phone, to fetch the live stream from the perspective of the foreground user. It detects objects through the customized object detection model. The customized template labels the objects that have been detected with their accuracy score. Then, the Directions API manager takes the step-by-step instructions from the source to the destination. The system will greatly improve the in-transit experience and will be very useful. Persons with visual disabilities will benefit from knowing what they have in front of them. To make it easier to navigate and help people get from one location to another, we have integrated the Directions API. The system uses a mobile phone camera to collect data. In order to offer the visually impaired an experience without delay, live video is treated in real time at a rate of 5 frames per second. The proposed SPECDUINO architectural schematic is shown in Fig. 1.

The system converts the formed AI model into a JSON model, and the JSON model is stored in the IBM Cloud Object Storage in a bucket. The system also allows for cross-origin resource sharing, allowing access to the JSON model hosted by authorized customers. The HTTP header mechanism is used for calling the API. The JSON model hosted in the cloud is assigned the HTTPS GET method. That way, the data is transferred on the HTTP call request from server to client. The system converts detectable objects to audio via Google text to speech via an API call and, thus, obtains the audio output of the peripheral speaker.

3.1 The Working Principle of the Proposed SPECDUINO

The simplest and most basic browsing method for persons who are blind or have a visual impairment uses the Google Maps API. The API where we integrate the

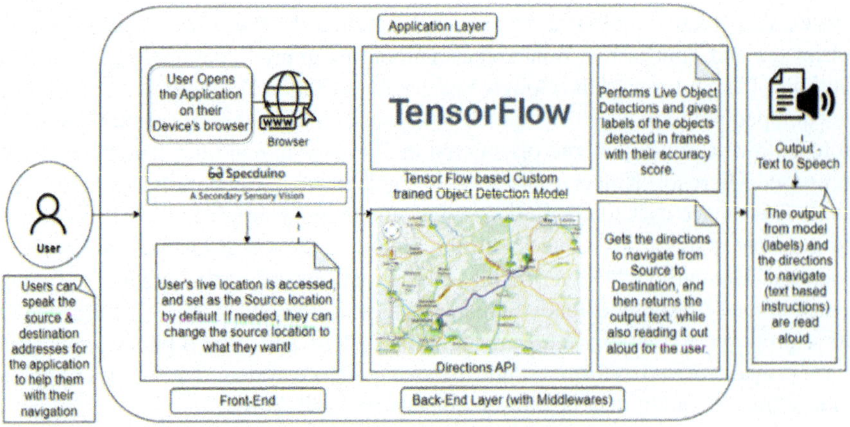

Fig. 1 SPECDUINO—framework

Web-based application will serve as the approach compass. A person who is visually impaired will be able to follow routes and navigation guided by AP maps. They can hear the understandable controls through the earphones or the mobile loudspeaker itself. Figure 2 demonstrates the working principles of the SPECDUINO.

At first, the app opens the camera and captures the video sequences in real time to process them. It requests the destination location from the user at the beginning itself, it is up to the user to use the Maps API to achieve its destination. After that, the data collected is collected and processed in the integration application. The model is formed by gathering images of the outside environment like vehicles, platforms, sewers, animals, etc. The TensorFlow model is converted into TensorFlow JavaScript (TFJS) for the processing of monitored data with labeling. Because the model is stored in the cloud storage bucket, the JSON template file will be easily accessible. As a follow-up, the Maps API will perform the correct operation to guide the user. Text-to-voice technology converts objects into audio output. This may be used in any device on any platform.

4 Experimental Results and Discussions

The proposed system has shown promising results in assisting blind persons to navigate unknown environments. By leveraging computer vision, sensor technologies, and hearing feedback, the system has provided real-time advice, improved safety, and improved independence for people with visual disabilities. As for the parameters of the model we developed, we used supervised learning. It was populated with entry images that had tags on them. The model currently detects the following objects: animal, splitter, platform, poles, sewer-cover-open, street-sign, and vehicle. Precise object detection samples and tags were shown in Figs. 3 and 4.

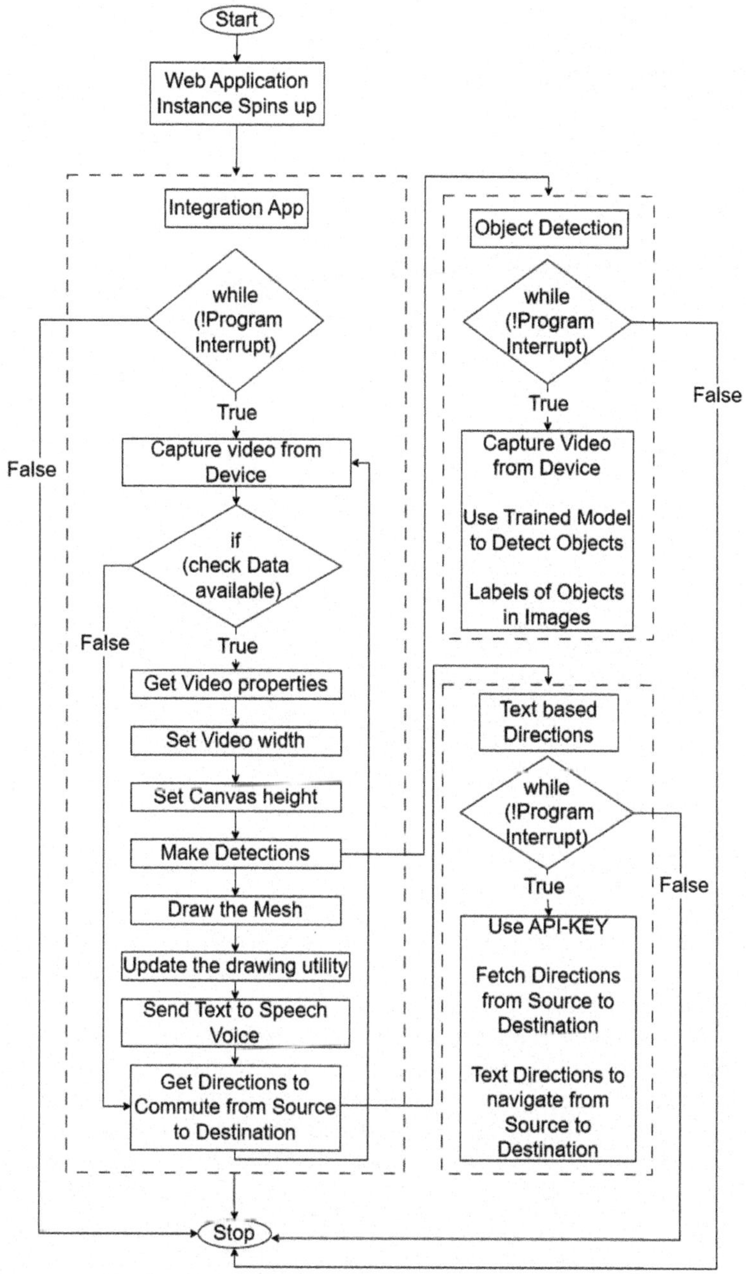

Fig. 2 SPECDUINO—working principle

Fig. 3 Platform and sewage hole depiction

Fig. 4 Custom trained model—sample output

The customized template has been trained to detect objects very well and has very low loss of location and loss of classification. We take these tags with their percent accuracy and then decide if the detected object should be informed to the user. More hazardous objects such as moving vehicles and open potholes are read aloud, even when the accuracy is lower than normal, because of its consequences.

In general, current technologies for blind navigation appear promising to improve the freedom and mobility of the visually impaired. Problems remain, such as better interaction with other assistive technologies and more accurate detection of barriers. Our custom model has been optimized to reduce losses, and has achieved greater loss reduction, with more steps. This ensures that the mean average accuracy (MAP) value of our model is high, and that our model is more accurate for detecting, locating, and classifying objects within an image frame. The Maps API gives the user instructions based on where they are traveling in relation to their target. For reading API routing, we use voice output from directions. The maps will provide the shortest route for the user to reach their destination quickly. An example of directional navigation can be found in Fig. 5.

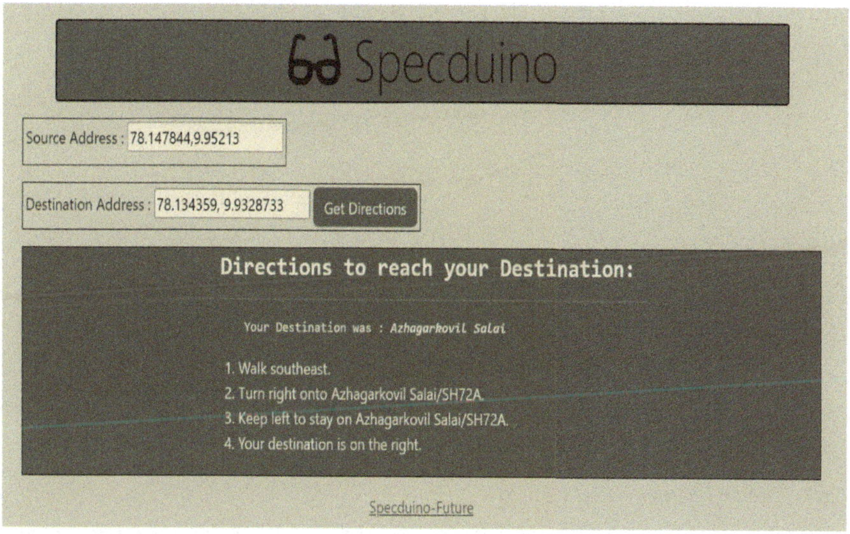

Fig. 5 SPECDUINO—directional navigation snapshot

Figures 6 and 7 depict the learning rate and the loss of the customized model. Figure 6 shows how the learning rate in the model has risen nearly twice over time and is clearly visible in the graph. The learning rate for the Stage 200 model is 0.04, while by the end of the 1600 cycle it is 0.08. This demonstrates that our model has learned to detect objects very rapidly. Figure 7 illustrates the overall loss that occurred in the model during training. The loss dramatically decreased by more than 2×, from 1 to 200 steps.

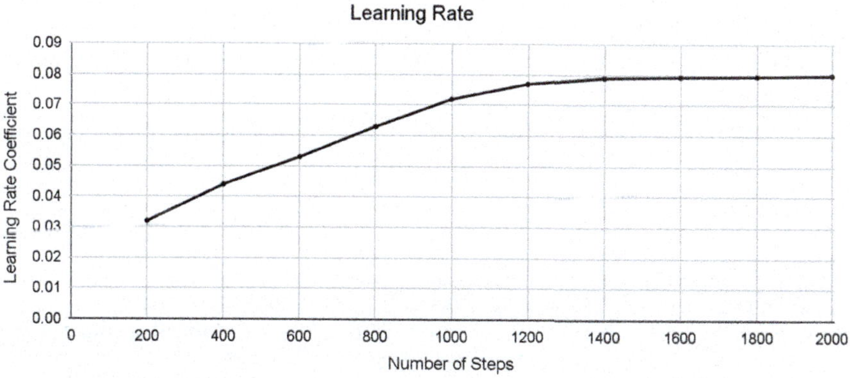

Fig. 6 Custom trained model—learning rate

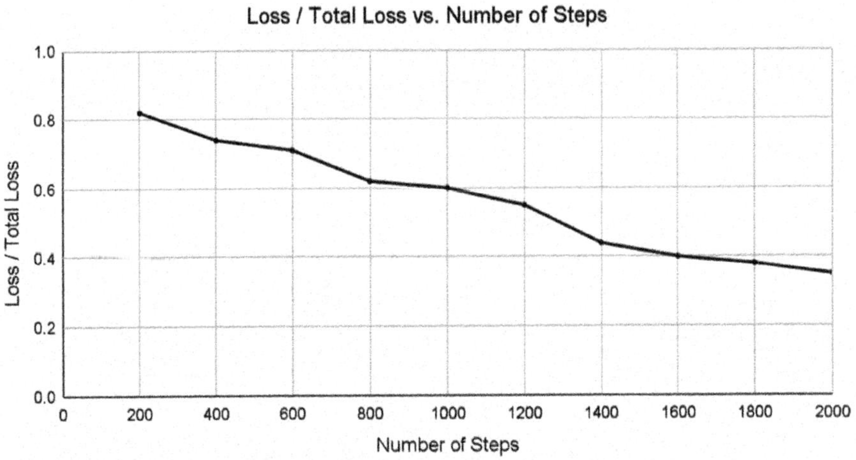

Fig. 7 Custom trained model—total loss

5 Conclusion

The need for guidance systems for persons with vision loss is addressed. The characteristics of the existing navigational guidance systems were compared to those of the proposed system: SPECDUINO. The SPECDUINO is a Web-based application that can be installed in the mobile and can be used as a navigation system for people with visual disabilities. This Web app can be used on Android, iOS, Windows, Linux, and macOS. A detailed discussion was held about the framework and working principles of the proposed application. Experimental results have been observed and discussed in depth. The proposed application gives blind people an accessible and convenient way to explore their surroundings using the Google Maps API. The initiative has been successful in achieving this objective by providing the blind with a practical and understandable tool.

In the future, the application can be significantly improved in terms of how the Internet can be used effectively like 5G and 6G. Thus the proposed method SPECDUINO has the potential to improve its performance by reducing response latency time.

References

1. Tapu R, Mocanu B, Zaharia T (2020) Wearable assistive devices for visually impaired: a state of the art survey. Pattern Recogn Lett 137:37–52
2. Elmannai W, Elleithy K (2017) Sensor-based assistive devices for visually-impaired people: current status, challenges, and future directions. Sensors 17(3):565

3. Grant P, Spencer L, Arnoldussen A, Hogle R, Nau A, Szlyk J, Nussdorf J, Fletcher DC, Gordon K, Seiple W (2016) The functional performance of the BrainPort V100 device in persons who are profoundly blind. J Vis Impairment Blindness 110(2):77–88
4. Sainarayanan G, Nagarajan R, Yaacob S (2007) Fuzzy image processing scheme for autonomous navigation of human blind. Appl Soft Comput 7(1):257–264
5. Ward J, Meijer P (2010) Visual experiences in the blind induced by an auditory sensory substitution device. Conscious Cogn 19(1):492–500
6. Hersh MA (2008) Assistive technology for visually impaired and blind people, vol 1 (ed: Johnson MA). Springer, London
7. Loomis JM (1985) Digital map and navigation system for the visually impaired. Unpublished manuscript, Department of Psychology, University of California, Santa Barbara
8. Crandall W, Bentzen BL, Myers L, Brabyn J (2001) New orientation and accessibility option for persons with visual impairment: transportation applications for remote infrared audible signage. Clin Exp Optom 84(3):120–131
9. Ran L, Helal S, Moore S (2004) Drishti: an integrated indoor/outdoor blind navigation system and service. In: Second IEEE annual conference on pervasive computing and communications proceedings. IEEE, pp 23–30
10. Harada T, Kaneko Y, Hirahara Y, Yanashima K, Magatani K (2004) Development of the navigation system for visually impaired. In: 26th annual international conference of the IEEE Engineering in Medicine and Biology Society, vol 2. IEEE, pp 4900–4903
11. Fernandes H, Costa P, Filipe V, Hadjileontiadis L, Barroso J (2010) Stereo vision in blind navigation assistance. In: 2010 world automation congress. IEEE, pp 1–6
12. Fernandes H, Costa P, Paredes H, Filipe V, Barroso J (2014) Integrating computer vision object recognition with location based services for the blind. In: Universal access in human-computer interaction, aging and assistive environments: 8th international conference, UAHCI 2014, held as part of HCI international 2014, Heraklion, Crete, Greece, proceedings, Part III, vol 8. Springer International Publishing, pp 493–500
13. Martinez-Sala AS, Losilla F, Sánchez-Aarnoutse JC, García-Haro J (2015) Design, implementation and evaluation of an indoor navigation system for visually impaired people. Sensors 15(12):32168–32187
14. Zöllner M, Huber S, Jetter HC, Reiterer H (2011) NAVI–a proof-of-concept of a mobile navigational aid for visually impaired based on the Microsoft Kinect. In: Human-computer interaction–INTERACT, 13th IFIP TC 13 international conference, Lisbon, Portugal, proceedings, Part IV, vol 13. Springer, Berlin, Heidelberg, pp 584–587
15. Li B, Munoz JP, Rong X, Chen Q, Xiao J, Tian Y, Arditi A, Yousuf M (2018) Vision-based mobile indoor assistive navigation aid for blind people. IEEE Trans Mob Comput 18(3):702–714
16. Zhang H, Ye C (2017) An indoor wayfinding system based on geometric features aided graph SLAM for the visually impaired. IEEE Trans Neural Syst Rehabil Eng 25(9):1592–1604
17. Jafri R, Campos RL, Ali SA, Arabnia HR (2017) Visual and infrared sensor data-based obstacle detection for the visually impaired using the Google project tango tablet development kit and the unity engine. IEEE Access 6:443–454
18. Neto LB, Grijalva F, Maike VRML, Martini LC, Florencio D, Baranauskas MCC, Rocha A, Goldenstein S (2016) A kinect-based wearable face recognition system to aid visually impaired users. IEEE Trans Hum-Mach Syst 47(1):52–64
19. Giudice NA, Klatzky RL, Bennett CR, Loomis JM (2013) Perception of 3-D location based on vision, touch, and extended touch. Exp Brain Res 224:141–153
20. Lin BS, Lee CC, Chiang PY (2017) Simple smartphone-based guiding system for visually impaired people. Sensors 17(6):1371

Student Behavior Analysis in Classroom Using Facial Emotion Recognitions and Graph Visualizations

J. Akil Arsath, N. Pathri Narayanan, A. M. Abirami, and P. Karthikeyan

Abstract Majority of the present generation of people is felt hard to handle their emotions due to many reasons. There is a drastic change in the behavior of students in the classroom environment which reflects in the emotional quotient (EQ). Few researchers contributed on emotions of learners. The novice faculty needs lot of skills to engage the class either online or offline mode and to ensure the learning happens effectively. There is a need for an automatic emotion recognition system in the classroom to facilitate the faculty. In this work, an automatic facial emotion recognition system is proposed to measure the students' facial expressions such as listening, joy, drowsy, and confusion during the class. The analysis dashboard of the proposed system indicates the student engagement index (SEI) that helps the faculty to improve their teaching plan and to enhance their class handling skill. The Viola–Jones algorithm is used along with the convolutional neural network (CNN) model to implement the proposed system. The model was trained using FER 2013 and MobileNetV2 datasets for facial features detection and for emotions recognitions. The proposed system is working with 88% accuracy. The change of emotional states of each student for each class is visualized graphically. Hence, it is proved that the proposed system will be highly suitable for both online and offline classes.

Keywords Emotions recognition · Facial expression · Student engagement · Student engagement index · Deep learning · Convolutional neural network

J. A. Arsath · N. P. Narayanan · A. M. Abirami (✉) · P. Karthikeyan
Thiagarajar College of Engineering, Madurai, India
e-mail: abiramiam@tce.edu

J. A. Arsath
e-mail: akil@student.tce.edu

N. P. Narayanan
e-mail: pathri@student.tce.edu

P. Karthikeyan
e-mail: karthikit@tce.edu

S. Bandyopadhyay et al. (eds.), *Intelligent Computing Systems and Applications*,
Lecture Notes in Networks and Systems 1010,
https://doi.org/10.1007/978-981-97-5412-0_16

1 Introduction

The present generation people fall under Gen Z people who are highly depressed than the other generations [1]. The learners of these generations will not be able to control their emotions due to many reasons such as environment and society. The education system is also changing all over the world based on the need of the learners and their behavior. In recent days, online classes have increased and most of the learners prefer that to acquire knowledge. The students' behaviors are not predictable in the online classes that make the environment really challenging. Hence, it is hard to engage such a class by a novice faculty. For the same situation in the real physical classroom environment, the faculty can easily monitor and control the behaviors of the student. However, the emotions of the students will be changing in the classroom due to many reasons such as theoretical topic, disturbance in the environment, post-lunch class, and low voice of the faculty. The facial emotion is the primary factor that gives more confident to the faculty who handled the class. More activity, interesting facts, and future scope of the topic need to be discussed in every class to maintain the classroom active. The emotions of learners are happy, sad, learning, drowsy, etc. Based on eyebrow, lip, and chin movements in the face, various emotions are coming.

In this paper, an automatic facial emotion recognition system has been proposed to improve the learning process in the classroom. The student's emotions can be captured by a series of video frames. To achieve the high accuracy in a real-time environment, more video frames need to be captured which results in high computation. On the other hand, less features can be used for analysis but accuracy would be lower. Here, a model is proposed with a goal to provide a solution for both accuracy and efficiency. The optimized solution in emotion recognition helps teachers in a real-time environment to adjust the strategies and methods to increase the teaching quality. CNN model is used with depth-wise convolutions layers which predict the output of four classes like listening, joy, confusion, and drowsy state of students. FER 2013 dataset is used for neutral and happy images and manual extraction of confused images for training the dataset. MobileNetV2 dataset is used in training for the detection of facial features.

The main objective of the work is to enhance the student engagement index (SEI) of the classroom. The proposed system supports to generate the SEI with the help of image vision technique such as convolutional neural network (CNN) model. The model has been trained with four major emotion faces and produced results with accuracy of 88%. Section 2 describes the motivating factor of the proposed system, Sect. 3 explains the literature review, Sect. 4 discusses the methodology of the proposed system, Sect. 5 discusses the experimental results, and Sect. 6 concludes the work.

2 Motivation

Student engagement is vital to the teaching learning process. If there is a system to monitor the student behavior inside the classrooms at regular intervals, then the effectiveness of teaching learning process can be measured and analyzed. In this paper, we have made an attempt to determine the student engagement index (SEI) based on the emotional state of each student in the class. The teachers can revisit their lecture plan based on this score and plan for the well-defined set of student engagement activities. Or the teachers can counsel the students if there is any deterioration in the attentiveness of students in the classrooms.

3 Related Work

The advancement in machine learning techniques greatly support the facial expression recognition and its related applications. A deep convolution neural network (DCNN) model is available along with transfer learning and pipeline tuning strategies for training facial data with pretrained VGC16 model [2]. This strategy helped them to get more accuracy by removing the overhead of initializing random weights to the model. Viola–Jones algorithm was the most predominant algorithm for the detection of emotions such as angry, disgust, fear, joy, sad, contempt, and surprise from eye and mouth regions with greater than 90% accuracy [3]. CNN model based on depth-wise separable convolution for efficient lightweight real-time emotion recognition was proposed to reduce the number of parameters while training and testing real-time images, for large datasets [4]. CNN model with an attention mechanism was used for increasing the accuracy on small datasets. It resulted in performance improvement as information is transmitted to the output layer directly [5].

Multichannel convolutional neural network with two channels of CNN was developed to improve training for smaller datasets, where the first channel acts like a usual CNN training model and the second one uses the pretrained parameters. It is done by convolutional auto-encoder (CAE), which learns Gabor-like filters and generates output for logistic classifiers [6]. Hybrid model with CNN-LSTM was proposed for efficient emotion recognition. It resulted in higher accuracy as CNN is effective in classification of images and LSTM is effective in handling the sequence of extracted features of images [7]. Student emotions monitoring system was developed for eLearning platforms using their head and eye movements [8]. Automated emotion recognition system for students was designed using MTCNN model and validated using public datasets [9].

Different public datasets like CK+, JAFFE, and BU-3DFE are available for image processing applications. Lopes et al. explored different image preprocessing techniques for determining various emotions from the facial expressions. They concluded that CK+ database gives better and faster results with greater than 95% accuracy [10].

Unique CNN-based FERC model was proposed for feature extraction and recognizing facial expressions for better performance. FERC uses two components of CNN, one is for background image removal and other for feature extraction. The expression vector (EV) is generated by tracking facial points of importance and applied to background removed faces. It works with different orientations due to the unique EV feature vector, and background removal added a great advantage in accurately determining the emotions [11]. Haar cascade method along with Viola–Jones algorithm was widely used by researchers for object recognition and features detection [12].

Convolution deep belief network (CDBN) model was used for learning the salient features of multimodal expressions using the signals from face, body gesture, voice, and physique [13, 14]. Teaching assessment model was proposed for analyzing the emotions of students in the classroom by using deep learning techniques supported by LBP encodings and mappings [15]. EmoNet model was developed using the CNN network which uses linear growing-shrinking filter method for extracting features robustly. It used progressive resizing of images and multiple filter sizes for training and validation [16]. Simulated framework for analyzing the facial expressions of online users of different courses was developed which classifies eight different emotions like angry, happy, sad, and so on [17]. Classroom emotions were explored through facial expressions by using multitask CNN (MTCNN) with masked dataset (RAF-DB) for training. The results were compared with VGG16 and ResNet18 and concluded that MTCNN performed better for classroom emotion recognitions with more than 90% accuracy [18].

From the literature survey, it is understood that many research focused on the facial features extraction and emotion recognition using eyes and head positions. To our knowledge, limited research work was carried out using expressions produced in the mouth region. This paper used both eyes and mouth expressions from the face of students to recognize their emotions at regular intervals of time during the class hours. It also generates visualizations like comparison charts, state transitions, along with student engagement index (SEI) score.

4 Methodology

This paper proposes a data visualization method for assessing the teaching leaning process happening inside the classroom. The proposed student behavior analysis system is shown in Fig. 1.

This research work helps the teachers to understand the emotional state of each student and thereby determines the student engagement index (SEI) for each class by analyzing their emotions at different instants of time.

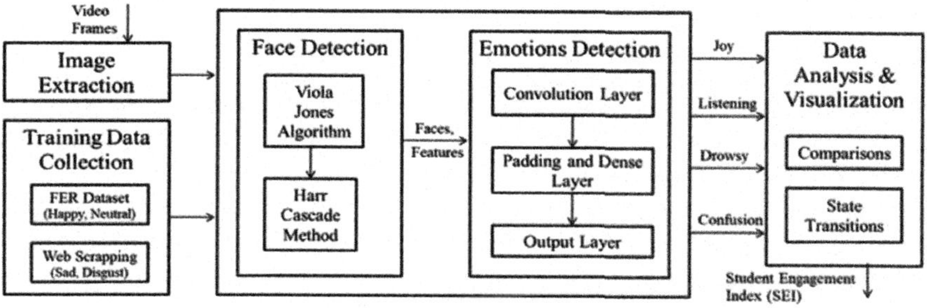

Fig. 1 Student behavior analysis system

4.1 *Image Extraction*

Videos of classroom are fed into this module at regular intervals for the extraction of frames. Initially, the frame per second (FPS) value may be set for the extraction of frames from the videos. For example, if video length is 30 min and FPS is set to 10, then 300 frames are extracted. If FPS is set to 0.5 for the same video, then 15 frames are extracted. These saved frames or images are then fed to the face detection phase.

4.2 *Training Data Collection*

This research work aims to categorize four different emotions like joy, listening, drowsy, and confusion. FER2013 dataset is used for training the model which includes 500 images for each class label [19]. The data consists of 48 × 48 pixel grayscale images of faces. *Happy*-faced images are considered for the class *joy* and *neutral* images for the class *listening*. For the class *confusion*, it does not have any dataset. So sad images are picked by Web scrapping; some images are selected from the disgust directory which look like confusion. OpenCV technique is used to find the emotion *drowsiness* of the students. Apart from these, the collection of images MobileNetv2 from ImageNet dataset is used for learning facial features from the faces [20]. MobileNetV2 is computationally efficient and well-suited for deployment on mobile devices with limited computing resources. It achieves greater efficiency by using a combination of depth-wise separable convolutions and linear bottlenecks. In the context of facial recognition, MobileNetV2 can be used as a feature extractor, where it is used to extract facial features from input images. These features can then be used for tasks such as face detection and emotions recognition.

4.3 Face Detection

During this phase, faces are detected from the input images. Viola–Jones algorithm is used for face detections. Though it is slow during the training phase, it is more powerful algorithm for detecting the faces from the real-time feed. This algorithms identifies multiple faces in the given images. It checks for the presence of facial features in sub-regions at different positions. Haar cascades method is used to identify the presence of the face in frame. If face exists, the region is cropped and fed to the subsequent emotion recognition phase.

4.4 Emotions Classification

Lightweight convolution neural network (CNN) model is used for the classification of emotions from the faces. The images are preprocessed and then augmented using different techniques such as flipping, zooming, and scaling. Eyes and mouth features are extracted from the faces for determining the emotions of students. OpenCV technique is used to find the eye aspect ratio (EAR) which helps to find the drowsiness state, depending on eye opening ratio. The forward propagation generates output patterns for the corresponding inputs. In the backward propagation, the weights are adjusted in the hidden layers using the error values calculated in the output layers so as to minimize loss values. The initial weights are the most important for model building and its performance. Else, the activation would be performed in different regions, resulting in reduced performance. These processes are repeated for different epochs until the error is reduced to a minimum value.

4.5 Data Analysis and Visualization

This phase analyzes the emotional classifications of the class. It prepares the comparison report and state transition of emotions at different occasions based on the number of students in the particular emotional state. It also calculates the student engagement index (SEI) for that specified class. Let N be the total number of students in the class. Let w_{jl} be the weight assigned for the states joy and listening; w_c be the weight assigned for the state confusion; and w_d be the weight assigned for the state drowsy. The SEI at an instant t is calculated as follows:

$$\mathrm{SEI}_t = \left(\left(n_j + n_l\right) * w_{jl} + n_c * w_c + n_d * w_d\right)/n \tag{1}$$

where n_j, n_l, n_c, and n_d are the percent of students in the emotional states joy, listening, confusion, and drowsy, respectively. The dashboard is developed and represented in the graphical forms, as shown in Fig. 2.

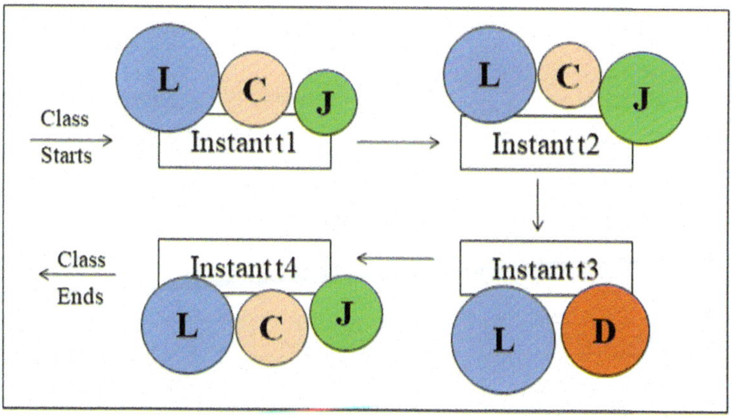

Fig. 2 Data visualization of class behavior

If 80–20 rule is applied for this scenario, it can be estimated that 80% of students have to actively participate in the class activities, whereas 20% may remain silent or not participated. The teacher may expect each student to be in the joy or listening state, so same weight is assigned for both the states. However, the teacher has to give special attention to the students who are in confused or drowsy state. If SEI is greater than 0.8, it means that the teacher has engaged the students in the class. The weights shall be assigned different values depending on the factors like morning first session, post-lunch session, mathematics, programming, or theory courses. Accordingly, the 80–20 rule may slightly be modified to 70–30 or any other ratio.

5 Experimental Results

The CNN architecture for the proposed student emotion recognition framework is shown in Fig. 3. The proposed model was trained using 500 images for four class labels.

and validation test was done for 100 images. A total of 50 epochs were run which include both forward and backward propagation to reduce the error value, as shown in Fig. 3. The trained model was used for the class consisting of 60 students. The model gave the accuracy of 88% and loss of 1.5%.

The flask application was developed for this application which captures the real-time videos and demonstrates the emotional ambience of classroom. The application also produces different visualizations for further analysis. The emotions of single student at different instants of time during the class were shown in Fig. 4.

The emotional state of one student captured in the classroom scenario is shown in Fig. 5. The x-axis uses the time, and the y-axis uses the emotional value. From Fig. 5, it is understood that the student was in the 'listening' state up to 20 min; the

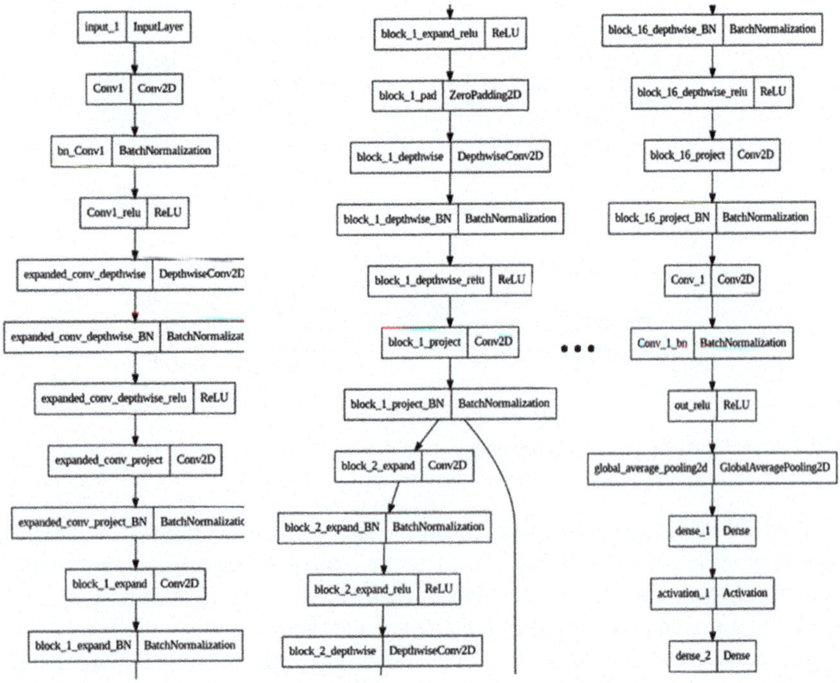

Fig. 3 CNN architecture for emotion recognition

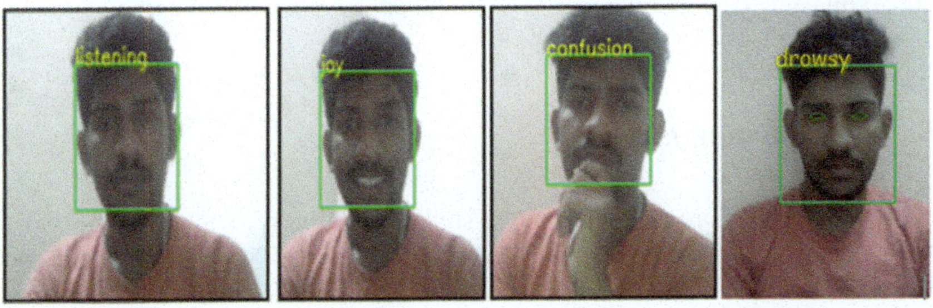

Fig. 4 Emotion states—listening, joy, confusion, and drowsy

teacher conducted some activity which made him to be in 'joy' state for 25–35 min; then the teacher would have taught continuously as the student was in 'listening' and 'confusion' states for the remaining period of time; he was never moved to the state 'drowsy'.

The application was used for capturing the emotions of different types of students in their classrooms and shown in Fig. 6. The model is so effective, and it captures multiple faces with different emotions. If the teacher wants to monitor the behavior

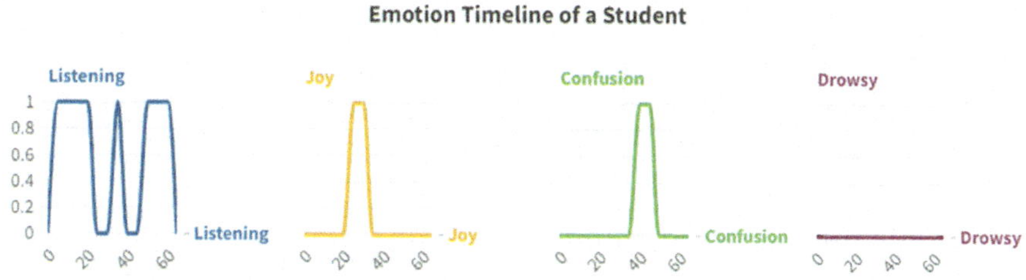

Fig. 5 Emotional timeline of a student in the classroom

of last bench students, multiple cameras may be placed in the classroom for capturing the videos. The corresponding video segment may be identified, processed, and then used for further behavior analysis.

Tables 1 and 2 present the confusion matrix and classification results of the proposed model.

From Table 2, it is seen that the proposed model correctly identified the class label 'joy', whereas it is not able to clearly differentiate 'listening-confusion', 'listening-drowsy', and 'confusion-drowsy' states. The model has to be trained further so as to classify confusion and drowsy states precisely. Table 3 compares the accuracy value of the proposed model with the existing work.

Fig. 6 Emotions recognitions of students in classrooms

Table 1 Confusion matrix of the proposed model (Instant t1)

Label (actual)	Listening (20)	Joy (14)	Confusion (14)	Drowsy (12)
Listening	19	0	2	1
Joy	0	14	0	0
Confusion	1	0	10	1
Drowsy	0	0	2	10

Table 2 Classification results of the proposed model

Time	Label	Precision (%)	Recall (%)	$F1$-score (%)	Accuracy (%)
Instant t1	Listening	86	95	90.4	88.3
	Joy	100	100	100	
	Confusion	83	62	71	
	Drowsy	75	90	82	
Instant t2	Listening	94	88	90	86.7
	Joy	100	100	100	
	Confusion	66	75	70	
	Drowsy	57	66	62	
Instant t3	Listening	92	92	92	90
	Joy	100	100	100	
	Confusion	80	80	80	
	Drowsy	83	83	83	
Instant t4	Listening	83	91	87	86.6
	Joy	100	100	100	
	Confusion	83	83	83	
	Drowsy	86	75	80	
Overall result (Average)		85.6	86.4	85.76	88

Table 3 Comparison of performance of emotion recognition models

Author(s)	Dataset used	Objective	Feature detection and classification algorithm	Accuracy (%)
Akhand et al. [2]	KDEF, JAFFE	Facial emotions classification	CNN, transfer learning	96
Baffour et al. [4]	FER 2013, JAFFE	Facial emotions classification	CNN (Depth-wise separable 2D convolution to reduce number of parameters)	88
Padilla et al. [12]	FEI face database	Face detection	Viola–Jones with Harr cascade method	98
Proposed model	FER 2013, MobileNetv2	Facial emotions classification	Harr cascade, CNN	88

The application generated the comparison chart as shown in Fig. 7. The bar chart was drawn for the class having 60 students for an instant t_1.

For the same instance, the emotion graph of the class is shown in Fig. 8. The bigger sized nodes in the middle represent the emotion node, and all other nodes represent the student nodes.

Few students are in the same state for the whole session, and few students have change of emotional states during the class hours. The change of state is visualized in Fig. 9 for the student S14. Within the instance t_1, he was in three different states,

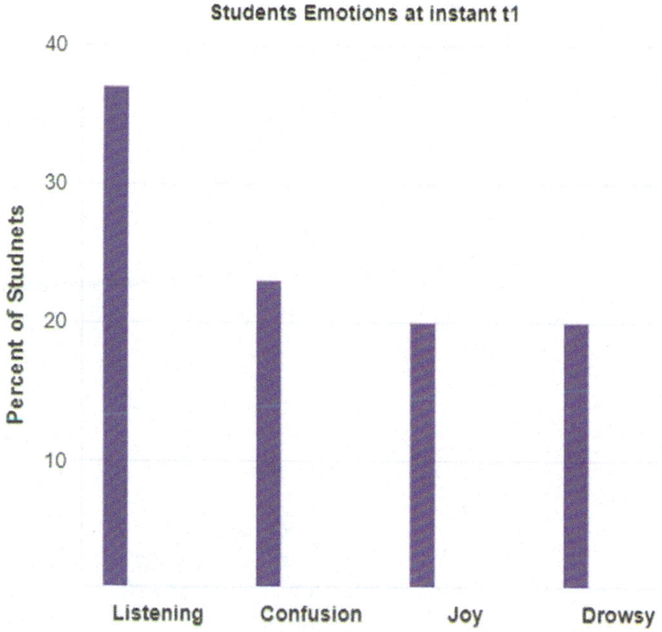

Fig. 7 Class emotions at instant t_1

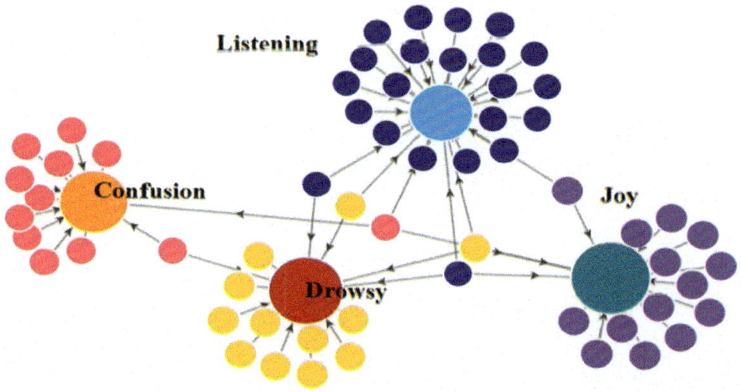

Fig. 8 Graph visualization of class emotions at instant t_1

namely joy, listening, and confusion. By placing the mouse cursor on any node, its details are visible, whereas the other links are blurred. This is visually shown in Figs. 10 and 11.

Assuming the weights w_{jl} as 0.8, w_c as 0.1, and w_d as 0.1, the SEI_{t1} is calculated as 0.52 (52%) using Eq. (1). If entire class time t is divided into four different instants

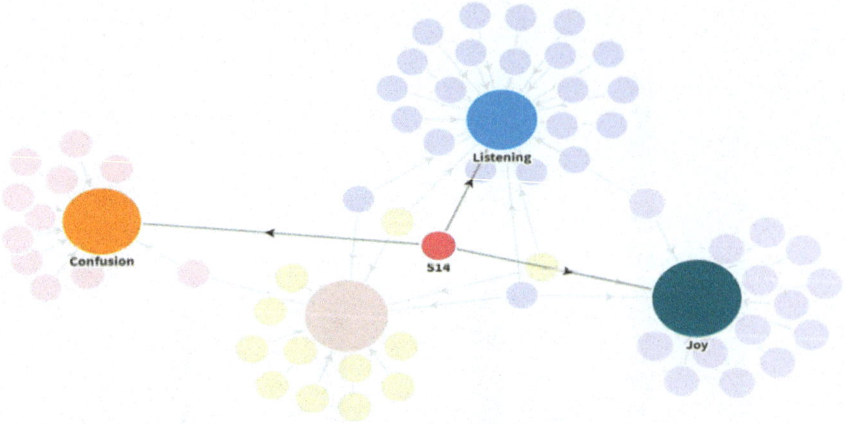

Fig. 9 Graph visualization of emotional change state of student (S14)

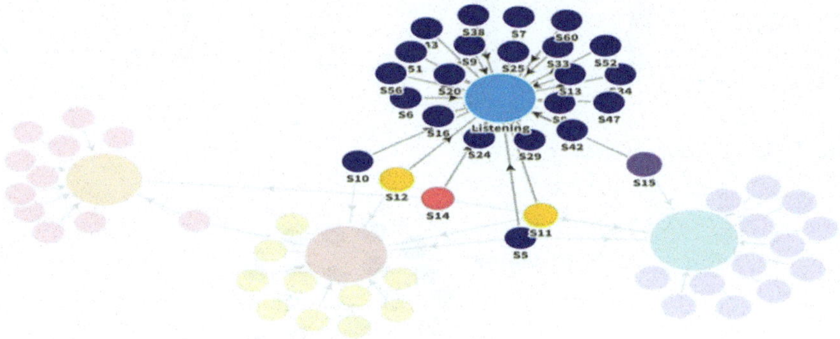

Fig. 10 Graph visualization of listening state of class at instance t_1

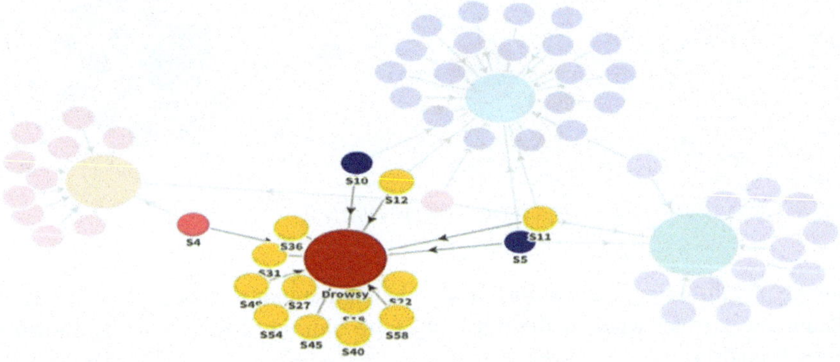

Fig. 11 Graph visualization of drowsy state of class at instance t_1

like t_1, t_2, t_3, and t_4, then the emotions of students at different instants are given in Table 4. The overall SEI is determined by taking the average of all SEI_ts. In the example shown, the overall SEI is calculated as 54%.

The emotional state of the entire class was shown in Fig. 12, with the gauge type of dashboard.

This type of dashboard helps the teachers to get overall emotional state of the class. The teacher can check the transition of emotional state of each student at different instants of time and give attention on them in subsequent classes. The teacher can judge himself whether the students actively engaged in the class activities or not. This helps him to either revisit his lecture plan activities or plan to have one-to-one discussion with the students.

Table 4 Student engagement index (SEI_t) of the class

Time	Listening	Joy	Confusion	Drowsy	SEIt
Instant t_1	22	14	12	12	0.52
Instant t_2	34	12	8	6	0.64
Instant t_3	26	12	10	12	0.54
Instant t_4	22	10	12	16	0.47

Fig. 12 Student engagement dashboard of the class

6 Conclusion

Classroom teaching has many advantages for both students and teachers. It is really very much essential to measure the teaching learning effectiveness of the classes. This type of emotions analysis of students would give better feedback to the teachers which in turn helps them to change their style of teaching methodologies. It also helps them to identify the complex topics for which the student feel difficult to understand. Hence, the proposed research work enables the teachers and the students to have better interaction, which in turn result in the active participation of students. Our work could lead to social isolation. If people are constantly being monitored for their emotions, they may feel like they have to censor themselves and hide their true feelings. This could make it difficult to form genuine relationships with others. Another drawback of our work is that in real-time scenarios, students may overlap, which leads to incorrect predictions. The future work includes the visual representation of emotional state change of individual student. It would definitely help the teacher to give personal attention to the students for their effective learning.

Acknowledgements This research work is supported by SERB Teachers Associateship for Research Excellence (TARE) fellowship (TAR/2022/000425).

References

1. https://www.aecf.org/blog/generation-z-and-mental-health. Accessed on Feb 2023
2. Akhand MAH, Roy S, Siddique N, Kamal AS, Shimamura T (2021) Facial emotion recognition using transfer learning in the deep CNN. Electronics 10(9). https://doi.org/10.3390/electronics10091036
3. Ali F, Khatun M, Turzo NA (2021) Facial emotion detection using neural network. Adv Res Comput Eng 2:33–52
4. Baffour AA, Guo J, Kusi GA (2018) Depth-wise separable convolution for real-time facial expression recognition. In: 2nd Borneo international conference on applied mathematics and engineering BICAME. IEEE Xplore, Borneo, pp 128–133
5. Daihong J, Zheng H, Lei D, Jin P (2021) Facial expression recognition based on attention mechanism. Sci Program. Article ID 6624251. https://doi.org/10.1155/2021/6624251
6. Hamester D, Barros P, Wermter S (2015) Face expression recognition with a 2-channelconvolutional neural network. In: International joint conference on neural networks IJCNN. IEEE Xplore, Ireland, pp. 1787–1794
7. Hung BT, Tien LM (2021) Facial expression recognition with CNN-LSTM. In: Research in intelligent and computing in engineering, pp 549–560. https://doi.org/10.1007/978-981-15-7527-3_52
8. Krithika LB, LakshmiPriya GG (2016) Student emotion recognition system (SERS) for e-learning improvement. Procedia Comput Sci 85:767–776
9. Li X, Yue R, Jia W, Wang H, Zheng Y (2021) Recognizing students' emotions based on facial expression analysis. In: 11th international conference on information technology in medicine and education (ITME), China, pp 96–100
10. Lopes AT, Aguiar E, DeSouza AF, Santos TO (2017) Facial expression recognition with convolutional neural networks: coping with few data and the training sample order. Pattern Recogn 61(12):610–628

11. Mehandale N (2020) Facial emotion recognition using convolutional neural networks (FERC). SN Appl Sci 2(446). https://doi.org/10.1007/s42452-020-2234-1
12. Padilla R, Fernandes CF, Filho C, Costa M (2012) Evaluation of Haar cascade classifiers for face detection. In: 4th international conference on digital image processing (ICDIP), 6, Malaysia (2012).
13. Bhupesh KRA, Akil JA, Abirami AM (2022) Multi-modal emotions detection service using deep learning techniques. In: International Conference on applied computational intelligence and analytics ACIA 2022, 2075(1), https://doi.org/10.1063/5.0133760
14. Ranganathan H, Chakraborty S, Panchanathan S (2016) Multimodal emotion recognition using deep learning architectures. In: 2016 IEEE Winter conference on applications of computer vision (WACV), New York, pp 1–9
15. Sahla KS, Senthil Kumar T (2016) Classroom teaching assessment based on student emotions. In: Corchado Rodriguez J, Mitra S, Thampi S, El-Alfy ES (eds) Intelligent systems technologies and applications 2016. ISTA 2016. Advances in intelligent systems and computing, vol 530. Springer, Cham. https://doi.org/10.1007/978-3-319-47952-1_37
16. Thiruthuvanathan MM, Krishnan B (2020) EMONET: a cross database progressive deep network for facial expression recognition. Int J Intell Eng Syst 13(6):31–41
17. Wang W, Xu K, Niu H, Miao X (2020) Emotion recognition of students based on facial expressions in online education based on the perspective of computer simulation. Complexity 1–9, Article ID 4065207
18. Yuan Q (2022) Research on classroom emotion recognition algorithm based on visual emotion classification. Comput Intell Neurosci, Article ID 6453499. https://doi.org/10.1155/2022/645 3499
19. https://www.kaggle.com/datasets/msambare/fer2013. Accessed on Aug 2022
20. https://www.kaggle.com/code/mgiraygokirmak/mobilenetv2. Accessed on June 2022

Requirement-Driven Developer Recommendation Framework Based on Github Developer Social Network

Saksham Sneh Mandal and Ananya Kanjilal

Abstract Developer recommendation is a field where focus is often made on the question and answer forums, bug tracking and resolutions and version control system commits. In this paper, a framework is proposed to utilize the huge information contained in GitHub repositories with respect to developer profiles and their activities. The proposed framework extracts and formalizes some of their features to create a Developer Network Graph (DNG). A novel ranking algorithm named Developer Experience and Requirement Attribute Algorithm (DERA) is proposed that uses developer experience and user requirements to rank and recommend developers. The proposed framework uses both feature weights of developers and their neighborhood scores to rank them, according to the requirements made by the user and display the result in real time. The model is experimentally validated on a dataset of 5000+ users and 200 languages. Results for different requirement scenarios are discussed for substantiation. The complexity of the proposed ranking algorithm (DERA) is identified as O(km) for k filtered users and m languages and is adaptable to real-life scenarios of finding developers that suit the business/project requirements.

Keywords Requirement-based developer recommendation · Github repository · Developer social network · Developer ranking

1 Introduction

In real life, developers use web forums, code storage systems or host their programming codes online, either for the benefit of others, or for requiring help from others, for storage purpose to be referenced later. Git, a version control system created for maintaining versions of code, is used widely by developer community nowadays. Over the years, multiple applications began hosting their own version of Git over Internet for users. GitHub, an Internet hosting service for a distributed version control

S. S. Mandal (✉) · A. Kanjilal
B. P. Poddar Institute of Management & Technology, Kolkata, India
e-mail: sakshamsneh@gmail.com

S. Bandyopadhyay et al. (eds.), *Intelligent Computing Systems and Applications*, Lecture Notes in Networks and Systems 1010,
https://doi.org/10.1007/978-981-97-5412-0_17

system is one such website based on Git, where users can store relevant profile details, create repositories, track their code issues, fork repositories created by other users, follow them, star/watch the repositories to keep an eye on the changes being made, etc. All of these leave a trace on the user, forming their influence over time. As GitHub data shows the actual work done by the users, it becomes a rich source of information that can be used for recommending users for specific needs.

Software development community require developers of particular skillset or experience from time to time to address specific project needs. Different users and requirements may have focus on different features or values the developer should bring in. For example, the developer should have 'n' years of experience in full stack programming, etc. The influence created in GitHub can be thus be utilized here to rank and recommend developers, according to their needs. In this work, a framework is proposed to capture project specific requirements (in the form of developer skillset and experience required). A directed graph-Developer Network Graph (DNG) comprising of developers as nodes with various attributes (based on profile information) and edges formed based on various kinds of relationships among the developers is presented A ranking algorithm, Developer Experience and Requirement Attributes (DERA) algorithm is proposed, that uses the normalized feature values, requirement-based feature weights, graph node ranking value and user details to create a requirement-based ranking of developers for recommendation. The approach is requirement driven and hence tunable by users to get the best suited developer for a specific project requirement.

2 Related Work and Background

Developer recommendation has become extremely relevant nowadays as a means to get the most suitable developer for a specific project role. As requirements and vacancies are created across different companies, developers are required, those who are specifically trained in some particular fields. The different approaches proposed in some recent are discussed in this section.

2.1 Developer Recommendation on Question and Answer Forums

Authors in [1] propose an approach for developer recommendation from stack exchange software engineering Q&A website based on K-Means clustering. Best cluster is found for test question and PageRank with time decay is applied on the directed network graph created using clusters' developer. In another work [2], authors propose recommendations based on the reputation of users, based on the satisfaction of the question asker of the answer. It is also based on past interactions, degree of knowledge on the particular topic, and availability to answer the question.

Authors propose bug-based recommenders in [3–5]. In [5], developers are recommended opportunities based on bug types and developer scores. It categorizes developers based on bug fix works, into core, active, and peripheral developers by their experiences. Then developer scoring provides a score to each bug-developer pair and thus recommends developers. In the work in [6], experts are identified using similarity of test query and user posts, which is then filtered using network algorithm and combined with user reputation. Authors in [7] uses both reputation of users like [2], and similarity of test question and user content like [1, 6] in the name of Exp-Rank.

2.2 Developer Recommendation on Source Code Repositories

The work in [8] presents an empirical study regarding the effectiveness and usefulness of historical commits for developer recommendation. They use four open-source repositories: jEdit, Spring Framework, Hadoop, and JDT-debug. Their results show a correlation of commit messages and developers' expertise.

In [9], authors have designed the DevRank algorithm using two features from GitHub data, the commit details, mapping developers to projects and follower details to map them within self. They assume that the more commits a developer refers to a project, the more influence scores propagate from the developer to the project. PageRank algorithm [10] is a well-known algorithm that is used to measure a significance of a node in a network using the formula given below and is used in several works including the proposed work in this paper.

The authors in [11] studied the GitHub data to figure the developer influence using follow, star, fork, and user activity. Multiple network graph metrics are used. H-Index an author analysis metric is used for star and fork analysis. Using a follower graph, Borda Count method and Spearman's Correlation coefficient is used to rank and recommend developers. In [12], authors use connections and language experience for analyzing productivity and contribution changes over projects. It reports prior connection as well as language experience increases the contribution to projects, while only connection implies lesser productivity. Work in [13] displays the actual influence/value produced, which is different between languages.

Our work in similar to [9, 11] in that GitHub repository information to recommend developers is also used. Similar to [9] a developer network is built, and in addition the graph is annotated with node weights and edge weights to encode more semantics. However, our approach is more adaptable as it considers user requirements and creates a ranking algorithm based on the user influence with respect to the requirements.

3 Scope of Work

As discussed in the earlier section, it is seen that few works have explored the rich information in GitHub [14] repository for developer recommendations that forms the basic premise of our work. In this work, a publicly available dataset of GitHub users is used to create an adaptable model of ranking developers for recommendations based on user requirements. It has two phases, the model creation and the ranking phase.

In the model creation phase, the dataset is used to generate a vector for each of the developers with their individual attribute details and a relationship pair of developers based on "follow" information on GitHub. Both of this information are now used to generate a network digraph with developers as nodes and relationships as edges. The developer network graph (DNG) is used to generate network metrics for the nodes for their neighborhood scores. The popular PageRank algorithm is used, as it utilizes the edge weights and is affected by nodes present in the vicinity of the node under consideration. The PageRank values are updated back in the vector for utilization in the ranking phase.

In the ranking phase, the experience and language proficiency selected by user requirements are used to find the effective influence of the calculated PageRank values. It is assumed that the developer has given equal time to each project. Time decay is applied to the influence as the influence decreases over the years. The developer's weight, decided by the user based on six different features, is added to the influence as the developer's actual value. The resultant experience is used to rank the developers.

Some assumptions are made in our model, like the name of the repository is unique, and if two users have similarly named repositories, they have worked on the same repository. Also, work done/time spent by any developer is equally divided between all of their repositories.

Our approach has been implemented as an application that allows the user to provide their specific values to different features that define the developer's work and influence.

4 Methodology

Figure 1 shows the different steps of the proposed approach.

4.1 Data Cleaning

In this step, the GitHub dataset is converted from JSON format to CSV format. The data thereafter is cleaned by removing suspicious user accounts; all accounts that do

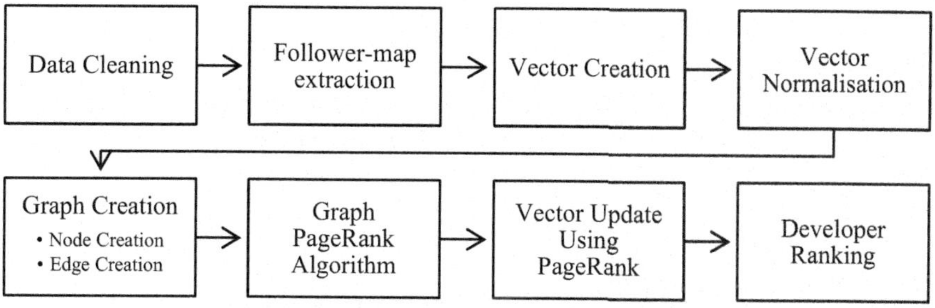

Fig. 1 Block diagram of proposed approach

not contribute the GitHub in any way; bot and organizational accounts, who cannot be recommended for user's requirements. The commit details are removed for each user, as commit details are not being referenced as part of the model.

4.2 Follower-Map Extraction Algorithm

Follower-map extraction is presented in Algorithm 1. Each user is set as key and its followers as value in follower-map. Next, for each user, all the following users are set as key and the user itself is added to their followers.

Algorithm 1 Follower-map extraction

```
function extract follower-map (dataframe: csv): {follower-map: dictionary}
follower-map ← empty dictionary/*dictionary is map of (id, followers) */
while row in dataframe then
  follower-map[row[id]] ← row[follower list]
  while f ← row[following list] then
    if f in follower-map then follower-map[f] += row[id]
    else follower-map[f] ← row[id]
    end if
  end while
end while
```

4.3 Language Extraction Algorithm

Languages are extracted from repository information for each row of dataset. A dictionary is created per row where unique language, respective counts create the key, value pair. They are assigned as a new column value of the dataset (Algorithm 2).

Algorithm 2 Language Extraction

```
function extract language (dataframe: csv): {langdict: dictionary}
 while row in dataframe then
  langdict ← empty dictionary /*dictionary is map of (language, count) */
  while repo ← row[repo] then
      if repo[language] in langdict then langdict[language] += 1
      else langdict [language] ← 1
      end if
  end while
  row[langdict] ← langdict
 end while
```

4.4 Conversion to Vector and Normalization

Dataset is used to create vector dataframe. For each row of the dataset, a row is created, containing the id of the user, name, repository count, gist count, follower count, following count, language count, commit/contribution count, repository list, language list, language dictionary and total days. The total experience is calculated as the difference between the max updated date of the dataset and created date of the user account in years. Min–Max scalar is used to normalize the numeric values in vector in a new dataframe.

4.5 Graph Creation

A Directed Weighted Graph is created, named Developer Network Graph (DNG). Each node represents one develop with node attributes like name, created date, repository count, gist count, commit count, follower count, following count, and language list.

$$\text{DNG} = (V, E) \tag{1}$$

where $V = \{v_i \in \text{vector}\},\ v_i$ is i - th node

$$E = \begin{cases} (u, v), (v, u) \text{ if common repo or language} \\ (u, v), v \text{ is a follower of } u \end{cases}, \quad u \in V, v \in V, u \neq v$$

Adjacency matrix,

$$A_{\text{DNG}} = \begin{cases} \text{repo weight} & \text{if } u \leftrightarrow v, \text{common repo} \\ \text{lang weight} & \text{if } u \leftrightarrow v, \text{common repo} \\ \text{follower weight} & \text{if } u \leftarrow v, v \text{ follows } u \\ 0 & \text{otherwise} \end{cases}$$

4.6 Developer Network Graph PageRank

PageRank algorithm is applied to the graph with all nodes and those users who are not available in the dataset are removed, as they were used only for ensuring the PageRank value aligned with the follower-map. The values are updated in the vector dataframe for each node.

Table 1 describes the complexity of each step of the framework based on the components.

Table 1 Algorithmic complexity of framework components

Component	Algorithm complexity
Data cleaning	$O(nm)$, n is the number of rows, m is number of repositories
Follower-Map Extraction	$O(nm)$, n is the number of rows, m is number of following ids
Language Extraction	$O(nm)$, n is the number of rows, m is number of repositories
Node Vector Conversion	$O(n)$, n is the number of rows
Vector Normalization	$O(n)$, n is the number of rows
DNG Node Creation	$O(n)$, n is the number of rows
DNG Edge Creation	$O(n^2)$, n is the number of rows
DNG PageRank	$O(E \cdot k)$, E is number of edges, k is iterations for convergence
Vector Update	$O(n)$, n is the number of rows

Requirement feature dictionary

Requirement formulization is done using a webapp UI implementation. User selects values for different features in the range [0,1] according to their requirements. This is stored in a feature weight dictionary with feature as key and weights as values. User weight is weighted average of normalized features using the requirement feature dictionary.

Developer Experience and Requirement Attributes (DERA) algorithm

DERA is the proposed ranking algorithm based on user attributes and requirements. Some assumptions in this regard are:

(i) A developer works equal time on all repositories. The developer experience is shared among all repositories in the ratio of their work om different languages. For example, if total work count on Language 1 is 3, Language 2 is 2; total time spent by the developer is 10 years, then the time spent on Language1 would be 10*3/(3 + 2) = 6 years and Language2 = 10*2/(3 + 2) = 4 years.

(ii) Measuring the user's influence on its neighborhood using network graph algorithms, the influence part caused by Language1 would be the product of the total time of Language1 and the total influence. A time decay factor is applied under the assumption the influence is decreasing over time.

(iii) Adding the calculated user's weight to the influence will result in the user's current experience.

Thus, the user's experience is,

$$e_i = \frac{\sum_j (f_{ij} * w_{ij})}{\sum_j (f_{ij})} + \prod\left(\frac{N(rl)}{N(r)}, t, pr_i, d\right) \tag{2}$$

where

e_i is the calculated experience of the i-th user, used to rank them;
f_{ij} is j-th feature weight of i-th user;
w_{ij} is j-th feature value (normalized) for i-th user;
$N(rl)$ is the number of repositories with selected languages for i-th user;
$N(r)$ is the total number of repositories of i-th user;
t is the experience of the i-th user in years;
pr_i is the PageRank value of i-th user from Network DiGraph; and
d is the time decay applied for influence decreasing over time.

Table 2 Algorithmic complexity of components

Component	Complexity
$\sum_j (f_{ij} * w_{ij})$	$O(n)$, where n is the number of features
$\sum_j (f_{ij})$	$O(n)$, where n is the number of features
$N(rl)$	$O(m)$, where m is the number of languages
$N(r)$	$O(1)$
pr_i	$O(1)$

Complexity analysis of DERA algorithm

The total time complexity is (Table 2):

$$(O(n) + O(n)) + (O(m) \cdot O(1) \cdot O(1)) = O(n) + O(m)$$

where n is the number of features and m is the number of languages.

Display algorithm (Generated output with recommended developer list)

The user is allowed to apply filters to language, and years of experience, requirement feature vector and get top "*n*" developers within the filtered data. Developer experience is calculated using the proposed DERA algorithm. The ranking is performed by descending sorting the filtered users by their new experience values. The top "*n*" users are selected and if the user has selected date sorting, the data is again sorted by date to get the oldest users first.

5 Experimental Setup

A representative user-centric dataset of 10 million GitHub Developers [15] is a dataset created/extracted from GitHub using GitHub APIs on Harvard Dataverse [16]. The same dataset has been used by [17] to identify malicious GitHub accounts, based on their profiles, and contribution activity. The same dataset available in JSON format is used, where the entry contains GitHub details of user, with details of commits and repositories the user has worked on. The dataset is collected from GitHub between July 20, 2018, and August 27, 2018. The entire dataset contains 10,649,574 user details, 118,602,740 commit details, and 20,999,258 repository details. Software configuration has Python3 as primary language, Python libraries including Pandas, NetworkX, Jupyter, Seaborn, Ipython, Concurrent Features; Gephi for network visualization; Google Colab as environment. Default CPU (Intel Xeon CPU and 13 GB of RAM) is selected from Colab as Hardware.

6 Results and Discussion

A part of the entire dataset of 4 million+ users has been extracted via the random sampling feature of the Pandas library. After cleaning 0.2% of data, the dataset contains 5248 users and 21 features. Table 3 gives feature correlation matrix.

Inferences from Matrix:

- As Langcount is based on the unique languages count of all repositories of a user, Public_repos can be observed to be highly correlated to Langcount at 0.53.
- Langcount also has a high correlation of 0.45 with commits as it displays the number of commits made by the user.
- Between public_repos and commits, the correlation value is 0.21, implying that despite having a lot of repositories, users do not commit as much.
- Public_gists does not show any high correlation to other values. As gists are often created by users for storing/displaying small pieces of code, that can be utilized by others, it doesn't seem to affect any other feature of the dataset.
- Followercount and Followingcount seem to have some correlation but very low, as the value of the correlation matrix between the two features is 0.12.
- Public_repos and following has a large correlation value of 0.49, implying users having more public repository links follow more users.

The dataset contains 200 unique languages, with the top 5 languages with their counts being JavaScript (1534), Java (1171), HTML (1127), Python (1006) and CSS (742). The vector dataframe contains 5248 rows and 13 columns. The generated weighted graph using dataframe and Gephi Tool, namely Developer Network Graph (DNG) contains 284,298 nodes and 6,068,052 edges. A subgraph of DNG is shown in Fig. 2.

The DNG is formed with all developers as nodes and individual attributes as node weights. The edges may be of three types with different weightages signifying separate significance levels; higher value indicates higher importance and vice versa.

(i) Repository edge: 0.2 (Developers working on same repository)
(ii) Language edge: 0.3 (Developers working on same programming language)

Table 3 Feature correlation matrix

	public_repos	public_gists	commits	followers	following	langcount
public_repos	1	0.13	0.21	0.12	0.49	0.53
public_gists	0.13	1	0.15	0.096	0.089	0.16
commits	0.21	0.15	1	0.19	0.043	0.45
followers	0.12	0.096	0.19	1	0.12	0.15
following	0.49	0.089	0.043	0.12	1	0.12
langcount	0.53	0.16	0.45	0.12	0.12	1

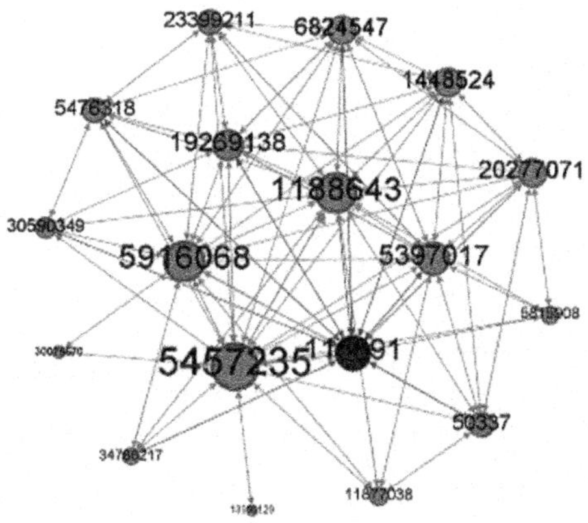

Fig. 2 Subgraph of DNG

(iii) Follower edge: 0.5 (Developers are connected by follower-followed relationship)

Figure 2 shows a subgraph of the graph generated using dataframe and Gephi Tool that has 18 nodes and 160 edges. Node size represents their relative PageRank values. The top 5 PageRank values for the total graph are noted in Table 4. The average computation time of running dataset to vector and graph process with a given number of nodes is shown in Table 5.

For the dataset, *n*, the number of features is fixed as 6 and *m* is the number of languages, based on user requirement. Thus, the time complexity for the filtered

Table 4 Developer with PageRank

Id	Login ID	PageRank
10704521	hongyangAndroid	0.001275
2352090	offensive-security	0.001011
13198583	hackedteam	0.000911
3346272	singwhatiwanna	0.000998
1061138	evilcos	0.000647

Table 5 Time taken for creation of DNG with different no. of users

No. of nodes	Time for follower-map extraction	Time for vector creation	Time for graph node creation	Time for graph edge creation	Total time taken
1000	2.8585	0.1403	0.0677	42.5708	51.1268
2000	3.7424	1.1998	0.1377	97.846	116.8661
5248	4.3877	2.58	0.3659	249.2546	286.6642

Table 6 Input-requirement feature set with weights

Repo count	Gist count	Commit count	Lang count	Follower count	Following count
1.0	0	0	0	0	0

Table 7 Output-recommended developer list

Id	Created	Repo	Gist	Commit	Follower	Following	Lang	Rank
1748769	17/5/12:10:59	3213	1	10	42	385	37	3.0
2128570	10/8/12:08:43	3313	5	1	198	6447	62	2.0
3011346	10/12/12:22:04	4203	30	201	79	2021	77	1.0

dataset of k rows is $O(k \cdot m)$, where k is the number of rows in the filtered dataset by selecting m languages. Time complexity of the DERA algorithm is $O(n) + O(m)$.

In the final step, the user can view the data and filter according to their requirements. A predefined time decay of 0.85 (similar to the damping factor in PageRank) is applied in DERA for ranking purposes.

A requirement scenario: The user wants 3 developers with a minimum experience of 5 years in HTML, JavaScript. Also, they want the more experienced developer to be recommended first. Their requirement has the developer should have worked on a lot of repositories/projects, while other features do not interest him (Tables 6 and 7).

The scatter plot of experience versus rank is also plotted to observe how language requirements and feature selection changes affect developers' ranking on GitHub data. When some specific feature is selected and all other features are ignored, the experience versus rank scatter plot depicts the changes. Figure 3 presents the scatter plot for exp versus rank for different features. For features repository count, Gists count, follower count, following count, the scatter plot drops sharply after nearly first 100 ranked users. This implies that the DERA calculated experience changes is very less for the given dataset after nearly the first 100 developers or so. Also, the top developers have a high value for these features and all developers after them have nearly consistent values in these features. However, the changes in experience for commit count and language count is not so steep. Thus, it can be implied that these features have a distributed value across all developers.

A comparison with related work with the proposed approach is presented in Table 8.

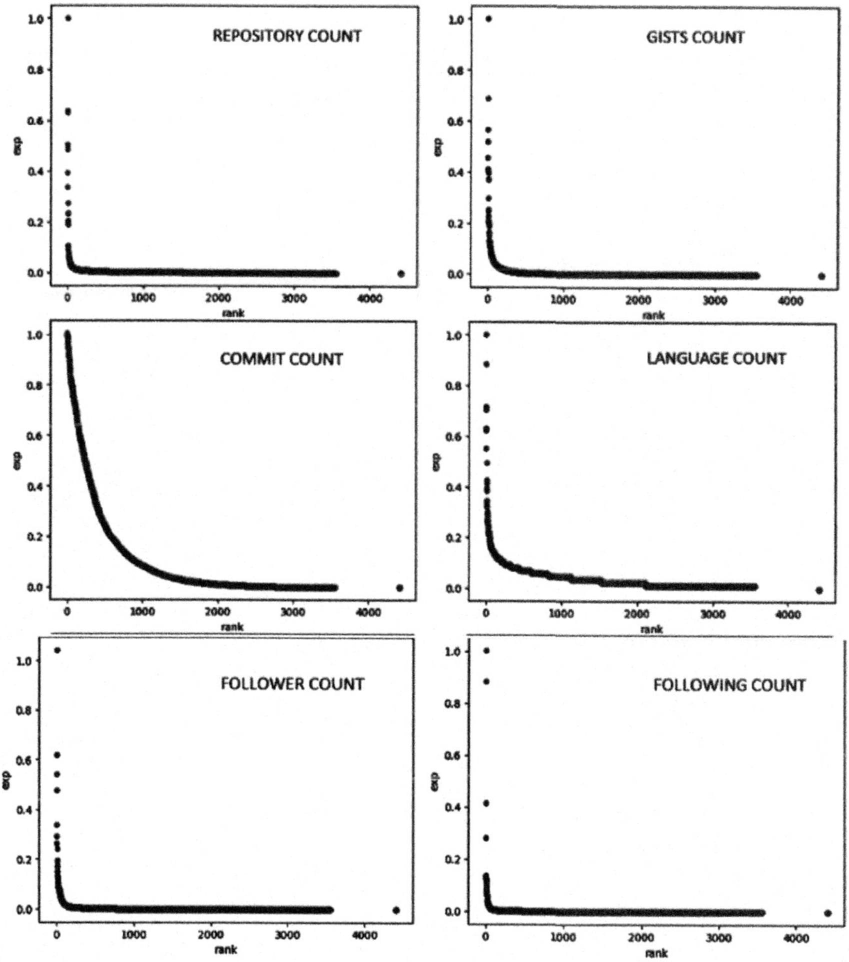

Fig. 3 Scatter plots for experience versus rank for selected features

Table 8 Comparison of related work

Ref. No	Dataset	Follower graph	Generic network graph	Page rank	HITs	Methodology
1	Stack Exchange Q&A Dataset	N	Y	Y	N	Clustering, Network Graph
2	Moodle	N	N	N	N	Reputation calculation based on knowledge
5	Eclipse, Mozilla Bug Repo	N	N	N	N	Using bug-developer pairing score
8	jEdit, Spring Framework, Hadoop, and JDT-debug Repo	N	N	N	N	NLP and cosine similarity of commit messages
9	GitHub Data	Y	Y	Y	Y	DevRank, PageRank and HITs for convergence
11	GitHub Data	Y	N	Y	Y	Multiple metrics for feature analysis
6	Stack Overflow Q&A Dataset	N	Y	Y	Y	NLP, cosine similarity, Network Graph
Proposed	Harvard Dataverse [16]	Y	Y	Y	N	DERA, DNG

7 Conclusion

In this work, an adaptable requirement driven graph-based framework is proposed for recommending developers from the information extracted from GitHub repositories. The framework allows users to capture requirements as custom feature weights like language, minimum experience (in years), etc. Custom edge weights have been applied while creating the graph, which affects the PageRank metrics used for the final ranking. The proposed Developer Network Graph is a weighted directed graph, where all the developers are nodes, and edges are created between nodes, if they have common repositories or language or if one node follows the other. For the first two conditions, bidirectional edges exist, and for the third, unidirectional. Also, the edge weights are dependent on the type of edge created. The proposed algorithm DERA is dependent upon the user's feature selection on GitHub data. It combines the weighted average of features using the user-selected weights with the experience of the developer and the influence calculated from the Developer Network Graph. At the same time, the follower details, repository details, and language details of users are also used in DERA to calculate the PageRank through the Graph created. The methodology allows users to input their feature values, language, and experience requirements to rank developers.

The model can be further extended to include other features of developers and consider similarity of developers as well. User contribution graphs can also be used in the future increasing the accuracy of the ranking when defining the user's experience in given languages.

References

1. Verma A, Sardana N, Lal S (2020) Developer recommendation for stack exchange software engineering Q&A Website based on K-means clustering and developer social network metric. Proc Comput Sci 167:1665–1674
2. Aciar S, Aciar G, Collazos C, González González C (2016) User recommender system based on knowledge, availability, and reputation from interactions in forums. IEEE Revista Iberoamericana de Tecnologias del Aprendizaje 11:1–1. https://doi.org/10.1109/RITA.2016.2518441
3. Chaitra BH, Swarnalatha KS (2022) Bug triaging: right developer recommendation for bug resolution using data mining technique. In: Emerging research in computing, information, communication and applications. Springer, Singapore, pp 609–618
4. Wang S, Wang L, Qu Y, Chen R, Guo S (2021) Toward more accurate developer recommendation via inference of development activities from interaction with bug repair process. J Softw Evolving Process 33(5). https://doi.org/10.1002/smr.2341
5. Zhang W, Zhao J, Peng R, Wang S, Yang Y (2022) SusRec: an approach to sustainable developer recommendation for bug resolution using multimodal ensemble learning. IEEE Trans Reliab 1–18. https://doi.org/10.1109/TR.2022.3176733
6. Elkorany A (2013) Integrated expert recommendation model for online communities. Int J Web Semantic Technol 4. https://doi.org/10.5121/ijwest.2013.4402
7. Zhao N, Cheng J, Chen N, Xiong F, Cheng P (2020) A novel expert finding system for community question answering complexity, pp 1–8. https://doi.org/10.1155/2020/5346085
8. Sun X, Yang H, Leung H, Bin L, Li H, Liao L (2018) Effectiveness of exploring historical commits for developer recommendation: an empirical study. Front Comput Sci 12. https://doi.org/10.1007/s11704-016-6023-3
9. Liao Z, Jin H, Li Y, Zhao B, Wu J, Liu S (2017) DevRank: mining influential developers in Github. In: GLOBECOM 2017—2017 IEEE global communications conference, Singapore, pp 1–6. https://doi.org/10.1109/GLOCOM.2017.8255005
10. Page L, Brin S, Motwani R, Winograd T (1999) The PageRank citation ranking: bringing order to the web. Technical Report. Stanford InfoLab
11. Hu Y, Wang S, Ren Y, Choo KKR (2018) User influence analysis for Github developer social networks. J Expert Syst Appl 108–118. https://doi.org/10.1016/j.eswa.2018.05.002
12. Casalnuovo C, Vasilescu B, Devanbu P, Filkov V (2015) Developer onboarding in GitHub: the role of prior social links and language experience. In: Proceedings of the 10th joint meeting on foundations of software engineering—ESEC/FSE. https://doi.org/10.1145/2786805.2786854

13. Sajedi Badashian A, Stroulia E (2016) Measuring user influence in Github: the million follower fallacy. In: IEEE/ACM 3rd international workshop on CrowdSourcing in Software Engineering (CSI-SE), Austin, TX, USA, pp 15–21. https://doi.org/10.1145/2897659.2897663
14. GitHub. https://github.com/. Last accessed 01 May 2023
15. Gong Q, Zhang J, Chen Y, Xiao Y, Fu X, Hui P, Li X, Wang X (2018) A representative user-centric dataset of 10 million GitHub developers. https://doi.org/10.7910/DVN/T6ZRJT
16. Harvard Dataverse. https://dataverse.harvard.edu/. Last accessed 01 May 2023
17. Gong Q, Zhang J, Chen Y, Li Q, Xiao Y, Wang X, Hui P (2019) Detecting malicious accounts in online developer communities using deep learning. In: Proceedings of the 28th ACM international Conference on Information and Knowledge Management (CIKM'19), Beijing, China

Development and Optimization of Automotive Testing Using Machine Vision

Kailash Jagadeesh

Abstract In this paper, a computer vision-based approach for optimizing component test benches in endurance testing of automotive components. As a use case, the paper explores testing of automotive throttle position sensor using fiducial markers and image-processing techniques. The proposed methodology is highly scalable and cost effective while maintaining the required testing accuracy. This research contributes to advancing computer vision techniques in optimizing testing methodologies in bulk manufacturing industries like the automotive industry.

Keywords Industry 4.0 · Computer vision · ArUco · Fiducial markers · Automotive testing

1 Introduction

1.1 Background and Motivation

The need for component testing in the automotive industry is crucial to ensure that vehicles meet certain standards of safety, efficiency, and effectiveness. Quality assurance [1] is critical to ensure that vehicles function correctly and safely before being released to the market. To achieve this, the components are tested for at least a few thousand endurance cycles during their development and production phases. This also requires monitoring and logging all these endurance tests for the traceability and serviceability of the part [2].

The rapid advancement in technology has led to emerging trends and developments which have played a significant role in different aspects of life. The field of automotive engineering is no exception, as the use of machine vision for automotive testing is gaining momentum as a valuable tool. Automotive manufacturers are increasingly using machine vision systems to test vehicle components and systems

K. Jagadeesh (✉)
Mechanical Engineering Department, NIT Trichy, Tiruchirapalli, India
e-mail: kailashjagadeesh2000@gmail.com

S. Bandyopadhyay et al. (eds.), *Intelligent Computing Systems and Applications*, Lecture Notes in Networks and Systems 1010,
https://doi.org/10.1007/978-981-97-5412-0_18

such as brakes, steering, suspension, and engine parts among others. This has revolutionized the field by providing accurate results while reducing testing timeframes within production processes substantially.

Traditionally used methods such as human observation has limited accuracy due to fatigue, subjectivity, and sensory limitations resulting from prolonged hours performing repetitive tasks. Machine vision tackles this by providing an automated process that uses digital cameras to capture images and processors with advanced algorithms to analyze the information [3, 4].

1.2 Problem Statement

One of the most critical components of any two-wheeler vehicle is the drive train system and the input to this system is the input given by the driver to the throttle. In conventional vehicles, throttles are usually linked with the engine carburetor using a cable controlling a butterfly valve [5] which in turn controls the fuel flow to the engine hence controlling the speed of the vehicle. However, modern trends have shifted this throttle into an e-throttle [6] which has a sensor converting the angular movements into electrical signals feeding it to the powertrain system. This e-throttle or throttle position sensor(TPS) (see Fig. 1) is more widely used due to the rise in electric and hybrid vehicles.

The TPS is very critical to ensure the proper operation of the vehicle which otherwise might introduce safety risks for the driver and other road users. Current TPS testing methodology relies on the use of human inspectors and encoder-based motors which are quite expensive. Therefore, there is a need to develop and optimize an automotive testing process using machine vision that can accurately and efficiently test the TPS component while cutting down the cost of these test benches and End Of Line setups (EOL).

Fig. 1 Electronic-throttle

1.3 Objective

The objective of this research paper is to develop and optimize an automotive testing process using machine vision for the throttle position sensor component. This is achieved by benchmarking the current EOL setups and test benches used by throttle manufacturers. This reference bench is used as a baseline to design and implement an efficient machine vision-based testing system. The vision-based testing system consists of a fiducial marker, a monocular camera[1], a laptop to capture and process the video feed, motor actuators (without encoders), and a power supply. The proposed machine vision-based testing system is designed to accurately and efficiently test the TPS component while also reducing the cost of EOL setups. The successful development of this system would enable manufacturers to monitor and evaluate the performance of TPS components throughout the production process, ensuring that only reliable and high-quality components are used in the vehicles.

2 Related Work

2.1 Overview of Machine Vision in the Automotive Industry

As part of Industry 4.0 [7], manufacturing automation has become an important area of research to optimize production processes and reduce costs. Many modern concepts like the Internet of Things (IoT) and artificial intelligence [8] are increasingly used to reduce human labor and increase process efficiency and implement multiple quality checks and implement traceability of production for effective quality control [9]. This has led to the widespread adoption of machine vision systems in automotive and other manufacturing sectors. Villalba-Diez et al. [8] work has introduced the usage of deep neural networks for industrial visual inspection with high-resolution cameras which increases the accuracy and has allowed the process to be fully automated. Zhou et al. [10] have proposed a system that uses multiple light sources and Charge Coupled Device (CCD) cameras which collect the surface details of a vehicle and classify any flaw or defect like dents and scratches to millimeter-level accuracy. Though many implementations are using deep learning techniques, the main issue is the lack of a proper dataset and labeling this dataset is a slow process. In these situations, many researchers have adopted image-processing techniques for better and faster results. El Hachem et al. [11] has implemented image-processing techniques for brick positioning for ceramic monoliths for the production of automotive exhaust systems using the Canny-Hough method. Anitha et al. [12] has presented an automated optical inspection system for identifying defects in the PCB assembly of SMT components. Duchoň et al. [13] implemented an inspection system to identify and verify the presence of various components of an automotive door using image processing.

[1] In the prototype, phone camera is used instead of webcam

(a) ArUco (b) AprilTag (c) ARTag (d) STag

Fig. 2 Types of Fiducial Markers

2.2 Fiducial Markers and Their Usage

Many researchers working on this domain focus on the detection and tracking of that particular component/object which is tricky and difficult due to factors like part-to-part variation, lighting conditions, absorption characteristics of the material, etc. This can be solved by adding a fiducial marker on the component which is relatively easier to track and provides consistent results. A fiducial marker is a simple tag that can be used as a reference object in image-processing techniques. There are many types of fiducial markers based on their shape, pattern, and method of detection. The most common ones are ArUco markers, AprilTag [14], ARTag and STag [15] (Fig. 2). Fiducial markers are used in various applications including robotics and 3D object tracking. Crivellaro et al. has presented a method that localizes an object in 3D space by using ArUco markers as control markers in their algorithm that works in real-time and under challenging lighting conditions. Garrido-Jurado et al. [16] has proposed an Augmented Reality setup where an array of ArUco markers are used to create an occlusion mask which is used for generating augmented characters in real-time. This shows that image-processing techniques can be used to provide the results in real-time as they use relatively fewer computations when compared to deep learning or machine learning-based methods.

3 Experimental Setup

To benchmark a conventional testing bench's performance, a test bench (Fig. 3) with a high torque rhino motor with encoder feedback was fabricated along with fixtures for fixing the TPS unit. This system has a 24V Switch Mode Power Supply (SMPS) Direct Current (DC) supply along with a motor driver with a Universal Asynchronous Receiver-Transmitter (UART) interface which is controlled using a laptop. The optimized test bench is built upon the conventional bench by adding a camera and an ArUco marker sticker stuck on the component side. The conventional bench is re-used to get a comparison and possible error measured between the systems. Many different fiducial markers can be used but for this first attempt, ArUco markers are

(a) (b)

Fig. 3 Test bench used

being used as they are well-supported by the OpenCV community and provide both position and orientation data. For the camera system, a simple webcam is used which is in turn connected to the laptop that controls the motor driver. The computation required was performed on the laptop to keep the setup simple and cost-effective.

4 Methodology

4.1 System Overview

The methodology presented in the paper aims to optimize component test benches for throttle position sensors using cameras and ArUco markers. the proposed approach involves a series of operations, camera calibration, data acquisition and preprocessing, marker detection, pose estimation and integration. The system follows a modular and scalable design which allows it to be commonized for multiple test benches operating in parallel (see Fig. 4).

4.2 Calibration of Camera System

Uncalibrated cameras can often introduce error or noise into the input image fed to the system causing significant distortion in the final detection. Monocular cameras have two types of distortion, namely, barrel distortion and pincushion distortion (see Fig. 5). Camera calibration [17] is performed to negate these distortions by calculating two types of parameters- intrinsic and extrinsic parameters. Intrinsic parameters are composed of the focal length in X and Y axes, principal points and lens distortion coefficients while extrinsic parameters are based on the position and orientation of the camera in the real-world coordinate system. The most common method of monocular camera calibration is by using the chessboard pattern technique. The chessboard pattern consists of a grid of alternating black and white squares on a

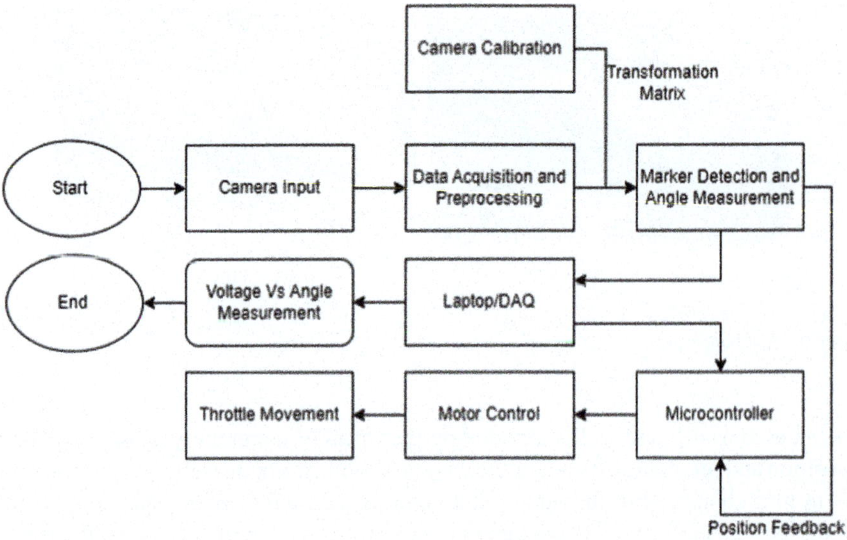

Fig. 4 System flow

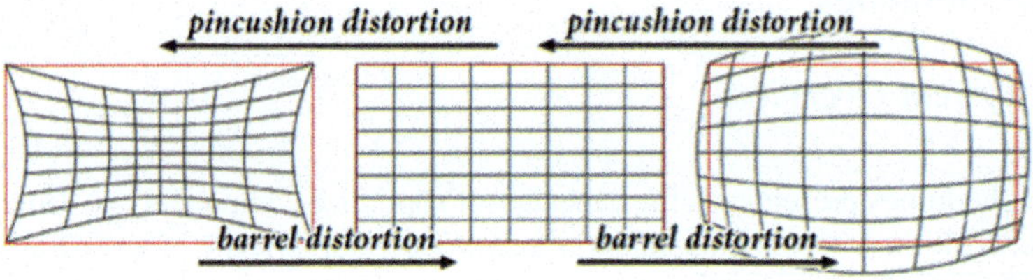

Fig. 5 Types of distortion

white background with a known number of rows and columns and known width for the squares which makes it a suitable pattern due to its sharp gradient change. The calibration process includes :

1. Capturing multiple images of the chessboard pattern from different angles, orientations and distances.
2. Define real-world coordinates by measuring the width of squares on the pattern.
3. Detect the corners of the squares detected in the images.
4. Compute the camera parameters by solving the system of equations using known real-world coordinates.

The equations that transform the 3D point (X_w, Y_w, Z_w) in real-world coordinates to the projection of 2D point (u, v) on the image plane can be represented as:

$$\begin{bmatrix} u' \\ v' \\ z' \end{bmatrix} = \mathbf{P} \begin{bmatrix} X_w \\ Y_w \\ Z_w \\ 1 \end{bmatrix} \tag{1}$$

where,

$$u = \frac{u'}{z'}; \quad v = \frac{v'}{z'} \tag{2}$$

where P is the projection matrix, $K$is the intrinsic parameters matrix and $[R|t]$ represents the extrinsic parameters matrix that includes rotation and translation vectors.

$$\mathbf{P} = \mathbf{K} \times [\mathbf{R|t}] \tag{3}$$

and,

$$\mathbf{K} = \begin{bmatrix} f_x & 0 & c_x \\ 0 & f_y & c_y \\ 0 & 0 & 1 \end{bmatrix} \tag{4}$$

where f_x,f_y are focal lengths in X and Y axes while c_x,c_y are coordinates of optical center in image plane, approximated to center of image. The intrinsic parameter matrix K is computed by solving a system of equations that relate the 3D coordinates of known points to their corresponding 2D image coordinates without any perspective transform. The distortion coefficients of the lens is calculated by:

$$\begin{bmatrix} x_{\text{corrected}} \\ y_{\text{corrected}} \end{bmatrix} = \begin{bmatrix} x(1 + k1 * r^2 + k2 * r^4 + k3 * r^6) \\ y(1 + k1 * r^2 + k2 * r^4 + k3 * r^6) \end{bmatrix} \tag{5}$$

where (x, y) are the coordinates of a point in the distorted image, $r^2 = x^2 + y^2$, and k_1, k_2, and k_3 are the radial distortion coefficients. The extrinsic parameters are the estimated using the detected chessboard corners and the known real-world coordinates using the function:

$$[R|t] = \text{solvePnP(objectpoints, imagepoints, K, distcoeff)} \tag{6}$$

This function estimates the pose of the object using the input object points, corresponding image points along with the intrinsic parameters and distortion coefficients [18].

The points expressed in the world frame (X_w, Y_w, Z_w) are projected into image plane $[u, v]$ using perspective transform model Π and the intrinsic parameter matrix:

$$[u, v, 1] = K\Pi^c T_w \begin{bmatrix} X_w \\ Y_w \\ Z_w \\ 1 \end{bmatrix} \tag{7}$$

(a) Input (b) Grayscale (c) Binary (d) Otsu Filter

Fig. 6 Preprocessing and marker detection

where ${}^{c}T_{w}$ is the extrinsic parameter matrix. Substituting the known values of intrinsic parameter matrix, the equation becomes:

$$\begin{bmatrix} u \\ v \\ 1 \end{bmatrix} = \begin{bmatrix} f_x & 0 & c_x \\ 0 & f_y & c_y \\ 0 & 0 & 1 \end{bmatrix} \begin{bmatrix} 1 & 0 & 0 & 0 \\ 0 & 1 & 0 & 0 \\ 0 & 0 & 1 & 0 \end{bmatrix} \begin{bmatrix} r_{11} & r_{12} & r_{13} & t_x \\ r_{21} & r_{22} & r_{23} & t_y \\ r_{31} & r_{32} & r_{33} & t_z \\ 0 & 0 & 0 & 1 \end{bmatrix} \begin{bmatrix} X_w \\ Y_w \\ Z_w \\ 1 \end{bmatrix} \tag{8}$$

This equation is solved for the values captured by multiple images in iteration thereby solving $[R|t]$.

4.3 Data Collection and Preprocessing

The input image captured by the camera is reduced to a standard size and converted to a grayscale image. After conversion to grayscale, an adaptive thresholding technique with a Gaussian filter and other image smoothening techniques are applied to remove noise and extract a binary image which is used for marker detection(see Fig. 6a–c).

4.4 Generation, Detection and Tracking of Fiducial Markers

The ArUco marker is made of a distinctive pattern used for identification and tracking purposes. It consists of a black border that surrounds an inner matrix with binary patterns. The black border ensures reliable detection even in different lighting conditions, while the binary matrix enables the marker's identification and the use of error detection and correction techniques. The marker size is directly proportional to the size of the inner matrix, for example, a marker with a 4 × 4 marker size has 16 bits. A dictionary of markers is used to match the marker based on binary codification. The marker can be generated by specifying parameters like marker ID, padding size, marker dictionary, and output image size.

Fig. 7 Angle measurement

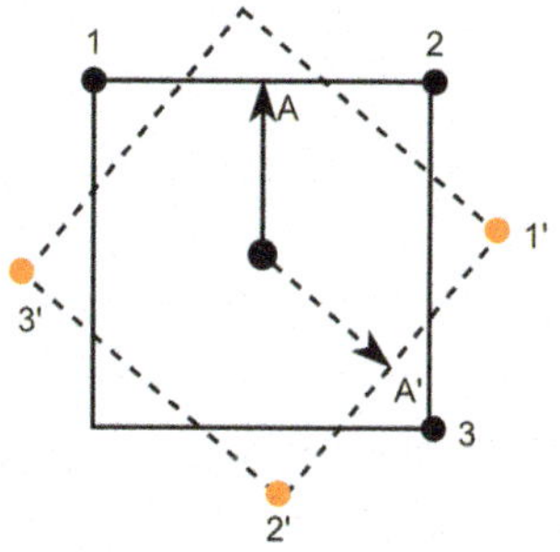

The marker detection is done in steps. First, the possible marker candidates are identified by analyzing the input binary image and calculating its convex hull and contours to detect all the squares present in the image. Some additional filtering ID applied to remove false positives. Next, the identified markers need to be decoded based on their inner codified matrix (see Fig. 6d). For this, every bit is extracted from the marker and identified. To achieve this, perspective transform is performed on the image to get the canonical form. This is followed by Otsu thresholding [19] to separate black and white bits. Based on the comparison of these bits with the dictionary, the marker is either identified or discarded. [image needed] The corners of the marker are finally detected using the Harris corner detection [20] and with the intrinsic and extrinsic parameter matrices, we can identify the 3D pose of the marker. This again involves solving the PnP problem (Perspective-n-Point) with these known parameters and comparing with the axes, the angular rotation of the marker is extracted. Once the corners are identified in order, we can calculate an imaginary axis that represents 0 deg orientation and the current orientation is the angle between this axis and the vector connecting the marker's center and top center which is between the first and second point of the marker. This can be calculated as follows:

Considering the rotation of ArUco marker as shown in Fig. 7, to find the angular rotation, the center of the marker and top center of the marker(center between corner 1 and 2) are computed and the slope of the vector connecting these two corners onto the reference plane is calculated which would be the angular rotation of the marker. The equations are shown in Eq. 9.

$$\theta_{\mathrm{ref}} = \tan^{-1}\left(\frac{Y_A - Y_C}{X_A - X_C}\right); \quad \theta_{\mathrm{moved}} = \tan^{-1}\left(\frac{Y_{A'} - Y_C}{X_{A'} - X_C}\right); \quad \theta_{\mathrm{final}} = \theta_{\mathrm{moved}} - \theta_{\mathrm{ref}} \tag{9}$$

4.5 *Integration with TPS*

The angular rotation of the ArUco marker that is measured is used as the feedback for the motor control system which controls the position of the throttle connected to the motor shaft thereby allowing position control of the system. Along with this, the throttle position sensor is provided with the input supply and the output is connected to the Analog to Digital Converter (ADC) of the microcontroller which acts as the input to the control algorithm which is in turn connected to the laptop which is the data acquisition unit of this system. This allows us to capture the voltage versus angle data for the TPS over the endurance tests consisting of more than one lakh test cycles. The laptop is also used to detect when the TPS output is beyond tolerance and measure the exact moment when the TPS unit failed.

5 Results and Discussion

5.1 *Evaluation of Fiducial Marker Detection and Tracking*

The ArUco marker detection algorithm and its feedback to the motor control system are capable of measuring angular position with 0.3 deg accuracy which is less than the current step measured by the manufacturers, which is 0.5 deg. Though the traditional setup which consists of a motor encoder can measure much more accurately, it is not needed for this testing methodology. Also due to this high resolution, the control loop needed to perform position control requires more tolerance in value comparison as the motor driver tries to reach the exact position causing jitter in the system and a measurement delay as well.

In Fig. 8, the output of the image-processing techniques is shown. Figure 8a shows the detected angular movement of the TPS as 2.73 deg when the motor was given the command to rotate 2.75 deg thereby demonstrating measurement with less than 0.05 deg of error. Figure 8b–d shows the algorithm's measurement when the throttle was continuously moving from 0 deg to 100 deg by the motor where the measured values are 40.82 deg, 61.70 deg, and 91.85 deg, respectively.

5.2 *Cost Comparison*

When both systems are compared in terms of cost, the camera-based system using ArUco markers is more cost-effective compared to traditional methods involving motor encoders. The bill of materials for both versions of test benches are shown

(a) (b) (c) (d)

Fig. 8 Results of machine vision system

in Table 1. On the cost comparison of one test bench we can see that we have achieved a cost reduction of 54% when compared to traditional testbench. One of the main advantages of this system is its scalability for multiple parallel testing. This is achieved by replacing the camera with a higher resolution camera and using multiple jigs which are arranged within the frame size of the camera. Assuming it is scaled up to 10 benches, we can see the cost comparison between the 2 systems from Table 2.

From the above table, we can see that upon scaling up for parallel testing, we can have over 65% of savings in the test bench cost by commonizing the camera system, microcontroller, and laptop/Data Acquisition(DAQ) system for the parallel test benches.

Table 1 Bill of materials

Traditional test bench		Optimized test bench	
Components	Cost	Components	Cost
24VDC Motor with encoder	Rs. 3000	12VDC motor	Rs. 500
High precision motor driver	Rs. 2000	H Bridge driver	Rs. 300
24V SMPS	Rs. 800	12 V Power adapter	Rs. 250
Microcontroller	Rs. 500	Microcontroller	Rs. 500
Jig fixture	Rs. 1000	Jig fixture	Rs. 1000
Laptop/DAQ system		Laptop/DAQ system	
		Camera	800
Total cost:	Rs. 7300	Total Cost:	Rs. 3350

Table 2 Comparison of scalability for 10 benches

Traditional test bench		Optimized test bench	
Components	Count	Components	Count
24VDC Motor with encoder	10 Nos	12VDC Motor	10 Nos
High precision motor driver	10 Nos	H Bridge driver	10 Nos
24V SMPS	10 Nos	12V Power Adapter	10 Nos
Microcontroller	10 No	Microcontroller	1 No
Jig fixture	10 Nos	Jig fixture	10 Nos
Laptop/DAQ system	10 Nos	Laptop/DAQ system	1 No
		Camera	1 No
Total cost: (based on Table 1)	Rs. 73,000	Total cost:	Rs. 24,000

6 Conclusion and Future Work

This paper has explored the possibility of optimizing conventional testing methodologies used in automotive industries with the concept of machine vision and Industry X.0. It has been shown that the use of machine vision in automotive testing can significantly improve accuracy, consistency and efficiency while reducing cost when compared to traditional methods. With a few more improvements, this proposal can be adopted in the industry as a viable alternative to conventional automotive testing methods. Due to its scalable and modular nature, this methodology can also be ported to any other component testing with a few minor modifications. Furthermore, this optimized testing methodology can be integrated with other Industry X.0 technologies to create a powerful framework for next-gen automotive manufacturing plants.

The first version of this test bench is promising, and there is a lot of potential for future enhancements. One potential area is to research tinyML models that are capable of running on microcontrollers and use very less compute yet have minimal latency. This would allow us to use a lower-resolution camera yet generate equally good or better results. Another potential area of improvement can be done on the type of fiducial marker used. In this version, the ArUco marker which is one of the most common fiducial markers was explored. However, there are a few advanced markers like ARTag, and AprilTag which are more robust in pose estimation and more resilient to dynamic lighting conditions. Adding suitable noise removal and signal filtering techniques like the application of the Kalman filter can also be explored to see the improvement in results, if any.

References

1. Magdalena M (2019) Quality assurance processes in series production of car elements. Conf Qual Prod Improv CQPI 1(1):610–617
2. Beckert SF, Paim WS (2017) Critical analysis of the acceptance criteria used in measurement systems evaluation. Int J Metrol Qual Eng 8:23
3. Ramshankar Y, Deivanathan R (2018) Development of machine vision system for automatic inspection of vehicle identification number. Int J Eng Manuf (IJEM) 8(2):21–32
4. Senthilkumar KS et al (2017) Low cost machine vision system for sorting of mechanical parts using MATLAB and ARDUINO embedded hardware. Int J Recent Trends Eng Res. Conference on Electronics, Information and Communication Systems (CELICS'17) Special Issue 11:1–2
5. Tsuchiya K et al (1980) Emission control of two-stroke motorcycle engines by the butterfly exhaust valve. SAE Trans 3056–3069
6. Kreß J, Rau J, Hebert H, Perez-Peña F, Schmidt K, Morgado-Estevez A (2023) Low-cost throttle-by-wire-system architecture for two-wheeler vehicles
7. Albers A, Gladysz B, Pinner T, Butenko V, Stürmlinger T (2016) Procedure for defining the system of objectives in the initial phase of an Industry 4.0 project focusing on intelligent quality control systems. Procedia CIRP 52:262–267. ISSN 2212-8271
8. Villalba-Diez J, Schmidt D, Gevers R, Ordieres-Meré J, Buchwitz M, Wellbrock W (2019) Deep learning for industrial computer Vision Quality Control in the Printing Industry 4.0. Sensors 19, 3987. https://doi.org/10.3390/s19183987

9. Dutta G, Kumar R, Sindhwani R et al (2021) Digitalization priorities of quality control processes for SMEs: a conceptual study in perspective of Industry 4.0 adoption. J Intell Manuf 32:1679–1698. https://doi.org/10.1007/s10845-021-01783-2
10. Zhou Q, Chen R, Huang B, Liu C, Yu J, Yu X (2019) An automatic surface defect inspection system for automobiles using machine vision methods. Sensors 19:644. https://doi.org/10.3390/s19030644
11. El Hachem C, Santiago R, Painvin L, Perrot G, Couturier R (2022) Brick orientation adjustment in the automotive industry using image processing techniques. In: 2022 8th International conference on control, decision and information technologies (CoDIT), Istanbul, Turkey, pp 729-733. https://doi.org/10.1109/CoDIT55151.2022.9803927
12. Anitha DB, Rao M (2020) Assembled PCB SMT component inspection. https://doi.org/10.35940/ijitee.L3422.1081219
13. Duchoň F et al (2016) Detection of door's components in automotive industry by simple image processing. In: ELEKTRO, Strbske Pleso, Slovakia, pp 137–142. https://doi.org/10.1109/ELEKTRO.2016.7512052
14. Olson E (2011) AprilTag: a robust and flexible visual fiducial system. In: 2011 IEEE International conference on robotics and automation, Shanghai, China, pp 3400–3407. https://doi.org/10.1109/ICRA.2011.5979561
15. Kalaitzakis M, Carroll S, Ambrosi A, Whitehead C, Vitzilaios N (2020) Experimental comparison of fiducial markers for pose estimation. In: 2020 International conference on unmanned aircraft systems (ICUAS), Athens, Greece, pp. 781–789. https://doi.org/10.1109/ICUAS48674.2020.9213977
16. Garrido-Jurado S, Muñoz-Salinas R, Madrid-Cuevas FJ, Marín-Jiménez MJ (2014) Automatic generation and detection of highly reliable fiducial markers under occlusion. Pattern Recogn 47(6):2280–2292. ISSN 0031-3203. https://doi.org/10.1016/j.patcog.2014.01.005
17. Zhang Z (1999) Flexible camera calibration by viewing a plane from unknown orientations. In: Proceedings of the Seventh IEEE International conference on computer vision, Kerkyra, Greece, vol 1, pp 666–673. https://doi.org/10.1109/ICCV.1999.791289
18. Perspective-N-point (PNP) pose computation. https://docs.opencv.org/4.x/d5/d1f/calib3d_solvePnP.html. Last accessed 14 Jun 2023
19. Otsu's method for image thresholding explained and implemented. https://muthu.co/otsus-method-for-image-thresholding-explained-and-implemented/. Last accessed 13 June 2023
20. Ye Z, Pei Y, Shi J (2009) An improved algorithm for Harris corner detection. In: 2nd International Congress on Image and Signal Processing. Tianjin, China, pp 1–4. https://doi.org/10.1109/CISP.2009.5304635

Analysis of Techniques and Methods for Health Informatics: A Quantum Leap in Health Care

B. Akoramurthy and B. Surendiran

Abstract If implemented properly, quantum blockchain has the complex strength to improve health informatics and aid in the fight against pandemic situations. The value of blockchain is examined here, and it is concluded that in the near future, blockchain-enabled quantum and its associated trials will be applied in health information networks aimed at gathering information from sensors, performing automatic patient monitoring, and storing information in a safe and reliable manner. The ability to store a large amount of data in a decentralized and secure manner while also allowing access at any time it is needed is a major benefit of this technology that greatly facilitates the ease with which activities may be carried out. Upon the application of quantum blockchain (QUAB), all of the advantages of probabilistic science may be utilized, including the capacity to gain non-visible imaging based on probabilistic science and the rapidity with which patients can be discovered and tracked. Data records may be kept private, legitimate, and easily accessible with the use of tools like quantum blockchain. Applying hybrid QUAB technologies to the processing of healthcare data has the prospect of increasing both processing speed and confidentiality. In this study, we explore the potential uses and advantages of QUAB in the medical, pharmaceutical, and health informatics infrastructure sectors. It is our hope that this article will serve as a resource for researchers of all levels working in the fields of QUAB and health care, allowing them to better visualize the breadth of ongoing study, identify potential pitfalls and opportunities, and make educated choices as they design fresh structures as well as applications for QUAB in medicine.

Keywords Health Informatics · Quantum Blockchain · Medical Images · Quantum Dots · Probabilistic computing

B. Akoramurthy (✉) · B. Surendiran
Department of CSE, National Institute of Technology Puducherry, Karaikal, India
e-mail: akor.theanchor@gmail.com

B. Surendiran
e-mail: surendiran@nitpy.ac.in

S. Bandyopadhyay et al. (eds.), *Intelligent Computing Systems and Applications*,
Lecture Notes in Networks and Systems 1010,
https://doi.org/10.1007/978-981-97-5412-0_19

1 Introduction

1.1 Health Informatics

The field of health informatics encompasses a wide range of topics, including the interplay between individuals, groups, diseases, and medical interventions. It is the study of how to best organize, share, and apply biological information and understanding for resolving issues and coming up with decisions. Contemporary technological advances, especially those pertaining to computers and communication, play an essential role in the subject, which encompasses both the theoretical and practical aspects of biomedical science. The field of health informatics investigates methods to improve medical care and health promotion by optimizing the development, dissemination, and use of clinical information. Significant advances in information and communication technologies, the realization that the body of knowledge of biomedical research is basically insurmountable through conventional printed techniques, and the increasing belief that the method of making educated choices is as essential to contemporary biomedical research as the accumulation of information on which therapeutic choices or studies are made, all contributed to the development of healthcare informatics as a new discipline.

1.2 Overview of Techniques and Methods in Health Informatics

Increasing amounts of healthcare data from sources like hospitals, clinics, patients, health insurers, and drug manufacturers have become accessible because of advancements in technology. This opens up a new window of potential for data science tools for enhancing medical care based on insights gleaned from collected data. Nevertheless, is challenging to get reliable findings across communities due to healthcare data's fragmented and private nature. For instance, due to the delicate nature of medical information, it can be challenging to communicate electronically stored medical records (EHR) between clinics that treat patients from various communities. A significant obstacle is thus erected to the development of productive, universal statistical methods, which necessitate various, "big data." Federated learning shows significant potential to integrate disparate medical sources of information while protecting individual privacy. This is accomplished by training an integrated universal framework on a centralized server while maintaining all personally identifiable information in the communities wherein the records originate. Table 1 exhibits the current research work on health informatics based federated learning.

1.3 Probabilistic or Non-deterministic or Quantum Computing

Grounded on the strange occurrences of quantum chemistry and physics, probabilistic, or non-deterministic or quantum technology (QT) stands an exciting new direction in computing. It is an impressive combination of many fields including math, physics, computing, and computational modeling. By controlling the actions of subatomic particles like atoms, electrons, and photons, it can do calculations at an exponentially faster rate than conventional computers while using much less energy. A broader structure of computation, quantum computing, has been developed thanks to quantum theory, which can solve issues that conventional computers couldn't. Quantum computers (QC) maintain and analyze information using quantum bits (or "Qubits") rather than the traditional binary bits 0 and 1 used by traditional computers. "Quantum computers" refer to devices capable of doing quantum computations. Even the most fantastic supercomputer we have at present would take hundreds of years to crack current cryptography methods, but QC could do it in minutes. It is expected that while QC will be able to crack a number of current encryption methods, they will also create secure alternatives. Miniaturized computers would not be able to make use of transistors, logic gates, or integrated circuits. Therefore, the rotation and state information of particles such as protons electrons and neutrons, and ions are utilized as qubits. It is possible to create new combinations by stacking them. This means they can potentially boost their power by running in parallel and making good use of memory. Unlike every other computing paradigm, QC can make full use of the hardware at its disposal since it contradicts the Church-Turing thesis.

In quantum theory, the fundamental unit is called a quantum bit (abbreviated as "qubit"), and it represents computer memory and processors in the form of elementary particles like atoms, electrons, and other subatomic particles. It is not limited to the numbers 0 or 1, but can even be both. It can perform a million times more calculations per second than the most powerful computers available today. Qubit production and management is a demanding engineering problem. The digital and analog components of the quantum computer are the source of its superior computational abilities. Due to their analog nature, quantum gates have no distortion limit, however the digital nature of gates provides a benchmark for overcoming this critical shortcoming. Thus, the mathematical models and logic gates of conventional computing are ineffective in quantum computing. Quantum computing can employ just traditional computer techniques. This calculation, however, calls for a novel approach to eliminate processing variations and noise. The same methodology for fixing bugs and fixing design faults is necessary.

In quantum theory, the fundamental unit is called a quantum bit (abbreviated as "qubit"), and it represents computer memory and processors in the form of elementary particles like atoms, electrons, and other subatomic particles. It is not limited to the numbers 0 or 1, but can even be both. It can perform a million times more calculations per second than the most powerful computers available today. Qubit production and management is a demanding engineering problem. The digital and

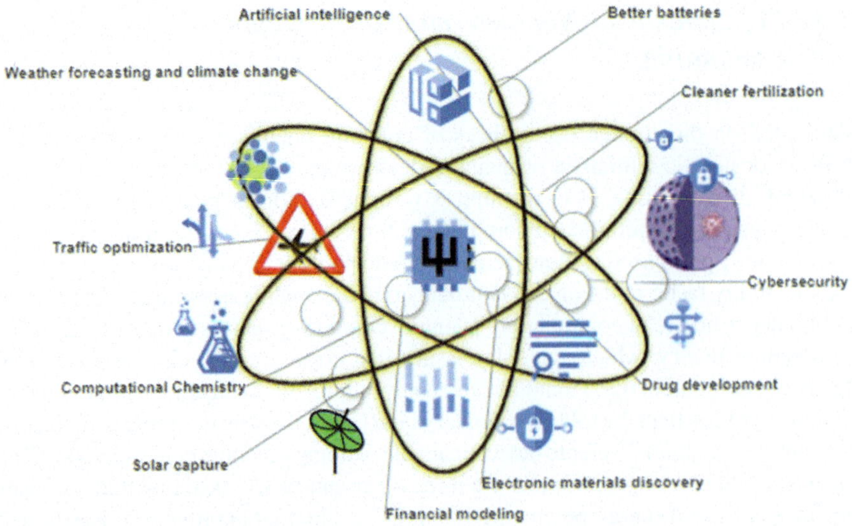

Fig. 1 Quantum computing applications

analog components of the quantum computer are the source of its superior computational abilities. Because of their analog nature, quantum gates lack a distortion limit, however, the digital nature of gates provides a benchmark for overcoming this critical shortcoming. Thus, the mathematical models and logic gates of conventional computing are ineffective in quantum computing. Quantum computing can employ just traditional computer techniques. This calculation, however, calls for a novel approach to eliminate processing variations and noise. The same methodology for fixing bugs and fixing design faults is necessary. Figure 1 exhibits the various possible and potential applications of quantum computing.

1.4 Introduction to Quantum Blockchain

When an event occurs in the digital world, a copy of it is stored on the blockchain and made available to everyone. All deals are guaranteed by the presence of several parties. It stores all financial dealings ever made. Blockchain, the game-changing technology that is disrupting many industries, made its debut in the market in the shape of Bitcoin, its most sophisticated iteration, under unexplained circumstances. In some contexts, virtual currencies like Bitcoin can function as an alternative to fiat currency. Bitcoins have seen explosive growth thanks to blockchain, a cutting-edge technology. In accordance with QC and quantum computation theory, quantum blockchain (QUAB) is inherently secure, dispersed, and decentralized.

Once entered into the QUAB, the data will not be changed in any way. Academic interest in QUAB research has increased in recent years as the fields of probabilistic computing and quantum information technology have matured. QUAB has begun an extensive program of study, progress, and outlay in the cutting-edge field of Blockchain Technology, which includes Bitcoin digging and other complex decentralized applications. Quantum's R&D will emphasize on cryptography, providing a fresh take on blockchain by fusing cutting-edge methods of implementation and functionalities with quantum computing and AI deep learning.

In May 2023, the market value of the cryptocurrency industry surpassed $1.14 trillion. Regardless of what occurs, bitcoins will become increasingly important in the global financial system. The most difficult part of digital currency is ensuring everyone uses it properly. In addition, blockchain technology looks to be a viable option. Utilizing cryptographic techniques widely regarded as unbreakable, even by brute force attacks, assures trustworthiness. The advantages of QUAB beyond regular blockchains are briefly explored in this work [12], which also analyses current developments in the field. We talk about the inside workings and design of the QUAB. The next step is to demonstrate the use of quantum technologies to a subset of the blockchain.

The advantages of QUAB over the standard blockchain and its endless possibilities for growth are also explored. According to the findings of this research [12], temporal entanglement, and not only spatial entanglement, provides a significant quantum advantage. Each component of the system has been successfully realized in experiments. Furthermore, our programming method may have non-classical effects on the past. Future-focused analyses of the underlying challenges, risks, and rewards of implementing these technologies are provided in the article which is based on UAV [13]. The research we conducted concludes with a discussion of the major unanswered questions, technical concerns, and potential futures for the discipline.

Blockchains are a type of distributed hyper ledger technology that is protected against tampering by cryptography. Blockchain-based systems have tremendous potential, but they are vulnerable to attacks from quantum computers since they depend on electronic certificates. Similarly, albeit to a smaller extent, cryptographic techniques employed to produce new blocks may be vulnerable to quantum computing, suggesting that parties with the ability to perform quantum computing might have an unfair advantage in the race for the benefits of mining. For the blockchain conundrum of the quantum era, authors [14] offer an experimental prototype of a post-quantum cryptographic blockchain platform for intellectual recognition of individuals that utilizes quantum cryptography across an inner-city fiber grid.

This article [15], discusses the current scientific advances in quantum-resistant cryptography, all of that is vital for quantum relationships; therefore, we additionally address the advantages, limitations, potential future advancements, and research issues associated with quantum technologies as applied to unmanned aerial vehicles. As part of this investigation, we are also developing a classification scheme for fields associated with quantum studies according to the reasoning behind our understanding and investigation of these fields.

In this research work [16], the authors take an overall look at distributed ledger technologies. Most widely used encryption methods are vulnerable to the advent of developed non-deterministic systems, whose security relies on the predictability of computing outcomes, as in the current systems. The research's scientists [17] presented a novel negotiation procedure for determining a block's legitimacy and allocating a genesis block to the blockchain network. To accomplish block validation and distribution, mediation procedures that consider an enhanced likelihood context are utilized. In this research [18], the authors propose a novel QB system as a means of strengthening blockchain security. The authors started by explaining what QuickBase was and how it came to be. In this work, we focus on the advantages. And second, we used the no-broadcast theorem to the development of a new cryptocurrency we named quantum coin. In addition, we utilize quantum entanglement and DPoS to develop a novel QUAB approach for Health Informatics.

1.5 Quantum Blockchain Different Elegances of Framework

The unification of modules that can be exploited in constructing quantum blockchain-based apps is made possible by the structure of a quantum blockchain (QUAB).

Using this framework, we may build domain-dependent features in a way that is dependable, tractable, flexible, and reasonably priced. The attributes of openness, reliability, accessibility, privacy, and efficiency must all be met by the framework.

1.5.1 Computable Components

Quantum blockchain computation is made up of fundamental building blocks called qubits and quantum gates, which are described in detail in the computing components.

Qubits

Quantum blockchain relies on blocks, data, and dates, times that are archived in a distributed network and processed using quantum techniques. In QUAB, qubits are used to store information.

Photons are another name for qubits. Light and radio waves emit electromagnetic radiation known as photons. A block is coded using superdense encoding. This technique is used to transmit a message consisting of two bits (00, 01, 10, or 11) using a single qubit.

Let's say User A wants to tell User B something, and they choose to convey it as two entangled qubits. The entanglement between the two qubits is set to a specific Bell state when User A applies a quantum gate to its qubit. The Bell state is intertwined and has other names as well. Two photons can be measured in their entangled quantum states using the Bell states [19]. Based on the spin of the qubit, there are four alternative Bell states (00, 01, 10, 11), each of which corresponds to a different pair of bits of information that User A must transmit to User B. One qubit is delivered from user A to user B, and the two-bit message is decoded by measuring the Bell state between the two qubits.

Using Quantum Gates for a Computational Purpose

One bit of logic is used to process data in traditional computers. It specifies a truth table in which the states 0 and 1 are interchangeable by defining them as 0 to 1 and 1 to 0. Primary gates such as AND, OR, and NOT are replaced by primary quantum gates in a quantum circuit (quantum array or quantum network). The one qubit information is transmitted via the phase flip gate Z, the bit flip gate X, and the Hadamard gate H.

The controlled-NOT (CNOT) gate operates on two qubits. If the first two bits are 1, then the third bit will be negated by the Toffoli gate (CCNOT controlled-controlled not) because it is a three-qubit gate. Using auxiliary wires and assigning values in accordance with the classical Boolean circuit, these CCNOT circuit gates can be simulated for any classical calculation. However, due to the nature of quantum measurement, only one bit of data can be extracted from each state transition recorded by the qubits. There are three types of algorithms—classical, quantum search, and quantum simulation—that can handle quantum computing.

In order to compute, it is necessary to use various models, each of which requires its own unique set of resources. Asymptotic notation is created to help comprehend an issue with the model and the prerequisites. A function's behavior in a problem can be summed up using asymptotic notation. If we are curious about how many gates are required to add two n-bit values, we won't be able to get an accurate count. The number of gates needed to carry out this algorithm is $24n + 2[\log n] + 16$. The algorithm's essential behavior is linear in the number of function calls.

1.6 *Quantum Blockchain and Its Importance in Health Informatics*

Medical insurance schemes and electronically stored medical information can help mankind lower the high intricacy and charge of the bionetwork. Information subjects have a legal right to be conversant about the storage and processing of individual private data under the umbrella of data protection regulation. The World Wide Web, however, is an unsecured route via which medical information passes, making it vulnerable to hacking attempts that might result in an attack on private information or a modification of the information stored, among other harms. Therefore, conventional healthcare systems have challenges in ensuring the safety and privacy of those involved [18]. As a service, blockchain has matured to improve the efficacy of the medical system of today without compromising the confidentiality or integrity of any participants. This paper [20] is motivated by these results as the contributors investigate a few security designs employed to protect medical information and add quality control to the standard encryption method. Next, we provide a blockchain-based health informatics system that restricts access to database records according to the user's declared duties.

As the state of medicine improves, the use of EMRs to increase efficiency and reliability in patient care has become standard practice. However, electronic medical records (EMRs) are stored in silos among hospitals and clinics [21], which creates difficulties when attempting to share data. Furthermore, especially delicate EMRs pose security and confidentiality problems because of their openness to manipulation and exploitation. The current developments in health care 5 Beyond, which employ Internet of Things (IoT) components [13] and cloud computing services to provide distant control over medical operations, captured the interest of the investigators from a philosophical city perspective. One of the main tenets of health care 5 Beyond was the regular monitoring, consolidation, transfer, exchange, and management of medical data. There exists a sum of encounters allied with keeping hackers out of patients' private information. Therefore, additional security measures are required to guarantee that only specified user gears of digital health systems have access to patient health information when it is being stored, retrieved, or sent in the cloud. To guarantee the well-being of sensitive medical data during its time on the cloud, a number of cryptographic methods have been developed. The analysis of relevant work in the field of QUAB is provided in Table 2.

Table 1 Synopsis of current research on health informatics federation-based learning

References	Challenge handled	Methods used
[1]	Privacy-preserving patient similarity	Homomorphic encryption
[2]	Patient similarity network	Federated data quality profiling
[3]	Mortality	LoAdaBoost
[4]	Malicious or faulty patients	Adaptive federated averaging
[5]	Personalized clinical decisions	Double deep Q-network
[6]	Medical data privacy and security	Split learning
[7]	Phenotype	Alternating direction method of multipliers
[8]	Patient information extraction	Decentralized FL approach
[9]	Representation learning for distributed biomedical data	Federated PCA
[10]	Binary supervised classification problem for predicting hospitalization	Cluster primal–dual splitting SVM
[11]	Diagnosis classification	Byzantine-robust aggregation techniques

Table 2 Evaluation of similar works on quantum blockchain

References	α	β	Υ	χ	ω	ψ	∂	ϵ	π	μ
[61]	✔	✖	✖	✖	✖	✔	✔	✖	✖	✖
[22]	✔	✖	✖	✖	✖	✖	✖	✖	✔	✔
[23]	✔	✔	✔	✔	✖	✖	✔	✖	✔	✖
[24]	✖	✔	✖	✔	✖	✖	✖	✔	✔	✖
[61]	✔	✔	✖	✖	✖	✖	✖	✖	✔	✔
[25]	✔	✖	✔	✔	✖	✔	✖	✔	✖	✖
[26]	✔	✖	✖	✖	✖	✔	✔	✖	✖	✖

α: blockchain, β: Computing Paradigms, Υ: Cybersecurity, χ: Data Management, ω: Financial Technology, ψ: Post-Quantum Cryptography, ∂: Quantum Blockchain ϵ: Quantum Resistance, π: Smart Society, μ: Tokens

2 Literature Survey

Privacy data includes not only patients' health but also financial status [27]. While examining the way that the real information is transmitted aspects, it is important to confirm that the freshly created database handles personal data properly and protects the confidentiality of patients throughout the dissemination of information [28]. Alkatheiri and Alghamdi [29] proposed a Blockchain-Assisted Cybersecurity (BCCS) for IoMT is created BCCS collects health data from the IoMT in conjunction with devices using blockchain and a traditional deep manner. Blockchain-enabled medical data stored in the cloud and exchange solution by [30]. It secures records by hashing them. The system in question fails to confirm user legitimacy, therefore a patient or doctor can regularly store phony records. The algorithm produces and maintains hash values for all data types, including photos, text, and integers.

Numerous blockchain-based approaches for health informatics data analysis, archiving, administration, and reformation have been presented [31–35].

In this [36], the researchers discussed protecting patient information and the integrity of the electronic health records system's infrastructure from prying eyes. Using blockchain technology and the decentralized IPFS file system, they proposed a smartphone-based cloud service for the sharing of medical data. The outcomes of their testing deployments are promising, and they offer a practical means of protecting data transmissions on portable infrastructures. Improvements in lightweight control of access, latency within the network, and data safeguarding with high confidentiality and security standards were uncovered during the security evaluation of the concept. A hypothetical architecture for a probabilistic blockchain based on spatial proximity was proposed in [37, 38].

3 Examining Current Quantum Computing-Based Health Informatics Methods for Competitive Effectiveness

Advantages of QUAB in Health Informatics

When applied to health informatics, probabilistic or quantum, or non-deterministic computing creates the ideal scenario to (i) make the system more resistant to simultaneous bouts, (ii) increase the speed at which data can be processed and made available to authorized users, (iii) make the healthcare system safer and more beneficial for all parties involved, (iv) make possible unified health informatics system that improves a management and system a patient-focused and eliminates information silo, and (v) be implemented in a matter of months (Table 3).

Disadvantages of QUAB in Health Informatics

One of the primary obstacles to using quantum computing in health informatics is that (i) widespread use may have unintended consequences for the natural world. (ii) It is difficult, particularly in underdeveloped countries, to organize non-deterministic computing resources for use in medical delivery in an everyday setting. Providing the necessary infrastructure for the immediate utilization of probabilistic or quantum or non-deterministic computing (particularly in medical or health informatics) is a complex and challenging endeavor. It's a somewhat pricey fix, too. Despite the fact that computing tools (such as Qiskit, Qism, Bluequbit, and silq) exist for low-volume solutions. More tools-supporting computers are needed to enable widespread integration.

4 Open Research Areas of Quantum Health Informatics

The many outstanding questions of quantum computing in health informatics are addressed here. We provide a categorization of these difficulties, an explanation for why they persist, and suggestions for where future research may go to find answers (Fig. 2).

5 Application of QC in Health Care

New investigations reveal that quantum computers outperform traditional ones in almost every measurable way. Malady early detection and therapy can be sped up with the use of quantum computing, which can substantially cut calculation timescales from decades to seconds [30]. It inspires novel approaches to realize advanced proficiency in certain jobs, as well as novel architectural designs and operational approaches. Thus, medical professionals have a great deal of opportunity to put quantum technology into action in an array of contexts, particularly in the

Table 3 Lists current and relevant quantum computing integrated medical care methods

Research work	A	B	C	D	E	F	G	H	I	Method used	The way forward
[39]	✔	✖		✔	✔	✔	✔	✔	✔	Concerns of quantum technology related to the protection of patient records are examined. Data encryption during the analysis of medical records is the topic at hand. Data security for medical institutions that employ quantum methods is also considered	Quantum computing-based methods for data security may be investigated at each of the three stages (Analysis, dissemination, and archiving) of the data lifecycle
[40]	✖	✔	✖	✖	✔	✖	✔	✔	✔	Quantum features associated with safety that can be managed are examined	The characteristics of other cryptographic fundamentals and methods can be addressed
[41]	✖	✔	✖	✖	✔	✔	✔	✔	✔	Quantum-supported cloud medical technology and its ecosystems are examined	It is possible to construct quant-based hardware as well as software design to safeguard medical cloud services
[42]	✔	✖	✖	✖	✖	✖	✔	✔	✖	Current and potential future uses of blockchain technology are examined	The scope of this study may be expanded to incorporate quantum technology powered by blockchain systems for the protection of emerging technologies
[43]	✔	✖	✖	✖	✖	✖	✔	✔	✖	This research demonstrates the immediate and long-term importance of blockchain and mobile apps in the healthcare sector. In this study, mobile gadgets are broken down into their component parts	Extending this study, we may examine the value of quantum technology for healthcare applications and data on handheld gadgets with limited resources

(continued)

Table 3 (continued)

Research work	A	B	C	D	E	F	G	H	I	Method used	The way forward
[44]	✖	✔	✖	✖	✔	✔	✔	✔	✖	Quantum technology and the protection of COVID-19 medical information and files are the topics of this article	Exploring a thorough architecture over the physical definition of quantum computing for medicine is possible
[45]	✖	✔	✖	✖	✔	✔	✔	✔	✖	Primitive concepts and algorithms from post-quantum cryptography are addressed for use in protecting medical records	In order to prevent assaults on conventional security measures, post-quantum encryption may be investigated for its potential application in the healthcare sector
[46]	✔	✖	✔	✖	✔	✖	✔	✖	✖	The safety of blockchain with non-fungible tokens (NFT) is discussed	The value of tokens that incorporate probabilistic computing may be further explained in the context of health care and other fields
[47]	✔	✔	✖	✖	✔	✖	✖	✖	✔	A comprehensive blockchain architecture for the healthcare sector is described	It is possible to investigate the viability of enhancing the adaptability of a distributed ledger system for health care
[44]	✖	✔	✖	✔	✔	✖	✔	✖	✔	Video conferencing encryption that makes use of quantum computers is addressed	In the future, this study can be expanded to cover situations in which physicians and patients utilize this capability to communicate confidential health information and related treatment plans

(continued)

Table 3 (continued)

Research work	A	B	C	D	E	F	G	H	I	Method used	The way forward
[48]	✖	✔	✖	✔	✔	✖	✔	✖	✔	Various situations in which digital technology may be useful in the pharmacy are described	Possible outcomes can be examined (mathematically or experimentally)
[45]	✖	✔	✖	✔	✔	✖	✔	✖	✔	The role of quantum technology for healthcare data analysis via ML/DL methods is discussed	Additional health information or data may be examined in this way to evaluate the effectiveness using quantum computing and to compare the results to those obtained using traditional computing methods
[45]	✖	✔	✖	✖	✔	✖	✔	✖	✔	This paper aim is to create a blockchain-based healthcare system that can safely store and provide access to patient records. In the field of security, this is accomplished through a searchable-encryption process	The widespread adoption of distributed ledger networks may be improved by further studying and assessing their scaling, efficiency, and safety
[43]	✔	✔	✖	✖	✔	✖	✔	✔	✔	Protecting patient health records in a connected city: a case study	The potential for incorporating rapid computing (including quantum processing) can be investigated

(continued)

Table 3 (continued)

Research work	A	B	C	D	E	F	G	H	I	Method used	The way forward
[49]	✖	✔	✖	✖	✔	✖	✔	✖	✔	It examines the beneficial effects of integrating medical care, quantum technology, and cloud services, and it discusses these topics in depth	The scope of this study may be broadened to incorporate security basics and procedures that combine three different methods of protection
[50]	✖	✔	✖	✖	✔	✖	✔	✖	✔	In this paper, the authors recommend employing blockchain technology powered by quantum computers to efficiently authenticate IoVs	In a formal setting, we may test the robustness of the authentication system against a variety of assaults
[51]	✖	✔	✖	✖	✔	✖	✔	✖	✔	Quantum computing's merits and downsides are weighed in light of its potential role in the smart healthcare system	An in-depth study based on tools and procedures may be included to help find and create optimal healthcare situations
[20]	✔	✖	✖	✔	✔	✔	✔	✖	✖	Emphasized blockchain and cryptography. Digital signature strength is highlighted here	Quantum gates and circuitry for cryptographic basics and blockchain protocols in actual time may be designed using this study

(continued)

Table 3 (continued)

Research work	A	B	C	D	E	F	G	H	I	Method used	The way forward
[17]	✔	✔	✔	✔	✔	✔	✖	✖	✖	Quantum networking and quantum UAVs are extensively discussed. This article elaborates on how quantum UAVs and networking could collaborate together to form a cutting-edge technological whole	Further development of this study may involve the creation of a computational framework and the exploration of the viability of designing circuitry for quantum systems
[52]	✔	✔	✖	✔	✔	✔	✔	✖	✖	Quantum computing's potential impact on global warming is investigated	Quantum computing's potential for pragmatic integration into ecological remedies is now being explored
[53]	✔	✖	✖	✔	✔	✔	✔	✖	✖	Conceptual debates of using the technology of blockchain in large-scale medical treatment	QUAB-based healthcare systems can be added to this effort
[54]	✔	✖	✖	✔	✔	✖	✖	✖	✖	Quantum, the use of cryptography, and blockchain integration are covered	Enhance implementation in practice and formal confirmation
[12]	✔	✖	✖	✖	✔	✖	✔	✖	✔	Talked about blockchain and medical care	Blockchain healthcare experiments are possible
[16]	✔	✔	✖	✖	✔	✔	✔	✖	✖	Proposed blockchain-quantum computing convergence	To assess practicality, official confirmation, and evaluation can be done

(continued)

Table 3 (continued)

Research work	A	B	C	D	E	F	G	H	I	Method used	The way forward
The way forward	✔	✖	✖	✖	✔	✔	✖	✖	✖	Broadly discussed quantum computing, encryption, and blockchain theory	To assess practicality, official proof and testing can be done

A: QUAB, B: Technologies that combine quantum effects (UAVs, satellites, ML/DL), C: Linear Optical QC, D: Q- Logic Gates, E: Q-Protocols, F: Q- simulation, G: Medical Records, H: Internet of Medical Things (IoMT), I: Additional Cutting-Edge Technologies (Robotics, Homorphic Encryption, AR/VR, Nano Technology and others)

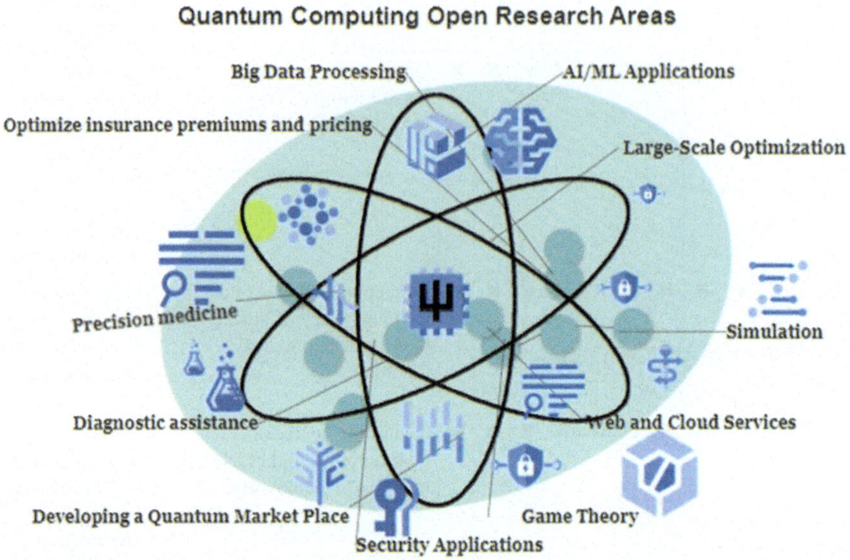

Fig. 2 Open research areas of quantum health informatics

fields of speedier diagnosis, individualized treatment plans, and cost minimization. According to a review of the pertinent research, the accessibility and readily available nature of medical data sources and knowledge from around the world have improved, leading to a rise in the usage of improved, leading to a rise in the usage of both conventional modeling and quantum-based techniques. In this part, we discuss some of the ways in which quantum technology may be used to improve medical practice, Figure provides an example of this.

5.1 Molecular Dynamics Simulations

When opposed to conventional computing, wherein the rate at which information is processed is determined by chips, the method by which quantum computers handle information employing qubits is radically different [17]. Quantum computers, which don't employ binary digits to store data, take advantage of quantum entanglement in order to create quantum algorithms that can compete with traditional machines [39]. Quantum systems can use ML/DL, optimization, and AI to do intricate models in the medical field [46]. Complicated molecular relationships and interconnected architectures of compounds with engaging ions are frequently at the heart of medical procedures. Duration continually serves as a constraint in models and other activities in this area, and its needs expand exponentially with the complexity of the issue [40]. As a result, researchers contend that systems built on quantum technology are an obvious choice for this application.

5.2 Precision Health

Precision health is the field that investigates how to best meet the unique healthcare requirements of each patient [34]. Because the human physiological system is so complicated, advanced, individualized, medical care will be necessary in the future [29]. Predictions of future illness risk using EHRs have been demonstrated to be accurate using conventional ML [41]. Classical ML techniques have come a long way, but they still have their constraints when it comes to things like accuracy and disturbances, variable size, and the intricacy of interactions between variables. As a result, the notion of applying quantum-enhanced ML arises [40], which may allow for more precise and nuanced early illness discovery. Care providers can then employ simulated diagnoses based on real-time data streams to learn how different scenarios affect patients' risks. Research on the genomic characteristics of cancer cells and their effect on drug susceptibility is underway at the cellular level. Chemical modeling with biochemical features that can be employed for finer efficacy prediction in cancer treatment is the subject of research at the moment. Pharmaceutical reasoning systems are one area where quantum-enhanced ML might speed up progress in medicine [48].

The purpose of precision health care is to determine and explain the connections between various factors and potential therapies, as well as to forecast the best next steps for each specific patient. Umbrella diagnoses are made when doctors rely only on a patient's stated symptoms, and the therapies for these conditions don't always work [51]. The use of quantum computing for illness prediction and therapy might facilitate the use of constant data flows with personalized therapies. Utilizing continuous monitoring, Q-enhanced predictive health maximizes efficiency and individualization in medical care [36]. Quantum-enhanced modeling may help patients stay on track with their therapies and actively participate in their care. Figure 3 depicts a potential use for precise healthcare enabled by quantum technology.

5.3 Assisting Doctors in Diagnosis

In addition to improving diagnosis and therapy options, early illness detection has the potential to reduce overall healthcare costs. For example, studies have shown that early detection of intestinal cancer reduces the cost of therapy by an amount of four, while simultaneously decreasing survival rates "by a ratio of nine" (?). However, existing methods of evaluation and therapy for the vast majority of ailments are expensive, time-consuming, and inaccurate (by a factor of 25%). Automated diagnostics have advanced rapidly in recent years, making imaging tests like X-rays, mammography, computed tomography scans, nuclear medicine tests and MRIs increasingly important. Noise, poor data quality, and a lack of replication create difficulties in both diagnosis and therapy.

Accordingly, quantum-based diagnostics has the potential to enhance image-assisted diagnosis by analyzing medical pictures and supervising the processes, such as boundary identification in health imageries. The existing methods rely on only one-cell approaches for diagnosis, though, there is a deficiency of investigation tools for data from individual cell sequencing experiments and the use of flow cytometry. In addition, these methodologies necessitate sophisticated data processing approaches, in particular when merging records from several sources. Classifying cells according to their biochemical and physical properties is seen as a significant obstacle in this setting. However, separating cancer cells from healthy ones needs a more expansive feature space, since the variable that predicts grows in importance and must be distinguished from the former. Such categorizations and only one-cell tests are made possible by quantum machine learning (QML) methods.

The examination of complex omics datasets including metabolite profiling, expression profiling, and large-scale study of proteins, and genomes is made possible by the identification and characterization of biomarkers. These procedures may result in a larger feature space, which may in turn generate complicated patterns and correlations that are difficult, if not unattainable, to analyze with traditional computational methods. Quantum dots shall play an imperative character in the detection of cancer. A great deal of phase I diseases that are diagnosed have a five-year survival rate of more than ninety percent when treated appropriately. Clinical resolution is aided greatly by the use of fluorescence naming or fluorescence imaging technology. In order to enhance the visualization of fluorescence, quantum dots are typically utilized due to their photochemical stability and high luminous production.

5.4 Investigation into New Medicines

Modifying non-covalent interactions at the nuclear level is essential for medicinal research, which is made possible by quantum computing. Important uses include analytics, drug development, diagnostics, and therapy. New quantum computing capabilities have made it feasible to encode hundreds of proteins and mimic their reactions with medications. Quantum computing aids the processing of this data by many orders of magnitude over traditional computing. With the use of quantum computing, medical professionals may quickly find the most useful patterns by comparing several datasets and their various permutations concurrently. Utilizing conventional techniques, such as a bio-barcode test, gold nanoparticles have made detecting disease-specific biomarkers in the blood feasible. One possible objective in this context is to make advantage of diagnostic correlations.

5.5 *Cancer Treatment Therapies*

Cancer is a major killer worldwide, second only to cardiovascular disease in terms of mortality rates [55, 56]. When the systems that control cell development and division don't work properly, it can lead to this illness. From a projected 7.9 million in 2021 and 7.7 million in 2020, cancer deaths in India reached an estimated 8.08 million in 2022. By 2025, the number of new cases of all cancers is projected to reach 15.7 million. Successful cancer therapy depends on lowering death rates, boosting patient survival rates, enhancing patient care, making accurate diagnoses quickly, and then administering targeted therapies [57]. The time and money spent on cancer therapy are both decreased when the disease is caught early. Furthermore, early detection might be especially crucial for aggressive and symptomless tumors. Regardless of offering suitable geomorphological information of cancer cells and tissue, clinical methods of cancer imaging such as positron emission tomography (PET) and computerized axial tomography (CAT) have a poor resolution as well as contrast in instances in which the cancer cells appear extremely identical to normal tissue. Thus, for widespread clinical deployment, it will be necessary to design probes that overcome the difficulties related to signal magnitude, equilibrium, and tissue penetration. Quantum Dots, sometimes known as "artificial atoms," are nanocrystals of semiconductors with a size of less than ten nm, placing them in the nanoscale zone across every dimension. There is significant therapeutic relevance to the widespread usage of Quantum Dots for the optical imaging of cancer cells at present [58].

5.6 *Reflections on What We've Learned*

Quantum computing might be used in a variety of diagnostic procedures and systems that rely on past data, imaging modalities like MRIs and CT scans, and more. DNA sequencing now takes around two to three months with conventional computers but might be greatly sped up with the aid of quantum computing. It might also speed up the process of analyzing DNA variations associated with myocardium. The expanding capabilities of quantum computing have the potential to improve medicine in surprising ways, but their widespread adoption might present security risks.

Thus, investing in quantum computing is necessary for improved delivery of medical care. Vaccine development might also benefit from increased levels of automation. Distributed quantum computing, a method in which a quantum supercomputer shares its resources through the cloud, also requires allocation.

6 What Can We Anticipate to Change with Medical Research and Care Delivery if Everyday Access to Quantum Computing Becomes a Reality?

With the arrival of practical quantum computers, X-analytics and ML/DL are projected to see rapid development. The spectrum of ML/DL ranges from completely data-driven to completely reliant on human input for making predictions [59]. Human judgment is required while developing a statistical framework, particularly when defining variables and establishing their interactions. ML/DL relies on the evaluation of trillions of data items and all of their possible directions of influence and interactions to create algorithms with little to no human intervention. An algorithm is more indicative of machine learning the more it can be constructed from unprocessed information with minimal human intervention. Real-time machine learning in the clinic requires ever-expanding databases of medical record data that are linked to other sources of data (like wearable technology). To solve this problem, future ML programs will need the speed and accuracy of quantum computing. Physicians may soon be able to choose a therapy precisely appropriate for each patient based on algorithms that are continually running and up-to-date to forecast response to therapy while taking into account the patient characteristics (race, age, sexual orientation, multiple chronic conditions, co-therapy/treatment, genetic genealogy, etc.). This is made possible by the combination of more data-telling predictive algorithms and the computational power of quantum computers.

Without a corresponding upsurge in the availability of medically useful data from a variety of sources, the benefits of quantum computing to medicine will remain theoretical. Merging data from disparate sources, such as prevailing patient information banks or clusters, bio-banks, medical wearables, etc., will help overcome obstacles to comprehending a medical system in which non-deterministic machine uses centuries of statistics from millions of medical experiences to suggest a particular medicine for a patient showcasing new-onset depressive symptoms, cancer, diabetes, or any medical issue.

Are virtual doctor appointments the wave of the future? Most people who go to the doctor anticipate more than just medicine. Especially when more is at stake, such with surgery, and when effective treatment demands a patient's commitment to intricate regimens, such as in the case of cancer. The doctor-patient connection, open communication, and trust inherent in team decision-making are not things quantum computers can replace. Nor can a computer replace a doctor's capacity to provide comfort through words of affirmation, compassion, social support, and even a

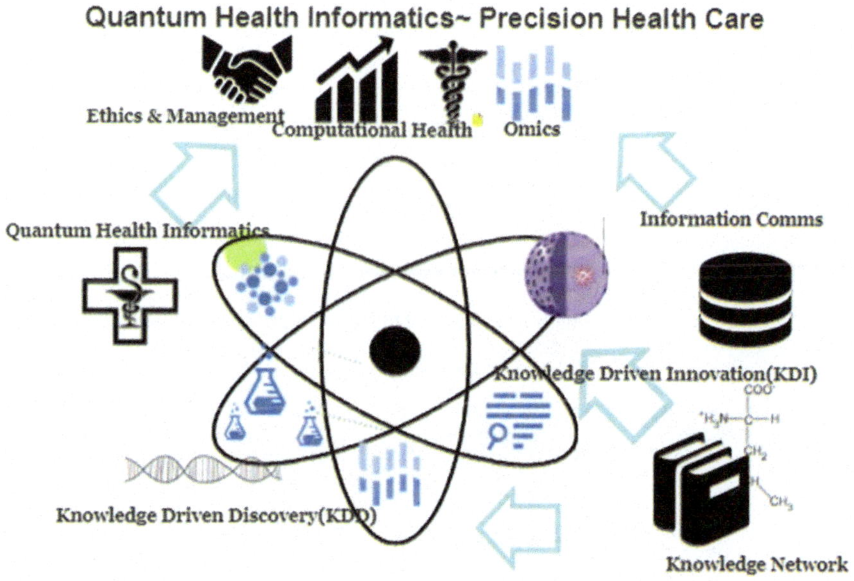

Fig. 3 Precise health care enabled by quantum technology

simple slap on the back, greeting, laugh or embrace. Never discount the importance of the trust patients have in their doctors and the role that medical professionals like nurses and administrative assistants play in a patient's recovery. An algorithm running on a "conventional machine" or the "next-generation quantum machine or non-deterministic machine" will never be able to replicate a doctor's promise "to obtain one with this method" or a family physician's encouragement to "take a stroll for half an hour regularly" to improve well-being (Fig. 3).

7 A Synopsis of the Primary Needs (P) of Quantum Computing in the Distribution of Medical Services, Together with the Many Problems (P) and Potential Solutions (P) to Those Needs

See Table 4.

Table 4 Summary of problems and its primary needs and its potential solutions (PPP) of QC in Health Informatics

Primary needs	Problems	Potential solutions
Quantum ML (QML)	Slower processing time Due to a paucity of means More complicated Higher costs of implementation	Quantum computing-based solutions
Quantum Gate (QG) Implementation	No limits on replication Linking topology presents several difficulties Finding Maxima and Minima Difficulties Insufficient error-checking code	Implementing Quantum Gate (QG)-Model Systems with Shor's factoring algorithm for processing factorization results
Computing capability	Lowest computing capability of conventional systems Complexity of computations is on high side Big scale issues Complicated in practice	Quantum computing's high-dimensional spaces Effectively depicting more complex issues Quantum constructive interference Quantum computing's lightning-fast speed
Access to Health Care in Quantum Form	Far and yet to born processing systems Problems in the flow of information systems Inadequate network and computer support Inadequate distribution of services	Progress of quantum services
Quantum computation in higher dimensions	Quantum states in increasing supply Old-style systems had less storage space available Inadequate means Rising processing demands	Hilbert Space Quantum Mechanics
Prerequisites for Actualization of Quantum	Costs to implement are higher. Reduced income, fewer available resources, and an absence of necessary skills	Affordable Q solutions. • Quantum HR development. • Creation of new Q physical systems
Super-Fast Internet (5G/B5G Networks)	Inadequate safety measures Insufficient scalability Both integrity and privacy have been compromised	Quantum walks and search algorithm computations

(continued)

Table 4 (continued)

Primary needs	Problems	Potential solutions
Risk Tolerance Property	Intolerance of error State of quantum entanglement Qubit errors A Quantum Error-Correction Code Is Missing	Keeping tabs on qubits with the help of a spare one Finding of logical mistakes The code used to identify the error We can control the spread of errors by:
Topologies with Dispersed Nodes	Quantum state separation in terms of space and time Delays in the operation of the quantum bus The need for a well-coordinated system of support facilities There is no wide-area network in place	Quantum computing research and development on a broad scale Effective use of a quantum bus system Quantum artificial neural networks (QUANN) operate on a feed-forward basis The effect of dipoles on one another
Quantum computation scalability	Challenges in scaling, recovering from failures, and accommodating increased processing loads. The absence of suitable emulation settings	Techniques of retaining information for later use. Employing Boltzmann machines with neural networks Transfer learning (TL) procedures are motivated by physical principles Use cases for quantum computing on an FPGA

8 Quantum Versus Other State-of-the-Art Technologies

8.1 Quantum Machine Learning

A primary focus of quantum machine learning studies is on discovering how to use quantum to accelerate the application of traditional machine learning methods. Machine learning with the help of quantum mechanics is called quantum machine learning, or QML. Data scientists and others are learning more about how quantum computing might be used for machine learning and other processes as the technology matures beyond its theoretical roots. A small number of pilot programs at large IT companies are currently testing quantum computing in practical contexts. The result may pave the way for further practical applications of quantum computing.

The foundation of quantum machine learning (QML) is a combination of quantum data and classical/quantum models.

8.1.1 Quantum Data

What we call "quantum data" can come from either a natural or artificial quantum system. For Google's demonstration of quantum supremacy, e.g., this can be data generated by a quantum computer, such as data collected from the Sycamore processor. As a result of superposition and entanglement, joint probability distributions can arise in quantum data that may need exponentially more processing resources in the classical realm to express or store.

Here are some types of quantum information that can be created or simulated by a quantum device:

- When it comes to quantum metrology, variational quantum models can be used for the construction or improvement of probes, which are small-scale quantum devices used for quantum-enhanced high-precision measurements like quantum sensing and quantum imaging.
- Materials research, mathematical chemistry, biological computation, and the development of medicines are just few of the fields that could benefit from the insights gained by chemical simulation.
- Employ machine learning to differentiate between non-orthogonal quantum states, then apply this knowledge to the development of organized quantum relays, recipients, and purifiers for use in quantum communication networks.

8.1.2 Quantum and Conventional Approaches Fused

If your data has a quantum mechanical origin, a quantum model can represent and generalize it. The quantum models cannot generalize quantum data using only quantum processors because these processors are still relatively tiny and noisy in the near future. For maximum efficiency, NISQ processors need to be paired with conventional co-processors. Because of TensorFlow's built-in support for computing environments that vary across CPUs, GPUs, and TPUs serves as the foundation upon which researchers can try out hybrid quantum–classical algorithms.

Quantum computational models with many parameters can be described using the term "quantum neural network" (QNN), which describes a model that runs most efficiently on a quantum computer. Parameterized quantum circuit (PQC) is a common synonym for this concept.

The field of QML also has applications in the following areas:

- Classical data modeling with quantum computers.
- Classical algorithms with quantum-like properties.
- Quantum classifiers for supervised learning
- QNN with adaptive layer-by-layer learning.
- The Study of Quantum Dynamics.
- Quantum state mixtures: a generative model
- Use of QNN for classification on future computers.

8.1.3 Quantum UAV

For the benefit of those who might not know, a "UAV" is a type of aircraft that can be flown without a human pilot on board but otherwise operates similarly to a plane. This word is frequently used in conversations about cars. The existence of a pilot who is human is one way to tell a piloted aircraft apart from a small aircraft or another unmanned aerial vehicle. Since the aircraft's autopilot function is indistinguishable during the takeoff process, flying an airplane requires a human pilot. UAVs equipped with quantum intelligence can be used for surveillance, monitoring, and gathering thermal imagery. Modifications to systems, distribution of food or medicine, sanitation services, and disaster aid are among the possibilities. Integrating quantum-based transmission of information, social separation assessment, and information harvesting with AI-based classification could considerably enhance quantum UAV-based platforms with quantum satellites. When combined with quantum communications satellites, quantum artificial intelligence (AI), and machine learning (ML), quantum drones can be used for environmental cleaning.

9 Quantum Blockchain's (QUAB) Limitations

Current blockchain security measures should be upgraded to be quantum-resistant before quantum computers become widespread. The blockchain can be made immune to quantum computers in two basic ways. There are two types of cryptography: quantum and post-quantum.

9.1 Post-Quantum Cryptography

Since post-quantum cryptography protects blockchain cryptography using mathematical techniques, it is sometimes referred to as quantum-resistant cryptography. Mathematical algorithms, instead of quantum features, are used in post-quantum cryptography. Post-quantum cryptography relies on the mathematical techniques of lattice cryptography, multivariate cryptography, hash cryptography, and code cryptography.

10 Case Studies and Success Stories in Health Informatics Using Quantum Blockchain

10.1 The Next Revolution Toward Safe and Smart Cities is Post-quantum Blockchain

Researchers have incorporated blockchains to quantum computing as part of the shift toward post-quantum cryptosystems. Other researchers have found ways to speed up the Bitcoin mining process by introducing a slight adjustment to the Grover's methods, which were introduced by the authors [60] Smart contracts are a practical application of quantum cryptography (QC). Quantum key distribution (QKD) refers to a broad field of study that draws on the physics of key establishment.

Traditional forms of cryptography and security protocols are vulnerable to quantum computers and quantum-based cryptography, including blockchain. Blockchain-based smart cities are an important application because they store sensitive information and data on users and their various components, making them vulnerable to future quantum attacks. In this essay, we discuss the potential risks that blockchain-based smart cities may face in the future. Highly intriguing post-quantum solutions, such as lattice-based cryptography, QKD, and entanglement in a time-based quantum blockchain, are described, and the conventional blockchain-based smart city framework is also on the way to revolute. These systems are suited for usage in a PQB-based smart city before the full potential of quantum computers is realized while being susceptible to quantum attacks in the lab.

10.2 Scalable Services for Post-quantum Smart Cities with Blockchain

As a result of the danger posed by quantum computing, the scientific community has been working to develop quantum-resistant protocols.

In addition, blockchain technology can be used to revolutionize several smart city applications. Decentralizing the smart city application with automatic safe data collecting and distribution, while defending against a failure at central server is made possible by blockchain technology, which eliminates the requirement for a trusted third party. In addition, any smart city application can benefit greatly from the combination of quantum computing and blockchain. As a result, this chapter does not merely focus on explaining the fundamentals of quantum computing, blockchain, and the hybrid of blockchain and quantum computing. Computing in addition, a post-quantum blockchain infrastructure for transport applications in smart cities is described. The detailed account includes the planning and development of an SRP platform with built-in smart contracts for safe storage and sharing of vehicle records between four parties: vehicle, toll booths, network management, and the local transit agency. In addition to facilitating easy administration, the work provided below of automobile registrations, but also strengthen coordination organization types all located in the same metropolitan area a smart city that is both strong and welcoming.

11 Conclusion

As with other health informatics systems, the implementation of blockchain-based technology in the battle during the pandemic has proven to be incredibly beneficial. Sensor data is a part of cutting-edge healthcare infrastructure that can assist keep tabs on patients without compromising their anonymity or the security of their personal health information. In a nutshell, this innovation allows for the secure and convenient long-term storage of massive amounts of information that can be retrieved at any time. This distinguishes it as a superior technology to all others. They can be utilized at any time and at all times they are needed since they are easy to get ahold of and can be retrieved quickly in an emergency. Leveraging the QUAB and associated system, locating a patient might take as little as a few seconds. Data can be hidden in a QUAB in a way that makes it both safe and not overly difficult to obtain. Potentially faster and more secure handling of medical records might be achieved with the help of QUAB. Numerous hypotheses have been proposed for both blockchain and quantum technologies. The purpose of this research was to establish the likelihood that these advancements have practical applications in the field of medicine at present and future times. Quantum mechanics, blockchain technology, deep learning, and even UAVs are just some of the advances that are being studied in this area.

Conventional systems have been rendered obsolete by QC, which provides inconceivable speeds, efficiencies, and dependabilities. We can create more computationally efficient healthcare applications using these fundamental aspects of non-deterministic computing. To that end, in this research, we conduct a systematic literature review of works that explore how to use probabilistic computers in the design of medical technologies. Particularly, we spoke about many areas where quantum computing can be useful in the medical field. Further, we gave a taxonomy of current QC architectures for healthcare systems and elucidated the most important prerequisites for creating quantum computing-enabled medical applications. We also spoke about how quantum technologies can guarantee the safety of QC used in healthcare applications, and we covered a range of quantum security concerns. Finally, we outlined the existing problems, what's causing them, and where future studies may go to take advantage of QC. To the scientific scholars attracted to the use and analyze diverse possibilities of QC in numerous health informatics applications, this innovative study may serve as a one-stop solution, outlining all the important areas of non-deterministic computing's insinuations inside the healthcare architype.

References

1. Lee J, Sun J, Wang F, Wang S, Jun CH, Jiang X (2018) Privacy-preserving patient similarity learning in a federated environment: development and analysis. JMIR Med Inform 6(2):e20. https://doi.org/10.2196/medinform.7744.PMID:29653917;PMCID:PMC5924379
2. Navaz AN, Serhani MA, El Kassabi HT, Taleb I (2023) Empowering patient similarity networks through innovative data-quality-aware federated profiling. Sensors (Basel) 23(14):6443. https://doi.org/10.3390/s23146443.PMID:37514736;PMCID:PMC10384464
3. Huang L, Yin Y, Fu Z, Zhang S, Deng H, Liu D (2020) LoAdaBoost: loss-based AdaBoost federated machine learning with reduced computational complexity on IID and non-IID intensive care data. PLoS ONE 15(4):e0230706. https://doi.org/10.1371/journal.pone.0230706.PMID:32302316;PMCID:PMC7164603
4. Muñoz-González L, Co KT, Lupu EC (2019) Byzantine-robust federated machine learning through adaptive model averaging. CoRR abs/1909.05125
5. Xue Z et al (2021) A resource-constrained and privacy-preserving edge-computing-enabled clinical decision system: a federated reinforcement learning approach. IEEE Internet Things J 8(11):9122–9138. https://doi.org/10.1109/JIOT.2021.3057653
6. Vepakomma P, Gupta O, Swedish T, Raskar R (2018) Split learning for health: distributed deep learning without sharing raw patient data. arXiv:1812.00564
7. Kim Y, Sun J, Yu H, Jiang X (2017) Federated tensor factorization for computational phenotyping. KDD 2017:887–895. https://doi.org/10.1145/3097983.3098118.PMID:29071165;PMCID:PMC5652331
8. Lu S, Zhang Y, Wang Y (2020) Decentralized federated learning for electronic health records. In: Proceedings of the 54th annual Conference on Information Sciences and Systems (CISS'20). IEEE, pp 1–5
9. Silva S, Gutman B, Romero E, Thompson PM, Altmann A, Lorenzi M (2018) Federated learning in distributed medical databases: meta-analysis of large-scale subcortical brain data. arXiv:1810.08553
10. Brisimi TS, Chen R, Mela T, Olshevsky A, Paschalidis IC, Shi W (2018) Federated learning of predictive models from federated electronic health records. Int J Med Inform 112:59–67
11. Grama M, Musat M, Muñoz-González L, Passerat-Palmbach J, Rueckert D, Alansary A (2020) Robust aggregation for adaptive privacy preserving federated learning in healthcare. arXiv:2009.08294
12. Chuntang L, Yinsong X, Jiahao T, Wenjie L (2019) Quantum blockchain: a decentralized, encrypted and distributed database based on quantum mechanics. J Quantum Comput 1(2):63. https://doi.org/10.32604/JQC.2019.06715
13. Kumar A, Augusto de Jesus Pacheco D, Kaushik K, Rodrigues JJPC (2022) Futuristic view of the internet of quantum drones: review, challenges and research agenda. In: Vehicular communications, p 100487. https://doi.org/10.1016/J.VEHCOM.2022.100487
14. Kiktenko EO et al (2018) Quantum-secured blockchain. Quantum Sci Technol 3(3):035004. https://doi.org/10.1088/2058-9565/AABC6B
15. Kumar A et al (2021) Survey of promising technologies for quantum drones and networks. IEEE Access, pp 1–1. https://doi.org/10.1109/ACCESS.2021.3109816
16. Ikeda K (2018) Security and privacy of blockchain and quantum computation. Adv Comput 111:199–228. https://doi.org/10.1016/BS.ADCOM.2018.03.003
17. Iovane G (2021) MuReQua chain: multiscale relativistic quantum blockchain. IEEE Access 9:39827–39838. https://doi.org/10.1109/ACCESS.2021.3064297
18. Singh K, Kaushik K, Ahatsham, Shahare V (2020) Role and impact of wearables in IoT healthcare. In: Advances in intelligent systems and computing, vol 1090, pp 735–742. https://doi.org/10.1007/978-981-15-1480-7_67
19. Latif S, Qadir J, Farooq S, Imran MA (2017) How 5G wireless (and concomitant technologies) will revolutionize healthcare? Future Internet 9(4):93

20. Bhavin M, Tanwar S, Sharma N, Tyagi S, Kumar N (2021) Blockchain and quantum blind signature-based hybrid scheme for healthcare 5.0 applications. J Inf Secur Appl 56:102673. https://doi.org/10.1016/J.JISA.2020.102673
21. Kumar A, Krishnamurthi R, Nayyar A, Sharma K, Grover V, Hossain E (2020) A novel smart healthcare design, simulation, and implementation using healthcare 4.0 processes. IEEE Access 8:118433–118471. https://doi.org/10.1109/ACCESS.2020.3004790
22. Kumar N, Goel V, Ranjan R, Altuwairiqi M, Alyami H, Asakipaam SA (2023) A blockchain-oriented framework for cloud-assisted system to countermeasure phishing for establishing secure smart city. In: Security and communication networks, Article ID 8168075, 13 pages. https://doi.org/10.1155/2023/8168075
23. Sharma P, Choi K, Krejcar O, Blazek P, Bhatia V, Prakash S (2023) Securing optical networks using quantum-secured blockchain: an overview. Sensors 23(3):1228. https://doi.org/10.3390/s23031228
24. Subhi D, Bacsardi L (2023) Using quantum nodes connected via the quantum cloud to perform IoT quantum network. Condensed Matter 8(1):24. https://doi.org/10.3390/condmat8010024
25. Li C-Y, Chen X-B, Chen Y-L, Hou Y-Y, Li J (2019) A new lattice-based signature scheme in post-quantum blockchain network. IEEE Access
26. Rajan D, Visser M (2019) Quantum blockchain using entanglement in time. Quantum Rep 1(1):3–11. https://doi.org/10.3390/quantum1010002
27. Gupta D, Rani S, Ahmed SH, Garg S, Piran MJ, Alrashoud M (2021) ICN-based enhanced cooperative caching for multimedia streaming in resource-constrained vehicular environment. IEEE Trans Intell Transp Syst 22:4588–4600
28. Srivastava AK, Grotjahn R, Ullrich PA (2019) A multimodel technique for estimating future changes in extreme precipitation. In: Proceedings of the AGU Fall Meeting Abstracts, San Francisco, CA, USA, 9–13 Dec 2019, volume 2019, p A51Q-2832
29. Alkatheiri MS, Alghamdi AS (1801) Blockchain-assisted cybersecurity for the internet of medical things in the healthcare industry. Electronics 2023:12. https://doi.org/10.3390/electronics12081801
30. Chen Y, Ding S, Xu Z, Zheng H, Yang S (2019) Blockchain-based medical records secure storage and medical service framework. J Med Syst 43:5
31. Rubí JNS, Gondim PRL (2020) Interoperable internet of medical things platform for e-health applications. Int J Distrib Sens Netw
32. Patel V (2019) A framework for secure and decentralized sharing of medical imaging data via blockchain consensus. Health Inform J 25:1398–1411
33. Zhang J, Xue N, Huang X (2016) A secure system for pervasive social network-based healthcare. IEEE Access 4:9239–9250
34. Alam T (2020) mHealth communication framework using blockchain and IoT technologies. Int J Sci Technol Res
35. Rahmadika S, Rhee K-H (2018) Blockchain technology for providing an architecture model of decentralized personal health information. Int J Eng Bus Manag
36. Nguyen DC, Pathirana PN, Ding M, Seneviratne A (2019) Blockchain for secure EHRs sharing of mobile cloud based E-Health systems. IEEE Access 7:66792–66806
37. Rajan D, Visser M (2019) Quantum blockchain using entanglement in time. Quantum Rep 1:3–11
38. Ringbauer M, Costa F, Goggin ME, White AG, Fedrizzi A (2018) Multi-time quantum correlations with no spatial analog. NPJ Quantum Inf 4:37
39. Chen X, Xu S, Qin T, Cui Y, Gao S, Kong W (2022) AQ–ABS: anti-quantum attribute-based signature for EMRs sharing with blockchain. In: 2022 IEEE Wireless Communications and Networking Conference (WCNC). IEEE, pp 1176–1181
40. Azzaoui AE, Sharma PK, Park JH (2022) Blockchain-based delegated quantum cloud architecture for medical big data security. J Netw Comput Appl 198:103304
41. Grosu GM, Nistor SE, Simion E (2022) A note on blockchain authentication methods for mobile devices in healthcare. IACR Cryptol ePrint Arch 2022:159

42. Ahmad Naqishbandi T, Syed Mohammed E, Venkatesan S, Sonya A, Cengiz K, Banday Y (2022) Secure blockchain-based mental healthcare framework:—a paradigm shift from traditional to advanced analytics. In: Quantum and blockchain for modern computing systems: vision and advancements. Springer, Cham, pp 341–364
43. Laxminarayana N, Mishra N, Tiwari P, Garg S, Behera BK, Farouk A (20220 Quantum-assisted activation for supervised learning in healthcare-based intrusion detection systems. In: IEEE transactions on artificial intelligence
44. Qu Z, Zhang Z, Zheng M (2022) A quantum blockchain-enabled framework for secure private electronic medical records in the Internet of Medical Things. In: Information sciences
45. Pandey SS, Dash T, Panigrahi PK, Farouk A (2022) Efficient quantum non-fungible tokens for blockchain. arXiv preprint arXiv:2209.02449
46. Trenfield SJ, Awad A, McCoubrey LE, Elbadawi M, Goyanes A, Gaisford S, Basit AW (2022) Advancing pharmacy and healthcare with virtual digital technologies. In: Advanced drug delivery reviews, p 114098
47. Malviya R, Sundaram S (2022) Exploring potential of quantum computing in creating smart healthcare. Open Biol J 9(1):56–57. https://doi.org/10.2174/1874196702109010056
48. Gupta S, Modgil S, Bhatt PC, Jabbour CJC, Kamble S (2022) Quantum computing led innovation for achieving a more sustainable Covid-19 healthcare industry. Technovation, p 102544
49. Li C, Dong M, Li J, Xu G, Chen XB, Liu W, Ota K (2022) Efficient medical big data management with keyword-searchable encryption in healthchain. IEEE Syst J
50. Tchagna Kouanou A, Tchito Tchapga C, Sone Ekonde M, Monthe V, Mezatio BA, Manga J, Simo GR, Muhozam Y (2022) Securing data in an internet of things network using blockchain technology: smart home case. SN Comput Sci 3(2):1–10
51. Charles WM (2022) The future of blockchain. In: Blockchain in life sciences. Springer, Singapore, pp 315–336
52. Yung KL, Xu L, Zhang C (2021) Spacecraft informatics. In: Enterprise information systems
53. Jeet R, Kang SS (2020) Investigating the progress of human E-healthcare systems with understanding the necessity of using emerging blockchain technology. In: Materials today: proceedings
54. Saif S, Biswas S, Chattopadhyay S (2019) Intelligent, secure big health data management using deep learning and blockchain technology: an overview
55. Fitzmaurice C et al (2017) Global, regional, and national cancer incidence, mortality, years of life lost, years lived with disability, and disability-adjusted life-years for 32 cancer groups, 1990 to 2015: a systematic analysis for the global burden of disease study. JAMA Oncol 3(4):524–548
56. Ahamed HA, Mohamed MJ, Arunachalam KD, Raiyaan GID, Musthafa MS, Begum SS et al (2020) Effects of azomite enriched diet on gonadal steroid hormone levels and milt quality indices in oreochromis mossambicus. Aquac Rep 17:100341. https://doi.org/10.1016/j.aqrep.2020.100341
57. Alshehri S et al (2020) Progress of cancer nanotechnology as diagnostics, therapeutics, and theranostics nanomedicine: preclinical promise and translational challenges. Pharmaceutics 13(1):24
58. Devi S, Kumar M, Tiwari A, Tiwari V, Kaushik D, Verma R, Bhatt S, Sahoo BM, Bhattacharya T, Alshehri S, Ghoneim MM, Babalghith AO, Batiha GE-S (2022) Quantum dots: an emerging approach for cancer therapy. Front Mater 8:798440. https://doi.org/10.3389/fmats.2021.798440
59. Beam AL, Kohane IS (2018) Big data and machine learning in health care. Jama-J Am Med Assoc 319(13):1317–1318
60. Berdik D, Otoum S, Schmidt N, Porter D, Jararweh Y (2021) A survey on blockchain for information systems management and security. Inf Process Manage 58(1):102397
61. Ren X, Xu M, Niyato D, Kang J, Xiong Z, Qiu C, Wang X (2023) Building resilient web 3.0 with quantum information technologies and blockchain: an ambilateral view. ARXIV-CS.CR

Model-Based Learning of Information Diffusion in Social Networks

T. R. Abijeeth Vasra, A. M. Abirami, and S. Sumitra

Abstract In recent years, the ability to predict how information will travel within a network has gained importance, especially in the setting of online social media platforms where the dissemination of information can have a big impact on social and political events. Researchers frequently employ a range of methods, such as network mapping and analysis, data mining, and machine learning algorithms, to study social networks and the connections between them. These methods can be used to locate important nodes and influential nodes within a network and to comprehend the underlying dynamics that govern the dissemination of knowledge. This research work attempts to study the use of three different epidemic models to a medium level twitter network of size 35,000 nodes. This paper discusses the impact of centrality measures, the use of hyperparameters and the results of models in detail. It recommends the suitable social network epidemic model for better information diffusion in the network, which can be adopted by businesses for digital marketing or advertising.

Keywords Social network analysis · Epidemic models · Digital marketing · Business intelligence · Twitter analysis

T. R. A. Vasra · A. M. Abirami (✉)
Department of Information Technology, Thiagarajar College of Engineering, Madurai, Tamil Nadu, India
e-mail: abiramiam@tce.edu

S. Sumitra
Department of Mathematics, Indian Institute of Space Science and Technology, Trivandrum, Kerala, India
e-mail: sumitra@iist.ac.in

S. Bandyopadhyay et al. (eds.), *Intelligent Computing Systems and Applications*, Lecture Notes in Networks and Systems 1010,
https://doi.org/10.1007/978-981-97-5412-0_20

1 Introduction

The Covid-19 outbreak has recently led to a rise in social media usage globally. Many businesses now do much of their business on social media, where they use digital marketing strategies to draw in new clients and keep existing ones [2]. The growth of social media analytics is well utilized by businesses for promotion of their brand values among the competitors. AllTheResearch report detailed the business growth in US and Asian countries due to social media marketing strategies. Western countries started using this in early 2000s, whereas the eastern countries in the recent past, and it is shown in Fig. 1. Social media includes both structured and unstructured data. These types of data are very much useful for business marketing analytics, opinion mining, and so forth. It serves as the powerful data source for researchers. Artificial intelligence (AI) algorithms including natural language processing (NLP) techniques and statistical methdos paly vital role in social media analytics. For instance, clustering and regression techniques are essential for forecasting analysis if a business wants to examine customer behavior to forecast their business. Social media firms analyze user profiles and behavior using AI and NLP algorithms. They create sophisticated algorithms that can identify persons in images and be used to create new content.

The practice of obtaining information from social networking platforms like Twitter, Facebook, LinkedIn, and other sites is known as social media analytics. Businesses may track their performance and use this analytics to make highly-informed decisions [18]. It is not just the total number of likes, dislikes, tweets, retweets, and other interactions. It needs the assistance of web crawlers, which gather information from many sources based on search queries, then pre-process it, model it, and analyze it for more in-depth interpretation.

Nowadays, cyberbullying on social media platforms has become a severe problem. Through these digital platforms, individuals send offensive or hazardous messages or attachments to one another, which has a psychological impact on the recipients. They can track out the senders of these harassments, remarks, or tweets with the aid of machine learning techniques like Bayesian analysis [6] and social media analytics can be tailored to these kinds of harassments or harmful messages [12].

Whenever there are elections, big sporting events, major movie releases and the like, the corresponding social media sites will become popular among users. During these days, users have access to various platforms to express their own thoughts and beliefs. It may either elicit favorable or unfavorable responses from followers. The social media network structure and the person disseminating the information play a role in the cascading effect or dissemination of information in the social network. Social media analytics must be performed on these data in order to educate the stakeholders at the appropriate time and assist them to take necessary preventive steps.

The purpose of social media analytics would be different for different stakeholders. The following are some of its objectives which help the analysts in different ways:

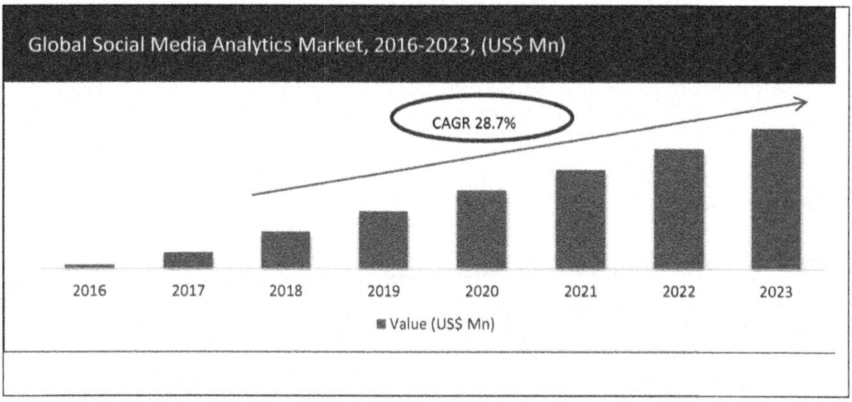

Fig. 1 Social media and global market analysis. *Source* www.alltheresearch.com/report/5/social-media-analytics-market

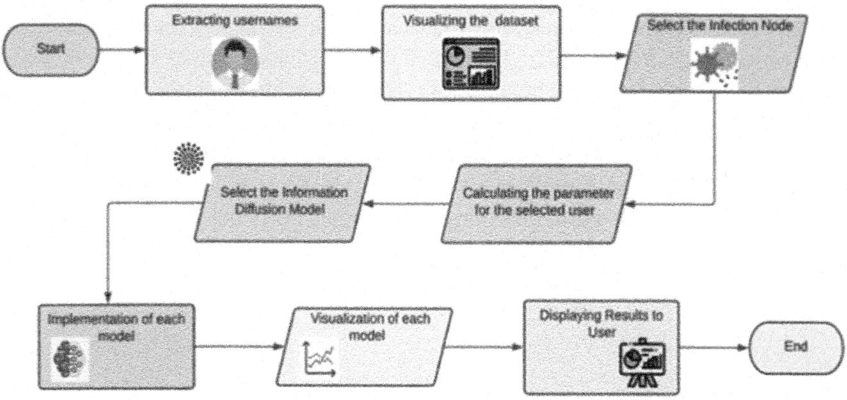

Fig. 2 Generic processes of social network analysis

- To perform opinion mining and to predict the market sustainability for the business
- To perform market study and to do benchmarking against the competitors' products
- To analyze the presence of external influential factors for the business
- To comprehend the present trend among the consumers so as to reach the right audience.

The dissemination of knowledge, concepts, and viewpoints within a network or community is referred to as the process of information diffusion. Researchers and professionals have created a variety of models and analytical tools that may be used to forecast and comprehend how information spreads within a specific context in order to assist this process. These models are created to capture the essential elements that affect how information spreads within a network. They are based on various theoretical frameworks and mathematical techniques. The generic processes involved in social network analysis is shown in Fig. 2. Overall, the study of social networks and the presence of types of relationships to forecast how information will spread in the network is a challenging task and a significant field of study with broad societal ramifications.

1.1 Motivation

In order to comprehend how information spreads in a particular society, it is crucial to analyze social networks and the connections between them. Predicting the information flow inside a network is made easier with the usage of social network models. These models are based on mathematical and computational methodologies that examine several aspects of social networks, such as the degree of connectedness between nodes and the strength of the ties between them. User profile data in the social media platforms can be represented mathematically using the epidemic social network models. The results of models can be powerfully visualized to help people understand and express complex information. Social media data analysts can determine which model can be utilized to spread the information in that specific network and give recommendations to the business. This research paper attempts to study the application of epidemic models to the twitter network of size 35,000 nodes.

The remaining section of the paper is organized as follows. Section 2 briefly explains the metrics used for social network analysis. Section 3 discusses the research work carried out in this particular domain. Section 4 describes the dataset used and the methodology in detail. Section 5 discusses the results of social network models using graphs. Section 6 concludes the paper with the future scope.

2 Background

Nowadays, social media is quite dynamic. It facilitates user interaction and content creation for users. To communicate their opinions, people use podcasts, blogs, wikis, online forums, and other media. These online sites directly record the pulse of humanity, which may then be examined to gain additional insights. With billions of users contributing text, photographs, and videos, social media data is vast and rich. With the aid of big data analytics, it is possible to foresee user behavior, preferences, and current trends from vast amounts of social media data.

Graph theory and network analysis are components of social media analytics. The fundamental ideas of graph theory, measures, and epidemic models are described in this section. In a graph, the dataset's elements typically serve as the nodes and the connections between them as the edges. If user A replies to user B's comments in a facebook network structure, then users A and B are linked by an edge from A to B and the sample structure is shown in Fig. 3.

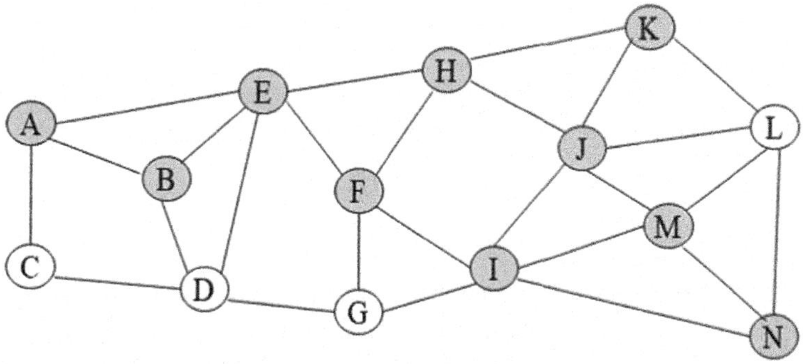

Fig. 3 Sample social media network structure

If the start node is A, then the information is passed on to neighbor nodes in a random fashion. Figure 3 shows that 10 out of 14 nodes received information. Nodes which received messages are gray colored. Message is passed deeper in this network. Nodes and edges have attributes like size, number of following nodes, number of follower nodes, weight, direction, strength of tie, and so on. Nodes and edges which are the basic elements of graph theory try to explain the digital marketing strategies, trending behvaior of consumer, the most popular post, and so on.

2.1 Centrality Measures

Measures of centrality are crucial for comprehending networks. Each measure operates differently and describes the significance of the graph's nodes.

1. **Degree Centrality** is a measure of node importance in a network that is widely used in network analysis. The degree centrality of a node is calculated as the number of edges (or connections) that it has with other nodes in the network. This measure is useful in identifying nodes that have a large number of direct connections and therefore potentially greater influence or control within the network. For example, in social media networks like Twitter, nodes with high degree centrality are likely to have a large number of followers or be following many other users.
2. **Betweenness Centrality** is a measure of node importance in a network. It captures the node which acts as an interface between other nodes. Specifically, it determines the total number of shortest paths between the every pair of nodes in the network that pass through a given node. This metric identifies the key influencers, nodes with high betweenness measure, within the network. They have a large impact on the flow of information or resources between other nodes. In a social network, a node with high betweenness centrality may be a bridge between different social groups or communities.

3. **Closeness Centrality** is a measure of node importance in a network that captures the average distance between a given node and all other nodes in the network. Nodes with high closeness centrality are often seen as more influential within the network, as they have a shorter average distance to other nodes and can more easily reach them. For example, in a social network, a node with high closeness centrality may be able to quickly reach and influence a large number of other users.
4. **Eigen Centrality** is a measure of node importance in a network that takes into account both the number of connections a node has and the importance of those connections. Specifically, it assigns a score to each node based on the scores of its neighbors, so that nodes that are connected to other highly connected nodes receive a higher score than nodes that are connected to less important nodes. Nodes with high eigenvector centrality are often seen as more influential within the network, as they are connected to other influential nodes.

2.2 *Epidemic Models*

Epidemic models have been widely used in network analysis to study the spread of infectious diseases. They have also been adapted for modeling the spread of information, ideas, and behaviors through social networks. A. G. McKendrick developed epidemic models by adopting the principles of Graph theory of mathematics in 1927. These partitioned models assume that there exists closed population with three categories, namely Susceptible (S), Infected (I), and Recovered (R). The model assumes that the population is constant and there is no birth, death, or migration of individuals in the network. The number of healthy people at time t is referred by $S(t)$, who are not yet infected but possible The number of infected people at time t is referred by $I(t)$, and they spread the disease to others. The number of people recovered from the disease at time t is referred by $R(t)$ [10]. The users of social networks are modeled based on their probability of being infected and their ability to infect others. The epidemic models can be enhanced by incorporating network properties such as centrality measures, which can help identify influential nodes in the network that may have a larger impact on the spread of diseases or information.

3 Related Work

Network analysis models are used in social network analysis to examine the relationships between the actors. The epidemic models explain how messages are spread throughout the network and how it is transmitted among neighbor nodes. The parameters of the network topology control the diffusion process rate and intensity. The volume of data flowing across the network makes it possible to use machine learning (ML) algorithms for information extraction. These techniques are employed in social media analytics such as spam detection, link analysis, fake user profile detection, trend setters, customer segmentation, cluster analysis, opinion mining, and other tasks.

A brief description of diffusion models, algorithmic perspective toward influential users, and the comparative study were performed [1, 15]. Preliminary study was done on the social network diffusion models and concluded that the diffusion depends on the centrality measure degree [13]. Different metrics like rate of adoption, quantum of exposure, and critical mass for the information diffusion models were explained based on different scenarios [10, 16]. Social media analytics was widely used by researchers [11] and for consumer behavior analysis [8].

General Influence Spread Rank (GISR) model, a parameterizable centrality measure was used for analyzing social networks by identifying influential actors through information spread models [14]. ML algorithms and causal regression methods were combined to study the application of influence models in social bots [3]. Independent Cascade model was used to retrieve a variable that holds a metric of how the information regarding current traffic has traveled through the network [9]. A candidate nodes-based random population initialization approach was used, and these candidate nodes aid in accelerating the convergence process for the entire population [7].

Exploratory study on the cost awareness model into the classic linear threshold opinion diffusion model was done to analyze the effect of variables like network topology, number of spreaders, spreader active time, injection strategies and the like [17]. Social media analytics was also useful during disaster events like heavy floods, cyclone and typhoons. This analysis supports for the vacating the affected people and for supplying the life materials using the spatial and temporal data received from social media websites [4]. Social media data with geo tags was used in the pattern analysis of user movements [5]. Agent-based techniques are used for determining the dynamics in social networks [20] and CNN algorithm for fake user profile identification [19].

4 Methodology

Social network analysis is performed in two different ways: measuring the amount of information diffusion in the network and determining the influential nodes which make the interactions stronger in the network.

Information gets diffused in the networks like viral or contagious diseases. This process of spread is dynamic, and it depends on the state and behavior of nodes in the network. This section details the application of epidemic models to social networks for the study of twitter dataset and described briefly in Fig. 4.

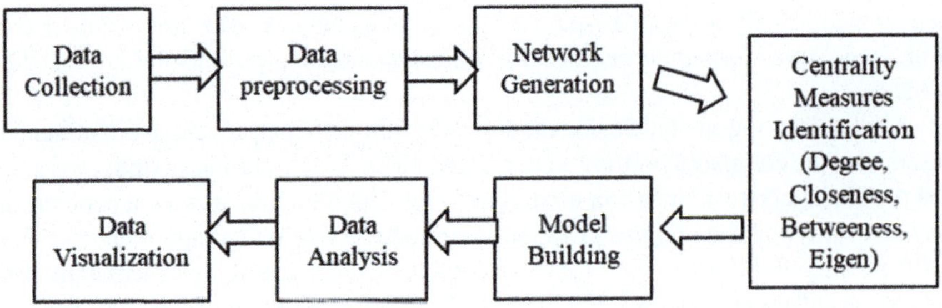

Fig. 4 Proposed methodology

4.1 Dataset Description

The dataset used in this analysis was collected from Twitter using the Tweepy Python library. The dataset includes information about a user named samdro1981 and his network. This username was randomly selected from the available list of networks. It consists of 35,000 rows, with each row representing an individual person in the network. It includes three columns: "id," which represents a unique identifier for each user; "source," which indicates the user who is following another user; and "target," which indicates the user who is being followed. With this dataset, it is possible to analyze the structure and connectivity of the network, as well as explore patterns of information diffusion and user behavior on Twitter.

4.2 Data Collection

Tweepy library is a powerful tool that enables developers to build innovative applications that leverage the vast amounts of data available on Twitter. With its ability to fetch user network data, Tweepy is an essential tool for businesses and organizations looking to make the most of their presence on Twitter.

4.3 Data Preprocessing

Preprocessing is an essential step in gaining insights into a target audience on Twitter. By cleaning and transforming the data, businesses and organizations can gain valuable insights that can be used to inform marketing strategies, improve customer engagement, and enhance overall brand awareness. Splitting the original dataset into edges and nodes datasets can be a useful step in analyzing the network. The edges dataset contains information about the connections between nodes, with each row representing a directed edge from a "source" node to a "target" node. This information can be used to create a graph or network visualization of the connections between users in the network.

Meanwhile, the nodes dataset contains information about each individual node in the network. This can include attributes such as the node's ID or username, as well as measures of the node's centrality or importance in the network, such as the node's degree (i.e., the number of connections it has to other nodes). This information can be used to gain a better understanding of the roles that different users play in the network, as well as to identify any particularly influential or connected nodes. Overall, by splitting the dataset into edges and nodes datasets, it is possible to gain a more detailed and nuanced understanding of the structure and dynamics of the Twitter network being analyzed.

4.4 Network Generation

Generating a network and visualizing it using NetworkX is the next step for gaining insights into a user network on Twitter. By visualizing the network, businesses and organizations can gain a better understanding of the relationships between users, identify key influencers or clusters of users, and make informed decisions about how to engage with their target audience.

4.5 Centrality Measures Identification

Centrality measures are a set of metrics that can be used to identify the most important nodes in a user network on Twitter. These measures can be used to gain valuable insights into the structure of the network and identify key influencers or clusters of users. One of the most common centrality measures used in network analysis is degree centrality. Degree centrality measures the number of edges (i.e., connections) that a node has in the network. Nodes with a high degree centrality are typically seen as more influential within the network, as they have a larger number of connections to other nodes. Another important centrality measure is betweenness centrality. Betweenness centrality measure identifies the node which acts as an interface between other nodes in the network. Nodes with a high betweenness centrality are often key influencers within the network, as they have a large impact on the flow of information between other nodes. Apart from these two, closeness and eigen centrality measures are also widely used in social network analysis.

4.6 Information Diffusion Models Building

Fitting data into a diffusion model is a process of estimating the parameters of a mathematical model that describes how information or behavior spreads through a network. The diffusion model is a popular model used in social network analysis to describe the spread of ideas, opinions, or behaviors through a network of interconnected individuals. Fitting the social media data into a diffusion model is an important tool for understanding how behaviors or ideas spread through a network. By estimating the parameters of the model and predicting the behavior of the network, businesses and organizations can gain valuable insights into the dynamics of the network.

4.7 Network Visualization

Bokeh is a powerful data visualization library in JavaScript that provides interactive and flexible visualizations. Bokeh provides a variety of graph types, such as line graphs, scatter plots, and bar charts, that can be used to analyze the results of network analysis. Using Bokeh for analyzing the results of network analysis can provide valuable insights into the structure of the network and the dynamics of the behavior being analyzed.

4.8 Results Analysis

Analyzing the results from visualizations is an important step in network analysis. Visualizations can provide valuable insights into the structure of the network and the behavior being analyzed. By analyzing the visualizations, businesses can make more informed decisions about how to engage with their target audience. The type of analysis includes identifying the present trends, detecting the outliers, and so on.

5 Results and Discussions

Twitter API, namely Tweepy was used to collect data from the twitter platform. It gives the dataset in the required format (nodes and edges) which can be readily used for analysis. Python libraries like Ndlib was used for building information diffusion models; matplotlib library was used for plotting the graphs; Streamlit was used to render the graphs in a web page. In addition, Flourish, an interactive networks visualizer tool was used in this application. It gets nodes and edges as inputs and produces colorful visualization in the form of graphs, as shown in Fig. 5.

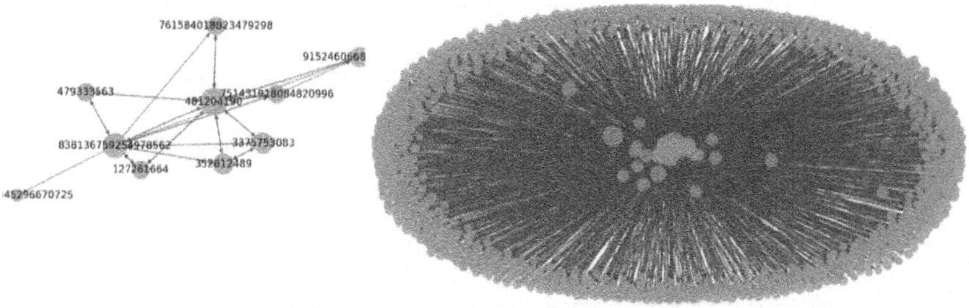

Fig. 5 Social network visualization—eigen centrality

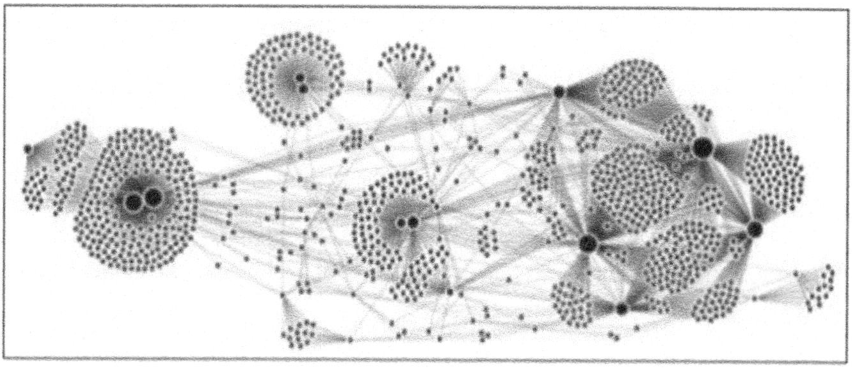

Fig. 6 Social network visualization—betweenness centrality

This graph projects the node with a bigger size whose eigen centrality measure is higher. Here, nodes are represented in blue whereas edges are in black color. Left side of Fig. 5 shows the top 10 nodes whose eigen values are higher and the right side of the figure shows the complex network structure. Figure 6 shows the social network structure where the betweenness centrality measure was used. Here red color denotes the node and blue color denotes the edges. This picture clearly depicts how many clusters are there in the network and which node is the interface between the clusters.

The web application was developed for the twitter dataset analysis using the centrality measures and the epidemic models like SIS and SIR. Figure 7 shows the front end design of the application. Top 10 nodes with high betweenness centrality are displayed on the left side panel along with the node's ID, centrality value, and username. Comparison of two models can be done by selecting the start node and the models from the given choice list; the iterations can also be specified in the sliding bar provided. If a mouse is placed over any node in the visualization, its closest neighbors are highlighted in the graph. On clicking the compare button, the models are built and the graph is plotted on the right side of the application window.

In terms of iterations, SIS and SIR models are iterative, while SI model is not. In SIS and SIR models, the population is divided into two groups: those who are susceptible and those who are infected. In the SIS model, individuals move back

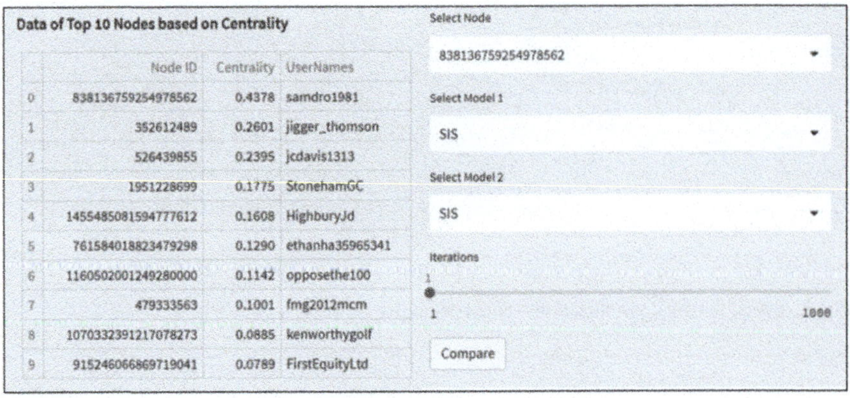

Fig. 7 Front end design—social network analysis

and forth between these two groups, while in the SIR model, individuals move from the susceptible group to the infected group and then to the recovered group. Each iteration of the model represents a time step in the spread of the disease or behavior. The parameter Beta refers to the rate of transmission of infection and lambda refers to the rate of recovery.

5.1 Discussion I

The following discussions compare the results of SIS and SIR models using the degree centrality measure. The models used the following parameters and values: Beta:0.01, Lambda:0.005, Gamma:0.01, Iterations:448.

Case 1: Highest Degree Centrality Node as the Start Node Highest degree centrality node was selected as the start node. From Fig. 8, it is clearly seen that the information diffusion occurs at a much faster rate in the SIR model than in the SIS model. This is found from the intersection point of the Susceptible and Infected lines in the graph. In the SIS model it took 350 iterations approximately, while in the SIR model it took only 50 iterations for the information spread. The information is reached to all the nodes in the network at a faster rate in the SIR model.

Case 2: Average Degree Centrality Node as the Start Node Average degree centrality node was selected as the start node. From Fig. 9, it is clearly seen, only in the SIR model there is a maximum spread of information in the network. It requires nearly 100 iterations approximately for the diffusion of information. In the SIS model, the Susceptible and Infected lines didn't even change from their default position which shows that it requires much more iterations or a different model to aid the diffusion in the network.

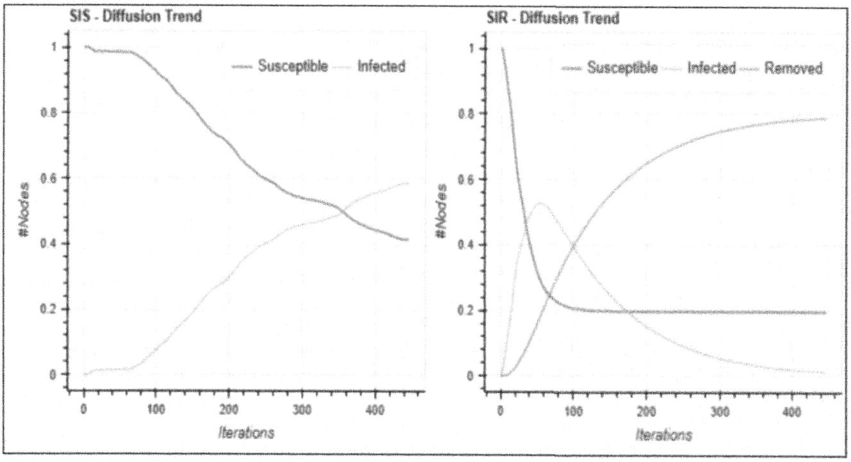

Fig. 8 Comparison of SIS and SIR models—highest degree centrality

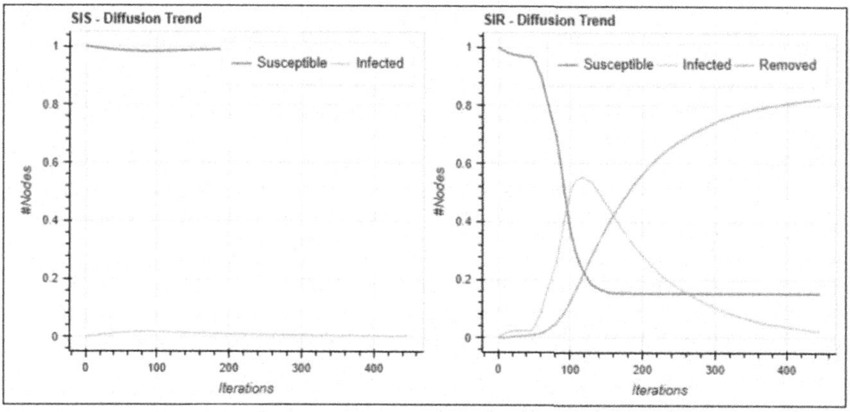

Fig. 9 Comparison of SIS and SIR models—average degree centrality

Case 3: Lowest Degree Centrality Node as the Start Node Lowest degree centrality node was selected as the start node. From Fig. 10, it is evident that in both the models there is no full information diffusion occurring in the network. It clearly shows that if an information is sent from a person with lowest centrality it takes more number of iterations for the spread of information.

Generally, the node with the highest degree centrality shall be used as the starting node for the information diffusion process. This is because the number of iterations needed for the process is comparatively less when node with higher centrality is chosen. The number of iterations required for each model can vary depending on the complexity of the model and the size of the population being studied.

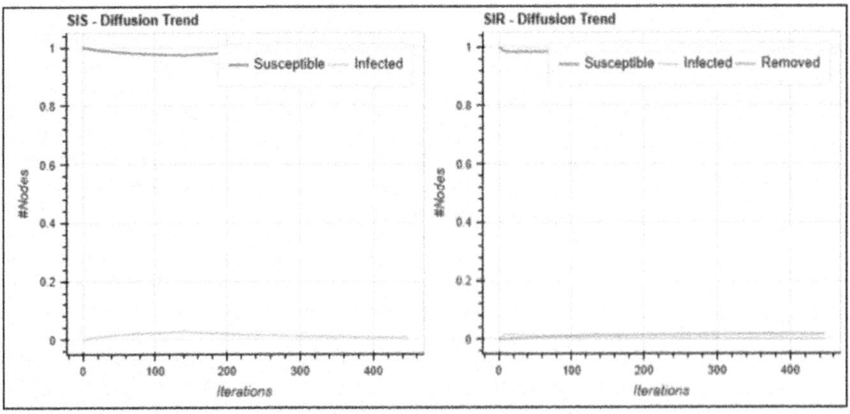

Fig. 10 Comparison of SIS and SIR models—lowest degree centrality

5.2 *Discussion II*

The hyperparameters beta and lambda decide the spread of information diffusion in the network. The comparison of SIS and SIR models is studied for various values of beta and lambda values.

Case 1: SIS Model Beta Versus Lambda The ratio of Beta and Lambda is the major factor of SIS model. Infection is grown exponentially during the initial period of time t and the rate of recovery is very small. As time grows, when Beta > Lambda, there exists small population which remains infected and the disease may prolong for certain time period. Here, Beta = 0.01 and Lambda = 0.005 were used. Figure 11 shows the rate of diffusion with the start node as highest degree centrality node and Fig. 12 with the start nodes as the second highest node. Though both the figures depict the same behavior for 100 iterations, the number of nodes infected by the former case was 5529 (out of 35,000), whereas it was 2548 in the latter case.

More number of people get recovered resulting in the end of epidemic after sometime, when Beta < Lambda. Here, Beta=0.005, Lambda=0.01 were used. The spread is slower in this case. The information is reached to a certain kind of population and there may be people who have to get information. In this SIS model, for both the scenarios, only limited population is reached with the information.

Case 2: SIR Model Beta Versus Lambda When Beta > lambda and t is small, the rate of infection is exponential. Here, Beta = 0.01 and Lambda = 0.005 were used. Nearly 18,000 nodes were infected when highest degree centrality node was selected as the start node at the 10th iteration; whereas nearly 9900 nodes were infected when the second highest degree was the start node. This scenarios is explained by Figs. 13 and 14, respectively. At the end of 100 iterations, more than 80% of nodes got infected and recovered by the former case, whereas 78% of nodes by the latter case. It means that information is reached nearly 80% of population.

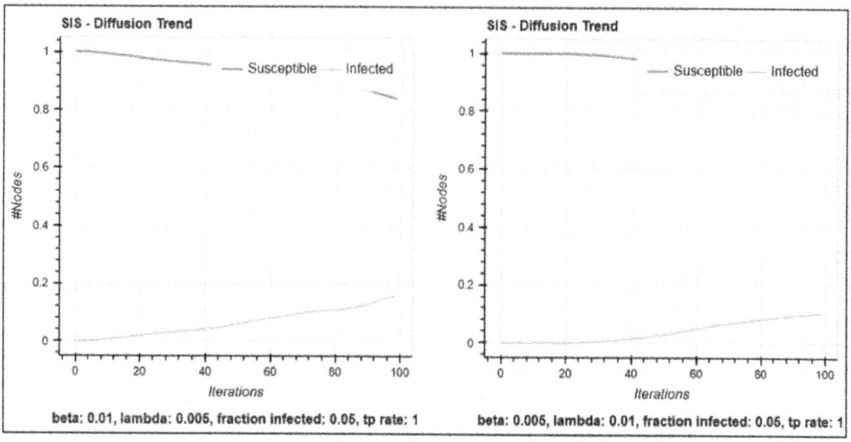

Fig. 11 Comparison of SIS model—highest degree centrality start node

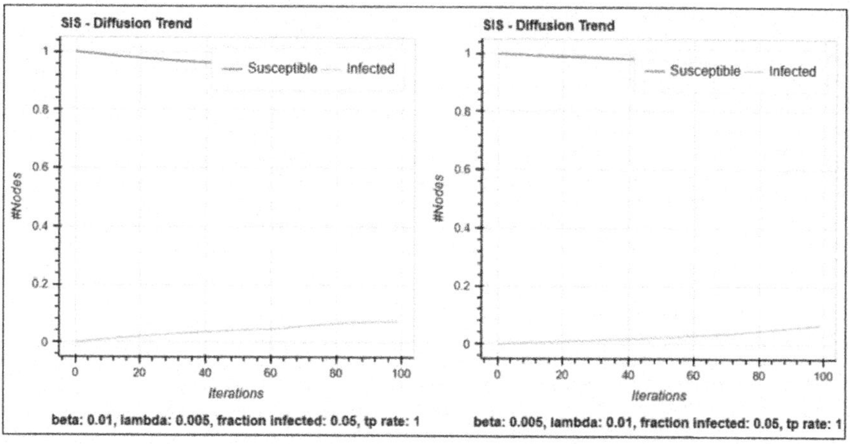

Fig. 12 Comparison of SIS model—second highest degree centrality start node

When Beta < Lambda, the information spread is very slow. Here, Beta = 0.005, Lambda = 0.01 were used. Only 2000 nodes were infected after 100 iterations. Generally, SIS models require more iterations than SIR models because of the constant movement of individuals between the susceptible and infected groups. SI models, on the other hand, do not require iterations because they assume that once an individual becomes infected, they remain infected for the duration of the simulation. If information has to be reached to a vast group of people, SIR model can be selected with suitable Beta and Lambda values; on the other hand, if information has to be reached in lower rate either SIS or SIR models can be selected with suitable Beta and Lambda values.

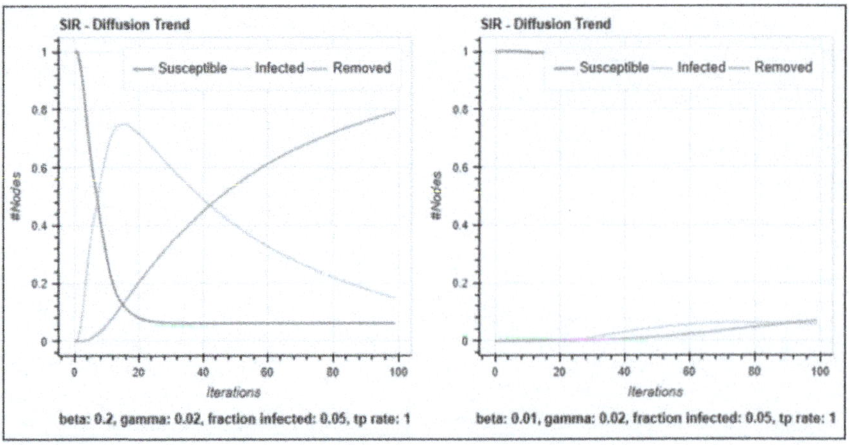

Fig. 13 Comparison of SIR model—highest degree centrality start node

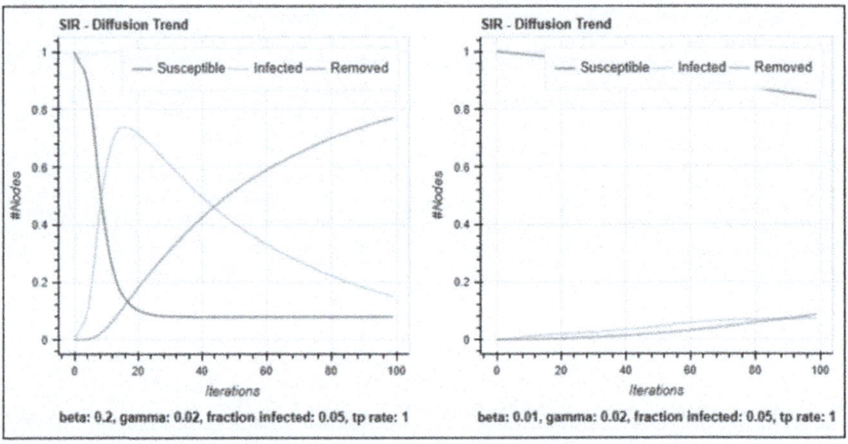

Fig. 14 Comparison of SIR model—second highest degree centrality start node

6 Conclusion

The social media platforms have completely changed the lifestyle of common men in their engagements and they use these platforms rigorously for communicating with each another. Social media platforms have made it possible for users to communicate with people all over the world and exchange information, ideas, and opinions, resulting in a massive global user network. These type of applications help them to connect with new clients, increase brand recognition, and boost sales. Businesses must comprehend social network characteristics and how they affect information dissemination in order to take full advantage of this potential.

Social network models can assist organizations in analyzing and comprehending the network's structure, user behavior, and the nature of shared material. This data can be utilized to produce content that is more likely to be shared and distributed extensively across the network or to construct efficient marketing campaigns that target particular network segments. Social network analysis applications would be helpful for companies, who can use it to change their business strategies for sustaining in the market. Future work shall include real time monitoring, content analysis and consumer behavior analysis and market prediction using their social media platforms for incorporating business intelligence.

Acknowledgements This research work is supported by SERB Teachers Associateship For Research Excellence (TARE) fellowship (TAR/2022/000425).

References

1. Abirami AM, Sarada S (2023) A comprehensive survey on information diffusion models for social media text: social media analytics. IGI Global 06. https://doi.org/10.4018/978-1-6684-8145-5
2. Abdelaziz Darwiesh KR (2022) Social media big data analysis: towards enhancing competitiveness of firms in a post-pandemic world. J Healthcare Eng 2022. https://doi.org/10.1155/2022/6967158
3. Cai M, Luo H, Meng X, Cui Y, Wang W (2023) Network distribution and sentiment interaction: Information diffusion mechanisms between social bots and human users on social media. Inform Process Manag 60(2):103197. https://doi.org/10.1016/j.ipm.2022.103197
4. Chae J, Thom D, Jang Y, Kim S, Ertl T, Ebert DS (2014) Public behavior response analysis in disaster events utilizing visual analytics of microblog data. Comput Graph 38:51–60. https://doi.org/10.1016/j.cag.2013.10.008
5. Chen S, Yuan X, Wang Z, Guo C, Liang J, Wang Z, Zhang X, Zhang J (2016) Interactive visual discovering of movement patterns from sparsely sampled geo-tagged social media data. IEEE Trans Vis Compu Graph 22(1):270–279. https://doi.org/10.1109/TVCG.2015.2467619
6. Christopher Perez SK (2023) An NLP-assisted Bayesian time-series analysis for prevalence of twitter cyberbullying during the covid-19 pandemic. Social Network Anal Mining 13 (2023). https://doi.org/10.1007/s13278-023-01053-4
7. Khatri I, Choudhry A, Rao A, Tyagi A, Vishwakarma DK, Prasad M (2022) Influcnce maximization in social networks using discretized Harris Hawks optimization algorithm and neighbour scout strategy
8. Chaudhary K, Mansaf Alam MS, Gumaei A (2021) Machine learning-based mathematical modelling for prediction of social media consumer behavior using big data analytics. J Big Data 8
9. Laynes Fiascunari V, Rabelo L (2022) Preliminary study for impact of social media networks on traffic prediction. In: Florez H, Gomez H (eds) Applied informatics. Springer International Publishing, Cham, pp 204–218
10. Li M, Wang X, Gao K, Zhang S (2017) A survey on information diffusion in online social networks: models and methods. Information 8(4). https://www.mdpi.com/2078-2489/8/4/118
11. Madila S, Dida M, Kaijage S (2021) A review of usage and applications of social media analytics. J Informa Syst Eng Manage 6 (2021). https://doi.org/10.21601/jisem/10958
12. Kauasar N, Asghar AliKhan MS (2022) Towards better representation learning using hybrid deep learning model for fake news detection. Soc Network Anal Mining 12. https://doi.org/10.1007/s13278-022-00986-6

13. Pawan Kumar AS (2021) Information diffusion modeling and analysis for socially interacting networks. Social Network Anal Mining 11 (2021). https://doi.org/10.1007/s13278-020-00719-7
14. Riquelme F, Vera JA (2022) A parameterizable influence spread-based centrality measure for influential users detection in social networks. Knowl Based Syst 257:109922. https://doi.org/10.1016/j.knosys.2022.109922
15. Singh SS, Srivastva D, Verma M, Singh J (2022) Influence maximization frameworks, performance, challenges and directions on social network: a theoretical study. J King Saud Univ Comput Inform Sci 34(9):7570–7603. https://doi.org/10.1016/j.jksuci.2021.08.009
16. Taie M, Kadry S (2017) Information diffusion in social networks. Springer, Berlin, pp 165–184 https://doi.org/10.1007/978-3-319-53004-8_8
17. Topirceanu A, Udrescu M (2022) On cost-effective strategies for opinion diffusion in complex networks. In: 2022 IEEE 16th International symposium on applied computational intelligence and informatics (SACI), pp 000179–000184. https://doi.org/10.1109/SACI55618.2022.9919571
18. Trifiro B, Clarke M, Huang S, Mills B, Ye Y, Zhang S, Zhou M, Chao C (2022) Media moments: how media events and business incentives drive Twitter engagement within the small business community. Soci Network Anal Mining 12. https://doi.org/10.1007/s13278-022-01003-6
19. Wanda P (2022) Runmax: fake profile classification using novel nonlinear activation in CNN. Soc Network Anal Mining 12. https://doi.org/10.1007/s13278-022-00983-9
20. Zakaria N (2021) Action network: a probabilistic graphical model for social simulation. Simulation 98. https://doi.org/10.1177/00375497211038759

Performance Prediction of 60 GHz Remote Optical Millimeter Wave Generation by Frequency 48 Tupling

Anu Sam and M. Ganesh Madhan

Abstract In this work, a novel scheme for frequency 48 tupling millimeter wave (mmw) generation using 2 dual parallel Mach–Zehnder modulator (DPMZM) through 2 stage modulation is investigated. When the first DPMZM is biased in maximum transmission point and adjusting the phase difference between the 2 MZM's, $(4n + 2)$ optical sidebands are generated. In the first stage, $\pm 6^{th}$ order optical sidebands beat at the detector and thereby 15 GHz signal for remodulating the second DPMZM is generated. In the second stage, DPMZM-2 acts as quadrupler which generates $\pm 4n$ order sidebands. By the suitable choice of modulation index of DPMZM-2 as 1.375, $\pm 2^{nd}$ order optical sidebands are enhanced and the other higher order optical sidebands are suppressed. The photodetector beats the $\pm 24^{th}$ order sidebands to generate frequency 48 tupling mm wave. Simulation results predict a radio frequency sideband suppression ratio (RFSSR) and optical sideband suppression ratio (OSSR) of 40 and 20 dB is obtained respectively, by considering the practical photodetector parameters.60 GHz mm wave is generated using a very low 1.25 GHz RF signal with a maximum power of − 40 dBm for a fiber length of 20 km when the optical amplifier gain is fixed as 21 dB.

Keywords 60 GHz · Dual parallel Mach–Zehnder modulator (DPMZM) · Frequency 12-tupling

A. Sam (✉) · M. Ganesh Madhan
Department of Electronics Engineering, Madras Institute of Technology, Anna University, Chennai, India
e-mail: anusam@mitindia.edu

M. Ganesh Madhan
e-mail: mganesh@annauniv.edu

S. Bandyopadhyay et al. (eds.), *Intelligent Computing Systems and Applications*, Lecture Notes in Networks and Systems 1010,
https://doi.org/10.1007/978-981-97-5412-0_21

1 Introduction

The integrated optical-wireless system provides a most promising solution for the high data rate traffic and huge demand for capacity, at a reasonable cost [1]. To meet the large demand of bandwidth, there is a need for high-frequency carriers in millimeter range. But the schemes of generating millimeter waves are quite challenging. One of the approaches is to generate it photonically, based on direct modulation, external modulation, optical heterodyning and optical up conversion [2]. Charan Teja V et al. reported the remote generation of 60 GHz millimeter wave by optical heterodyning technique [3]. M. Mohamed et al. investigated three modulation techniques for frequency quadrupling using a dual electrode MZM [4]. Reference [5] demonstrated the generation of 80 and 160 GHz from a 10 GHz RF signal by 2MZMs with large tunability. P. Shih et al. demonstrated the generation of millimeter wave using frequency quadrupling and four wave mixing [6]. Different frequency multiplication techniques are implemented for the generation of 60 GHz and also the transmission performance is studied [7]. Baskaran M et al. reported the generation of 60 GHz by utilizing the stimulated Brillouin scattering which needs an additional pump source. Moreover, it does not require any RF source [8]. For PON-based wireless transmission, a 60 GHz millimeter wave is generated using a simple digital circuit [9].

In this work, generation of 60 GHz millimeter wave by two stage modulation is proposed. A 15 GHz signal generated at the first stage is used as a modulation signal to drive second DPMZM, whereas first DPMZM generates frequency 12 tupled signal, when the modulation index is chosen as 5.136. The modulation index of second DPMZM is selected as 1.375, by considering the maximum RF voltage derived from the photodetector, for frequency quadrupling. The laser power delivered to DPMZM-1 and DPMZM-2 are fixed as 22 dBm and 19 dBm respectively, to ensure the modulation index of 1.375. A 60 GHz mm wave is generated from 1.25 GHz RF signal by, frequency 48 tupling. The generation of frequency 12 tupled and quadrupled mm wave signal are based on the references [10, 11]. We obtained a 48 tupled mm wave signal with a RFSSR of 40 dB using a lower modulation index and considering the practical photodetector parameters, compared with [12].

1.1 Principle of Operation

The schematic of 60 GHz remote mm wave generation using 48 tupling is shown in Fig. 1. The input laser light is fed to two DPMZMs. The DPMZMs are assigned with fixed laser power. The amplified output from DPMZM-1 is detected by a high power photodetector and is used to drive the second DPMZM. The output from second DPMZM is heterodyned at the detector and frequency 48 tupled mm wave signal is obtained. The first DPMZM consists of 2 MZM's (MZM1 and MZM2) connected in parallel. The two MZM's are separated by a phase shift of 90°. The first MZM output is given by

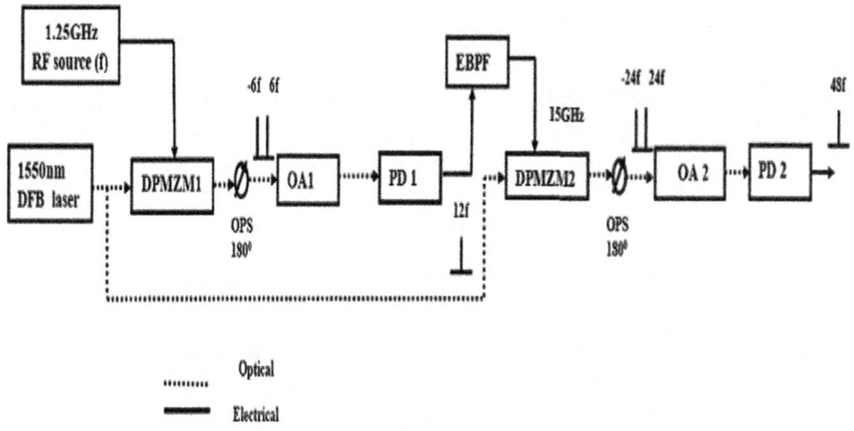

Fig. 1 Schematic diagram of 60 GHz remote mm wave generation using 48 tupling; RF: radio frequency source; DPMZM: Dual Parallel Mach–Zehnder modulator; OA: optical amplifier;OPS: Optical Phase shifter; PD: photo diode; EBPF: Electrical Bandpass Filter

$$E_{\text{MZM}-1} = \alpha E_i[\gamma e^{jnV_{21}/V_\pi} + (1-\gamma)e^{j\Pi V_{11}/V_\pi}] \tag{1}$$

where α is the insertion loss, γ is the power splitting ratio, V_Π is the halfwave voltage and E_i=$E_0e^{jw_c t}$ represents the light wave. Where E_0 and ω_c are the amplitude and angular frequency of DFB laser. V_{21} and V_{11} are the modulating signals fed to the 2 arms of the MZM 1.

The second MZM output is given by [10]

$$E_{\text{MZM}-2} = \alpha E_i[\gamma e^{j\Pi V_{22}/V_\Pi} + (1-\gamma)e^{j\Pi V_{12}/V_\Pi}] \tag{2}$$

V_{22} and V_{12} are the modulating signals fed to the 2 arms of the MZM 2.

Substituting $(1-\gamma) = \beta$, the combined output signals from both MZMs are given by

$$\begin{aligned}
E_0 &= E_{\text{MZM}-1} + e^{j\Pi}E_{\text{MZM}-2}\\
E_0 &= E_{\text{MZM}-1} - E_{\text{MZM}-2}\\
E_0 &= \alpha E_i\left[\gamma e^{j\Pi V_{21}/V_\Pi} + \beta e^{j\Pi V_{11}/V_\Pi} - \gamma e^{j\Pi V_{22}/V_\Pi} - \beta e^{j\Pi V_{12}/V_\Pi}\right]\\
E_0 &= \alpha E_i\left[\gamma e^{jm\sin(wt+\Pi)} + \beta e^{jm\sin wt} - \gamma e^{jm\sin(wt+3\Pi/2)} - \beta e^{jm\sin(wt+\Pi/2)}\right]
\end{aligned} \tag{3}$$

where $m = \Pi V_m/V_\Pi$ is termed as modulation index.

Expanding equation (3) using Jacobi-Anger expansions, we get

$$\begin{aligned}
E_0 = \alpha E_i\Big[&\gamma \sum J_n(m)e^{j(wt+n\Pi)} + \beta \sum J_n(m)e^{jnwt}\\
&-\gamma \sum J_n(m)e^{j(wt+n3\Pi/2)} - \beta \sum J_n(m)e^{j(wt+n\Pi/2)}\Big]
\end{aligned} \tag{4}$$

$$= \alpha E_i\Big[\gamma \sum J_n(m)e^{jnwt}\{[(1 - e^{jn\Pi/2})(\gamma(-1)^n + \beta)]\}\tag{5}$$

where J_n (m) is the Bessel function of first kind of order 'n'

Neglecting (β–γ) and after simplification, Equation (5) can be written as

$$E_0 = \alpha E_i\Big[\sum J_{4n+2}(m)e^{j(4n+2)wt}\Big]\tag{6}$$

$$E_0 = \alpha E_i[J_2(m)\cos(2wt) + J_6(m)\cos(6wt) + J_{10}(m)\cos(10wt)]\tag{7}$$

Since Bessel functions are of decreasing nature and modulation index is chosen as 5.136, so that all higher order other than 6th order sidebands are neglected. All the odd order sidebands are suppressed, as the two MZMs are biased at maximum transmission point. $J_2(5.136) = 1.28 * E{-}4$, $J_6(5.136) = 0.145503$, $J_{10}(5.136) = 0.00186$. At the receiving end, the ± 6th order sidebands are heterodyned at the photodetector.

$$\begin{aligned} I_{\text{out}} &= R|E_o|^2 \\ &= RE_i(t)J^2(\beta)\cos(12wt) \end{aligned}\tag{8}$$

The photocurrent obtained is given as the modulating signal to drive the second DPMZM. For quadrupling, the modulation index, $m = 1.375$ is chosen, such that all the higher order sidebands are neglected. In literature, the suitable modulation index for quadrupling is selected as 2.4048 [11]. In practice, to obtain amodulation index as 2.4048, the modulating signal amplitude should be 2.29 V. This mandates an optical amplifier of gain 30 dB along with a larger optical power. However, photodetector saturates at higher power levels. Moreover higher magnitudes of photocurrent is not feasible. Hence, we consider a typical photodetector, Finisar HPDV2120R, whose saturation current is 34 mA. To operate within the saturation current limit, the modulation index is fixed as 1.375, for which the input current to the first photodetector is calculated as 25 mA, which is less than the saturation current of the photodetector.

In this scheme, all the odd order sidebands are suppressed, as MZM3 and MZM4 are biased at maximum transmission point. $J_0(1.375) = 0.5803$, $J_2(1.375) = 0.201$, $J_6(1.375) = 1.37 * E{-}4$. Further, the laser carrier is also suppressed by introducing an optical phase shift of 180°. The higher order sidebands are neglected as they have minimal power levels. At the photodetector, the ± 24th order sidebands are heterodyned and frequency 48 tupled mm wave signal is obtained

$$I_{\text{out}} = RE_i(t)\, J_2(\beta)\cos(4w\alpha t)\tag{9}$$

Substituting $w\alpha = 12w_m$

$$I_{\text{out}} = RE_i(t)\, J_2(\beta)\cos(48w_m t)\tag{10}$$

Simulation Results and Discussion. The proposed system for frequency 48 tupling is simulated on optisystem16 platform. A DFB laser operating at 193.1 THz with a linewidth of 1 MHz provides input to 2 DPMZMs. The two DPMZMs are provided with laser powers of 22 dBm and 19 dBm, respectively. The half wave voltages of the two dual parallel Mach–Zehnder are 3.5 V and 3 V, respectively. The gain of the 2 optical amplifiers is fixed as 21 dB and 20 dB, respectively. The insertion loss and extinction ratio of each MZM is assigned as 5 and 30 dB. The modulation index for the frequency 12-tupling in the first stage is chosen as 5.136. The modulating signal amplitude is calculated as 5.725 V. The $\pm 6^{th}$ order optical sidebands are heterodyned in high power photodetector, HPDV2120R, with responsivity of 0.52 A/W.

In the second stage, the generated 15 GHz RF signal is used as a modulating signal to modulate the amplified optical carrier using second DPMZM. For frequency quadrupling to be achieved, the required modulation index is 2.4048. However, considering the practical limitations of photodetector into account, the modulation index is fixed as 1.375. The quality of generated mm wave is not compromised even when a low modulation index is selected.

In order to attain the modulating signal amplitude of 1.314 V, the laser power delivered to the first DPMZM has to be fixed as 22 dBm [13]. Figure 2 shows the variation of modulating signal amplitude with the laser power provided to the first DPMZM. The modulating signal amplitude is directly consonant to the laser power. The optical and RF spectrum of frequency 48 tupling is shown in Fig. 3. The optical spectrum reveals first and second sidebands alone.

The $\pm 1^{st}$ order sidebands are suppressed as the DPMZM-2 is biased in maximum transmission point. The frequency 48 tupling achieved an OSSR and RFSSR of 20 dB and 40 dB, respectively. Figure 4 shows the modulating signal of amplitude 1.314 V

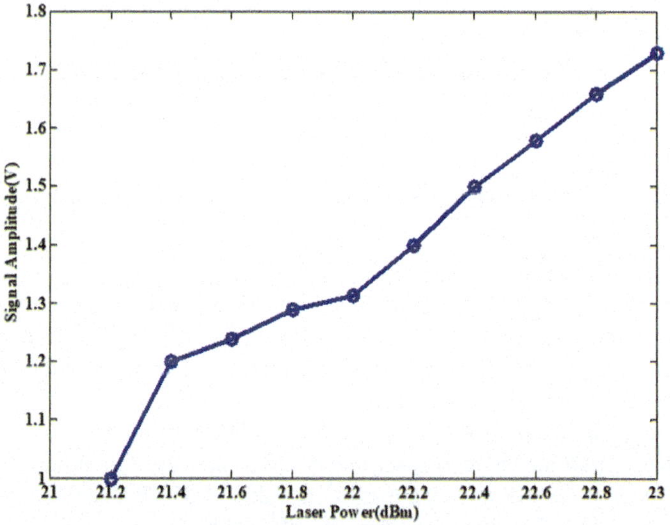

Fig. 2 Relation of detected RF voltage with laser power delivered to DPMZM-1

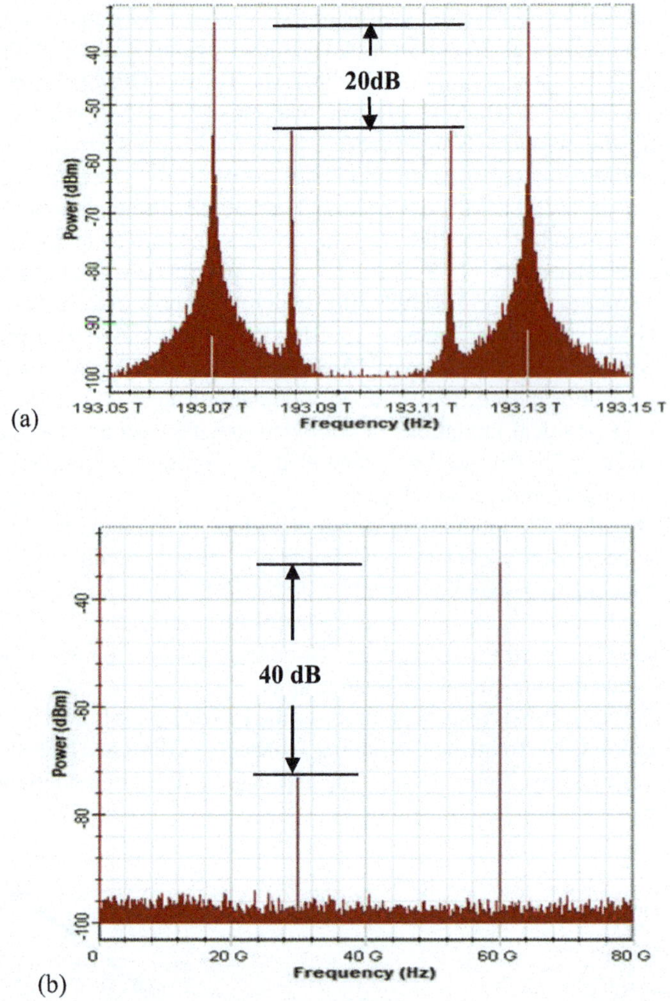

Fig. 3 **a** Optical spectrum and **b** radio frequency spectrum of frequency 48 tupling at modulation index, $m = 1.375$

when the laser power delivered to the first DPMZM is 22 dBm. Figure 5 shows the effect of laser power delivered to first DPMZM on OSSR and RFSSR of 48 tupling. Simulations predict a maximum value of OSSR around 42 dB, when the laser power delivered to first DPMZM is 31.5 dBm.

However at this condition, photodetector (HPDV2120R) saturates due to high power. Hence, a lower value of laser power, 22dBm is used to attain an OSSR of 20 dB. Similarly, RFSSR shows a maximum value of 65 dB when the laser power delivered to the first DPMZM is 28dBm. The effect of laser power delivered to second DPMZM on OSSR and RFSSR is shown in Fig. 6. The slope of OSSR with laser

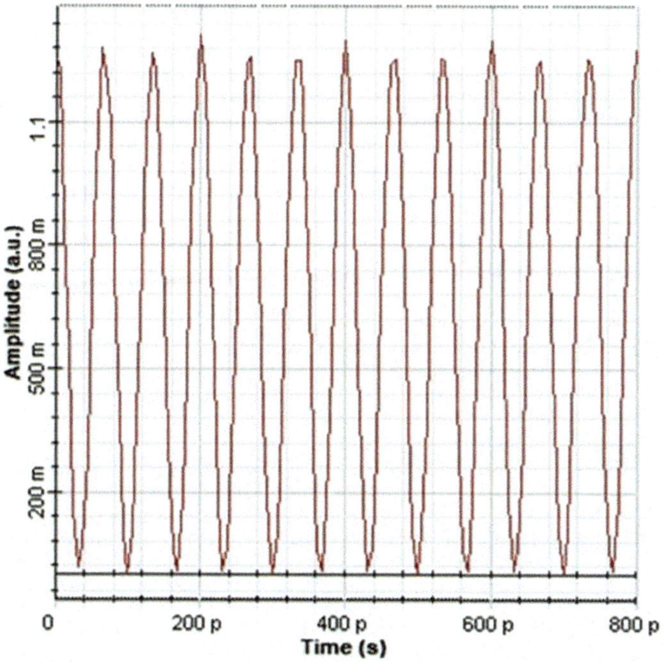

Fig. 4 15 GHz RF signal, when the laser power delivered to DPMZM-1 is 22 dBm

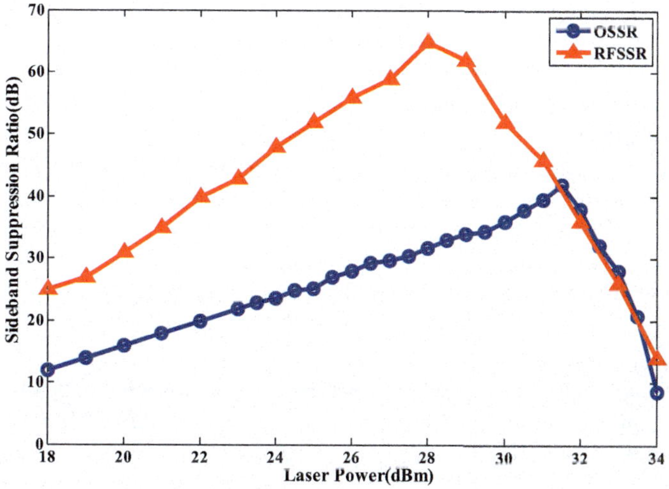

Fig. 5 Effect of laser power delivered to first DPMZM on OSSR and RFSSR

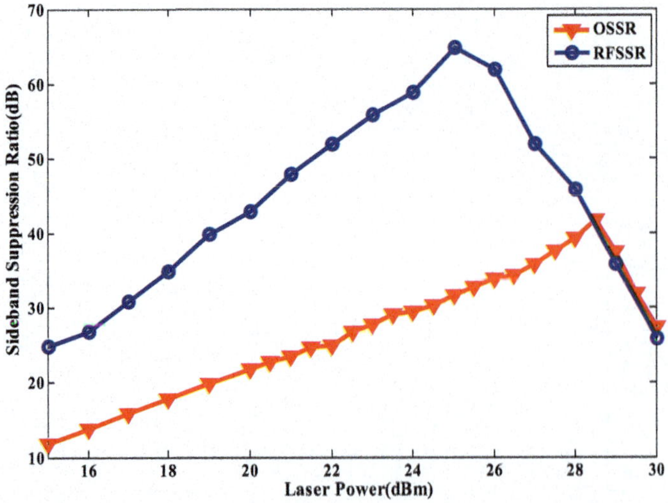

Fig. 6 Effect of laser power delivered to second DPMZM on OSSR and RFSSR

power is lower than RFSSR. The optical power delivered at second DPMZM is fixed as 19dBm and the corresponding OSSR and RFSSR values are 20 dB and 40 dB, respectively. But the maximum OSSR is 42 dB for a laser power of 28.5 dBm.

Figure 7 shows the effect of optical amplifier gain (OA1) on OSSR and RFSSR of frequency 48 tupling. It is evident from the figure that, the RFSSR and OSSR attains its maximum value of 64 and 35 dB when the optical amplifier gain is within the range of 27–30 dB. For practical realization of available devices, the optical amplifier gain is fixed as 21 dB. So the optimum OSSR and RFSSR is achieved as 20 dB and 40 dB, respectively.

The effect of optical amplifier gain on 60 GHz RF power for different fiber lengths is illustrated in Fig. 8. The amplified signal from the optical amplifier (OA2)is transported through 20 km optical fiber. 60 GHz mm wave is generated by frequency 48 tupling with a maximum power of – 40 dBm for a fiber length of 20 km when the optical amplifier (OA1, OA2) gain are fixed as 21 dB and 20 dB respectively, even though a high RF power is generated at a gain of 25 dB. This is due to the practical realizations of devices.

Conclusion. This work suggests a novel straight forward and practically realizable system to generate frequency 48 tupling mm wave using 2 DPMZMs. 60 GHz mm wave is generated from a low frequency 1.25 GHz RF signal by frequency 48 tupling. This system is simulated and optimized by adjusting the modulation index and optical amplifier gain. The selected parameter values are fitted into the system by considering the practical device constraints. The proposed work attained an OSSR and RFSSR of 20 dB and 40 dB, respectively. The OSSR and RFSSR of frequency 48 tupling is analyzed with the effect of laser power and optical amplifier gain.

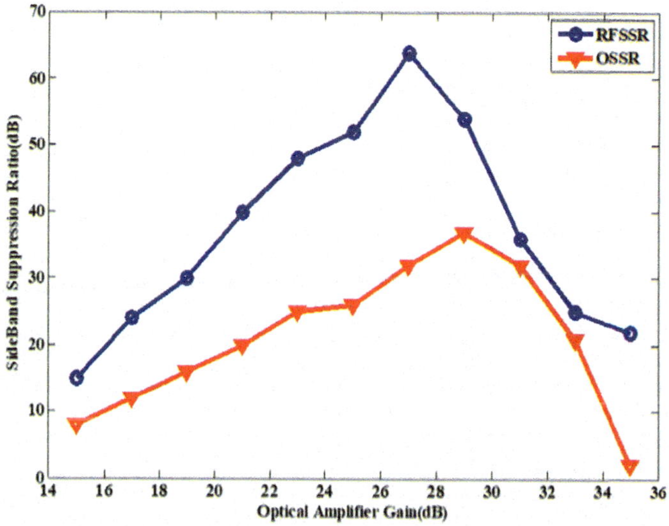

Fig. 7 Effect of optical amplifier gain (OA1) on OSSR and RFSSR of frequency 48 tupling

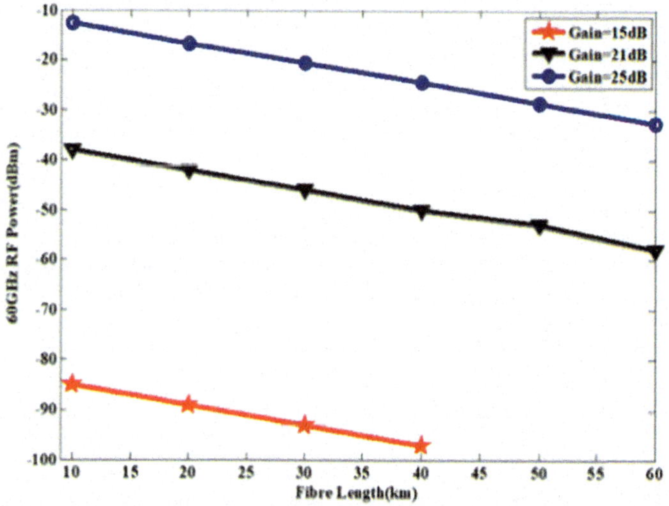

Fig. 8 Effect of optical amplifier gain on 60 GHz RF power

Acknowledgements The authors acknowledge Anna University, Chennai, for granting financial support to carry out the research work under Anna Centenary Research Fellowship (ACRF) scheme. One of the authors, Anu Sam is grateful to Anna University, Chennai for the award of Anna Centenary Research Fellowship [Ref No: CFR/ACRF/ 21254391260/AR1].

References

1. Jia Z, Yu J, Ellinas G, Chang G-K (2007) Key enabling technologies for optical–wireless networks: optical millimeter-wave generation, wavelength reuse, and architecture. J Lightwave Technol 25:3452–3471
2. Dar AB, Ahmad F (2022) Optical millimeter-wave generation techniques: an overview. Optik 258:168858
3. Charan Teja V, Ganesh Madhan M (2016) Numerical simulation of millimeter wave generation in a digital fiber optic link. In: International conference on communication and signal processing, ICCSP, pp 351–355
4. Mohamed M, Zhang X, Hraimel B et al (2008) Analysis of frequency quadrupling using a single Mach-Zehnder modulator for millimeter-wave generation and distribution over fiber systems. Optics Express 16:10786–10802
5. Huang Z, Cao C, Feng Z et al (2021) A filterless radio-over-fiber system that generates 80 and 160 GHz millimeter waves based on two MZMs. In: Applied optics
6. Shih P, Chen J et al (2010) Optical millimeter-wave signal generation via frequency 12-tupling. IEEE/OSA J Lightwave Technol 28(1):71–78
7. Chithra S, Rimmya C, Madhan MG (2020) Photonic generation of 60 GHz mm wave based on frequency multiplication techniques using DP- MZM. In: International conference on Emerging Trends in Information Technology and Engineering, ic-ETITE
8. Baskaran M, Ganesh Madhan M (2014) A novel approach for simultaneous millimeter wave generation and high bit rate transmission for radio over fiber (RoF) systems. Optik 125:6347–6351
9. Rimmya C, Ganesh Madhan M (2019) An efficient scheme for simultaneous 60 GHz MM wave generation and CSRZ-DQPSK transmission in a passive optical network. Microwave Optical Technol Lett 1–8
10. Dar AB, Ahmad F (2019) Filterless optical millimeter-wave generation using cascaded-parallel Mach–Zehnder modulators with tunable frequency multiplication factor. Optical Quantum Electronics 53:56
11. Jha RK, Al-Sharee NA, Hassan SIS, Malek F, Ngah R, Abbas SA, Aljunid SA (2013) A cost-effective method for high-quality 60GHz optical millimeter wave signal generation based on frequency quadrupling. Progr Electromagnetics Res 137:255–274
12. Shang Y, Feng Z, Cao C, Huang Z, Wu Z, Xu X, Geng J (2022) A using remodulation filterless scheme of generating frequency 32-tupling millimeter-wave based on two DPMZMs. Optics Laser Technol 148:107793
13. Ralston JD, Mathur A, Welch DF (1996) High-power 1550-nm distributed feedback lasers. Conf Lasers Electro-Optics 9

Highly Secure Lsb-based Image Steganography with Four-factor Security

A. Sreenidhi, B. Shruti, Ambati Divya, N. Subhashini, and S. Muthulakshmi

Abstract In view of rapidly evolving technology, there comes a fair share of vulnerabilities, and a greater need for protection of confidential data. Although encryption techniques are popular to make sensitive data indecipherable, the same can be extracted if the appropriate key is obtained. Steganography, however, aims to keep the very existence of the message hidden from prying eyes. It is the art of secretly writing sensitive text inside an image or an audio file in such a way, that anyone viewing the image would not see any difference between the original and modified image. However, neither technique is perfect or secures enough on sensitive data. The four factors in focus are the Rail Fence Cipher, RSA Encryption, Aztec Code, and LSB Steganography. The data will be passed one by one through all the processes, and the output from one will be passed into the next, thus providing the user with four-fold security. Once the final encrypted data in the form of an Aztec code is hidden inside the cover image, it will be visually indistinguishable from the original image, since only the least bits of the image has been altered to hide the image. This method is more secure than conventional methodologies since even if one factor is compromised, the sensitive data remains encrypted and undecipherable to all malicious intent.

Keywords Image Steganography · RSA Encryption · Rail Fence Cipher · Aztec code · LSB steganography

1 Introduction

Our rapidly digitizing world has opened doors to an entire world of opportunities. There are many methods to convert the data into something incomprehensible, such as public-key cryptography, symmetric encryption, and watermarking. Hiding the data in images or videos provides access to data to the user with well-enhanced

A. Sreenidhi · B. Shruti · A. Divya · N. Subhashini (✉) · S. Muthulakshmi
School of Electronics Engineering, Vellore Institute of Technology, Chennai, India
e-mail: subhashini.n@vit.ac.in

S. Bandyopadhyay et al. (eds.), *Intelligent Computing Systems and Applications*,
Lecture Notes in Networks and Systems 1010,
https://doi.org/10.1007/978-981-97-5412-0_22

accuracy and flexibility. But, on the other side, also endangers the possibility of tampering with the content and copyright violation. The main intent of data hiding is to protect the data from unauthorized parties. The practice must be inaudible or invisible to the observers.

Steganography involves hiding secret or confidential information within various media formats that serve as a cover for the same. The secret information is incorporated into the cover image in a way that it becomes invisible to the naked eye; making it non-suspicious. In this modern world, where technology changes faster than humans, security requirements need to be constantly updated to be in step with the fast-paced world. Therefore, to challenge the attackers, different cryptography algorithms can be combined making the hidden data more secure and flexible.

Rail Fence Cipher is a further development method of Caesar Cipher. It is an arrangement of the position of the characters, instead of requiring calculation of a heavy mathematical operation. The input text is written on continuous rails of a hypothetical matrix in a downward-slanting manner. On reaching the bottom, we traverse upwards moving diagonally, and this is repeated until the entire string is covered. The cipher algorithm depends on the shift key value, which is to be provided by the user.

Now, the ciphered text is encrypted using the RSA algorithm to double the security. The RSA algorithm brings one of the significant improvements in the field of cryptography. Until a certain period, only symmetric key encryption was in effect which used only one key for both encryption as well as decryption. If that key is disclosed, the entire communication stands at a risk to be compromised, irrespective of the complexity of conversion applied. Hence making the algorithm sensitive and data vulnerable. This is the primary reason for the proposal of asymmetric methods. RSA is an asymmetric cryptography algorithm that is widely used for secure data transmission.

Asymmetric algorithms involve the usage of different keys for the sender and the receiver. It involves the usage of 2 different keys for encryption and decryption, namely the public key and the private key. The public key is utilized for encryption and is accessible publicly. It cannot be used for the decryption of encrypted data. On the other hand, the Private Key used for decryption is kept private, and only the intended recipient has access to the same. It is a well-known fact that it is not easy to factorize a large integer. Unless the unauthorized person is able to pinpoint the exact co-prime utilized, it is not possible to obtain the Private Key. The encryption strength also depends on the key size. The larger the key size, the greater the strength of the algorithm. The strength of encryption increases exponentially which makes the decryption hard, opening another window in the security providing arena. Considering the complexity of Elliptic Curve Cryptography (ECC) over the RSA algorithm, the emphasis of this study is on RSA.

Further, RSA encrypted string is converted into an Aztec code, adding to the security level. The Aztec Code is a two-dimensional (2D) scannable matrix barcode designed to have much greater accuracy than its counterparts. Furthermore, it utilizes less space than other codes due to the lack of requirement of a 'quiet zone' since it is read from the inside out; thereby making the overall process more efficient. The Aztec code is then stored in the Least Significant Bit (LSB) plane of the image.

The main idea behind LSB embedding is that even if changes are made to the last bit of any given pixel, there will not be any significant difference in the outward appearance of the image in question. Hence, by changing the LSB of any given image, we can store secret information in the same without changing the external appearance. Performing bit plane slicing on the output image will show where the Aztec code is hidden. In [1], authors focus on encrypting secret message into QR codes, where the QR code is used as the container where the secret message is stored. The container is used to hide the payload and also provide false information to the attacker. Only a QR reader has the ability to read the QR code's message.

In [2], authors throw light on cryptography, its basic concepts, and compression. In [3] the authors have used Fresnelet transform to offer a frequency-domain steganography-based information camouflaging strategy (FT). The source/cover/stego image is pre-processed before getting decomposed using the Fresnelet transform. Furthermore, in this study, a QR code is employed as cryptic data, which adds an extra layer of creativity to the presented methodology.

Several symmetric and asymmetric encryption algorithms have been reviewed in [4] to determine the effectiveness of encryption techniques. The techniques are compared, analyzed, and tested using cryptographic tools. With an aim to combine the concepts of steganography and QR codes for implementing a new method for secure transmission, the authors have presented a methodology that splits into two phases: Encryption of the message using key values obtained by a heuristic algorithm and concealing the cipher text inside the QR code[5].

In [6], the authors provide a literature review on a variety of different methods, algorithms, and schemes in image steganography to analyze and investigate them. The authors discuss the types of images and types of host documents in [7]. Techniques of digital watermarking and steganography in frequency and spatial domains are studied. LSB can be changed in each pixel with information without compromising the quality of an image and the intensity of the color is not affected.

Four popular secret key encryption algorithms DES, 3DES, AES, and blowfish have been implemented and their performance is compared by encrypting input files of varying contents in [8]. In [9], RSA algorithm is reviewed and analyzed. Its benefits and drawbacks have been discussed, and solutions to the problem have been discussed. The authors discuss RSA as the most reliable cryptographic method now in use for securing data network communication.

A system was proposed in [10] in which, RSA and enhanced RSA are compared based on execution time. It is proven that RSA provides the highest security to business applications. This system can be used for encrypting long messages without employing hybrid and symmetric algorithms. In [11], an approach is suggested, that uses Discrete Wavelet Transform and Quick Response (QR code) to create a fusion steganography scheme (DWT). This method incorporates frequency-domain encoding and decoding actions. The QR code graphic conceals the written content. The discrete wavelet transform conceals the picture of the QR code.

A method that integrates the principles of Steganography with QR codes is used in [12]. The suggested procedure is divided into two stages: Using a QR code encoder to encrypt the message and generate a QR code followed by using a color picture to hide the QR code. An adaptive method that can select the most optimal pattern to minimize the error ratio due to message embedding is proposed in [13]. This adaptive pattern can optimize the performance of the inverted LSB substitution method, based on the two-bit + least-significant-bit (LSB) pattern in the container image.

Encrypting messages inside valid QR codes is the focus in [14]. The QR code is used as the container where the secret message is stored. It uses the container not only to hide the payload but also to give misleading information to the adversary. Only a QR reader has the ability to read the QR code's message.

In [15], authors examine and analyze the steganography algorithm of incorporating the QR code comprising secret information into the cover picture, which would be based on the JSteg algorithm, to assure information security as well as ease and convenience of transmission. Many cover pictures have been utilized to examine secret messages of variable size, and standard parameters have been set to validate, evaluate and verify the efficiency. Hiding the message in the Least Significant Bit (LSB) method and also using the LSB method used for various file formats has been discussed in [16].

2 Methods and Materials

This paper aims to develop a system that provides four-factor security to secure and incorporate confidential information within images. The driving idea behind the same is that even if one of the methods were to be compromised somehow, the data would still have three more levels of security to protect the hidden information, making it extremely hard to decipher or tamper with. The four methodologies utilized in this paper include Rail Fence Cipher, RSA Encryption, Aztec Code, and LSB Substitution.

Initially, the user is prompted to enter the secret message needed to be hidden, along with a shift value for the first stage of the process, i.e., Rail Fence Cipher. The Rail Fence Cipher is a simple, transposition cipher with the main ideology to jumble the arrangement of the characters of a message to make it incomprehensible. It works by positioning the characters of the message diagonally across a matrix created using the shift value as the number of rows, and the length of the string as the number of columns. After the completion of the string, each line is read horizontally, and the characters are collected to form the cipher text, as depicted in Fig. 1.

In the next stage, the cipher text is encrypted using the RSA algorithm. RSA is an asymmetric algorithm, meaning that it utilizes two separate keys for encryption and decryption; namely the public and private key. As the name suggests, the public key is accessible to everyone to use for encryption, however only the intended recipient of the message will have access to the private key. In this paper, we will be using inbuilt

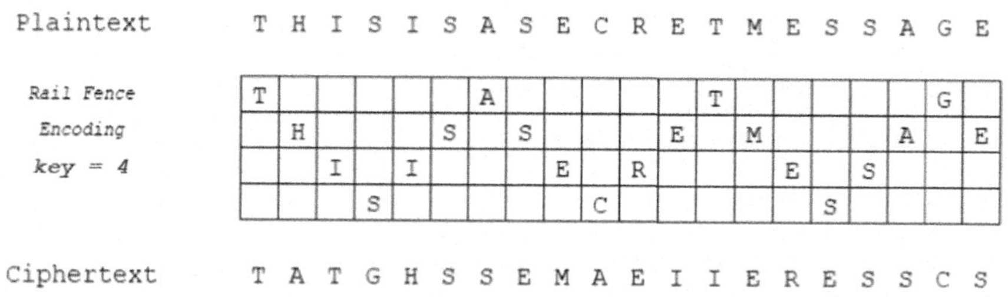

Fig. 1 Overview of Rail Fence Cipher

functions in Python to generate the same. RSA offers much greater key lengths than its counterparts, increasing by multiples of 256, up to 4096 bits. A comparison of popular encryption methods has been presented in Table 1.

It is known that the greater the key value, the greater the strength of encryption. However, there always exists a trade-off between performance and security, hence a key length that offers sufficient encryption strength as well as an acceptable execution time is required. Keeping in mind the same, an optimum key length of 512 bits was chosen for this work.

In the third stage, the RSA encrypted data is converted into an Aztec code. The Aztec code is a two-dimensional matrix symbology designed to be more accurate than other 2D matrices. The name of the code is derived from the central finder pattern's resemblance to an Aztec pyramid when viewed from above. One of the main benefits of the code is that it utilizes less space as compared to other symbologies like QR codes, due to the lack of requirement of a 'quiet zone'. Since Aztec codes are read from the inside out, it eliminates the requirement of an external delimiter boundary, thereby requiring lesser storage space.

In the fourth and final stage, the Aztec code is hidden in the least bit plane of the chosen cover image. The main idea behind LSB substitution is that even if changes are made to the least significant bits of any pixel of an image, there will not be any significant visual change in the overall appearance of the same. Hence, the least bit plane of the image is replaced by the image of the Aztec code and exported as the output image, which looks identically the same as the original image, thereby successfully completing the process. The entire process of hiding and retrieving the secret message has been depicted in the form of a flowchart, in Figs. 2 and 3.

Table 1 Comparison of popular encryption algorithms

Size (bits)	DES	Triple DES	AES	Blowfish	RSA
Plaintext block size	64	64	128	64	176
Cipher text block size	64	64	128	64	176
Key size	56	112 or 168	128, 192 or 256	448	512

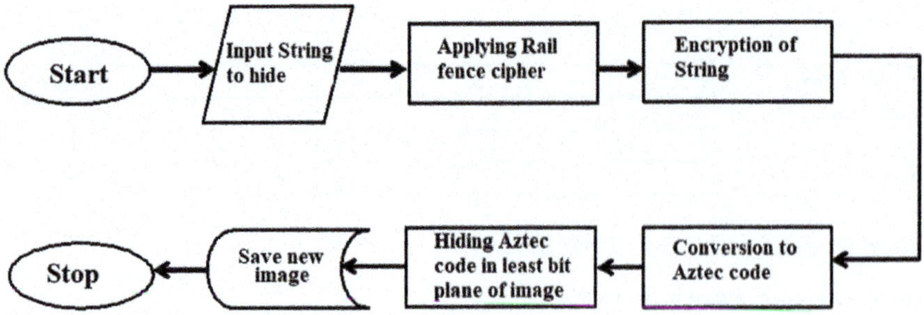

Fig. 2 Block diagram for hiding secret message

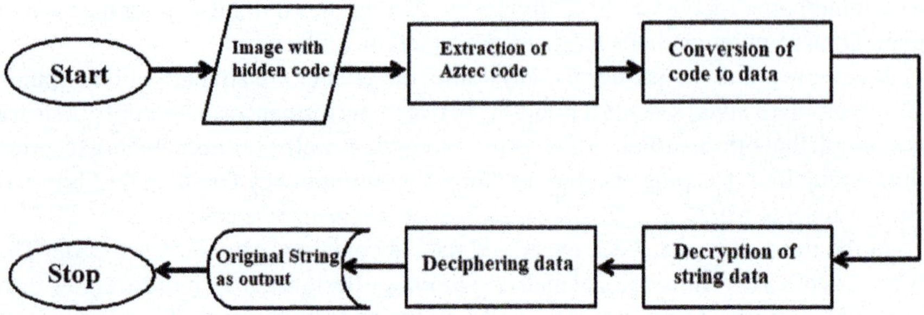

Fig. 3 Block diagram for retrieving secret message

3 Results and Discussion

The methodology discussed above was implemented in the web-based coding platform 'Google Colab' using the Python language. The process for achieving four-factor security of sensitive data was successfully implemented and tested. The secret message provided by the user is ciphered, encrypted, and concealed in the cover image in the form of an Aztec code without any changes to outward appearance; and further is also retrieved successfully. Figure 4 shows the comparison of the original image and the output image. It is clear that there is no visual difference between the two images; hence, the Aztec code is hidden successfully in the least significant bit plane without raising suspicion.

Any grayscale image consists of 8 bit planes where the first-bit plane is the most significant, and the 8th-bit plane is the least significant. The Aztec code image replaces the 8th-bit plane of the cover image. The same can be visualized using bit plane slicing. Figure 5 displays all the bit planes of the output image, and as expected the 8th-bit plane contains the Aztec code. During decryption, the same is extracted to decrypt the secret message.

Further, the performance of the same can be evaluated using a few metrics suited for the cryptographic algorithms, namely execution time, brute force decryption time, and Mean Square Error.

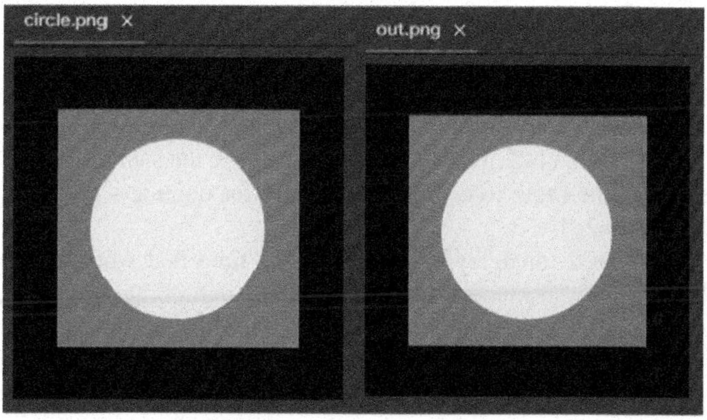

Fig. 4 Cover image versus output image

Fig. 5 Bit planes of output image

RSA has the highest execution time out of all other encryption algorithms [17], mainly owing to the large key size. Hence, to strike a balance between execution time as well as security, a key size of 512 bits was utilized which provided optimum results. Even though execution time is mildly greater (343 ms) as compared to the AES Fernet algorithm (148 ms), RSA compensates for the same with exponentially greater security. The brute force decryption time for popular algorithms has been presented in Table 2.

The time required for decryption denotes the time that would be required to obtain the correct key if random keys were tried using a brute force approach. As seen above, the RSA algorithm is the strongest since it takes the longest time for forceful malicious decryption of the cipher text. Table 3 records the decryption time for the most commonly used shift values of Rail Fence Cipher, combined with RSA.

As seen from Table 3, the more the options available for shift values, the longer it will take to decrypt the message using a Brute force attack.

It was observed that the quality of the output image is not drastically altered. Mean Square Error is one of the most common methods used to compare image quality. It measures deviation by calculating the average of the squared difference between the observed and predicted values; i.e., it can be used to calculate the deviation of the output image from the original image. Values closer to zero signify that the quality distortion is minimal. MSE value between the cover and output image was calculated to be ~ 0.54; hence there is a very marginal change in the image quality.

Table 2 Time taken for brute force decryption of different encryption algorithms

Key size (bits)	Cipher	No. of alternative keys	Decryption time
56	DES	$2^{56} = 7.2 \times 10^{16}$	2^{55} ns = 1.125 years
128	AES	$2^{128} = 3.4 \times 10^{38}$	2^{127} ns = 5.3×10^{21} years
168	3DES	$2^{168} = 3.7 \times 10^{50}$	2^{167} ns = 5.3×10^{33} years
192	AES	$2^{192} = 6.3 \times 10^{57}$	2^{191} ns = 5.3×10^{40} years
256	AES	$2^{256} = 1.2 \times 10^{77}$	2^{255} ns = 5.3×10^{60} years
512	RSA	$2^{512} = 1.3 \times 10^{134}$	2^{511} ns = 5.3×10^{90} years

Table 3 Time taken for Brute force decryption for different shift values

Shift key for Rail Fence Cipher	Alternate keys	Decryption time
3	2	$15.9 * 10^{90}$ years
4	3	$21.2 * 10^{90}$ years
5	4	$26.5 * 10^{90}$ years
6	5	$31.8 * 10^{90}$ years
7	6	$37.1 * 10^{90}$ years

Further, the CPU process time can be measured, i.e., the time that the CPU required to dedicate to the particular process for calculations. The load of the CPU is reflected, hence greater the CPU time used in the encryption process, the higher the CPU load. On average, the CPU time for encoding and decoding is around ~ 40 s.

4 Conclusion

In this work, a methodology to achieve four-factor security of sensitive data was successfully developed, with the help of an improved LSB-based Steganography. Rail Fence Cipher and asymmetric RSA encryption algorithm were used to make the sensitive message indecipherable, further converting the encrypted data into an Aztec code.

The code was hidden in the least bit plane of the cover image, making it virtually indistinguishable. Hence, sensitive data was concealed within an image file without changing the outward appearance, and successfully retrieved as well. To verify the same, bit plane slicing was executed to visualize the Aztec code stored in the LSB of the image. Also, the merits of the system were discussed through means of execution time and strength of the same. Further research can be conducted on how to increase the security even further by improving encryption algorithms or using different algorithms for Steganography and comparing their performance.

References

1. Alajmi M, Elashry I, El-Sayed HS, Allah OSF (2020) Steganography of encrypted messages inside valid QR codes. IEEE Access
2. Kumari (2017) A research paper on cryptography encryption and compression techniques. Int J Eng Comput Sci
3. Maheswari J, Hemanth D (2015) Frequency domain QR code-based image steganography using Fresnelet transform. AEU Int J Electronics Commun 69(2):539–544
4. Mandal ARD (2019) A review paper on encryption techniques. IJRAR
5. Manimekalai RB (2017) Hide and seek: a new way to hide encrypted data in QR code using the concepts steganography and cryptography. Int J Adv Res Comput Commun Eng
6. Mohammad AR (2018) Image steganography techniques. IJARCCE Int J Adv Res Comput Commun Eng
7. Mstafa R, Bach C (2013) Information hiding in images using steganography techniques. In: ASEE Northeast section conference
8. Nadeem MYJ (2005) A performance comparison of data encryption algorithms
9. Nisha MF (2017) RSA public key cryptography algorithm—a review Shireen. Int J Sci Technol Res
10. Preetha MN (2013) A study and performance analysis of RSA algorithm. Int J Comput Sci Mobile Comput IJCSMC
11. Ramesh M, Prabakaran G, Bhavani R (2013) QR-DWTcode image steganography. Int J Computational Intell Inf
12. Rani, K.R.E: Data security through QR Code encryption and steganography. Adv Comput Int J

13. Rustad S, Syukur A, Andono PN (2020) InvertedLSB image steganography using adaptive pattern to improve imperceptibility. J King Saud Univ Comput Inf Sci
14. Saravanan M, Priya A (2019) An algorithm for security enhancement in image transmission using steganography. J Inst Electronics Comput
15. Sun Y, Yu M, Wang J (2020) Research and development of QR code steganography based on JSteg algorithm in DCT domain. In: IEEE 15th International Conference on Solid-State& Integrated Circuit Technology (ICSICT)
16. Thangadurai GSD (2013) Evaluation of LSB based image steganography technique for various file formats. Int J Comput Intell Inf
17. Nair JB, Raghavendra SP (2019) A novel technique for steganography using LSB method. JNNCE J Eng Manage Peer Rev Bi-annual 3

VASEC: Visual Attention-Driven Key Shot Extraction and Video Enhancement with Parallelly Decoded Dense Captions

P. Priyadharshini, E. Vasudha, and K. Haripriya

Abstract In today's information era, vast amount of multimedia content has surged the necessity for techniques to detect significant and relevant content. We introduce visual attention-driven model (VAFF) for extracting the prominent shots of a video which is followed by enhancement and event-wise video captioning where parallel computation is exploited. The pre-trained GoogleNet model is utilized to obtain the visual features (object/static) used for key shot extraction and pre-trained I3D model is utilized to obtain the kinetic features (RGB and Flow). The modified version of 0/1 Knapsack Algorithm used the predicted importance score for key shot extraction. $F1$ score for the Visual Attention Feature Fusion (VAFF) showed improvement for SumMe and considerable value for TVSum due to their respective lengths. The shortened video is enhanced using pre-trained ESRGAN. Dense video captions are generated using the Temporally Sensitive Pre-trained model (TSP) features extracted from the enhanced video. Thus, research aims to provide a robust pipeline for static and real-time video analysis.

Keywords Video summarization · Enhancement · Dense video captioning · Visual attention · Keyshots · ESRGAN · Parallel decoding · Event-wise captions

1 Introduction

The lightning proliferation of video content has created a demand for efficient techniques to extract key shots from videos for various applications like video summarization, video browsing, content-based video obtainment, and interdisciplinary research. Visual attention mechanisms, which selectively focus on salient regions or frames, have proven to be effective in identifying important shots and guiding the extraction process. By prioritizing and combining different visual cues, such as color,

P. Priyadharshini · E. Vasudha (✉) · K. Haripriya
Department of Computer Science and Engineering, Anna University, Chennai, India
e-mail: vasudhae38@gmail.com

S. Bandyopadhyay et al. (eds.), *Intelligent Computing Systems and Applications*, Lecture Notes in Networks and Systems 1010,
https://doi.org/10.1007/978-981-97-5412-0_23

motion, and texture, attention-based models enhance the robustness and richness of extracted shots.

Improving the quality of shortened videos is crucial, especially for applications like closed circuit television (CCTV) where distortions and blurring can affect interpretation. Enhancing the video quality before captioning helps reduce false captions and improves the dense captioning process. Furthermore, generating event-wise captions for the extracted key shots enhances the understanding and interpretability of the summarized video content. These captions have various applications, including assisting visually impaired individuals, facilitating efficient video browsing and retrieval, and enabling automated video analysis.

The combination of efficient shot extraction, enhanced video quality, and informative event-wise captions revolutionizes the way users consume video content. It enables more accurate and efficient video retrieval, improves the browsing experience, and enhances the interpretability of the summarized content. Moreover, these advancements have implications in fields such as surveillance, multimedia analysis, and video recommendation systems, where the ability to efficiently extract key shots and provide informative captions can greatly impact decision-making processes.

Thus, there is a need to develop a robust pipeline aiding both static and real-time video analysis by exploiting attention and parallelism for localization of events and captioning. Hence, in this work, we investigate how visual attention and the fusion of multiple features help in improving the extraction of prominent events, and how parallelizing the dense captioning helps in increasing performance of the system.

2 Related Work

Ghauri et al. [1] proposed a supervised video summarization technique that extracts visual and motion-related features using pre-trained models, applies separate attention mechanisms, fuses the feature vectors, and predicts importance scores for frame selection. Feng et al. [2] proposed a technique for video summarization and shot segmentation that involves extracting RGB features, deep features and light flow features, utilizing a Bi-LSTM encoder to capture contextual information, and generating importance scores for shots using an LSTM decoder based on the representation vector, output from the encoder. Apostolidis et al. [3] proposed a supervised video summarization method that combines global and local multi-head attention mechanisms, incorporating absolute positional encoding and feature addition, to generate importance scores for selecting key fragments in the video summary within a specified duration constraint. Fajtl et al. [4] proposed a novel approach for supervised video summarization, replacing the LSTM encoder-decoder architecture with attention. By utilizing Kernel Temporal Segmentation (KTS) algorithm for shot identification and a modified Knapsack algorithm for keyshot selection, the keyshots obtained from the algorithm are concatenated to form the summarized video. Ji et al. [5] proposed a video summarization technique that employs attention-based encoder-decoder architecture, utilizing Bi-LSTM as the encoder and LSTM decoder to generate frame-level importance scores.

The limitations of the existing work include the result analysis on splits of the benchmark datasets (TvSum and SumMe) which included some repetitive videos and some videos were excluded. So, the model's performance wouldn't have been measured correctly. Hence, we provide non-overlapping splits of two benchmark datasets and hence evaluate the accurate model performance. The other limitation of the existing work is their model complexity majority uses encoders and decoders, while we use parallel attention mechanism for capturing contextual information from features extracted using pre-trained models and a simple stack of linear layers for prediction.

Olivastri et al. [6] introduced an encoder-decoder architecture for an end-to-end video captioning, optimizing both components simultaneously. They adopted a two-stage training approach, initializing the architecture with pre-trained encoders and decoders, and then fine-tuning the whole network to learn pertinent features for the video caption generation. Li et al. [7] proposed a Multi-level Attention Model-based Recurrent Neural Network (MAM-RNN) for video captioning, utilizing region-level and frame-level attention mechanisms to encode visual features and generate captions using LSTM as the decoder. Subhashini et al. [8] proposed a sequence-to-sequence captioning method utilizing two-layer LSTM architecture, with the top layer responsible for encoding the visual frame sequence using latent vector representation and the lower layer for decoding the output word sequence to generate captions. Lashin and Rathu [9] proposed a multi-modal dense video captioning model that consists of two phases: temporal event localization and caption generation. Pre-trained Bidirectional Single Stream Temporal action proposals network (Bi-SST) is used to generate temporal regions potentially containing events. In the captioning module, audio and visual features are extracted using the VGGish and I3D models, and decoder architecture combines intermediate feature representations with GloVe embeddings to generate captions for the proposed events. Mun et al. [10] proposed a streamlined dense video captioning framework that leverages visual and linguistic contexts to generate coherent captions by explicitly modeling temporal dependencies across events in a video.

Andrei et al. [11] proposed SUPERVEGAN, a Super Resolution Video Enhancement GAN that enhances the quality of low-bitrate compressed video streams through joint super resolution and artifact removal. Ledig et al. [12] introduced a naturalistic single image super resolution method using generative adversarial network (GAN), demonstrating superior photo-realism for large upscaling factors. Ma et al. [13] introduced CVEGAN, perceptually-inspired GAN for compressed video enhancement, enhanced residual non-local blocks (ERNB), and enhanced convolutional block attention modules (ECBAM), and a discriminator based on SRGAN with additional ERNBs. Cherian et al. [14] proposed a method for optimizing image resolution on surveillance images using SRGAN, where an input image is processed through SRGAN to enhance resolution, stitched together, and further enhanced using the Contrast Limited Adaptive Histogram Equalization (CLAHE) filter, resulting in higher resolution images with improved contrasts and readability.

3 Proposed System

The system comprises of a video processing pipeline with three stages: Video Key Shots Extraction, i.e., Video Summarization, Video Enhancement, and Dense Video Captioning (Fig. 1).

3.1 Video Key Shots Extraction

In our research, we consider the video as a sequential collection of frames, with a frame rate of 24 fps (frames per second). To reduce redundancy and computational complexity, we perform subsampling by selecting every other frame, resulting in a subsampling rate of 1 out of 2 frames. This subsampling does not compromise the overall performance or the resulting output since alternating frame subsampling does not significantly affect the overall perception of the video (Fig. 2).

Feature Extraction

To generate a collective representation of the video's content and motion-oriented information, we employ two distinct models. Firstly, for content-based features, a GoogleNet model pre-trained on the ImageNet dataset is used. By leveraging the network's middle layers, we extract feature maps that encode visual information at various levels of abstraction. These feature maps yield high-dimensional feature vectors that effectively represent the content of the input image.

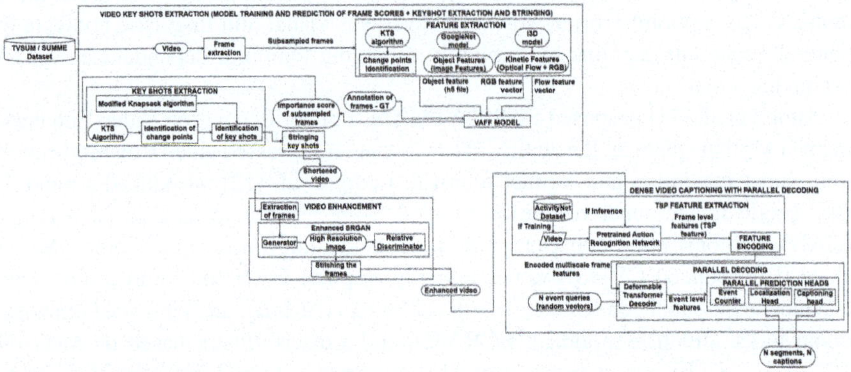

Fig. 1 The outline of the proposed system

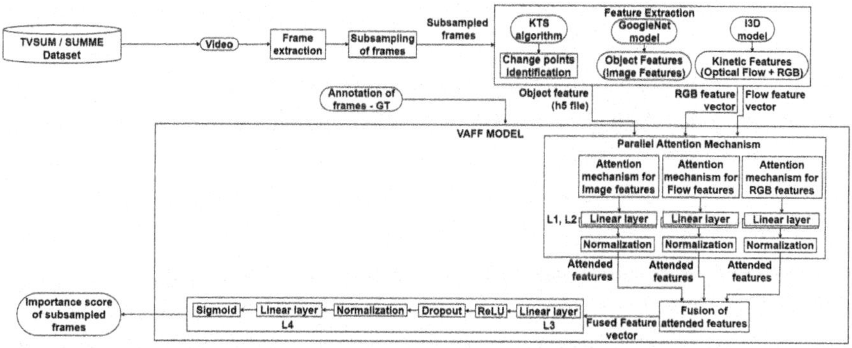

Fig. 2 Video key shots extraction

For motion-related features, specifically kinetic features, we employ a pre-trained I3D (Inflated 3D) model, which has been trained on the Kinetics 400 dataset. We extract both RGB and Flow features from the penultimate layer of the I3D model, before performing a summation operation. This extraction process results in two tensors, corresponding to RGB and flow streams.

Visual Attention Fusion Framework (VAFF)

In our approach, visual features are attentively fused in parallel using a Visual Attention Fusion Framework (VAFF). This framework assigns higher weights to relevant portions of the feature vectors, thereby emphasizing their importance. A separate attention mechanism is provided for each feature type and the attended features are processed by two linear layers and a normalization layer. Attention weights enables selective focus on specific regions, enhancing the overall performance of the key shot extraction process [14]. The attended features are then passed for intermediate fusion resulting in a single fused feature vector. The vector is transformed using the Linear layer and ReLU activation function before performing dropout and normalization to avoid overfitting. The latent representation is fed into a final linear layer after which the sigmoid activation function provides a target between 0 and 1 signifying the importance of the frame in the video.

Keyshot Extraction

Change points in the video are identified using Kernel Temporal Segmentation (KTS) [15] to identify the shots in the video. The modified version of the 0/1 Knapsack Algorithm where input is the shots of the video which is identified using change points detected by KTS, and importance scores of the shots which is the sum of the predicted importance scores of the frames in the shot is used to pick the keyshots such that the total video length will amount to 15% of the original video length. The optimization problem here is to maximize the sum of shot scores.

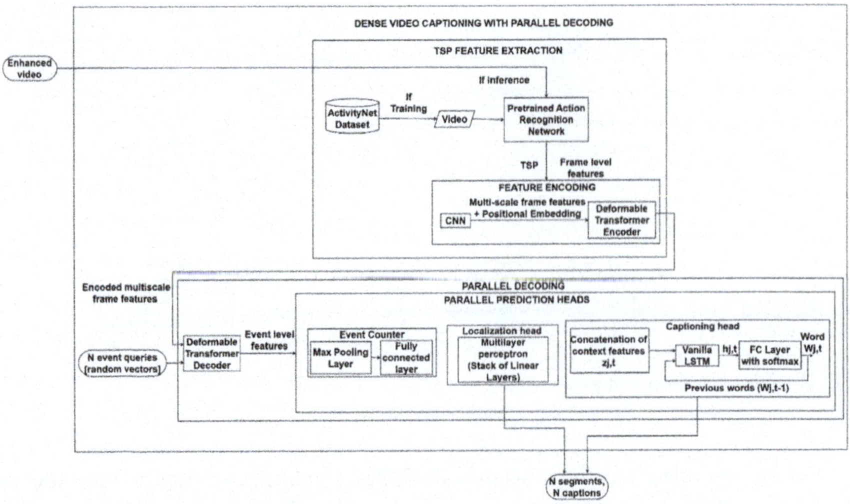

Fig. 3 Dense video captioning with parallel decoding

3.2 *Video Enhancement*

Pre-trained Enhanced Super resolution GAN model weights are loaded for enhancement. The video is extracted into frames upon which the ESRGAN is used for enhancement. The enhanced images are stitched to form the video.

3.3 *Dense Video Captioning*

Dense video captioning is generating corresponding captions (text descriptions) for a set of events in an unclipped video. The features of the video are extracted using a Temporally Sensitive Pre-trained (TSP) model. The Deformable Detection Transformer (DETR) encoder is used for encoding the features. The decoding network consists of a Deformable DETR decoder and three parallel heads–Event counter, Localization head, and Captioning head. Therefore, localization, selection, and captioning tasks are parallelized in a single end-to-end framework (Fig. 3).

Dense Video Captioning With Parallel Decoding
See Fig. 3.

TSP Feature Extraction

What Is TSP Feature?
TSP model is a pre-trained model which is trained using raw (unclipped) videos along with temporal annotations. The ResNet-based encoder is learned from the raw

video input. Sample video clips of fixed size are provided as input to the encoder. The clips are assigned two labels—a binary temporal region label indicating whether the clip is from the foreground (activity) or background (no action) of the video, and an action class label if the clip is from an action segment. Thus temporally-sensitive features helps localization algorithms to distinguish exactly between action and no action context in extensive unclipped videos.

TSP Feature Extraction Methodology
Input: Video frames
Methodology

1. The input frame is passed through a CNN model and gets transformed to a smaller feature map.
2. Then, an additional 1×1 convolution is used to minimize channel dimension from C to d.
3. After that, the spatial dimensions will be flattened because the encoder expects a sequence as input, resulting in a $d \times HW$ feature map.
4. Then, this newly generated feature map will be added by positional encoding. Positional encoding is used to recover spatial information which is lost during flattening.
5. This newly generated feature map along with its positional encoding is provided as input to the Deformable Detection Transformer's encoder where a deformable attention module is used to extract the frame-frame relationship across multiple scales.

Output: Encoded multi-scale frame features.

Parallel Decoding
The following section describes about the components of the parallel decoding module. It comprises of an Event Queries Generation module and Parallel Decoding Network.

Event Queries Generation
Input: N event queries (random vectors) and the encoded multi scale features
Description

1. Event queries are K vectors with length d (same as input token), working as inputs to decoder. Each event query stores information for one event of the video. Event queries are initialized randomly and gets refined after each decoder layer.
2. There are two attention types: self-attention and cross-attention. In the self-attention layer, query, key, and value are taken from event queries. This layer can be considered as to learn the dependencies between different events.
3. In the cross-attention layer, only query is taken from event queries, whereas key and value are from encoder output, which is unchanged after feature extraction. Each event query is searching information for corresponding events, assigning high attention weight to highly-correlated tokens.

4. After multiple layers' refinement, the decoders have collected enough information about events and can be used for final prediction.

Output: The refined *N* event queries

Parallel Decoding Network
Parallel Decoding network comprises of Event Counter, Localization Head and Captioning Head.

Event Counter
Input: Event level features
Description

1. The event counter detects the video's number of events.
2. It consists of a max pooling layer and a fully connected layer with the softmax activation function which initially squeezes the critical data of the event queries qj to a global feature vector, and then predict a vector rlen of fixed size, where each value indicates the probability of a particular event number.
3. *N*set = argmax(rlen) is used to obtain the predicted event number.
4. The top *N*set events with exact boundaries and proper captions are selected from *N* event queries to obtain the final output.

Output: Event number

Localization Head
Input: Event level features
Description

1. All event queries use and share a Multi-Layer Perceptron (MLP).
2. Its input is a (*d*), vector (event level features) and output is the binary class and bounding box of one event. At last, we accomplish event detection of one video by combining output of all event queries (Fig. 4).
3. For training, we need to match *N* ground truths and *K* predictions for computing loss. We pad *N* events into *K*, excluding *N*–*K* '0' class denoting background (no action) segments.
4. There are two sets of *K* elements and we need to match them one-by-one. It can be viewed as a Bipartite Matching (BM) problem where we need to use edges to link two sets and each node can have only an edge, and we need to maximize the total of edge weights. Here an edge weight is defined as an IoU (Intersection over Union) value of the ground truth and the predicted bounding box.
5. The output from the localization head is the beginning and finishing time of the events in the video with its confidence scores.

Output: The bounding box of events, binary class labels indicating action/no action segments.

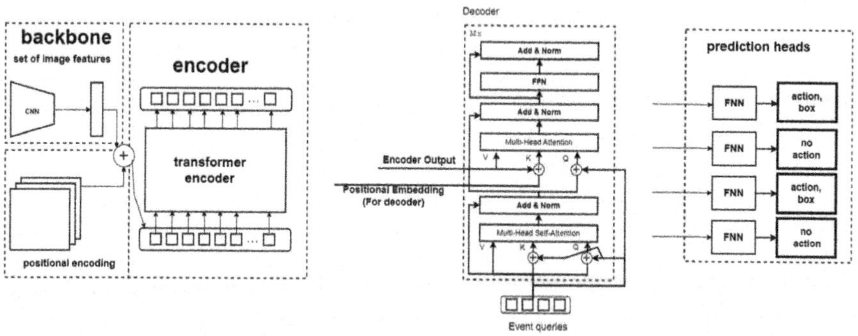

Fig. 4 Deformable detection transformer

Deformable Detection Transformer
See Fig. 4.

Captioning Head
Input: Event level features
Description

1. The captioning head consists of vanilla LSTM which takes event level representation q_j at each timestamp as input. The FC (Fully Connected) layer with the softmax activation over the hidden state $h_{j,t}$ of LSTM predicts the word $w_{j,t}$.
2. The LSTM takes in the concatenation of the context features $z_{j,t}$, event query features q_j and previous words $w_{j,t-1}$.
3. The FC layer with softmax activation over $h_{j,t}$ provides the probability of the next word $w_{j,t}$.
4. With the evolution of LSTM, we obtain a sentence $S_j = w_{j1}, ..., w_{jMj}$, where M_j is the sentence length.

Output: Caption sentence

4 Experimental Setup

4.1 Dataset

The benchmark datasets used are:
Video Summarization:
SumMe: 25 videos of length 1–6 min with 18 users' annotations.
TVSum: 50 videos of length 2–10 min, annotated by 20 users.
Dense Video Captioning
ActivityNet: 20,000 YouTube unclipped videos with 100,000 caption annotations.

The length of videos is 120 s on average. Mostly, videos contain 3 events with corresponding begin, finish time and annotated captions. The train split contains 10,024 videos, validation split contains 4926 and test split contains 5044.

4.2 Training Setup

Training details for Video Summarization: When reproducing the results of the previous work, some issues were identified in evaluation for both benchmark datasets. On observing, we found that the splits which were used for evaluation contained some repetitive videos while some were excluded. Hence, we provide 5 non-overlapping splits in which videos are equally divided across splits without any video getting repeated or excluded.

Fivefold cross validation is used. Each fold is trained for 300 epochs with 50 epochs as the early stopping criterion. The provided results are $F1$ scores averaged across the 5 non-overlapping splits of the corresponding datasets.

Training details for Dense Video Captioning: Parallel Dense Video Captioning model is trained for 30 epochs.

5 Results and Discussion

5.1 Performance Metrics

Video Summarization:
F1 Score: $F1$ score is defined as the harmonic mean of precision and recall.

$$F1 \text{ Score} = \frac{2 * P * R}{P + R} \tag{1}$$

where P is Precision and R is Recall of the model.
Spearman's correlation coefficient: Spearman's correlation measures the robustness and direction of the monotonic relationship between two variables.

$$r = \frac{n\left(\sum xy\right) - \left(\sum x\right)\left(\sum y\right)}{\sqrt{\left[n\sum x^2 - \left(\sum x\right)^2\right]\left[n\sum y^2 - \left(\sum y\right)^2\right]}} \tag{2}$$

Where x is ground truth, y is predicted value, the number of samples is denoted as n.

Kendall's correlation coefficient: Kendall's Tau is a nonparametric measure of the robustness and direction of the relationship that exists between two variables.

$$\text{Kendall's Tau} = (C - D/C + D)$$

where the number of concordant pairs is denoted as C and the number of discordant pairs as D.

Video Enhancement

PSNR: The peak signal-to-noise ratio (PSNR) between two pictures is measured in decibels. The quality of the compressed or rebuilt image improves with increasing PSNR.

$$\text{PSNR} = \frac{20 * \log_{10}(\text{Max Pixel})}{\sqrt{\text{MSE}}} \tag{3}$$

where MaxPixel is the maximum pixel value in the image and MSE corresponds to Mean Square Error.

Dense Video Captioning

METEOR: A metric for assessing the output of machine translation is called METEOR (Metric for Evaluation of Translation with Explicit Ordering).

$$\text{METEOR} = (\text{alpha} * \text{precision}) + ((1 - \text{alpha}) * \text{recall} * (1 - \text{penalty})) \tag{4}$$

where

$$\text{penalty} = \text{beta} * \text{e}^{-\text{gamma}*\text{sim}} \tag{5}$$

sim is the unigram similarity between the ground truth and prediction calculated using a WordNet based cosine similarity metric.

SODA: To efficiently capture the story of a video using generated and ground truth captions, SODA first determines the best temporal matching between them. Then, in order to penalize redundant captions, it computes METEOR ratings for matching captions and derives an optimized score from those METEOR values.

5.2 *Result Analysis*

Our model VAFF which takes in fused attended features for prediction has been evaluated on the five non-overlapping splits. On comparing the results for $F1$ score, as shown in Table 2, an improved $F1$ score was obtained for SumMe, whereas TVSum showed comparable results because of its length. Correlation coefficients (Spearman's Rho, Kendall's Tau) displayed in Table 1 show an improvement in model performance for both SumMe and TVSum. Thus, non-overlapping splits show an overall improvement in model performance and provides a fair comparison.

Table 1 Video summarization metrics values obtained

Dataset	$F1$ score	Spearman's rho	Kendall's tau
SumMe	52.3	0.185	0.166
TVSum	60.48	0.64	0.584

Table 2 Collation of various methods on standard datasets

Dataset	Method	$F1$ score	Spearman's rho	Kendall's tau
SumMe	MAVS [16]	43.1	–	
SumMe	M-AVS [5]	44.4	–	–
SumMe	reSEQ2SEQ [17]	44.9	–	–
SumMe	VASNET [4]	49.7	0.16	0.17
SumMe	MC-VSA [18]	51.6	–	–
SumMe	VAFF (**ours**)	**52.3**	**0.185**	**0.166**
TVSum	M-AVS [5]	61.0	–	–
TVSum	VASNET [4]	61.4	0.16	0.17
TVSum	MC-VSA [18]	63.7	0.116	0.142
TVSum	reSEQ2SEQ [17]	63.9	–	–
TVSum	MAVS [16]	67.5	–	–
TVSum	VAFF (**ours**)	**60.48**	**0.64**	**0.584**

6 Conclusion and Future Work

In this work, we have put forward Visual Attention-Driven Key Shot Extraction to get the important scores of subsampled frames. Using those important scores, we are extracting video key shots to get the summarized video using Kernel Temporal Segmentation Algorithm and Modified Knapsack Algorithm. An Enhanced SRGAN is used to enhance the summarized video. Then TSP features are extracted to generate dense captions of the summarized video. For future work, topic modeling techniques can be applied on the generated captions for video analysis.

References

1. Ghauri JA, Hakimov S, Ewerth R (2021) Supervised video summarization via multiple feature sets with parallel attention. In: 2021 IEEE international conference on multimedia and expo (ICME). IEEE, pp 1–6
2. Feng X, Zhu Y, Yang C (2021) Video summarization based on fusing features and shot segmentation. In: 2021 IEEE international conference on network intelligence and digital content (NIDC2021). IEEE, pp 1–5

3. Apostolidis E, Balaouras G, Mezaris V, Patras I (2021) Combining global and local attention with positional encoding for video summarization. In: 2021 IEEE international symposium on multimedia (ISM). IEEE, pp 1–6
4. Fajtl J, Sadeghi Sokeh H, Argyriou V, Monekosso D, Remagnino P (2019) Summarizing videos with attention. In: Computer vision-ACCV 2018 workshops: 14th Asian conference on computer vision, Perth, Australia, 2–6 Dec 2018, Revised Selected Papers 14. Springer International Publishing, pp 165–180
5. Ji Z, Xiong K, Pang Y, Li X (2019) Video summarization with attention-based encoder-decoder networks. IEEE Trans Circ Syst Video Technol 30(6):1709–1717
6. Olivastri S, Singh G, Cuzzolin F (2019) End-to-end video captioning. In: Proceedings of the IEEE/CVF international conference on computer vision workshops
7. Li X, Zhao B, Lu X (2017) MAM-RNN: multi-level attention model based RNN for video captioning. IJCAI 2017:3403–3409
8. Venugopalan S, Rohrbach M, Donahue J, Mooney R, Darrell T, Saenko K (2015) Sequence to sequence-video to text. In: Proceedings of the IEEE international conference on computer vision. IEEE, pp 4534–4542
9. Iashin V, Rahtu E (2020) Multi-modal dense video captioning. In: Proceedings of the IEEE/CVF conference on computer vision and pattern recognition workshops, pp 491–492
10. Mun J, Yang L, Ren Z, Xu N, Han B (2019) Streamlined dense video captioning. In: Proceedings of the IEEE/CVF conference on computer vision and pattern recognition, pp 11449–11458
11. Andrei SS, Shapovalova N, Mayol-Cuevas W (2021) SUPERVEGAN: super resolution video enhancement GAN for perceptually improving low bitrate streams. IEEE Access 9:91160–91174
12. Ledig C, Theis L, Huszár F, Caballero J, Cunningham A, Acosta A, Aitken A, Tejani A, Totz J, Wang Z, Shi W (2017) Photo-realistic single image super-resolution using a generative adversarial network. In: Proceedings of the IEEE conference on computer vision and pattern recognition (CVPR), pp 105–114
13. Ma D, Zhang F, Bull DR (2020) CVEGAN: a perceptually-inspired GAN for compressed video enhancement. arXiv preprint arXiv:2011.09190
14. Cherian AK, Poovammal E, Rathi Y (2021) Improving image resolution on surveillance images using SRGAN. In: Inventive systems and control: proceedings of ICISC 2021. Springer, Singapore
15. Archaoui Z, Moulines E, Bach FR (2009) Kernel change-point analysis. In: Advances in neural information processing systems, pp 609–616
16. Feng L, Li Z, Kuang Z, Zhang W (2018) Extractive video summarizer with memory augmented neural networks. In: ACM multimedia conference on multimedia conference. ACM, pp 976–983
17. Zhang K, Grauman K, Sha F (2018) Retrospective encoders for video summarization. In: ECCV—15th European conference on computer vision, vol 11212. Springer, pp 391–408
18. Liu Y-T, Li Y-J, Frank Wang Y-C (2020) Transforming multi-concept attention into video summarization. In: ACCV—15th Asian conference on computer vision, vol 12626. Springer, pp 498–513

Prognosticating COVID-19 Pneumonia Versus Common Pneumonia for Pediatrics Using Chest Radiographic Images

D. Suganya and R. Kalpana

Abstract There is currently no authorized medication or vaccination for the recently identified coronavirus, COVID-19, which is extremely contagious and causes illnesses. Omicron variant is the most transmissible variant for all age groups especially for children. The most typical COVID-19 symptoms include a coughing, fever, and sore throat. When symptoms intensify into a severe form of pneumonia, acute breathing crisis septic shock, pulmonary edema, and multiple organ failure may happen. Identifying the different pneumonias is difficult for children. To identify the most precise, crucial element of learning, deep convolutional neural networks, DenseNet, ResNet, MobileNet, Xception, InceptionResNetV2, VGGNet, InceptionV3, and NASNet were selected from deep CNN. The collected features are then used to classify participants as COVID-19 modules or controls using different machine learning separators. To develop a robust capacity for the normalization of anonymous data, this strategy has avoided methods of preprocessing specific work data. The effectiveness of the proposed method is evaluated based on publicly available Chest X-ray (CXR) images of COVID-19. The deep CNN is used to extract the features, which are integrated with the addition of Generative Adversarial Network (CYCLE-GAN). The binary classification will be made into a hybrid Deep CNN-LSTM proposed model. This model classifies the images into COVID-19 pneumonia, common pneumonia, and no pneumonia. It achieves an accuracy of 96.3% and precision of 94.5% and F_1-score of 91.3%. This effective classification helps the medical practitioner to detect the type pneumonia and its severity clearly and further go for the appropriate medical treatment.

Keywords COVID-19 · Deep Convolutional Neural Networks · DCNN · LSTM · Long short-term memory · Cycle Generative Adversarial Network

---

D. Suganya (✉) · R. Kalpana
Department of Computer Science and Engineering, Puducherry Technological University, Puducherry 605014, India
e-mail: suganyajagathguru@ptuniv.edu.in

R. Kalpana
e-mail: rkalpana@ptuniv.edu.in

S. Bandyopadhyay et al. (eds.), *Intelligent Computing Systems and Applications*, Lecture Notes in Networks and Systems 1010,
https://doi.org/10.1007/978-981-97-5412-0_24

1 Introduction

SARS-CoV-2 is the virus that causes the highly contagious infection referred to as COVID-19 (coronavirus disease).Since January 2022, a new wave of pandemic spreads all over the world. Omicron mutant like B.1.1.529 and XBB is considered to be more dangerous infection which leads to respiratory failure. An estimated death occurs is ~ 6 million worldwide. It has a devastating impact on the global population. COVID-19 appears to be the world's most important health catastrophe since the 1918 flu pandemic. The first cases were identified at the end of 2019 in China, SARS-CoV-2 spread rapidly throughout the world, and the World Health Organization declared it as a worldwide pandemic on March 2020.

Throughout the world, the virus harms billions of people's physical, emotional, and social well-being. In comparison with MERS and SARS, COVID-19 has a substantially higher rate of transmission, severe side effects that can last even after recovery, and frequent mutations (out from the original SARS-CoV-2 model) that cause high mortality and unrestrained violence [1]. Other organ systems other than the respiratory system (the main target organ) have been identified as harming the clinical manifestations of the coronavirus, including the kidneys, brain, hematological system, endocrine system, kidneys, etc., without any solutions till date [2]. The shortage of emergency medicine and life-saving medicine has lead scientists and industrialists to re-use available treatments and develop vaccines in a shorter period [3].

The main contributions of the proposed work are as follows:

1. A novel SARS-CoV-2 Omicron variant is identified for children using a hybrid Deep CNN-LSTM model with the help of CXR images.
2. Data preprocessing and augmentation is processed using CYCLE-GAN which helps to create synthetic images that exactly match the real image to solve the over-fitting problem in the datasets.
3. A hybrid DCNN-LSTM approach is used to detect whether the children is affect by Omicron variant pneumonia or Common pneumonia and no pneumonia (Normal).
4. Publicly available kaggle datasets were taken and fused with newly created pediatric dataset to prognosticate various pneumococcal disorders for pediatrics and the performance is validated.

The structure of the paper is organized in this way; the background of the study is discussed in Sect. 2. The methods and materials used for the detection of covid-19 are explained in Sect. 3, and the experimental setup and result can be analyzed in the Sect. 4, conclusion places the final session.

2 Backgrounds

Kassania et al. [4] proposed a COVID-19 identification using chest X-ray and CT-images by concatenating convolution neural network and RESNET to categorize COVID-19 images. The dataset used is Kaggle CXR and CT-images. The leading advantage of the model is less detection time for clinical applications. The presence of artifacts and low contrast boundaries of the images are the major limitations. Malghdid et al. [5] demonstrated of COVID-19 pneumonia diagnosis from CXR and CT-images using ALexnet. The dataset set contains of 230 CXR images and 412 CT-images. The main advantage of the study is it predicts fever using temperature sensor, fatigue prediction using inertial sensor and microphone sensor to detect cough.

Hall et al. [6] suggested finding of COVID-19 in CXR on a small dataset using deep learning through VGG16 CNN and RESNET50. The dataset comprises 135 COVID-19 cases and the information is lack at the stage of the disease is the major drawback. A COVID-19screening using deep learning method has been discussed by Farooq et al. [7] through VGG16 CNN and RESNET. The work was done on the COVIDx dataset that consists of 5941 radiography images. The model has an advantage of higher PPV and sensitivity leading to no false classification. The test case of the COVID-19 cases is smaller is the major con of the work.

A modified deep CNN for coronavirus detection was discussed by Rahimzadeh et al. [8] using Kaggle CXR images. The model works using Xception and ResNet50V2 and imbalance in the dataset is the main drawback. Coronavirus disease is screened using deep learning method is discussed by Xu et al. [9] using CNN, VNet20 and classical ResNet. The dataset consists of 618 CT-images. The model produces reliable results by digitizing and standardizing information of the images. The limited model samples count results in low detection rate.

3 Proposed Methodologies

The whole COVID-19 detection system consists of numerous components (see Fig. 1). Initially, the input images of CXR images need to be preprocessed. The preprocessing pipeline includes two steps, namely data resizing followed by data normalization. The preprocessed datasets is further breaks into training and testing dataset. Data augmentation is carried out through CYCLE-GAN to generate high-resolution synthetic images.

The training dataset is further used to train the convolution neural network and CNN-long short-term memory hybrid architectures. Drive loss and accuracy are evaluated after each epoch. Confusion matrix, accuracy, specificity, recall, sensitivity, and F_1-score evaluation metrics were used for evaluating the performance of the proposed system.

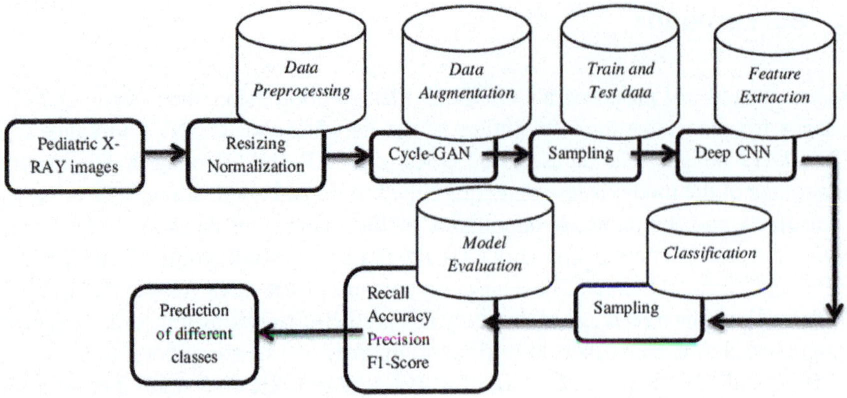

Fig. 1 Workflow of proposed architecture

3.1 Dataset Description

To measure the efficiency of the feature extraction and classification method, the publicly accessible dataset of chest X-ray images is used, which is accessible through a Kaggle repository [10]. 454 images of chest X-ray belonging to 3 classes, namely COVID-19 pneumonia, no pneumonia, and normal pneumonia have been taken from Kaggle. Fusing multiclass dataset, initially the Kaggle pneumonias detection dataset were taken and then fuse few pediatric chest X-ray images [11]. Both dataset images are fused or added to balance it with both + ve and healthy occurrences.

3.2 Data Preprocessing

Images within the database are obtained from various sources; therefore, there is a great variation in the intensity of the images. Therefore, the program does not take any widespread preventative efforts to enhance CNN architecture efficiency. During the removal process, this enables the model to withstand the noise, artifact, and flexibility of the input images. In order to enhance the training process, two common preprocessing stages were applied to model training. Resizing: The quality and size of the images in this database range from 356 * 456 to 1126 * 869 pixels, by resizing all realistic images to 64 * 64 pixels to provide the same size for all inputs [12].

Normalization: As a precautionary measure, we used ImageNet mean subtraction to re-measure the pixel density values. ImageNet means regularly generated on the ImageNet website. Tightening is another important pre-procedure step. The values of all photos from [0, 255] were normalized using min–max normalization in the power range [0, 1] to get this, and the equation used for normalization (X_{norm}) is as follows.

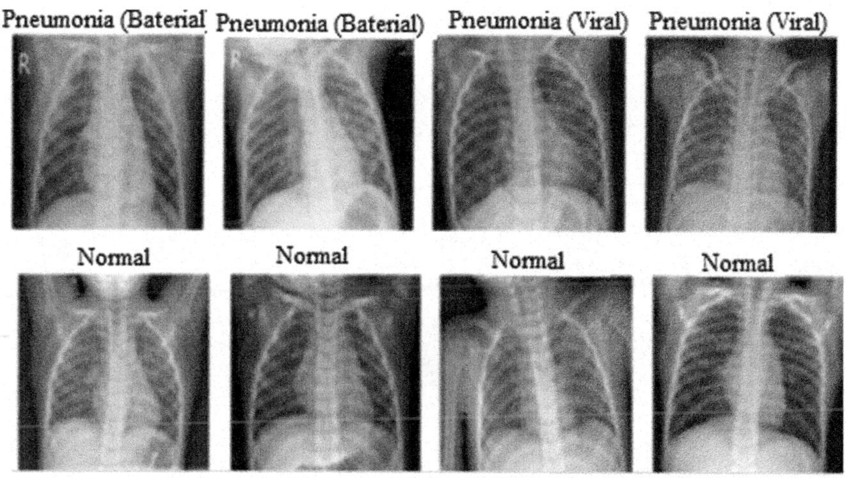

Fig. 2 Pediatric X-ray images with pneumonia and normal cases

$$X_{\text{norm}} = \frac{X - X_{\min}}{X_{\max} - X_{\min}} \quad (1)$$

The pixel intensity is defined as x, the minimum value is defined as x-min and maximum values of the input image is defined as x-max in Eq. (1). By eliminating bias and gaining uniform distribution throughout the database, this process facilitates model integration. Figure 2 depicts the sample images of pneumonia and normal cases of pediatric chest X-ray images which need to be preprocessed.

3.3 Data Augmentation

The main challenge in COVID-19 variants detection for pediatrics is unbalanced datasets. Here Deep Conditional-Generative Adversial Network is used to generate more images which exactly match the original to balance the datasets.

Data Generation by GAN

Two networks are simultaneously trained in GANs, the first focuses on image production and the second focuses on discrimination as shown in Fig. 3. Due to the inefficiency of image production and measures to combat background change, it gains power in education and industry. Many image-creating functions, such as advanced editing, text-to-text integration, and image-to-image translation, have shown that GANs can do well. If diagnostic images are to be published in public places, patient consent is required by law [13]. GANs are widely used to create private and sufficient images for analysis. Another obstacle to interpreting medical images in supervised learning is the lack of technology. However, many healthcare organizations are

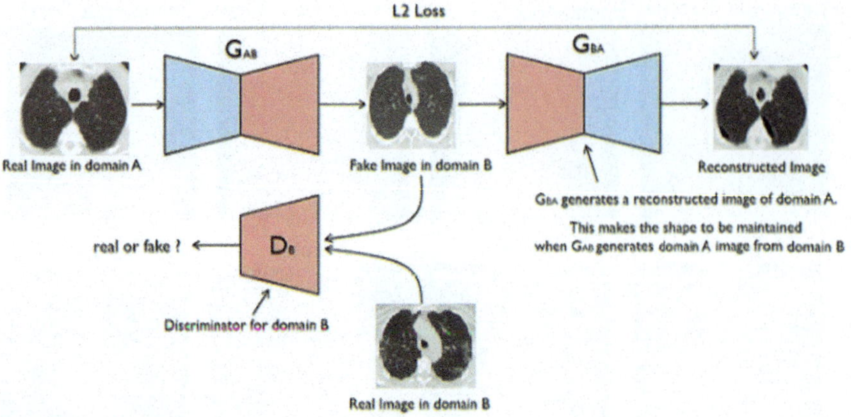

Fig. 3 CYCLE-GAN architecture

working to create large datasets that are publicly accessible. The issue remains a major issue for organizations such as cancer photography archives, Biobank, the North American radiologist community, and the national history of biomedical imaging. Rotation, probe, measurement, and expandable degradation are common ways to enlarge training samples. This, however, does not provide sufficient variability to be found in actual samples.

GANs provide highly structured data samples that are similar to real data. It has been used successfully in various articles to enrich photographic data sets. Radford et al. promoted the first variation of Deep Convolutional GAN (CYCLE-GAN), where both producers and non-producers deepen CNN. Goodfellow et al. introduced CYCLE-GAN as a sustainable version of GAN training and preparation and is the basis of many modern GANs. Both neural networks in the model are simultaneously trained. Figure 3 shows the standard GAN with both networks [14]. The first is racism (called D) which differentiates real and fake images. Accept input x and output $D(x)$. The generator (also known as G) is the second network and includes images that D declared as the most realistic. G puts Gz in the Pg image space by taking input images $z(1) \ldots z(m)$ from a simple Pz stream, which is the same distribution in Eq. (2). The main goal of G is to successfully retrieve P data. The next loss function can be enhanced by training networks in this way.

$$(G)(D)Ex \sim pdata(x) + Ez \sim pz[\log(1 - ((z)))] \tag{2}$$

The detector is trained to increase $D(x)$ in samples with x P data and reduce D $(x)! \sim P$ *data* in a sample. The $G(z)$ generator takes the samples in such a way that D sees them as real images. The generator enhances its ability to create more realistic images as a result of this training, while the developer improves his or her ability to distinguish real images from composite image samples [15].

Generator

Network input is a variant of 100 random values taken from the same distribution, producing a result of 64 * 64 * 1. To make an example of a 5 * 5 kernel size, the network has a fully integrated layer built to be 4 * 4 * 1024 and four layers with straight lines. Devolution (deconvolution) enhances the image by enlarging the pixels by adding a zero pixel in the center. In addition to the output layer, each layer incorporates bulk familiarity, which stabilizes the GAN network and prevents the generation from rolling to a single location. Except for the final layer, which uses the tanh activation function, all levels use ReLU as the activation function.

Discriminator

It has a standard CNN that accepts 64 64 1 (X-ray) as input and makes two judgments of whether the X-ray is real or mobile. This network contains 4 Conv layers and a fully connected layer with a kernel size 5 5. Instead of assembling layers, stridden convolution is used to reduce the size of the space. In addition to the input and output layers, each layer used batch-normalization. The ReLU rewarding function $f(x) =$ max (x, leakx) is used in each layer, while the Sigmoid function is used for the probability (0, 1) image sampling school in the final layer.

Training CYCLE-GAN

The collection of real-world images comes from a variety of sources. As a result, their decisions differed. We resized the images to 64 × 64 pixels to reduce GPU processing. The network was then trained separately for each class. All training tests, however, use the same parameter settings. All tests were performed 500 times, but CYCLE-GAN began producing virtual X-rays after 50 sessions in all classes during the training phase.

3.4 *Feature Extraction*

Convolutional neural network is a form of multilayer perceptron; however, a simple neural network cannot interpret complex features. In numerous implementations including classification of images, acquisition of object and analyzing medical images, CNN has achieved great performance [16]. The fundamental principle of CNN is to identify local features in the input from the upper layer and then send it to the layers below to get the sophisticated features as shown in Fig. 4. Conv layers, pooling layers, and fully integrated layers are all covered by CNN.

A collection of letters is used in the convolutional layer to identify the feature tensor maps. These kernels use a "stride (s)" to integrate all inputs, resulting in a numerical output volume size [17]. After applying the convolutional layer to start the pulse process, the input volume decreased significantly. In order to attach the input to 0 and maintain the input size using low-level features, zero padding is necessary. The convolutional layer is given with the following functionality:

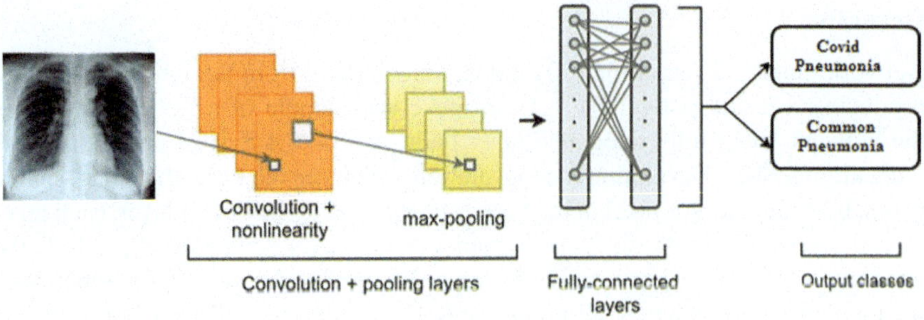

Fig. 4 CNN architecture

$$(a, b) = (J * K)(a, b) = \sum \sum I(a + m, b + n)K(m, n) \tag{3}$$

When referring to the input matrix, K-represents $m \times n$ matrix with 2D filters and F represents 2D feature map as output in Eq. (3). The effectiveness of convolution layer is described by $J * K$. For limiting the number of parameters, the pooling layer takes a small sample of the input size. The method that produces the highest value in the input area is high integration. Using factors n from the convolutional and pooling layers as a basis, the FC layer makes a decision.

3.5 Classification

Three different types of X-ray images were build an integrated method for automatically detecting COVID-19 instances. This structure was created by merging the CNN and LSTM networks, where CNN extracts the details from the images and LSTM serves as a separator. The 20 layers of the network are divided into 12 flexible layers, 5 integration layers, 1 FC layer, 1 LSTM layer, and 1 output layer with a Softmax function. Each conv block consists of 2 or 3 2D CNNs, one layer of integration, and a dropout layer with a 25% dropout rate [18]. Extract features using ReLU function, a flexible layer with 3 * 3 character size is used. By using a max-pooling layer with 2×2 sizes, the input image size is reduced. The architecture's final step involves sending a working map to the LSTM layer in order to retrieve time data. The outgoing shape behind the convolutional block is discovered to be (none, 7, 7, 512) [19]. Utilizing the technique of re-shaping, the LSTM layer's input size is decreased (50, 152). This model organizes the CXR images to a fully linked layer after examining the time characteristics to determine which of the three categories each image falls into (COVID-19 pneumonia, no pneumonia, and normal pneumonia).

Activation Functions used in LSTM

ReLU-Activation Function: ReLU is the corrective line activation function which is called as breakdown or indirect function. If the input is positive, produces a direct output; otherwise, it produces zero. The opening function in sensory networks is most frequently utilized, particularly in convolutional neural networks (CNNs) and multilayer perceptrons. Compared to its predecessors, such as sigmoid or tanh, it is simpler yet more powerful.

SoftMax Activation Function: The neural network models' outgoing layer, which forecasts the distribution of various possibilities, is activated using the SoftMax function [20]. In situations where class membership is necessary for more than two class labels, SoftMax is utilized as a facilitator for multiple class division challenges.

4 Experimental Setup and Result Analysis

First, a previous processing pipe was used to process the raw X-ray pictures. To reduce the data size and increase its commonality, preprocessing pipeline is used. Use high-resolution CYCLE-GAN photos after that. After that, the database was split into a set of testing and training data with 20:80 ratios, CNN and CNN-LSTM used to train the training samples in a particular formats. The training accuracy and training loss is being calculated after each period. Following that, a series of image tests are used to acquire the images.

4.1 Evaluation Metrics

Various performances can be evaluated with different metrics for the classification models, each metrics were calculated with different ways. Accuracy, F1-score, precision, and recalls were evaluated by calculating true positive, true negative, false positive, and false negative values.

$$\text{Precision} = \frac{\text{Tp}}{\text{Tp} + \text{Fp}} \tag{4}$$

$$\text{Recall} = \frac{\text{Tp}}{\text{Tp} + \text{Fn}} \tag{5}$$

$$\text{Precision} = \frac{2 \times (\text{precision} \times \text{recall})}{\text{precision} + \text{recall}} \tag{6}$$

$$\text{Accuracy} = \frac{\text{Tp} + \text{Tn}}{\text{Tp} + \text{Fn} + \text{Tn} + \text{Fp}} \tag{7}$$

$$\text{Specificity} = \frac{\text{Tn}}{\text{Tn} + \text{Fp}} \tag{8}$$

$$\text{MAE} = \frac{1}{N}\sum_{i=1}^{N}|yi - y\prime| \tag{9}$$

$$\text{MSE} = \frac{1}{N}\sum_{i=1}^{N}(yi - y'')^2 \tag{10}$$

$$R2 = 1 - \frac{\sum_{i=1}^{N}(yi - y\prime)^2}{\sum_{i=1}^{N}(yi - y\prime\prime)^2} \tag{11}$$

where y' is the predicted value of y and y'' is the mean value of y. The error value can be calculated with the help of metrics such as mean absolute value, mean absolute error, mean squared error can be evaluated using mean values as mentioned in the above equations.

4.2 Results Analysis

The proposed Deep CNN-LSTM hybrid model features for pediatric COVID-19 illness classification are shown in the below diagrams. The suggested DCNN-LSTM network outperforms other models, according to research, since it has smaller negative and false positive values, better realistic and consistently hostile values, and better realistic and consistent positive values (see Fig. 5). The suggested approach can efficiently distinguish COVID-19 instances as a result.

The training and validation graph for accuracy and loss can be plotted as shown in Fig. 6. The X-ray image of chest has been taken as input to produce three different outputs. Initially, it converts single images into batch.

Various metrics can be evaluated to show the output as given in Table 1. This CNN-LSTM model produces the accuracy of ~ 89%, F_1-score of ~ 91%, precision is ~ 91%, and recall is ~ 92% and Kappa-Score of ~ 90%. By comparing the existing bagging tree classification model, this proposed CNN-LSTM methods show better performance. The performance chart for comparing different classifier is mapped and shown in Fig. 7.

The performance metrics of the proposed method are given in Table 2, which contains Mean Absolute Value (MAE) and Mean Squared Error (MSE) with the values of 0.151515 and 0.242424. The R-Square value is 0.648000.

The AUC value is calculated for pediatric pneumonia classification. The ROC curve is drawn with the help of true positive rate and false positive rate, which achieves the area value of 0.935 as shown in Fig. 8.

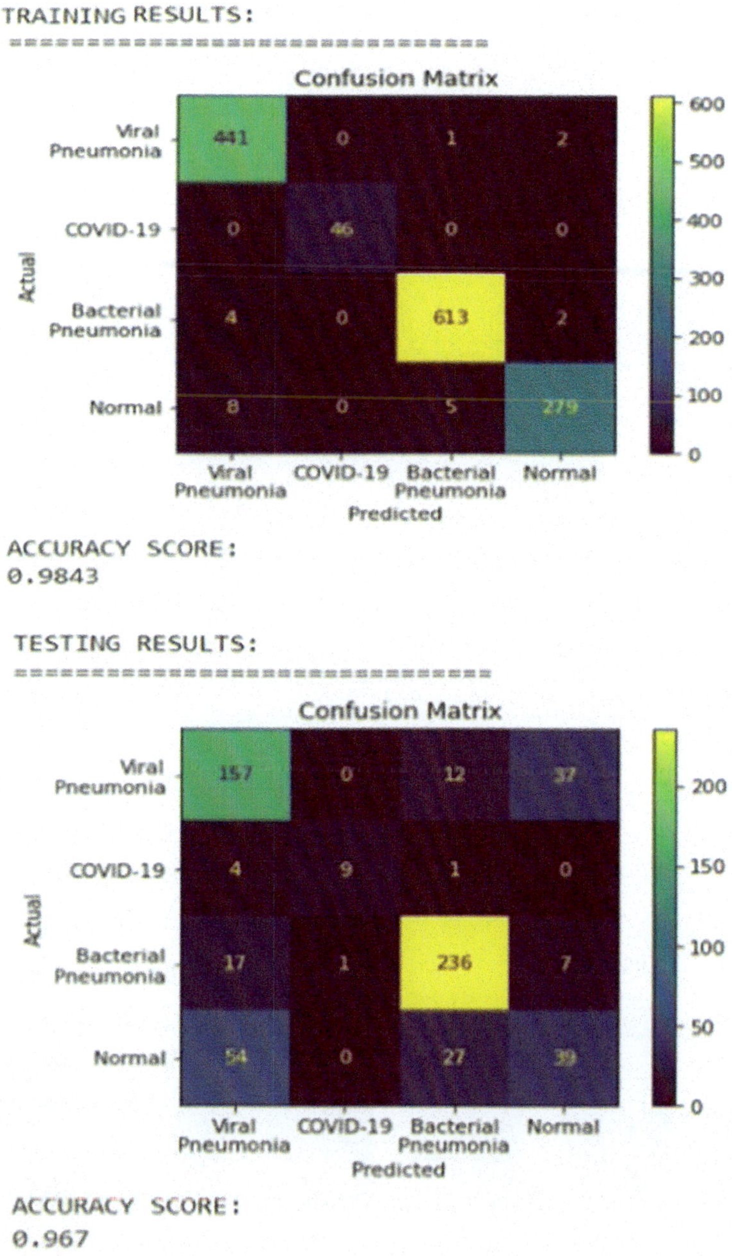

Fig. 5 Confusion matrix for training and testing using DCNN-LSTM model

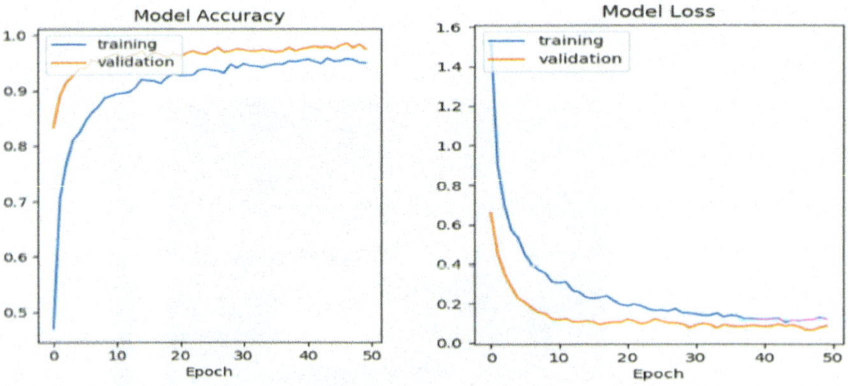

Fig. 6 Training and validation graph

Table 1 Comparison of performance metrics

Metrics	Methods	
	CNN Classifier (%)	Deep CNN-LSTM with CYCLE-GAN Classifier (%)
Accuracy	92.8	96.3
F1-Score	90.3	91.3
Precision	91.4	94.5
Recall	89.3	92.4
Kappa_Score	90.4	93.3

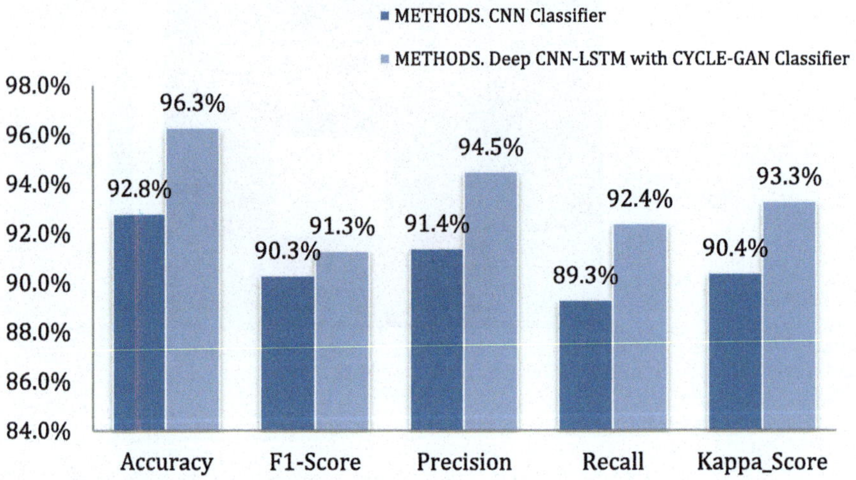

Fig. 7 Performance chart for different classifiers

Table 2 Performance metrics

S. No	Measurements	Values
1	MAE	0.151515
2	MSE	0.242424
3	R-Square	0.648000
4	AUC-ROC	0.935

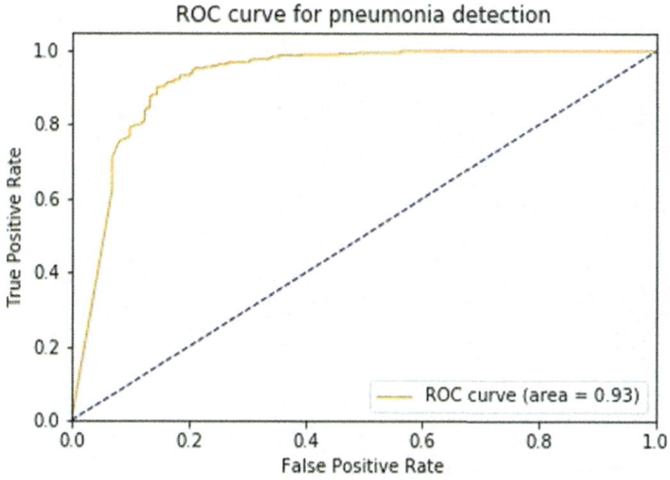

Fig. 8 AUC-ROC curve for pediatric pneumonia detection

5 Conclusion

Many nations are experiencing a resource deficit as COVID-19 instances rise regularly. Therefore, it is important to note all the good things about this health emergency especially for kids and adolescents. Here, the LSTM network serves as a coronavirus detection phase while CNN serves as a feature detector. In the same database, both the proposed Deep CNN-LSTM with CYCLE-GAN and alternative CNN structures are employed. The identification of COVID using CXR, a thorough integrated study strategy and feature reduction approach is proposed in this paper. The chest's X-ray characteristics were eliminated using a defect approach. The effectiveness of CNN network training has significantly increased. The CYCLE-GAN extensive LSTM research model utilized for COVID-19 acquisition in the suggested method was employed to address the issue of unequal data. The database of chest X-rays from Kaggle was fused with few pediatric chest X-ray images for testing. According to experimental findings, the proposed approach has an accuracy rate of 96.3% when using CXR data. The suggested approach efficiency for automatically obtaining different variants of COVID-19 was assessed using advanced reading models. The COVID-19 treatment is expedited by the proposed method's high rate of success with a few number of COVID-19 samples. This suggested strategy aids in the treatment of serious illnesses and helps doctors and radiologists diagnose COVID-19 for pediatrics effectively. This study has significant shortcomings that could be addressed in more detail in further studies.

References

1. Chen Y, Gou X, Feng X, Liu Y, Qin G, Feng Q, Yang W, Chen W (2019) Bone suppression of chest radiographs with cascaded convolutional networks in the wavelet domain. IEEE Access 7:8346–8357. https://doi.org/10.1109/ACCESS.2018.2890300
2. Wang Y, Wu Q, Dey N, Fong S, Ashour AS (2020) Deep back propagation–long short-term memory network-based upper-limbs EMG signal classification for automated rehabilitation. Biocybern Biomed Eng 40:987–1001. https://doi.org/10.1016/j.bbe.2020.05.003
3. Wang W, Xu Y, Gao R, Lu R, Han K, Wu G, Tan W (2020) Detection of SARSCoV-2 in different types of clinical specimens. JAMA
4. Kassaniaa SH, Kassanib PH, Wesolowskic MJ, Schneidera KA, Detersa R (2021) Automatic detection of coronavirus disease (COVID-19) in chest X-ray and CT images: a machine learning-based approach, vol 41
5. Maghdid HS, Asaad AT, Ghafoor KZ, Sadiq AS, Khan, MK (2021) Diagnosing covid-19 pneumonia from X-ray and CT images using deep learning and transfer learning algorithms, vol 117340E
6. Hall LO, Paul R, Goldgof DB, Goldgof GM (2020) Finding covid-19 from chest x-rays using deep learning on a small dataset. arXiv:2004.02060
7. Farooq M, Hafeez A (2020) Covid-resnet: a deep learning framework for screening of covid19 from radiographs. arXiv:2003.14395
8. Rahimzadeh M, Attar A (2020) A new modified deep convolutional neural network for detecting COVID-19 from X-ray images. http://arxiv.org/abs/2004.08052
9. Xu X et al (2020) Deep learning system to screen coronavirus disease 2019 pneumonia
10. Kermany D, Goldbaum M, Cai W et al (2018) Identifying medical diagnoses and treatable diseases by image-based deep learning. Cell 172(5):1122–1131. https://doi.org/10.1016/j.cell.2018.02.010
11. Kermany D, Zhang K, Goldbaum M (2018) Labeled optical coherence tomography and chest X-ray images for classification. Mendeley Data, v2. https://doi.org/10.17632/rscbjbr9sj.2
12. Wang L, Wong A (2020) COVID-Net: a tailored deep convolutional neural network design for detection of COVID19 cases from chest radiography images. arXiv200309871 Cs Eess. Accessed 29 Mar 2020 [Online]. Available http://arxiv.org/abs/2003.09871
13. Smith LN (2017) Cyclical learning rates for training neural networks. arXiv150601186 Cs. Accessed 29 Mar 2020 [Online]. Available http://arxiv.org/abs/1506.01186
14. Howard J, Ruder S (2018) Universal language model fine-tuning for text classification. ArXiv180106146 Cs Stat. Accessed 29 Mar 2020 [Online]. Available http://arxiv.org/abs/1801.06146
15. Zhang J, Xie Y, Li Y, Shen C, Xia Y (2020) Covid-19 screening on chest x-ray images using deeplearning based anomaly detection
16. Li L, Qin L, Xu Z, Yin Y, Wang X, Kong B, Bai J, Lu Y, Fang Z, Song Q, Cao K, Liu D, Wang G, Xu Q, Fang X, Zhang S, Xia J, Xia J (2020) Artificial intelligence distinguishes covid-19 from community acquired pneumonia on chest CT. Radiology 0(00:200905, 0. PMID: 32191588
17. Suganya D, Kalpana R (2022) Automated COVID-19 diagnosis using deep multiple instance learning with CycleGAN. In: 2022 International Conference on Smart Technologies and Systems for Next Generation Computing (ICSTSN), pp 1–6. https://doi.org/10.1109/ICSTSN53084.2022.9761334

18. Suganya D, Kalpana R (2023) Automated detection of Covid-19 waves with computerized tomography scan using deep learning. In: Awasthi S, Sanyal G, Travieso-Gonzalez CM, Kumar Srivastava P, Singh DK, Kant R (eds) Sustainable computing. Springer, Cham. https://doi.org/10.1007/978-3-031-13577-4_3
19. Suganya D, Kalpana R (2023) Prognosticating various acute covid lung disorders from COVID-19 patient using chest CT images. In: Engineering applications of artificial intelligence, 105820. ISSN 0952-1976. https://doi.org/10.1016/j.engappai.2023.105820. https://www.sciencedirect.com/science/article/pii/S0952197623000040
20. Apostolopoulos ID, Mpesiana TA (2020) Covid-19: automatic detection from x-ray images utilizing transfer learning with convolutional neural networks. In: Physical and engineering sciences in medicine

Recognition of Logo of Pirated Content Using Deep Learning-Based Regression Classification Algorithm

Kiran Kumar Jakkur Patalappa and Supriya Maganahalli Chandramouli

Abstract With the current state of online infrastructure, the phenomenon of content piracy is seeing rapid expansion and dissemination across several nations. The primary objective of internet infrastructure is to provide a platform for the transmission of authorized and legal content from the service provider to the recipient. Over the course of time, pirates have utilized the digital internet infrastructure system to replicate the original content and subsequently redistribute it using the same infrastructure. The utilization of visual analytics pertaining to the broadcast logo serves as a method for ascertaining the presence of stolen content. This study will focus on the development of a comprehensive collection of TV broadcast channel logos that are scalable and encompass a wide range of geographical locations and genres. Additionally, it will use publicly accessible statistics specifically pertaining to TV broadcast channel logos of Indian stations. A comprehensive collection of 450 television broadcast channel logos in many regional languages has been compiled, encompassing a range of categories such as sports, movies, kids and cartoons, entertainment, and more. To enhance the deep learning logo classification and expand the logo corpus, each logo undergoes exposure to a diverse range of data enrichment techniques. This study further investigates the identification of different logo classes using the use of the state-of-the-art object recognition approach known as YOLO. The experimental findings are recorded for several inference algorithms using different pixel contexts.

Keywords Deep learning · Data enrichment · Object detection · YOLO · TV broadcast channel logo corpus

---

K. K. J. Patalappa (✉) · S. M. Chandramouli
Sri Siddhartha Institute of Technology (SSIT), SSAHE, Tumkur, India
e-mail: jpkiran@gmail.com

S. Bandyopadhyay et al. (eds.), *Intelligent Computing Systems and Applications*,
Lecture Notes in Networks and Systems 1010,
https://doi.org/10.1007/978-981-97-5412-0_25

1 Introduction

The primary focal points of investigation within the domain of computer vision and pattern recognition encompass the identification and categorization of logos. The phrase "logo detection" pertains to the procedure of identifying the specific logo inside a given input picture. The task at hand presents a challenge due to the presence of several diverse things inside the given input image, hence necessitating the need to refine the focus toward the desired logo of interest. The concept of "logo classification" pertains to the categorization of a recognized symbol or emblem within a certain designation. Logo detection and classification have found extensive use across several sectors. The need for real-time brand recognition and identification has become a crucial necessity in several applications across multiple industries. An example of a novel IPTV framework is shown, which focuses on the automated detection, labeling, recognition, and replacement of television advertisements [1]. Additionally, the framework includes the identification and recognition of vehicle logos to facilitate smooth traffic flow. In the realm of pavement engineering, there has been significant research conducted on the identification of pavement cracks [2–7]. Additionally, studies have been conducted to explore the concept of brand recognition [8]. According to Ref. [9], the topic of discussion is integrated circuits (IC) authentication, as mentioned in Ref. [10]. In the field of brand management, several techniques have been developed to address important aspects such as brand recognition and reputation management on social media platforms, mark recognition and identification of counterfeits logo detection and localization [11–13]. Logo items will be created by combining text and visual symbols. Depending on factors including size, shape, texture, transformations, rotation, contrast, and occlusion, the logo in the real picture may be altered. As a result, it's likely that the process of recognizing and classifying logos in such contexts will get more challenging over time.

Recent machine learning (ML) research has shown that neural networks are quite accurate at performing video recognition tasks. Neural networks can be constructed and taught to perform a variety of visual recognition tasks, such as picture categorization, localization, and detection. A neural network may learn to distinguish between data elements without human interaction. Ibragimov et al. [14] proposes a method based on convolutional neural networks rather than outmoded key point-based methodologies. When identifying logo regions, segmentation methods are applied and for classification, SVM and CNN-derived features are used. Sahel et al. [15] uses a method in which pre-trained CNN is used to identify the regions and SVM is used to classify them. For the purpose of identifying logos, Iandola et al. [16] looked at various DCNN (deep convolutional neural network) designs. Oliveira et al. [17] describes the use of a quick region-based convolutional neural network to identify and detect logos (FRCNN). This method makes use of CNN models with ILSVRC Image Net data as input. Li et al. [18] suggests a simple and reliable object detection and identification system. Zhao et al. [19] highlights on different object detection methods using deep learning wherein the approach is broken down into two parts: the first makes use of CNN to find and identify objects, and the second makes use of categorization based on supervised pre-training for a specific job.

The ramifications of content piracy on broadcasters and content suppliers are substantial, resulting in a decrease in the perceived worth of live telecasting, particularly in the domains of sports and entertainment. Users now have more options than ever before when it comes to selecting the specific shows they want to watch online. This trend is a direct response to the growing demand for such online material. Because of the significant demand, this will afford content pirates the opportunity to engage in and perpetrate acts of content piracy, therefore unlawfully disseminating the content. Figure 1 illustrates a schematic representation of the pipeline utilized by unauthorized entities for the transfer of illicit material. The technique involves establishing a hosting site that has the URL for streaming transcoded media portions, as well as creating a connected site that promotes and attracts those seeking unauthorized access to copyrighted information. Pirates disseminate their hyperlinks on the prospective social networking platform. The use of Google advertising might potentially serve as an additional avenue for the dissemination of these illicit websites. The presence of illicit connections facilitates the cost-effective streaming of unlawfully altered pirated material, so presenting a significant peril to the industry involved in the development and dissemination of content.

Identifying the presence of pirated content constitutes the initial and pivotal measure in halting its dissemination. The procedure may be categorized into four steps: detection, collecting, calculating, and updating. The initial stage should be identifying public links that point to unlawfully obtained content. This is achieved through the process of indexing web hyperlinks and social media advertisements. The subsequent phase entails the collection of streaming data that necessitates scrutiny to determine the presence of a watermark, a forensic fingerprint, or visual analytics. The third phase involves doing computations on the collected data. The use of diverse

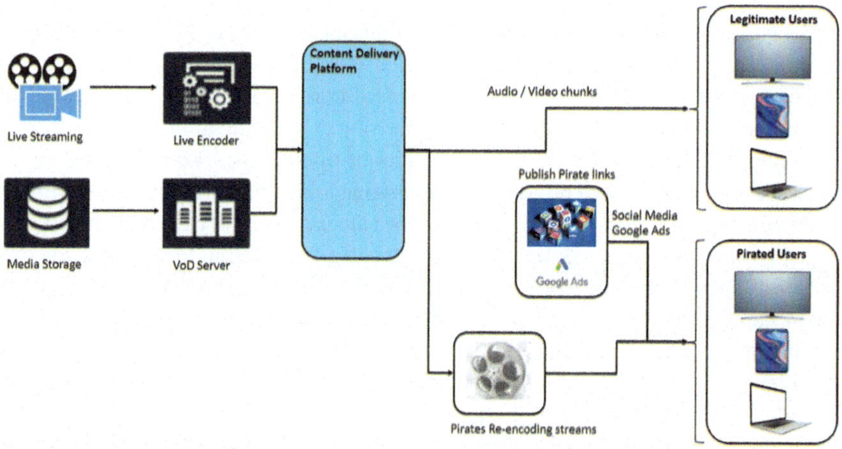

Fig. 1 Schematic representation of pipeline utilized by unauthorized entities

video analytics approaches might be employed to accomplish this task. The ultimate phase entails using the insights derived from step 3 and implementing the requisite measures.

In order to effectively identify instances of content piracy in real-time, it is imperative to employ visual analytics techniques to compare original material with unlawfully retransmitted content. The inclusion of a broadcast logo would serve as a significant indicator. The broadcast logo is visible in every frame of the original footage. The independent management of the broadcast logo's identification can be achieved without necessitating interface with the systems responsible for content distribution and delivery. The examination of video frames is vital in order to locate the broadcast logo placeholder and afterward ascertain the identity of the logo. The material retransmitted by the pirates may have had broadcast logos that were modified, distorted, inadequately apparent, or obscured.

The primary goal of this research is to construct the broadcast logo dataset, which would include 450 unique TV logo categories from various parts of India. The process of generating this extensive information is characterized by high costs, complexity, and time requirements. In addition to the provided dataset, our study delves into contemporary deep learning methodologies for the purpose of generic object detection. Subsequently, similar strategies are employed to identify and categorize broadcast logos, followed by the implementation of comprehensive experimentation.

2 Related Work

Various methodologies for broad object detection are available. The primary objectives of these techniques are the identification and categorization of the item inside the given picture, the delineation of a rectangular bounding box to denote the object's existence, and the assignment of a confidence score to assess the accuracy of the classification. The categorization of generic object identification approaches may be generally classified into two distinct sorts of operations. The first kind encompasses a pipeline for detecting and categorizing objects. This pipeline begins by extracting region suggestions, followed by the categorization of each proposed region based on several item categories. The second category, referred to as type 2, encompasses a cohesive methodology for identifying and classifying objects. This methodology categorizes the object detection problem as either a regression or classification problem. The group referred to as type 1 (two pass) encompasses a range of techniques that are built upon region proposal networks. These techniques include RCNN [20], Fast RCNN [21], FasterRCNN [22], SPP-Net [23], RFCN [24], FPN [25], and Mask RCNN [26]. The methods YOLO [27], YOLOv2 [28], SSD [29], MultiBox [30], AttentionNet [31], GCNN [32], DSOD [33], and DSSD [34] are classified as belonging to the type 2 (single pass) group in the context of regression and classification.

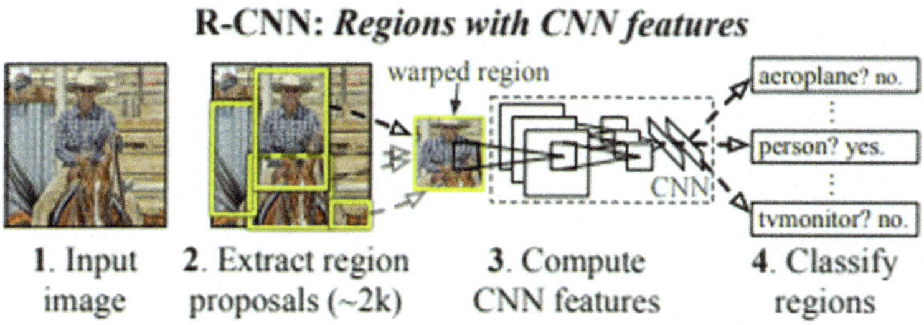

Fig. 2 Schematic representation of R-CNN [20] pipeline

2.1 Two Pass Proposal-Based Method

The R-CNN data flow diagram is shown in Fig. 2. The system includes feature generation, categorization, region extraction, and input image processing. R-CNN uses the selective search [35] method to derive 2000 region suggestions for a given image. The selective search method employs many criteria such as color, shape, texture, size, and other relevant aspects to effectively cluster regions that are connected to one another.

2.2 Single Pass Classification-Based Method

Region proposal-based approaches have several associated phases in their pipeline, from the extraction of region proposals through the construction of bounding box regression. Because of this, each stage takes its own amount of time to produce the results, which is a bottleneck for real-time applications. Regression–classification strategies aim to optimize the efficiency of classification processes by generating bounding box coordinates and class probabilities using pixel data from images. This method significantly reduces the amount of time needed to complete categorization tasks. The pipeline of a solid-state drive (SSD) is depicted in Fig. 3. For each object in the supplied image, the SSD method will employ the current ground truth bounding box. The analysis focuses on differences in aspect ratio inside each ground truth box. This functionality will facilitate the estimation of form coordinates and confidence levels for all sorts of objects contained inside the default box. As seen in Fig. 3, the process of training involves the comparison of ground truth boxes with the default boxes. The items that are congruent will yield favorable outcomes, whereas those that are not congruent will yield negative outcomes.

The identification and classification of pavement distress (PD) is essential for the effective management and rehabilitation of road infrastructure. Ongoing research is being conducted on the identification and monitoring of automation in the context of process discovery. However, it is imperative for road management to prioritize

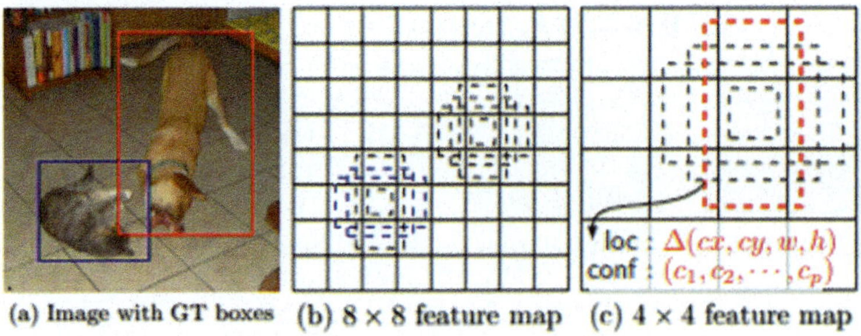

Fig. 3 Schematic representation of SSD [29] pipeline

various forms of professional development in order to effectively implement significant actions. Furthermore, the lack of a standardized Parkinson's disease (PD) dataset exacerbates the absence of a benchmark for evaluating different methodologies. The concerns are discussed in Ref. [36], and this research paper introduces two novel perspectives on the subject matter. Initially, a substantial dataset pertaining to professional development (PD) is generated. The collection consists of a total of 45,788 photographs, which were acquired using a high-resolution industrial camera mounted on a mobile vehicle. These photographs were collected under diverse weather and lighting circumstances. Each picture is enclosed by a bounding box that serves to delineate and characterize the specific region and type of pain being depicted. Furthermore, the YOLO network, which is a deep learning-based framework for object identification, is employed to predict potential locations and categories of concern. The detection accuracy has an overall value of 73.64%.

The ability to recognize and classify objects in real-time from video recordings serves as a fundamental component for the development of various analytical functionalities. These functionalities may include estimating the overall population in a certain location or tracking the volume of traffic in a particular area over a period. The work often encounters challenges such as delayed processing in classification and detection, as well as the incidence of inaccurate detection, which can be attributed to the use of tiny and lightweight datasets. The YOLO (You Only Look Once)-based detection and classification technique (YOLO) was developed in [37] to address these challenges. The utilization of this technology enhances computational efficiency and processing speed, while simultaneously upholding a high level of precision in the identification of objects inside video recordings. Each item class that the classification algorithm has been trained on is assigned a bounding box and an annotation that includes specific information about that class.

Tao et al. [36] suggests a technique for finding things in pictures of traffic scenes. It is dependable, swift, and accurate. Typical object detectors start by making recommendations. After that, the characteristics are extracted. Then, these proposals are subjected to a classifier. However, the speed is slow and the accuracy is subpar.

YOLO, a wonderful deep learning-based object recognition system, presents one convolutional neural network for both location and categorization. All YOLO's fully connected layers are exchanged for an average pool layer in order to form a new network.

The study described in Ref. [38] utilizes a radar sensor operating at a frequency of 77 GHz, specifically a frequency modulated continuous wave radar sensor. The purpose of this sensor is to develop a system for detecting and classifying vehicles in urban environments, using a single shot approach. The identification and categorization of individuals utilizing roadways relies on the use of the YOLO (You Only Look Once) real-time object detection method, which is implemented on the pre-processed radar range doppler angle power spectrum. To conduct training and testing of the proposed system, data is gathered by deploying a test car in urban areas. A mean average accuracy of 70.64% was achieved when evaluating a distinct test data set.

This section focused on regression/classification proposal approaches using special techniques like you look only once (YOLO) algorithms to construct regions of interest, detect, and categorize Indian TV broadcast channel logos. The corpus of logos for Indian TV broadcast channels is subject to these algorithms.

3 Dataset Collection

The focus of this section will be on compiling a dataset from readily available sources and then applying data augmentation techniques to increase the dataset.

Publicly accessible TV broadcast channel logo databases [39] include logos from all over the world. These pictures are scarce, and they do not even have logo classes. The logo class has a high frequency of occurrence, often with little differences in size and contrast. For the purpose of this study, a collection of prominent television broadcast channel logos has been assembled. The logos were captured using broadcast video television from digital television (DTV) broadcast and cable operators. The dataset manually edited comprises 450 unique television broadcast channel logos, encompassing a diverse range of genres such as entertainment/drama, kids/children, films, news, science fiction, spirituality, sports, and travel/documentary. Considerable attention has been devoted to the deliberate selection of well-recognized broadcast channel logos that correspond to each distinct type of television entertainment. Figure 4 presents a graphic representation of the logos that were captured and classified according to different television genres. Further, Table 1 provides information on the mode used to capture the logo. Table 1 also indicates some sample channel names for which the logos has been captured.

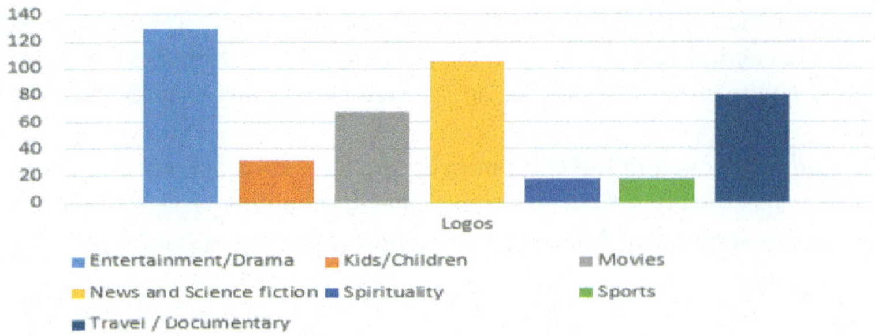

Fig. 4 Pictures arranged by different genre of television

Table 1 Few examples of logos captured from videos as part of digital television (DTV) broadcast and cable operators

TV viewing channels category	Channels	Service providers	Duration (min)
Entertainment/drama	Comedy Central, Sony, …	Cable	3
Kids/children	Cartoon Network, Disney Juniors	Cable	3
Movies	Star Movies, Rishtey Cineplex	Cable	3
News and science faction	Animal HD, NGC	Cable	3
Spirituality	Sanskara, Sri Venkateshwara Bhakthi	Cable	3
Sports	ESPN, Star Sports	Cable	3
Travel/documentary	India Today, Times Now	Cable	3

3.1 Method for Data Enrichment

Data enrichment is a technique for unintentionally expanding the size of the data collection while keeping the data variety required for model training constant. This method seeks to expand and improve the information by adjusting like cropping, titling, zooming, etc. Data enrichment, as seen in Fig. 5, involves altering the original image's intrinsic pixel values to generate similar-looking but otherwise distinct pictures.

The fastai [40] image processing technique was used to expand the dataset shown in Table 1. Modifications such as (brightness/contrast/crop/dihedral/flip/jitter/perspectivewrap/rotate/RGBrandomize/skew/squish/symmetric wrap/tilt/zoom/cutouts) are among the most notable. Since the broadcast channel logo images are copyrighted, actual broadcast images are not shown, instead sample reference image [41] is used to show the concept.

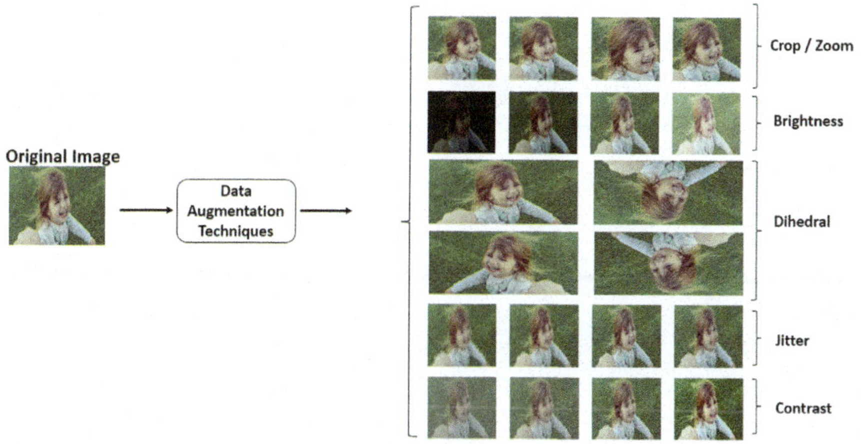

Fig. 5 Some of data augmentation techniques. Image used here is taken from [41]

4 Methodology

Figure 6 shows a diagram of the whole setup. During the training phase, the supplied input picture is exposed to data enrichment techniques to supplement data of various sorts, enhancing the efficacy and reliability of the classification model. The accuracy of the classification model will increase even when the logo is obscured on purpose. The real data improved images will be sent into the block object detection method, which uses a regression classification technique. The model for block object detection will be trained using the annotated, data-rich images. The given image is processed by a regression classification approach (YOLO [28]) (single pass method) to isolate several item classes for testing purposes. Figure 6 depicts a block network for object identification and classification that can identify items inside an image, generate location coordinates for the object, and predict its class.

Figure 7 depicts the regression classification process as it applies to a video picture used in broadcast. In contrast to region proposal-based object detection methods (two pass method), which search each image for region of interest that identifies and localizes the object before classifying object, regression classification methods (single pass method) may aim to provide bounding box coordinates and class availability probabilities from picture pixels in a single algorithm call. A square grid of size $S \times S$ is created from the supplied picture. Each of the image's many items may occupy a single $S \times S$ grid cell, or they may be spread out across many cells. An object's anticipated grid is the cell that contains both the object's center and its corresponding grid. Objects are recognized and classified by a single CNN network. The probability of the C_p class and the bounding boxes for the B class must be predicted for each grid cell. There are five inputs into the bounding box prediction: (B_x, B_y, B_w, B_h, and C). B_x, By represents the coordinates of the center of the box in reference to a certain grid cell. The width and height of the bounding box are recorded as a pair (B_w, B_h).

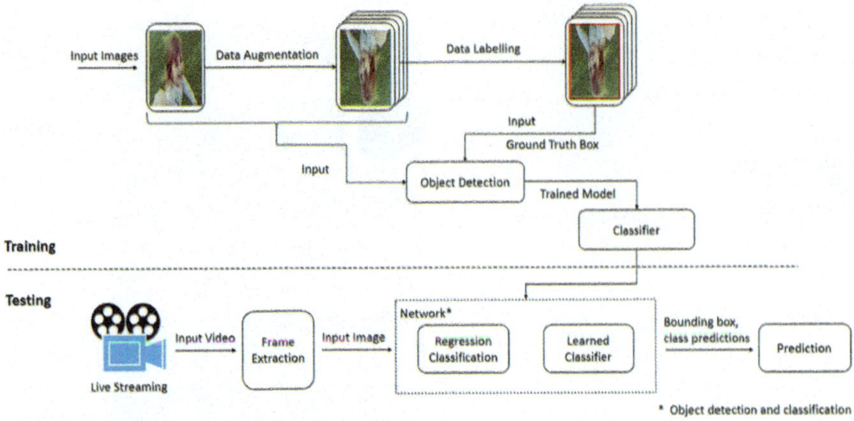

Fig. 6 Schematic representation of proposed pipeline for detecting, localizing, and classifying. Image used here is taken from [41]

Parameter C specifies the degree of certainty with which we can access the object. Object existence in a specific cell is not proven if the score is 0. When the forecast box and the ground truth box have the same intersection of union (IoU), confidence is at its highest.

Since the broadcast channel logo images are copyrighted, actual broadcast images are not shown, instead sample reference image [41] is used to show the proposed pipeline for detection, localization, and classifying.

4.1 *Object Detection and Feature Extraction Approach*

For object detection and feature extraction, a network architecture in the VGG fashion is employed. Highlights of the underlying network architecture are shown in Fig. 8.

The input size for this network will be $416 \times 416 \times 3$. The final feature maps are 13 by 13. As a result, there is a higher chance that the object will be in the middle of the feature map. The 13×13 feature map design generates a lower mAP but an improved recall of 88%.

The network employs several crucial parameters to improve mean average precision (mAP) and recall, such as:

- Batch normalization accelerates model convergence during training. This also limits the use of other normalization methods without overfitting, including dropout.
- Using the high-resolution classifier, images with greater definition by training with 416×416, which leads in a 4% increase in mAP, the network can modify the filters.

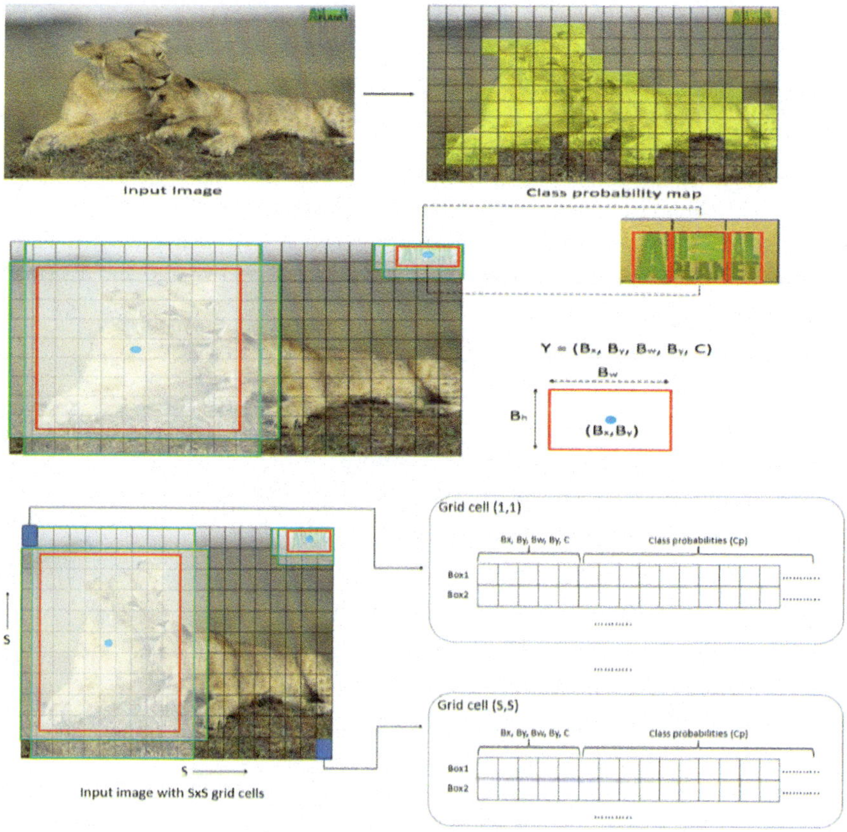

Fig. 7 Schematic diagram representing classification and pixel-based object detection

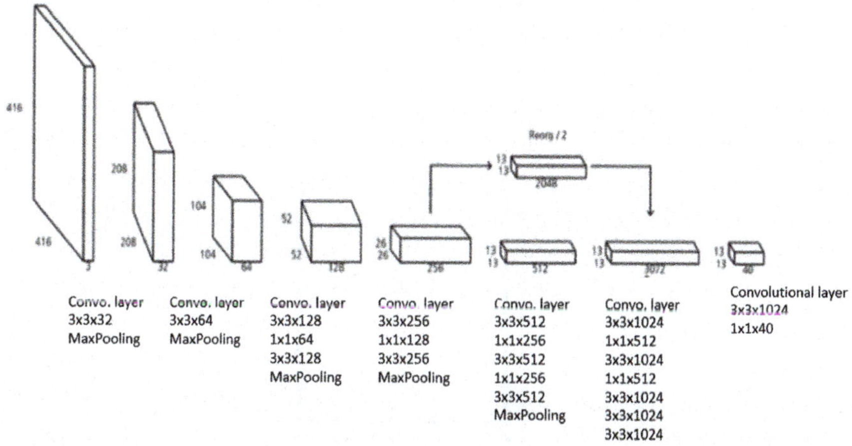

Fig. 8 VGG network architecture

- Instead of employing anchor boxes for bounding boxes, replace the fully linked layer in the region proposal network with anchor bounding boxes. To anticipate boundary boxes, utilize anchor.
- The top-K bounding boxes for images are identified with the highest level of accuracy using the K-means clustering algorithm. The maximization of IOU, as opposed to a reduction in Euclidean distance, is the aim of this strategy. In YOLO, $K = 5$ is employed to enhance the algorithmic trade-off.
- The sigma function is used to confine the value of the five parameters that YOLO predicts (B_x, B_y, B_w, B_h, C (objectness score)) to fall between 0 and 1. This is described in Fig. 9 and Eq. 1 [27].

$$
\begin{aligned}
b_x &= \sigma(t_x) + c_x \\
b_y &= \sigma\left(t_y\right) + c_y \\
b_w &= p_w e^{t_w} \\
b_h &= p_h e^{t_h} \\
p_r(\text{object}) &* \text{IOU}(b, \text{object}) = \sigma t_o
\end{aligned} \tag{1}
$$

t_x t_y t_w t_h are predictions made by YOLO
c_x c_y is the top left corner of the grid cell of the anchor
p_w p_h are the width and height of the anchor
c_x c_y p_w p_h are normalized by the image width and height
b_x b_y b_w b_h are the predicted boundary box
$\sigma(t_o)$ is the box confidence score

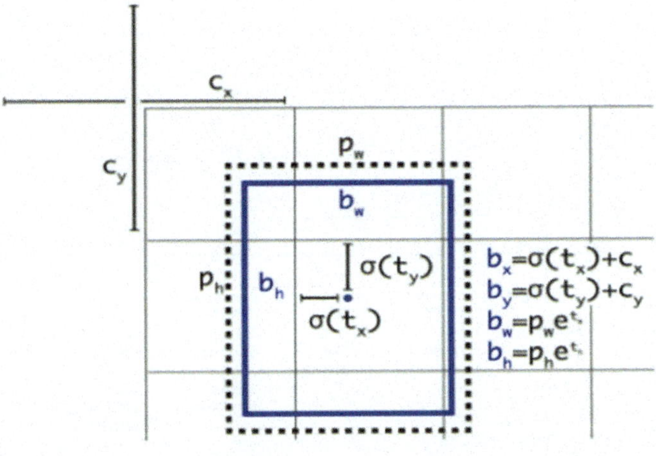

Fig. 9 Prediction of 5 parameters and narrowing the result between 0 and 1, YOLO [27]

5 Experimental Results

Many data transformations are applied to each of the 450 distinct broadcast TV station logos that were collected (Table 1). The technique depicted in Fig. 6 is also utilized to train the model using the data set. Each TV broadcast logo picture is altered using many approaches to provide 125 authentic data enriched images. Table 2 displays the data enrichment techniques, the TV genre category, the number of photographs with authentic data enhanced logos, and the total number of photos. There are a total of 56,250 photos in the entire collection. There are 54,000 photos in the training set and a total of 2250 photos in the testing set.

The outcomes of the experiments are shown in Table 3. The trained model is used to validate a sample of randomly selected images from each TV genre group. There are 130 distinct channel logo classes in the "Entertainment/Drama" TV genre category. Each of the 130 channel logo classes has a set number of photos selected from it for verification purposes. The same may be said about the many categories of television shows. For each of these categories of television shows, we calculated their precision, recall, and accuracy. The following experiments are planned to ensure that the desired confidence score of more than 0.5 is achieved by comparing the projected box to the ground truth using the intersection of union (IoU) method.

Pirates may go to great lengths to conceal the emblem by, for example, changing it or making it difficult to see. Data enrichment methods, such as random erasing/cutout and jitter, have been used to purposely mimic this condition. These fabricated visuals will simulate genuine user reactions to the stolen content once the broadcast logo has been altered. A sample of our trained model's output is shown in Fig. 10.

Since the broadcast channel logo images are copyrighted, actual broadcast images are not shown, instead sample reference image [41, 42] is used to show the experimental results.

Table 2 Depicting overall logos which have been collected

TV viewing channels category	Different classes	Logo enrichment per TV channel	Total images
Entertainment/drama	130	125	16,250
Kids/children	31	125	3875
Movies	68	125	8500
News and science faction	105	125	13,125
Spirituality	18	125	2250
Sports	18	125	2250
Travel/documentary	80	125	10,000

Table 3 Depicting results on the experiments carried out

TV viewing channels category	Different classes used	Pictures for testing	Precision	Recall	Accuracy
Entertainment/ drama	130	650	0.825	0.811	0.822
Kids/children	31	155	0.833	0.794	0.813
Movies	68	340	0.817	0.791	0.811
News and science faction	105	525	0.852	0.816	0.852
Spirituality	18	90	0.863	0.827	0.859
Sports	18	90	0.852	0.793	0.848
Travel/ documentary	80	400	0.842	0.816	0.841

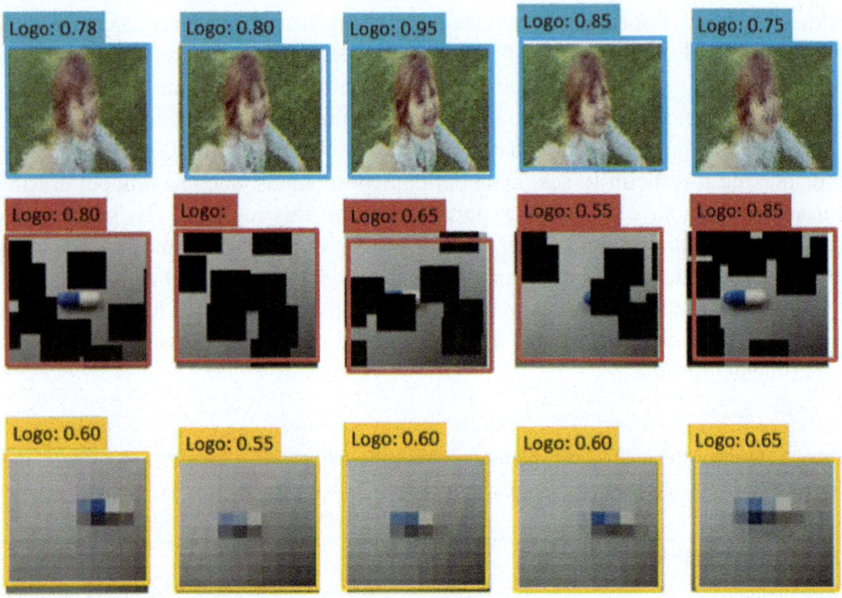

Fig. 10 Depicting experimental result images. Image used here is taken from [41]

6 Comparison with Other State-of-Arts Classification Methods

In this part, the results of the proposed YOLO model are compared to those of the SSD method of object recognition. According to Sect. 3 of this article, the obtained test data is analyzed using the SSD method to determine an evaluation. The results are summarized in Table 4. Significant modifications were made to the resulting

Table 4 Depicting comparison results

TV viewing channels category	Different classes used	Pictures for testing	Using our proposed system			Using SSD method		
			Precision	Recall	Accuracy	Precision	Recall	Accuracy
Entertainment/drama	130	650	0.825	0.811	0.822	0.835	0.812	0.832
Kids/children	31	155	0.833	0.794	0.813	0.857	0.832	0.812
Movies	68	340	0.817	0.791	0.811	0.814	0.798	0.803
News and science fiction	105	525	0.852	0.816	0.852	0.805	0.792	0.810
Spirituality	18	90	0.863	0.827	0.859	0.867	0.825	0.853
Sports	18	90	0.852	0.793	0.848	0.833	0.790	0.823
Travel/documentary	80	400	0.842	0.816	0.841	0.823	0.801	0.815

dataset to account for a wide range of circumstances including subtle shifts in color, brightness, and object orientation. The results demonstrate the efficacy of You Only Look Once (YOLO) in combating content piracy in real-world datasets.

7 Conclusion

The main topics of this paper included a discussion and assessment of state-of-the-art YOLO object detection algorithms and a system pipeline that employs regression classification methods for the detection and classification of broadcast logos in the context of broadcast and broadband content piracy. The primary originality of this work comes in how effectively it tackles the issue of identifying content piracy in video analytics using the YOLO technique. A key advantage of applying the YOLO technique is the ability to find, recognize, and classify the broadcast logo in a single pipeline step. In terms of systematically examining the effects of jitter and random erasing/cutout on the performance of object recognition, the results of studies carried out on a variety of TV genre categories were encouraging and successful.

The direction to future research work can be comprised of looking on different steps to combat content piracy. Step 1: Searching the social media to gather information that talks about illegal streaming content URL. Step 2: Collecting data using illegal streaming content URL identified in step #1. Step 3: Visualizing the illegal streaming content website data to identify the malicious tracking elements embedded and how user is exposed to risk their sensitive information. Step 4: Using video frame analysis to find and identify the desired objects. Step 5: Initiating the necessary action against unauthorized rebroadcasters.

References

1. Arsan T, Bulut EE, Eren B et al (2022) A novel IPTV framework for automatic TV commercials detection, labeling, recognition and replacement. Multimed Tools Appl. https://doi.org/10.1007/s11042-021-11563-y
2. Zhang J, Yang S, Bo C, Zhang Z (2021) Vehicle logo detection based on deep convolutional networks. Comput Electr Eng 90:107004. ISSN 0045-7906. https://doi.org/10.1016/j.compeleceng.2021.107004
3. Liu R, Han Q, Min W, Zhou L, Xu J (2019) Vehicle logo recognition based on enhanced matching for small objects, constrained region and SSFPD network. Sensors 19(20)
4. Jiang X, Sun K, Ma L, Qu Z, Ren C (2022) Vehicle logo detection method based on improved YOLO. Electronics 11:3400. https://doi.org/10.3390/electronics11203400
5. Lu W, Zhao H, He Q, Huang H, Jin X (2021) Category-consistent deep network learning for accurate vehicle logo recognition. Neurocomputing 463:623–636. ISSN 0925-2312. https://doi.org/10.1016/j.neucom.2021.08.030
6. Li Y, Chen Y, Yuan S, Liu J, Zhao X, Yang Y, Liu Y (2021) Vehicle detection from road image sequences for intelligent traffic scheduling. Comput Electr Eng 95:107406. ISSN 0045-7906. https://doi.org/10.1016/j.compeleceng.2021.107406
7. Hou Y et al (2022) A deep learning method for pavement crack identification based on limited field images. IEEE Trans Intell Transp Syst 23(11):22156–22165. https://doi.org/10.1109/TITS.2022.3160524
8. Yousaf W et al (2021) Patch-CNN: deep learning for logo detection and brand recognition. J Intell Fuzzy Syst 40(3):3849–3862
9. Hu B, Zhou N, Zhou Q, Wang X, Liu W (2020) DiffNet: a learning to compare deep network for product recognition. IEEE Access 8
10. Mallaiyan Sathiaseelan MA, Paradis OP, Taheri S, Asadizanjani N (2021) Why is deep learning challenging for printed circuit board (PCB) component recognition and how can we address it? Cryptography 5(1). [Online] Available: https://www.mdpi.com/2410-387X/5/1/9
11. Mallaiyan Sathiaseelan MA, Vutukuru MY, Pandurangi SV, Taheri S, Asadizanjani N (2021) Ic synthlogo: A synthetic logo image dataset for counterfeit and recycled ic detection. In: 2021 IEEE international symposium on the physical and failure analysis of integrated circuits (IPFA), pp 1–8
12. Sathiaseelan MAM et al (2022) Logo detection and localization for IC authentication, marking recognition, and counterfeit detection. In: 2022 IEEE international symposium on the physical and failure analysis of integrated circuits (IPFA), pp 1–7. https://doi.org/10.1109/IPFA55383.2022.9915713
13. Liu L, Dzyabura D, Mizik N (2018) Visual listening. In: Extracting brand image portrayed on social media. AAAI
14. Ibragimov E, Lee H-J, Lee J-J, Kim N (2022) Automated pavement distress detection using region based convolutional neural networks. Int J Pavement Eng 23:6. https://doi.org/10.1080/10298436.2020.1833204
15. Sahel S, Alsahafi M, Alghamdi M, Alsubait T (2021) Logo detection using deep learning with pretrained CNN models. Eng Technol Appl Sci Res 11(1):6724–6729
16. Iandola FN, Shen A, Gao P, Keutzer K (2015) Deep logo: hitting logo recognition with the deep neural network hammer. arXiv preprint arXiv: 1510.02131
17. Oliveira G, Frazão X, Pimentel A, Ribeiro B (2016) Automatic graphic logo detection via fast region-based convolutional networks. In: International joint conference on neural networks (IJCNN). IEEE, pp 985–991
18. Li Y, He X, Dong Y, Kong Q, Zeng Y (2022) Spike calibration: fast and accurate conversion of spiking neural network for object detection and segmentation. arXiv:2207.02702
19. Zhao Z-Q, Zheng P, Xu S-t, Wu X (2019) Object detection with deep learning: a review. arXiv: 1807.05511v2 [cs.CV]
20. Girshick R, Donahue J, Darrell T, Malik J (2014) Rich feature hierarchies for accurate object detection and semantic segmentation. In: CVPR

21. Girshick R (2015) Fast R-CNN. In: ICCV
22. Ren S, He K, Girshick R, Sun J (2015) Faster R-CNN: towards realtime object detection with region proposal networks. In: NIPS, pp 91–99
23. He K, Zhang X, Ren S, Sun J (2015) Spatial pyramid pooling in deep convolutional networks for visual recognition. IEEE Trans Pattern Anal Mach Intell 37(9):1904–1916
24. Li Y, He K, Sun J et al (2016) R-FCN: object detection via region-based fully convolutional networks. In: NIPS, pp 379–387
25. Lin T-Y, Dollar P, Girshick RB, He K, Hariharan B, Belongie SJ (2017) Feature pyramid networks for object detection. In: CVPR
26. He K, Gkioxari G, Dollar P, Girshick RB (2017) Mask R-CNN. In: ICCV
27. Redmon J, Divvala S, Girshick, Farhadi A (2016) You only look once: unified, real-time object detection. In: CVPR
28. Redmon J, Farhadi A (2016) Yolo9000: better, faster, stronger. arXiv:1612.08242
29. Liu W, Anguelov D, Erhan D, Szegedy C, Reed S, Fu C-Y, Berg AC (2016) SSD: Single shot multibox detector. In: ECCV
30. Erhan D, Szegedy C, Toshev A, Anguelov D (2014) Scalable object detection using deep neural networks. In: CVPR
31. Yoo D, Park S, Lee J-Y, Paek AS, So Kweon I (2015) Attentionnet: aggregating weak directions for accurate object detection. In: CVPR
32. Najibi M, Rastegari M, Davis LS (2016) G-CNN: an iterative grid-based object detector. In: CVPR
33. Shen Z, Liu Z, Li J, Jiang YG, Chen Y, Xue X (2017) DSOD: learning deeply supervised object detectors from scratch. In: ICCV
34. Fu CY, Liu W, Ranga A, Tyagi A, Berg AC (2017) DSSD: deconvolutional single shot detector. arXiv:1701.06659
35. Uijlings JR, Van De Sande KE, Gevers T, Smeulders AW (2013) Selective search for object recognition. Int J Comput Vision 104(2):154–171
36. Tao J et al (2017) An object detection system based on YOLO in traffic scene. In: 2017 6th international conference on computer science and network technology (ICCSNT). IEEE
37. Jana AP, Biswas A (2018) YOLO based detection and classification of objects in video records. In: 2018 3rd IEEE international conference on recent trends in electronics, information & communication technology (RTEICT). IEEE
38. Pérez R et al (2019) Deep learning radar object detection and classification for urban automotive scenarios. In: 2019 Kleinheubach conference. IEEE
39. Publicly available TV Broadcast channel logo. https://zenodo.org/record/2557474#.XzPcxCgzaUk
40. FastAI vision transformation. https://docs.fast.ai/vision.transform.html#vision.transform
41. Data augmentation example. https://towardsdatascience.com/data-augmentations-in-fastai-84979bbcefaa
42. Data augmentation example. https://becominghuman.ai/data-augmentation-using-fastai-aefa88ca03f1

Evaluation of Various Machine Learning Algorithms for Seed Prediction Based on the Soil Nutrient Quantification

Balakrishnan Ramprakash and Sankayya Muthuramalingam

Abstract The current condition of the soil is an important consideration in crop yield projections. An analysis of the soil's nutrient content can help farmers and soil analysts obtain a higher yield of seedlings suited to the cultivation process by facilitating the necessary preparations. This study presents a number of machine learning techniques that have been implemented in order to predict seedlings based on soil nutrient measurements. The Department of Agriculture Department in South Tamil Nadu provided the data that was used in this research experimental design. This data includes soil nutrient level samples from a variety of districts located throughout the South Tamil Nadu region for a selection of districts. The evaluation of crop yields is becoming an increasingly important area of research, including machine learning. The challenge of accurately predicting yields is an extremely significant one in the agricultural industry. Any farmer worth his salt is going to want to know how much yield he may anticipate receiving from his next harvest. In the past, predictions of yield were made by factoring in the years of experience that farmers had gained working with a certain crop and field. On the basis of the data that is now available, the prediction of the yield is a significant problem that has not yet been resolved. The implementation of methods that are driven by machine learning is the approach that is going to be most successful in achieving this objective. Five supervised machine learning methods were used to analyze the study's data: DT, NB, SVM, RF, and LR are codes for the following: Follow these ML classification techniques to plainly categorize the results. Regression is a form of supervised learning algorithm used to predict a continuous output variable given one or more predictors or features, also known as input variables. Experiments have been carried out in order to discover the method that is the most accurate in terms of seed prediction for the purpose of

B. Ramprakash (✉)
Department of Applied Mathematics and Computational Science, Thiagarajar College of Engineering, Madurai, India
e-mail: raam.prakash729@gmail.com

S. Muthuramalingam
Department of Information Technology, Thiagarajar College of Engineering, Madurai, India

S. Bandyopadhyay et al. (eds.), *Intelligent Computing Systems and Applications*, Lecture Notes in Networks and Systems 1010,
https://doi.org/10.1007/978-981-97-5412-0_26

continuing the farming process. In light of the findings of the experiments, it has been hypothesized that the machine learning technique known as random forest, which was among those used in the course of the research, will prove to be the most accurate method for seed prediction at a percentage level of 99.9%.

Keywords Soil nutrient · Seed · Machine learning · Naïve Bayes · Random forest · Decision tree · Support vector machine · Logistic regression · Prediction

1 Introduction

Agriculture has been considered India's primary and most significant cultural activity since the country's earliest times. This viewpoint dates back to India's Vedic period. The ancient people cultivated the crops on their own land, therefore over time, the plants have changed to accommodate the needs of those ancient people. Modern agriculture is slowly declining as a result of improvements in efficiency and productivity made possible by new technologies. People are devoting the majority of their time and energy, as a result of these fruitful inventions, to the production of artificial objects, which are hybrid products, which in turn leads to an unhealthy way of living. People living in our day and age typically lack knowledge regarding the proper planting and harvesting of crops at the appropriate times and locations. Because of these methods of farming, seasonal climatic conditions are also being altered, which has a negative impact on key resources like soil, water, and air and increases the risk of food insecurity.

Over the course of the past few years, a number of researchers have begun to apply the principles of machine learning to the sector of agriculture. Machine learning is a subfield of computer science that focuses on the development of software that enables computers to educate themselves automatically through experimentation and repetition. It is composed of a collection of clearly outlined models that, in order to achieve the desired outcomes, engage in the collection of particular data and the application of particular algorithms. Beneficial to the industry, the incorporation of machine learning techniques in agriculture has increased agricultural yield and quality. It is possible to determine, for a particular crop, under what circumstances it would generate the best yield by applying the algorithms of machine learning. Several factors, such as the composition of the type of soil, the climate, the topography of the region, and the presence of diseases or insect infestations, contribute to the ability to predict production. These elements individual chemical and physical properties are the result of the contributions made by each of these elements. It is essential to provide proper care for the soil in order to maintain fertility, boost productivity, and safeguard the environment. When farmers want to know what kind of crop they can grow successfully on a certain piece of land, it is incredibly important for them to conduct an analysis of the nutrients in the soil. Both the establishment of nutrient deficiency limits and the monitoring of soil fertility changes are necessary for the management of soil fertility. This allows for the anticipation of nutrient deficiencies.

This can be accomplished by maintaining a high level of agricultural productivity. According to Singhatiya and Ghosh [1], one of the most significant variables in obtaining high and sustainable output is maximizing the soil's fertility. The seeds used to cultivate new crops must be selected with great attention and be of high quality. The selection of the to-be-planted seeds may be the most important step in the agriculture procedure. This crop will produce well due to the high quality of its seeds. Agriculture requires both extensive knowledge and labor. The agricultural process requires the following cultivation steps: Soil preparation, seed selection, seed planting, irrigation, manure and fertilization, crop protection, crop harvesting, and crop storage.

This work is to improve the better analysis of cultivation time in order to select the better seed for producing more output of crops, which should help overcome the economic condition of farmers. The purpose of this work is to improve the better analysis of cultivation time. The format of this research work as described in the following: Sect. 2 contains a representation of the literature review, Sect. 3 has a discussion of the materials and technique, Sect. 4 contains a presentation of the results and discussion, and Sect. 5 contains a conclusion and prospective applications for the future.

2 Literature Review

Sujatha and Isakki [2] proposed a method that would use data mining techniques to estimate agricultural output in an effort to increase the value and profit of agricultural land. This methodology would utilize data mining techniques to estimate agricultural output. Singh et al. [3] used a variety of machine learning techniques to predict the category of rice crop production based on the presence of macro- and micronutrients in the dataset. This was done to maximize the accuracy of their predictions. These two models are also superior when it comes to predicting the yield based on the nutrient status of the soil. Supriya [4] presented a data mining-based strategy for predicting the category of the investigated soil datasets agricultural yield. This method was demonstrated for predicting the category. Both the Naive Bayes and the K-nearest neighbor techniques are used in this process. In addition to this, the system architecture is constructed using data mining technology and then reviewed. In addition, Pandith et al. [5] proposed a method for predicting wheat crops by applying fuzzy-c means clustering in conjunction with neural networks. This was done in order to make the prediction. Priya et al. [6] made their predictions for crop yields in the state of Tamil Nadu, India, by employing the random forest methodology. This work was finished up in the same year it was started. In order to conduct out the experiments, the program R Studio was applied; the dataset for the study is comprised of a range of factors including things like rainfall, temperature, crop production, and so on. Sellam and Poovammal [7] assessed environmental factors that influence crop yield, all of these environmental variables affect crop yield, the dependent variable. Weather conditions and soil parameters have a greater impact on crop production than cultivation area and the Food Price Index. Ghadge et al. [8] made a tool that uses data mining methods to help farmers keep track of the quality of their soil. The system gives inspection of soil quality top priority so that it can figure out which crops are best suited for a certain type of soil. Singh et al. [3] proposed a method for classifying crop yields based on the macro- and micronutrient content of the soil by employing a number of machine learning techniques. This technology analyzes crop data and suggests high-yield, high-profit commodities. By monitoring environmental and agronomic factors, Shirsath et al. [9] designed a decision support system to increase crop yields. The testing soil was furnished by the Agra region. Despite a large number of datasets, this system demanded a great deal of memory and processing capacity.

Renuka [10] used a number of machine learning methods, such as *K*NN, SVM, and decision tree, to estimate the sugarcane crop yield. For the purpose of this study, the Python platform was utilized to carry out the research. The decision tree was expected to have an accuracy of 99%, which was the highest conceivable, and it also had a smaller mean square error. Using Naive Bayes, decision tree, and a hybrid technique that combines Naive Bayes and decision tree, Manjula and Djodiltachoumy [11] prescribed to choose nutrients for evaluating the soil supplements. The nutrients consisted of a wide variety of elements, some of which were nitrogen, phosphorus, potassium, calcium, magnesium, sulfur, iron, and zinc. When evaluating the efficacy of various categorization algorithms, both the degree to which they are accurate and the amount of time it takes for them to complete their assignments are taken into consideration. In their research, to forecast soil fertility, Jayalakshmi and Devi [12] employed ML methods. The authors created a model that, given a set of soil parameters, can accurately and efficiently predict whether soil conditions are optimal for crop production using random Forest and *K*-nearest neighbor. Kumar et al. [13] suggested that machine learning provides an alternative inspection method with the advantages of automated inspection, convenience, and a reduction in cost. The visual patterns on rice plants are analyzed with machine learning classifiers such as support vector machine, logistic regression, decision tree, Nave Bayes, random forest, linear discriminant analysis, and principal component analysis, and the health of the plants is determined based on the classification outcomes. Sarijaloo et al. [14] use various machine learning algorithms, such as decision tree, gradient boosting machine, random forest, adaptive boosting, XGBoost, and neural network, to predict the yield of maize hybrids using data from the 2020 Syngenta Crop Challenge. Deepa and Gokila [15] explained that in order to predict precipitation and crop yield, scientists and statisticians use machine learning and statistical methods. Kharif and Rabi are examined as two distinct seasons. To forecast seasonal precipitation and crop yield, regressor prediction models are constructed. This research focuses predominantly on the rainfall-dependent seasonal crop production forecasting.

This proposed research's primary objective is to implement and evaluate a variety of machine learning algorithms using Python libraries in order to generate the most accurate predictive model which is also feasible. This model will be used to assist farmers in finding a better way to increase their crop yield, which will depend on the level of required soil nutrient and will enable us to select the type of seed that will be most suitable for the subsequent prediction process.

3 Materials and Methodology

3.1 Study Area and Dataset Description

The main goal of the current study is to try to predict the amount of nutrients in seeds by measuring the nutrients in the soil. Experiments were done on the Colab platform so that the goal of this study could be realized. The study challenge has been framed out by doing in-depth analysis of the problem domain, including conversations with agriculturalists and soil chemists, and reviewing relevant published material. Real data have been acquired for the purpose of the present research challenge from the Soil Testing Lab, the Directorate of Agriculture Department, South Tamil Nadu. The information was gathered from seed-growing regions located in a number of districts located in the South Tamil Nadu region. This dataset contains 5000 instances, 12 input factors indicating the soil nutrient state of the South region, and one output attribute, which is seed name. Table 1 represents the soil nutrients of various ranges and collection of seed names.

Following the conclusion of data collection, preprocessing was performed to convert nominal values into numeric form, impute missing data, and identify data outliers using statistical methods. This was done in order to ensure that the data were accurate and complete. The min–max normalization that was carried out was

Table 1 Soil nutrient dataset

N	P	K	PH	EC	S	CU	FE	MN	ZN	B	Duration	Seed name
143	69	217	5.9	0.58	0.23	10.2	116.35	59.96	54.85	21.29	Short	IR 64
170	36	216	5.9	0.15	0.28	15.69	114.2	56.87	31.28	28.62	Short	CO 47
158	66	219	6.8	0.34	0.2	15.29	65.87	51.81	57.12	27.59	Short	ADT 36
133	45	207	6.4	0.94	0.21	8.48	103.1	43.81	68.5	47.29	Short	ADT 37
132	48	218	6.7	0.54	0.19	5.59	63.4	56.4	46.71	31.04	Short	ADT 43
137	36	187	6.5	0.27	0.19	14.19	103.63	55.62	62.27	22.35	Short	ADT 45
173	62	194	6.3	0.42	0.16	5.9	106.01	47.26	38.41	14.54	Short	ADT 47
152	56	218	6.4	0.37	0.27	14.24	107.55	40.8	65.93	14.53	Short	ADT 48
143	71	173	6.9	0.19	0.25	7.79	70.8	68.69	51.48	20.04	Short	ASD 16
141	20	208	6.6	0.39	0.26	7.49	104.46	29.38	68.91	18.31	Short	ASD 17

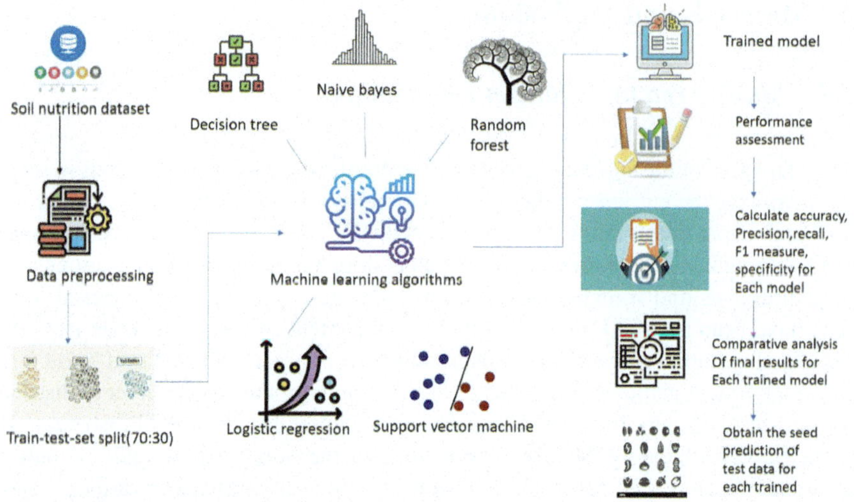

Fig. 1 Proposed framework

used to perform transformations on each of the input attributes that were within the range [16]. The overall approach that will be taken to complete the task that has been proposed is outlined in detail in Fig. 1.

Figure 1 depicts the proposed system architecture for predicting seeds using machine learning techniques and includes a copy of the proposed framework. The entire configuration can be broken down into three primary components: (i) data collection and preliminary processing, (ii) the model was trained using machine learning, and (iii) the results of the predictions. The technology is versatile, easily modifiable, and has the potential to be utilized for the production of seed that is suitable for improved crop yields on agricultural land. The next step is data gathering, which includes the collection of information on the varying levels of soil nutrients found in different districts throughout South Tamil Nadu. We are accumulating datasets from a variety of sources and saving those datasets in the data storage as we obtain them. After that, the data are put through a preprocessing approach, which consists of two processes: data cleaning and data reduction. These actions are performed on the data before the preprocessing technique is applied. In the process of data cleansing, erroneous and incomplete data are deleted from the database. After that, a simplified version of the digital information, either in numerical or in alphabetical form, is produced through the process of data reduction. The outputs of the phases of obtaining data and data preprocessing will be sent into the machine learning algorithm, which will then generate prediction results based on those outputs. These analyses allow this overall data to be split into train data consisting of 70% and test data consisting of 30% to be processed with five different ML models to train and produce the various levels of accuracy with better seed for given good yields, and the proposed technique presented can be used to predict absolute seed to be given growth rate in an efficient manner for increasing yield production with the aid of this suggested framework.

3.2 Machine Learning Techniques

Decision Tree

One of the algorithms that sees the most use in the context of supervised learning is called a decision tree. In most cases, the models for regression or classification that are produced by a decision tree take the form of a tree. It does this by breaking a dataset down into smaller subsets and then generating a decision tree based on those parts. The end result is a tree that has nodes that are either decision nodes or leaf nodes. A decision node will often consist of two or more branches, with each branch indicating a distinct value for the property that is being validated. The leaf node is an alternative to the overall numerical goal that can be pursued. The root node, also known as the most refined node, corresponds to the node that is the highest in the tree.

Naïve Bayes Classifier

The Bayes theorem serves as the basis for it, and it possesses autonomous properties. The estimation used for each class designation is determined by the instance's likelihood. A minimal quantity of training data is sufficient for accurately predicting the required class label for classification. When applied to data sets containing multiclass predictors, the Naive Bayes algorithm is especially useful.

Logistic Regression

A classification algorithm is a method for organizing data whose purpose is to designate observations to one of a limited number of categories. The logistic regression hypothesis tends to restrict the cost function to values between 0 and 1 as the admissible range of values. This is because these are the only values that are considered to be acceptable. It is one of the types of analysis that can be employed in order to make predictions about the future.

Support Vector Machine

Support vector machine is a supervised machine learning method or model that can be applied to classification and regression problems. In contrast, its primary application consists of categorization exams. The training data for an SVM are frequently represented as points in space, and these points are then divided into groups based on the maximum understandable distance between them. The SVM algorithm represents each item of data as a point in an n-dimensional space, and the value of each feature corresponds to the value of a given coordinate in that space. Finding the hyperplane that most effectively separates the two classes enables the classification process to be completed.

Random Forest

The random forest method and related random choice forest classifiers are agricultural research strategies for supervised learning that are applicable to a variety of tasks, such as division, association, and regression. Several trees are manipulated during the training and testing phases in order to determine which class most accurately represents the classification or predictive regression mode of decision trees. This process is repeated until all assignments have been produced. The random decision forest classifiers work better with individual trees than they do with their training and testing samples. Learning in this manner is referred to as supervised learning. After that, the regression is performed with the random forest classifier. The random forest method will continue to use the numbers that are in the wrong order. In the forest, the decision tree can be found a greater percentage of the time than the unfit classifier. The advantages of the random forest classifier, which were discussed previously, are also applicable to this performance indicator, for which our system makes use of the RFC algorithm.

3.3 Required Level of Soil Nutrients Statistical Analysis

Important nutrients for plant are required for profitable and sustainable agricultural production. A lack of any important nutrient will result in poor growth of the crop or pasture, which will in turn limit production and bring about a reduction in profit for the growers. To overcome this problem, this proposed work focuses on the majority of the nutrients that plants require for growth can be found in the soil. Nitrogen, phosphorous, and potassium are considered to be the three primary nutrients. The three of them together are known collectively as NPK. Other critical nutrients are PH, EC, S, CU, MN, and Zn, B, which are listed in Table 1. The quality of the soil is the most essential factor to consider when calculating the potential yield of a crop. As a result, the nutritional quality of the soil in Tamil Nadu needs to be evaluated. Plants are only able to extract and make use of a small part of the total quantity of nutrients present in the soil. This percentage differs depending on the nutrient and is regulated by a wide variety of soil, plant, and environmental factors. In this figure, statistical methods for assessing nutrient availability and the various nutrient ratios required for analysis of the seed to growth are explained. Also included are the nutrient levels that require the most and the least of each nutrient.

4 Result and Discussion

4.1 Experiment Analysis

In this section, the results that were acquired after applying the machine learning algorithms to the gathered dataset are shown in Table 2. These results were achieved by executing the machine learning algorithms. On a dataset that has been previously trained, the machine learning algorithms DT, LR, SVM, NB, and RF have been applied. This study has made use of the Colab notebook, which is written in Python. The database of soil nutrients serves as the foundation for the model that has been proposed. The above statistical report produces the most and least required soil nutrient ratio in Fig. 2, which indicates whether or not the seed will be suitable for further yield generation. This report is generated based on the soil nutrient level that was checked out. As a consequence of the experiments, we can observe that RF has achieved the highest possible level of precision. The accuracy of classification is summarized in the table below.

These matrices are also displayed in the bar chart located in Fig. 3. The performance metrics are displayed here in terms of accuracy, recall, specificity, and precision, as well as the *F*-score. On the test dataset, the accuracies of RF, NB, SVM, LR, and DT were, respectively, 0.99, 0.0.93, 0.89, 0.87, and 0.94; RF demonstrates superior performance in comparison with other machine learning approaches. In a manner parallel to the accuracy measure, all of the other metrics were won by RF, demonstrating that it is the superior algorithm. Additionally, DT did an excellent job. RF, on the other hand, demonstrated greater accuracy than DT. On the other hand, LR did not do particularly well when applied to our dataset. Since this is the case, it should come as no surprise that RF and DT are both effective classifiers for seed prediction based on the soil nutrient information.

Table 2 Comparison of accuracy rate in ML algorithms

ML algorithm	Accuracy (%)	Precision (%)	Recall (%)	Specificity (%)	*F*-score
Decision trees	94.52	89.72	75.33	80.17	0.8293
Logistic regression	87.91	70.61	76.34	85.43	0.7163
SVM	89.97	75.34	80.11	85.37	0.6724
Naïve Bayes	93.73	80.43	90.77	83.25	0.8213
Random forest	99.95	94.54	88.71	91.56	0.9357

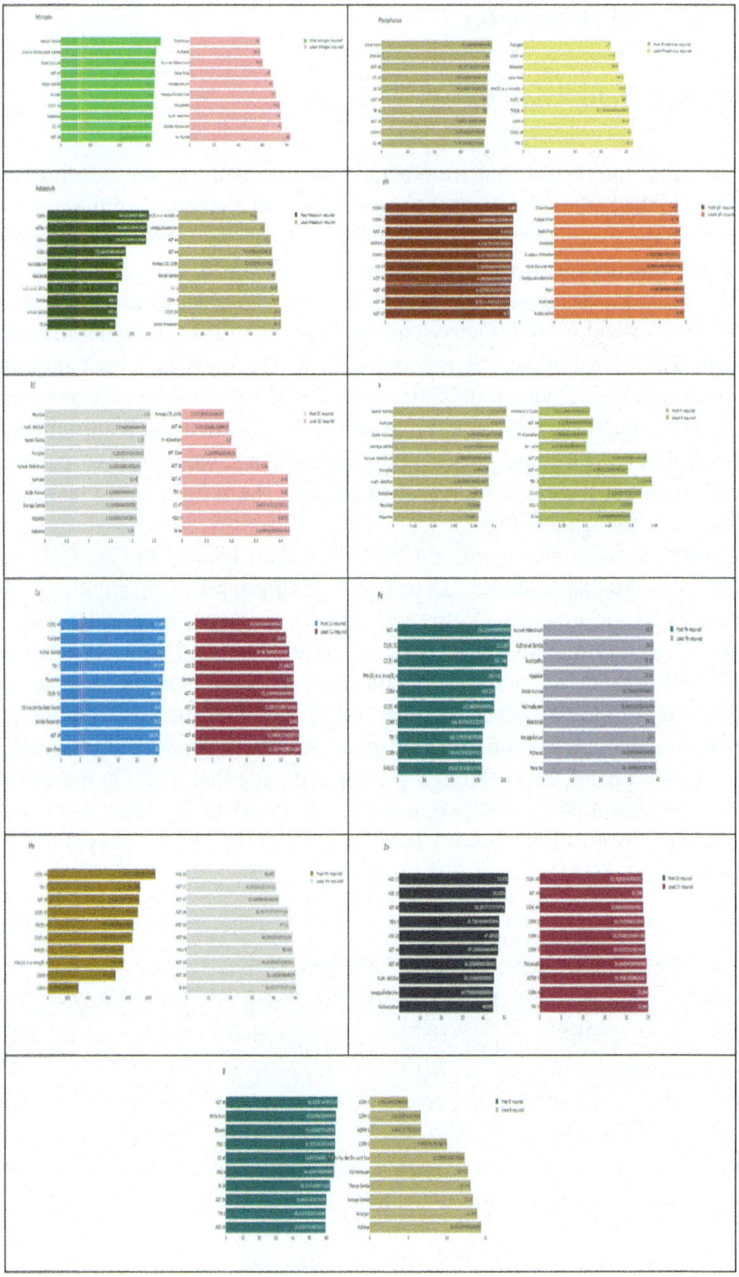

Fig. 2 Statistical analysis needed to determine the required amount of soil nutrients

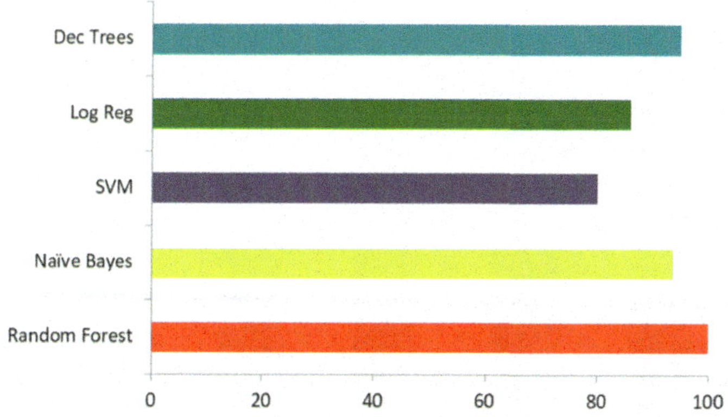

Fig. 3 Accuracy rate analysis

4.2 *Training and Testing Accuracy Comparison of ML Algorithms*

The properties of the training data have a significant impact on the classification accuracy determined by comparing the training and testing accuracy ranges for each ML algorithm Fig. 4. It is essential that the number of classes should be sufficient to explain the soil nutrient levels for this research area, and it is also essential that the data obtained from training and testing provide a description that is representative of each instance. In this part, the results attained by the random forest in classifying the test data given in Figure 4 are compared with those produced by other machine learning classifiers. With the use of this comparison, we hope to find out whether the high accuracy values that we attained using the RF are dependent on the classifier. Furthermore, the research comes to the conclusion that RF classifiers are highly suggested for high-level accuracy of soil nutrients data sets.

4.3 *Prediction Using Random Forest*

The random forest method is capable of processing some of the input data, particularly those shown in Fig. 5. However, these data must be analyzed in conjunction with the nutrient content of the soil. Because of this, it will be feasible to develop seeds in the subsequent procedure that have the appropriate properties for the increase of yield. The software was designed to learn from the dataset that was sent to it, with each and every nutrient's data from the soil being used. The random forest learning algorithm, which is a member of the family of supervised machine learning algorithms, is applied to the real-time data, and then the algorithm processes the data and generates an output of the appropriate seed. For this reason, the forecast result

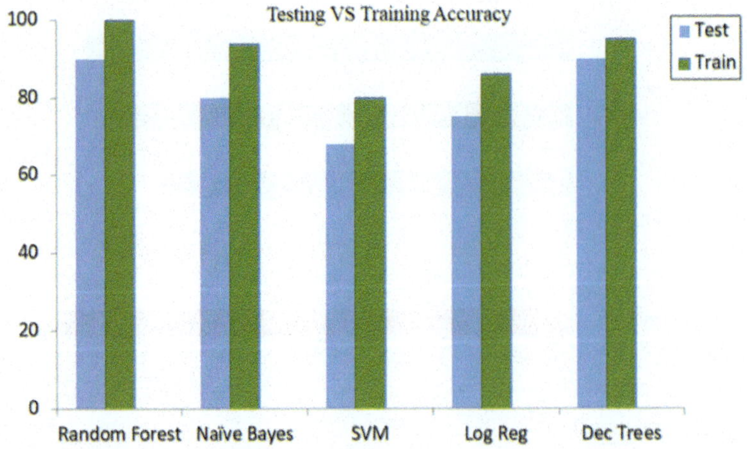

Fig. 4 Comparison ranges of training and testing split up in ML algorithms

```
Test input1: [143, 69, 217, 5.9, 0.58, 0.23, 10.2, 116.35, 59.96, 54.85, 21.29]
Test Output:  ['IR 64']

Test input1: [170, 36, 200, 6, 0.19, 0.3, 20, 116, 65, 60, 30]
Test Output:  ['CO 47']

Test input1: [130, 75, 207, 6.8, 0.9, 0.58, 0.323, 15.56, 100, 67, 55]
Test Output:  ['ADT 38']

Test input1: [150, 90, 220, 7, 0.58, 0.23, 10.2, 116.35, 59.96, 54.85, 21.29]
Test Output:  ['IR 64']

Test input1: [100, 49, 230, 5.9, 0.58, 0.23, 10.2, 116.35, 59.96, 54.85, 21.29]
Test Output:  ['IR 64']
```

Fig. 5 Result of predicted seed

provides a better choice for the farmer, who may choose for himself to select the ideal seed with the yield only when it is necessary. This helps to avoid the waste of irrelevant seeds that could have been used for growing.

4.4 Accuracy Analysis for Prediction of Seed

Due to the fact that the predicted accuracy is so high throughout both the training and testing processes, only selected handful of the samples that were used to construct the predictive models were able to be used for seed accuracy prediction perfectly.

This is seen in Table 3, which can be found below. Therefore, selecting the appropriate values for the model parameters is of the utmost significance if one wishes to avoid the learning process from improving model performance. The number of training samples was also raised, which helped improve the accuracy of the predictions. Therefore, the quantity of the dataset is a crucial component to consider while developing an improved prediction model. The following tables, which are organized according to the target variable, provide a summary of the prediction accuracies that were created during the process of model training and evaluation.

5 Conclusion and Future Scope

After classifying the soil based on its soil type, land type, and macro-nutrients (N, P, K, PH, EC, S, CU, MN, ZN, B) as well as its duration period, agricultural stakeholders are given advice on the best time to grow crops and the best seeds to use to improve the soil and increase crop yield. This is done in order to provide the agricultural stakeholder with the most up-to-date information regarding the soil. Additionally, the farmer can receive a projection of the crop's production in response to the crop he or she has ordered. When giving suggestions to the farmer, they take into account other crops that can be grown during that cropping season and are also good for the ratio of nutrients in the soil. The module for alternating crop rotation provides an additional three different plant species that are capable of being grown in that soil environment. The crop suggestion method being considered offers an accuracy rate of 99.9%. Following the results of the experimental study, one may get the conclusion that ML approaches are capable of being utilized effectively for the purpose of yield prediction of seed. In contrast, the results of this research indicate that random forest is the method that provides the best accurate predictions regarding seeds. These powerful machine learning algorithms will be of assistance to farmers in the process of yield prediction in advance based on the properties of the soil. In the not-too-distant future, it will be possible to employ big data techniques for crop production prediction using massive soil data sets. If there is a high crop yield projection, seed suggestions can be adopted based on the results of the yield prediction to assist soil analyzers and farmers in making decisions in a manner that is appropriate to the situation. This study examines the possibility of developing a seed selection for predicting a superior crop for cultivation based on big data and machine learning. Random forest and additional supervised learning techniques are more appropriate for seed selection. This study provides future work with a reference for crop prediction based on the random forest which is superior to the ML algorithm.

Table 3 Seed with accuracy range

S. No.	Type	Accuracy (%)	S. No.	Type	Accuracy (%)
1	ADT36	85	31	Kadaikazhuthan	89
2	ADT 37	90	32	Kalar palai	74
3	ADT 39	89	33	Kaliyan samba	90
4	ADT 43	74	34	Karuppukavuni	76
5	ADT 44	88	35	Kichedi samba	89
6	ADT 45	92	36	Kumvalai	78
7	ADT 46	79	37	Kuzhi vedichan	92
8	ADT 47	90	38	MDU 4	79
9	ADT 48	93	39	Mara Nel	85
10	ADT 49	78	40	Murugungaikar	90
11	ADTRH 1	94	41	PMK (R) 4 or Anna(R) 4	76
12	ASD 16	81	42	PY 4 (Jawahar)	89
13	ASD 17	72	43	Perunkar	74
14	ASD 19	80	44	Pichavari	84
15	ASD 20	95	45	Ponmani	89
16	BPT 5204	83	46	Poongkar	74
17	CO 43	79	47	RMD(R) 1	80
18	CO 46	85	48	Sadaikkar	76
19	CO 47	90	49	Seeraga Semba	74
20	CO(R) 48	73	50	TPS 2	92
21	CO(R) 49	95	51	TPS 3	79
22	CO(R) 50	89	52	TPS(R)4	92
23	CORH 1	74	53	TRY 1	85
24	CORH 2	87	54	TRY 3	90
25	CORH 3	70	55	Thuyamalli	76
26	CORH 4	76	56	Vadan samba	89
27	Chitraka	92	57	Vai kunda	74
28	IR 20	79	58	Val sivappu	95
29	IR 36	85	59	Vellai kuruvai kar	85
30	IR 64	90	60	White Ponni	90

References

1. Singhatiya S, Ghosh DS (2018) A review on soil property detection using machine learning approach. Smart Moves J IJOSci 4(8):6–9. Smriti (2015) A review on soil property detection using machine learning approach. Int J Online Sci 4
2. Sujatha R, Isakki P (2016) A study on crop yield forecasting using classification techniques. In: 2016 international conference on computing technologies and intelligent data engineering (ICCTIDE'16). IEEE, pp 1–4

3. Singh V, Sarwar A, Sharma V (2017) Analysis of soil and prediction of crop yield (rice) using machine learning approach. Int J Adv Res Comput Sci 8(5)
4. Supriya DM (2017) Analysis of soil behavior and prediction of crop yield using data mining approach. Int J Innov Res Comput Commun Eng
5. Pandith V, Kour H, Singh S, Manhas J, Sharma V (2020) Performance evaluation of machine learning techniques for mustard crop yield prediction from soil analysis. J Sci Res 64(2):394–398
6. Priya P, Muthaiah U, Balamurugan M (2018) Predicting yield of the crop using machine learning algorithm. Int J Eng Sci Res Technol 7(1):1–7
7. Sellam V, Poovammal E (2016) Prediction of crop yield using regression analysis. Indian J Sci Technol 9(38):1–5
8. Ghadge R, Kulkarni J, More P, Nene S, Priya RL (2018) Prediction of crop yield using machine learning. Int Res J Eng Technol (IRJET) 5
9. Shirsath R, Khadke N, More D, Patil P, Patil H (2017) Agriculture decision support system using data mining. In: 2017 international conference on intelligent computing and control (I2C2). IEEE, pp 1–5
10. Renuka ST (2019) Evaluation of machine learning algorithms for crop yield prediction. Int J Eng Adv Technol (IJEAT) 8(6)
11. Manjula E, Djodiltachoumy S (2017) Data mining technique to analyze soil nutrients based on hybrid classification. Int J Adv Res Comput Sci 8(8)
12. Jayalakshmi R, Devi MS (2019) Relevance of machine learning algorithms on soil fertility prediction using R. Int J Comput Intell Inform 8(4):193–199
13. Kumar R, Baloch G, Buriro AB, Bhatti J (2021) Fungal blast disease detection in rice seed using machine learning. Int J Adv Comput Sci Appl 12(2)
14. Sarijaloo FB, Porta M, Taslimi B, Pardalos PM (2021) Yield performance estimation of corn hybrids using machine learning algorithms. Artif Intell Agricult 5:82–89
15. Deepa S, Gokila S (2022) Comparative analysis of different machine learning prediction models for seasonal rainfall and crop production in cultivation. In: Computational methods and data engineering: proceedings of ICCMDE 2021. Springer Nature Singapore, Singapore, pp 277–289
16. Brownlee J (2016) *K*-nearest neighbors for machine learning. Machine Learning Mastery, p 15

Hybrid Deep Feature Optimization Methodology for Plant Leaf Classification

Ariyaman Goswami, Mridu Sahu, Maroi Agrebi, Saurav Mallik, and Pawan Kumar Singh

Abstract Research on plant leaf classification is still ongoing which holds high value in different sectors. Applications of plant leaf classification range from finding new medicines for diseases and tracking the impact of climate change and global warming, and for that we need to have advances on the accuracy of existing algorithms. Therefore, we develop a hybrid feature selection algorithm approach that uses the concept of deep feature optimization which accomplishes well both in terms of computational complexity and classification accuracy. Our model pipeline produces self-learned deep features using pre-trained transfer learning models such as MobileNetV3-Large and EfficientNet from the images of plant leaves. The features sets are merged and the feature dimensionality is reduced using a hybridization of ReliefF and Genetic algorithm. Lastly, a support vector machine (SVM) classifier has been applied to identify the selected feature subset into its corresponding accent classes. The proposed framework has been evaluated on two different widely available plant leaf datasets

A. Goswami · P. K. Singh (✉)
Department of Information Technology, Jadavpur University, Jadavpur University Second Campus, Plot No. 8, Salt Lake Bypass, LB Block, Sector III, Salt Lake City, Kolkata, West Bengal 700106, India
e-mail: pawansingh.ju@gmail.com; pawankrsingh@ieee.org

A. Goswami
e-mail: ariyamangoswami@gmail.com

M. Sahu
Department of Information Technology, National Institute of Technology, Raipur, India
e-mail: mrisahu.it@nitrr.ac.in

M. Agrebi
Department of Computer Science, LAMIH UMR CNRS 8201, Université Polytechnique Hauts-de-France, 59313 Valenciennes, France
e-mail: maroi.agrebi@gmail.com

S. Mallik
Department of Pharmacology and Toxicology, R. Ken Coit College of Pharmacy, The University of Arizona, Tucson, AZ, USA
e-mail: sauravmtech2@gmail.com

S. Bandyopadhyay et al. (eds.), *Intelligent Computing Systems and Applications*,
Lecture Notes in Networks and Systems 1010,
https://doi.org/10.1007/978-981-97-5412-0_27

namely, PlantVillage and Flavia. The proposed pipeline achieves 97.94% and 98.52% classification accuracies, respectively, which is comparable to all the latest models.

Keywords Plant leaf classification · MobileNetV3-Large · EfficientNet · PlantVillage dataset · Flavia dataset

1 Introduction

A total of 390,000 plant species are currently known to science [1]. Because of this, it is challenging to recognize and categorize all of these species. A lot can be discovered about a plant merely from its leaves, as it has been observed that leaves of particular plant species hold a significant quantity of information. Many plants are currently in danger of going extinct because of global warming and other factors. Both non-endangered and endangered plant species need to be effectively protected and safeguarded. Hence, in order to accurately identify and classify plants, an automated or computerized system must be developed.

Based on the end-to-end benefit of deep learning algorithms, it is possible to automatically understand the distinctive characteristics that each plant possesses as shown by its leaves. As a result, a key method for plant leaf recognition is one that is based on deep learning. Convolutional neural networks (CNNs) are particularly well adapted for solving computer vision issues like plant classification. Hard-core feature extraction and subject expertise are no longer required because of this. Consecutive convolution operations between the input image and convolutional filters can be used to create a feature map, and discriminative patterns can be retrieved from certain plant leaves [2].

Feature selection [3] is a crucial aspect of plant leaf classification as it enables the identification of the most relevant and discriminative features that distinguish one plant species from another. By choosing the most useful characteristics from a huge pool of features retrieved from plant leaves, this procedure reduces the number of dimensions in the data and enhances the precision and effectiveness of the classification model. The performance of machine learning algorithms can be greatly improved by using efficient feature selection techniques, especially when there are many more characteristics than samples. The classification model can minimize overfitting, improve interpretability, and lower the computational cost of both training and testing by concentrating on the most informative features.

Another important reason for feature selection in plant leaf classification task is to reduce the impact of noise and irrelevant features on the accuracy of the model. In real-world scenarios, plant leaves can exhibit various types of noise, such as variations in lighting conditions, camera angles, and image quality, which can significantly affect the performance of the classification model. Feature selection helps to eliminate the noise and irrelevant features, which can improve the strength as well as generality of the model. Additionally, by selecting only the most informative features, the classification model can focus on the essential characteristics of the plant leaves, which can improve the interpretability and reliability of the classification results. In a nutshell, the development of precise and successful plant leaf categorization systems can greatly benefit from the use of effective feature selection strategies.

1.1 Motivation and Contributions

The use of deep learning along with techniques for feature selection together with CNNs has especially demonstrated excellent results in the domain of image classification due to their superior capacity of identifying and learning lower-level characteristics and variances in images, as is evident from the works. This information is then used for classifying the images. Hence in our work, in order to choose the final optimal subset of features, we have attempted to construct a hybrid feature optimization strategy using a filter-based feature selection algorithm namely, ReliefF along with the Genetic algorithm (GA), a wrapper-based algorithm. When there is a lack of available data, researchers frequently employ transfer learning, wherein the values for the weights of the deep neural network are obtained from the prior training cycle of an advanced model, but in a different domain.

In the present work, the contributions made are highlighted below as follows:

- We utilize the strategy of two pre-trained transfer learning-based CNN models such as MobileNetV3-Large [4] and EfficientNet [5] to extract the feature vectors.
- We use the feature selection methodology to choose between the best optimizing features in order to generate more accurate classification accuracy.
- We achieve impressive performance on two benchmark datasets such as the PlantVillage dataset [6] consisting of 54,000 samples and Flavia dataset [7] consisting of 4320 samples.

2 Literature Survey

Many researchers have attempted to use various models on plant leaves throughout the years, and some of them have had great success raising the classification accuracy of their developed model.

On the PlantVillage dataset, Siddiqui et al.'s [8] use of deep learning algorithms for disease recognition of plants achieves 96.81% recognition accuracy. Cruz et al. [9] introduces a multi-task deep CNN called DeepSoy for the classification of soybean plants. Li et al. [10] presents a model for the plant leaf disease recognition by means of a semi-supervised few-shot learning-based approach. Their model shows the best accuracy of 96.8% on the subset of the PlantVillage dataset. Turkoglu et al. [11] introduces a novel plant identification system based on a multi-division CNN which produces an accuracy of 97.3% on the Flavia dataset. Figueroa-Mata et al. [12] presents an unique methodology for plant species identification using a Convolutional Siamese Network (CSN) which gives the best accuracy of 93.7% on Flavia dataset.

Saleem et al. [13] presents a model for plant leaf classification by applying the graphical analysis of leaf shape features. The aim of this work is to develop an automated system that can correctly classify plant species on the basis of their leaf shapes. This model gives the best accuracy of 96.8% on the Flavia dataset.

Some researchers have also implemented the concept of feature selection for plant leaf classification. Mane et al. [14] presents a novel approach to feature selection in plant leaf classification. The present work employs a metaheuristic algorithm to address the limitations of traditional feature selection techniques. Swarm behavior, simulated annealing, and evolution are examples of natural processes that serve as motivation for metaheuristic algorithms. These algorithms provide a flexible and robust framework for exploring large search spaces efficiently. Results on the PlantVillage dataset show that the suggested strategy is effective, with the highest classification accuracy of 96.15%. A hybrid metaheuristic feature selection strategy is suggested in the work by Pham et al. [15] to determine the most pertinent and instructive aspects for disease categorization. A model for the categorization of tomato plant diseases is presented by Gadekallu et al. [16] and integrates many methodologies, including deep neural network (DNN) followed by principal component analysis (PCA) and whale optimization algorithm (WOA). The classification accuracy achieved by their model is found to be high, indicating its ability to accurately classify different tomato plant diseases with the best accuracy of 97.8%.

Feature selection plays a significant part in enhancing the overall classification accuracy and reducing the computational complexity of the model. This motivates us to propose a hybrid feature selection model for solving on plant leaf classification problem.

3 Proposed Hybrid Feature Optimization Methodology

Our proposed model involves a hybrid feature optimization methodology for plant leaf classification. Figure 1 illustrates the overall block diagram of our proposed hybrid feature optimization approach.

3.1 Data Acquisition

Two standard benchmark plant leaf classification datasets are used in this work. These two datasets are described below.

3.1.1 PlantVillage Dataset

This dataset [6] is a comprehensive and widely used dataset in the arena of agricultural research and computer vision. Over 54,000 images of both healthy as well

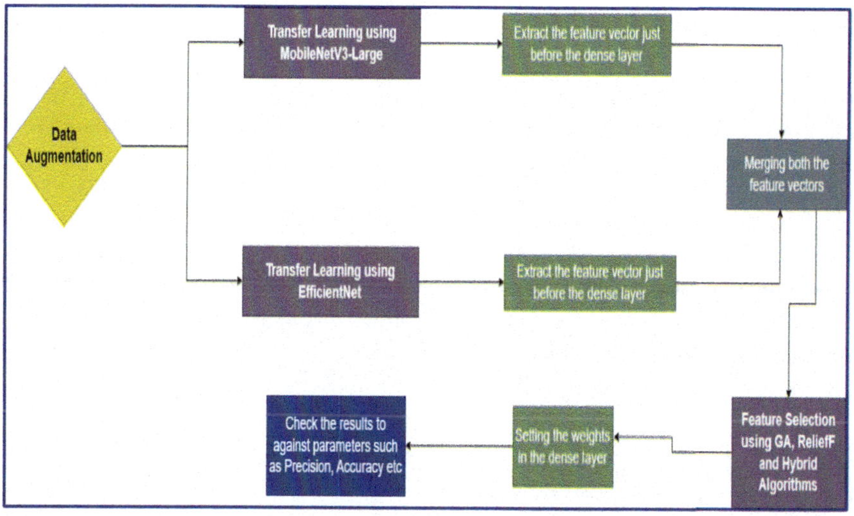

Fig. 1 Pictorial illustration of our proposed hybrid feature optimization-based methodology for solving plant leaf classification problem

as diseased plant leaves, stems, and roots from 14 different crop species, such as tomatoes, potatoes, grapes, and apples, make up this dataset. It was created to address the challenges of plant disease identification and crop health monitoring. Researchers have utilized the PlantVillage dataset for tasks such as binary classification (disease vs. healthy), multi class classification (disease type identification), and severity estimation (quantifying the extent of disease damage).

3.1.2 Flavia Dataset

This dataset [7] is a commonly used benchmark dataset in the field of plant leaf classification. It was created to facilitate research on automated plant species identification and classification. The dataset consists of images of leaves from four different plant species: Fagus sylvatica (beech), Quercus suber (cork oak), Populus nigra (black poplar), and Alnus sp. (alder). Researchers have developed and evaluated multiple machine learning and computer vision methods for plant species detection using this dataset. The dataset comprises 1907 image samples of leaves, with each species having approximately 476 samples. The images in the Flavia dataset were captured under controlled conditions to ensure consistent lighting and background. Each leaf was placed on a white background and photographed using a digital camera. The images were then manually segmented to isolate the leaf region from the background, resulting in grayscale images of the leaves.

3.2 Data Augmentation

It is always desirable to utilize augmentation to enrich the datasets even if both the Flavia and PlantVillage datasets used in this study both contain thousands of annotated photos because it reduces the chance of overfitting as well as increases the ability of the proposed model to generalize. Also, data augmentation enhances the model's robustness to variations in real-world scenarios. By applying transformations such as rotations, translations, and scaling, the model becomes more adept at recognizing objects from different perspectives and orientations. In this work, we have implemented geometric transformations on the datasets including flipping, rotating and cropping the images.

3.3 Transfer Learning

We used the pre-trained models of MobileNetV3-Large [4] as well as EfficientNet [5] and fine-tune these models on our dataset in order to get the feature vectors.

3.3.1 MobileNetV3-Large

This architecture is a CNN model specifically designed for efficient and lightweight mobile and embedded vision applications. It is an extension of the MobileNet family of models that aims to provide high accuracy with low computational complexity and memory footprint. MobileNetV3-Large builds upon the successes of its predecessors, MobileNetV1 and MobileNetV2, and incorporates several architectural changes to further improve performance. Figure 2 shows the diagrammatic representation of the MobileNetv3-Large model applied in the present model.

3.3.2 EfficientNet

The EfficientNet architecture is a contemporary CNN model designed to achieve high accuracy with improved efficiency. Researchers developed it due to the growing need for models that can successfully balance computational expense and model performance. The architecture introduces a novel approach by considering scaling across multiple dimensions simultaneously. This compound scaling ensures that the model achieves optimal performance with minimal computational cost.

EfficientNet introduces a family of models, denoted as EfficientNet-B0 to EfficientNet-B7, with each variant having a different scaling coefficient. The successive models (B1 through B7) are obtained by raising the scaling coefficient from

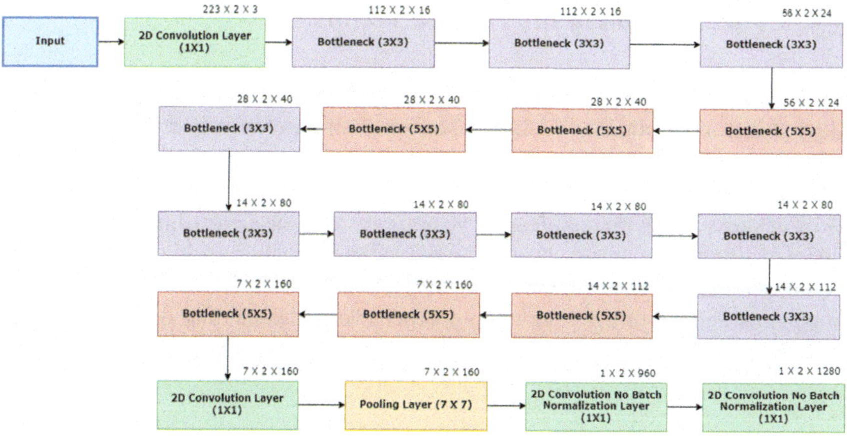

Fig. 2 Pictorial representation of MobileNetV3-Large model used in the present work

EfficientNet-B0, which serves as the baseline model. As the scaling coefficient increases, the model becomes deeper, wider, and has a higher input resolution. Figure 3 displays the diagrammatic representation of the EfficientNet network applied in the present model.

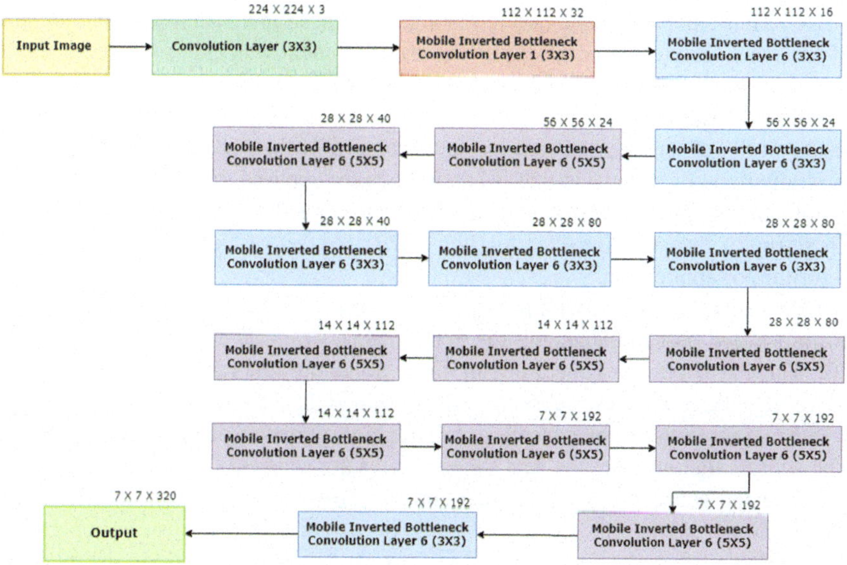

Fig. 3 Diagrammatic representation of the EfficientNet model used in the present work

3.4 Concatenation of the Feature Vectors

We merge the feature vectors that we got from the two separate models, namely EfficientNet and MobileNetV3-Large. Merging feature vectors from different models offers several advantages. Firstly, it can enhance model performance by leveraging the strengths of multiple models. By combining models with complementary strengths and weaknesses, the merged feature vector can compensate for limitations in each model, resulting in a more robust and reliable representation. Additionally, merging feature vectors can provide a broader perspective on the data, incorporating diverse viewpoints and capturing a more comprehensive understanding of the underlying patterns.

3.5 Hybrid Feature Optimization Methodology

In this work, two distinct feature selection algorithms—the ReliefF method and the GA—have been implemented. These are the algorithms we are implementing as we are trying different approaches of feature selection namely filter-based and wrapper-based. This helps us in getting the most relevant features, in our case, the top 500 features to fit into the dense layer for inferencing. We have taken the help of Py_FS package [17] in order to implement the feature selection algorithms.

ReliefF is a popular feature selection algorithm used in machine learning to identify relevant features in a dataset [18]. One important advantage of ReliefF is that it is computationally efficient and can handle datasets having large feature dimensionality. It is also robust to noisy and irrelevant features, as it updates the feature weights based on the difference between instances instead of their absolute values. One of the limitations of ReliefF is that it is a filter-based method, meaning that it selects features independently of the machine learning algorithm used to train the model. Therefore, it may not always select the superlative set of features for a particular algorithm or application.

GA, which is a metaheuristic algorithm, is used for feature selection and plays an imperative part in enhancing the performance of classification algorithms [19, 20]. The GA begins by generating an initial population of subsets of features from the original dataset. These subsets are then evaluated based on their fitness function, which measures the performance of a classification algorithm on the selected features. The subsets that produce the best fitness scores are selected for further processing. The selected subsets of features then undergo crossover and mutation operations, which mimic the natural selection process in genetics. In the crossover operation, two subsets of features are combined to create a new subset, while in the mutation operation, a randomly selected feature is replaced with another feature from the original dataset. Until a stopping requirement is satisfied, like as a pre-determined number of generations or a specified level of fitness, the GA algorithm iterates through the processes of selection, crossover, followed by mutation. The optimized feature subset selected by the GA is then used as input for the classification algorithm.

There are several limitations of just using GA on a very large dataset; hence, the time taken to do feature selection using GA is really high. So we make use of a hybrid feature selection approach where we first select features based on the ReliefF algorithm and then use GA to get the final feature vectors to be put in the dense layer. We took out the top 700 relevant features using the ReliefF algorithm and then used GA algorithm to finally get the top 300 most relevant features. After getting the most relevant features, it is passed through the support vector machine (SVM) classifier [21] to get the labels.

4 Results Analysis

In the present work, we have calculated three evaluation metrices such as $F1$-score, precision, and accuracy for measuring the effectiveness of the proposed hybrid feature optimization methodology for the task of plant leaf classification. Table 1 displays the values of the metrices for both PlantVillage and Flavia datasets before using the hybrid feature selection methodology, whereas Table 2 shows the values of the metrices for both datasets after using the hybrid feature selection methodology.

Table 3 shows the performance comparison of our proposed hybrid plant leaf classification method with some recently proposed contemporary pipelines. As shown in Table 3, the developed model may classify plant leaves similarly to a number of existing approaches already proposed, but no open comparisons can be made because there is no common standard dataset.

Overall, it can be concluded from the above performance comparison that the results are reasonable which makes our proposed method theoretically comprehensive and robust.

Table 1 Results attained (with respect to $F1$-score, precision, and accuracy) before using the hybrid feature selection methodology on both PlantVillage and Flavia datasets

Metrics (%)	PlantVillage	Flavia
$F1$-score	95.40	95.22
Precision	94.67	93.72
Accuracy	95.22	96.84

Table 2 Results obtained (with respect to $F1$-score, precision, and accuracy) after using the proposed hybrid feature selection methodology on both PlantVillage and Flavia datasets

Metrics (%)	PlantVillage	Flavia
$F1$-score	97.42	98.33
Precision	97.35	98.41
Accuracy	97.94	98.52

Table 3 Comparison of results obtained by our proposed hybrid plant leaf classification method with some recent existing methods

Authors	Year of publication	Method	Datasets used	Accuracy (%)
Mane et al. [14]	2020	Ensemble of CNN + GA, particle swarm optimization, ant colony optimization	PlantVillage	96.15
Siddiqui et al. [8]	2021	CNN along with data augmentation		96.2
Li et al. [10]	2021	Semi-supervised few-shot learning approach on CNN		96.8
Goswami et al.	**2023**	**Proposed hybrid feature selection methodology**		**97.94**
Saleem et al. [13]	2019	Transfer learning and KNN	Flavia	96.8
Figueroa-Mata et al. [12]	2020	CSN		93.7
Pham et al. [15]	2020	Feed-forward neural network + Hybrid metaheuristic feature selection		95.54
Turkoglu et al. [11]	2021	A multi-division CNN		97.3
Gadekallu et al. [16]	2021	PCA + WOA-based DNN model		97.8
Goswami et al.	**2023**	**Proposed hybrid feature selection methodology**		**98.52**

Bold signifies the highest classification accuracy

5 Conclusion and Future Works

In conclusion, our present work has revolved around the development and application of a hybrid feature optimization methodology for plant leaf classification. We utilized two benchmark datasets like the Flavia dataset and the PlantVillage dataset, along with various data augmentation techniques, to conduct extensive experimentation and analysis to investigate the effectiveness and efficiency of our proposed algorithm. One of the key strengths of our research project lies in the utilization of different state-of-the-art transfer learning models such as MobileNetV3-Large and EfficientNet along with different feature extraction and selection algorithms to extract important features, reduce complexity, and improve accuracy. Additionally, our work also demonstrated that the feature optimization process significantly improved the accuracy of our algorithm while requiring a reduced number of epochs for training. By leveraging the extracted and selected features, our algorithm achieved better discrimination power and generalization capabilities.

Finding the best hyperparameters for our proposed model is necessary because the outcomes of our methodology are experimental in nature. A potential drawback of our strategy could be that this empirical approach might require some human involvement (i.e., some experimentation) in order to refine the parameters to perfection. There are several different metaheuristic optimization algorithms such as harmony search algorithm [22], particle swarm optimization [23], ant colony optimization [24] which may result in better accuracy with lesser epochs.

References

1. Christenhusz MJ, Byng JW (2016) The number of known plants species in the world and its annual increase. Phytotaxa 261(3):201–217
2. LeCun Y, Bengio Y, Hinton G (1998) Convolutional networks for images, speech, and time series. In: The handbook of brain theory and neural networks, vol 3361(10), pp 1995–2012
3. Guyon I, Elisseeff A (2003) An introduction to variable and feature selection. J Mach Learn Res 3:1157–1182
4. Howard A, Sandler M, Chu G, Chen L, Chen B, Tan M, Wang W, Zhu Y, Pang R, Vasudevan V, Le QV, Adam H (2019) Searching for mobileNetV3. In: 2019 IEEE/CVF international conference on computer vision (ICCV). Seoul, Korea (South), pp 1314–1324. https://doi.org/10.1109/ICCV.2019.00140
5. Tan M, Le Q (2019) EfficientNet: rethinking model scaling for convolutional neural networks. Retrieved from arXiv preprint arXiv:1905.11946v5
6. Hughes DP, Salathe M, Schulte Hostedde A (2015) Data from: Honey bee nest thermoregulation: diversity promotes stability. Dryad Digit Repository. https://doi.org/10.5061/dryad.5pt92
7. Nilsback ME, Zisserman A (2008) Machine learning techniques for plant leaf recognition. In: The international conference on pattern recognition (ICPR)
8. Siddiqui FH, Shaukat J (2021) Deep learning-based recognition of plant diseases. Comput Electron Agric 183:105015
9. Cruz JA, Neves DM, Leite DC, Torres GA, do Nascimento VM, Torres RD (2018) DeepSoy: a multi-task deep convolutional neural network for soybean plant classification. Plant Methods 14(1):56
10. Li Y, Chao X (2021) Semi-supervised few-shot learning approach for plant diseases recognition. Plant Methods 17:68. https://doi.org/10.1186/s13007-021-00770-1
11. Turkoglu M, Aslan M, Arı A, Alçin ZM, Hanbay D (2021) A multi-division convolutional neural network-based plant identification system. PeerJ Comput Sci 7:e572. https://doi.org/10.7717/peerj-cs.572
12. Figueroa-Mata G, Mata-Montero E (2020) Using a convolutional Siamese network for image-based plant species identification with small datasets. Biomimetics 5(1):8. https://doi.org/10.3390/biomimetics5010008
13. Saleem G, Akhtar M, Ahmed N, Qureshi WS (2019) Automated analysis of visual leaf shape features for plant classification. Comput Electron Agric 157:270–280. https://doi.org/10.1016/j.compag.2018.12.038
14. Mane SB (2020) Meta-heuristic feature selection for plant leaf classification. Plant Methods 12(1):58
15. Pham TN, Tran LV, Dao SVT (2020) Early disease classification of mango leaves using feed-forward neural network and hybrid metaheuristic feature selection. IEEE Access 8:189960–189973. https://doi.org/10.1109/ACCESS.2020.3031914
16. Gadekallu TR, Rajput DS, Reddy MPK et al (2021) A novel PCA–whale optimization-based deep neural network model for classification of tomato plant diseases using GPU. J Real-Time Image Proc 18:1383–1396. https://doi.org/10.1007/s11554-020-00987-8

17. Guha R, Chatterjee B, Khalid Hassan SK, Ahmed S, Bhattacharyya T, Sarkar R (2022) Py_FS: a Python package for feature selection using meta-heuristic optimization algorithms. In: Computational intelligence in pattern recognition. Springer, Singapore, pp 495–504
18. Kira K, Rendell L (1992) A practical approach to feature selection. In: Proceedings of the ninth international workshop on machine learning, pp 249–256
19. Holland JH (1975) Adaptation in natural and artificial systems. University of Michigan Press
20. Scholz J (2018) Genetic algorithms and the traveling salesman problem: a historical review. https://doi.org/10.13140/RG.2.2.22632.78088/1. Retrieved from https://arxiv.org/abs/1901.05737
21. Cortes C, Vapnik V (1995) Support-vector networks. Mach Learn 20(3):273–297
22. Geem ZW, Kim JH, Loganathan GV (2001) Harmony search algorithm for solving Sudoku. Ind Eng Chem Res 44(20):7541–7551
23. Kennedy J, Eberhart RC (1995) Particle swarm optimization. In: Proceedings of the IEEE international conference on neural networks, vol 4, pp 1942–1948
24. Dorigo M, Gambardella LM (1997) Ant system: optimization by a colony of cooperating agents. IEEE Trans Syst Man Cybernet Part B Cybernet 26(1):29–41

Performance Investigation of Principal Component Analysis for Diabetic Detection Using SVM with Kernel Functions

P. Jagdish Kumar, P. Mamatha, P. Meena Kumari, and Gadicharla Sahiti Sriya

Abstract Diabetes mellitus has been increasing in a day-to-day scenario where this deadly disease increases the blood sugar level. In earlier days, only very few people were affected by this disease. In the present years, our immobilized activity, unhealthy eating, and mental stress lead to millions of adults who are mostly affected by this. In some instances, even the existence of this disease is unseen or unknown until they undergo a medical examination. Before all these things could have happened, the disease had sharpened itself by thriving in society; even some people ended up dying at very early ages. If we fail to diagnose, it leads to severe complications. This paper aims to create a model that can detect diabetes early to avoid the risk of diabetes. The proposed is an investigation model that selects the features of the diabetes dataset using a principal component analysis (PCA) and support vector machine (SVM) to classify the selected features. In this, SVM has been analyzed with different support vector kernels such as linear, polynomial, RBF, and sigmoid. Experimentation has been carried out in PIMA Indian diabetes dataset and the diabetes dataset to evaluate the performances of various principal components with SVM kernels. In the experimentation, it has been analyzed that the principal components 1 and 2 with SVM RBF will be the best suitable for predictive modeling of diabetes detection.

Keywords Diabetes mellitus · Principal components · Support vector machines · PIMA · Early diabetes

P. J. Kumar (✉) · P. M. Kumari
Department of CSE, IcfaiTech, IFHE, Hyderabad 501203, India
e-mail: jagdish.info123@gmail.com

P. Mamatha
Department of Computer Science and Engineering, AVN Institute of Engineering and Technology, Hyderabad, Telangana, India

G. S. Sriya
Bridgetown International University, Bridgetown, India

S. Bandyopadhyay et al. (eds.), *Intelligent Computing Systems and Applications*, Lecture Notes in Networks and Systems 1010,
https://doi.org/10.1007/978-981-97-5412-0_28

1 Introduction

Diabetes is a long-term medical illness that impacts how the body handles blood sugar (glucose) [1]. For humans, glucose is a vital energy source obtained from food. However, pancreas generates a hormone named insulin that controls the blood sugar levels, has certain difficulty in controlling diabetes. Type 1, 2 and gestational diabetes are three main diabetic types. If a human insulin has been unintentionally targeted and if cant produce insulin it terms to be Type 1 diabetes. Type 2 is the most common where the human body cant produce enough insulin to the body to maintain the blood sugar level. Obesity, poor food, and physical inactivity are are lifestyle factors associated with type 2 diabetes. Some women are affected by gestational diabetes, which arises during pregnancy. It often emerges between the 24th and 28th week of pregnancy and goes away after delivery. The risk of problems during pregnancy and delivery increases with gestational diabetes, in the future type 2 diabetes is more likely to affect both the mother and the kid [2].

Machine Learning (ML) techniques have been used in detecting and managing diabetes by analyzing large datasets to identify patterns, correlations, and predictive models related to the disease [3, 4]. It has contributed to diabetic detection in predictive modeling, risk stratification, early detection, decision support systems, and personalized medicine [5]. Early detection of diabetes has been carried out by identifying subtle patterns or anomalies in health data that may indicate the onset of the disease. By analyzing symptoms, physiological measurements, and lifestyle factors, data mining algorithms, especially the ML algorithm, can help healthcare providers detect diabetes at an earlier stage when interventions may be more effective [6]. In recent years, there have been works based on ML used for early detection and diagnosis of diabetes [5–7]. Every research work has its own merits and demerits in its performance. This paper proposes an investigation model for diabetic detection based on Principal Component Analysis (PCA) and Support Vector Machine (SVM). In this process, PCA has been used for feature selection, and the different support vector kernels, such as Gaussian radial basis function, linear, sigmoid and polynomial are investigated to improve the performance of diabetic detection.

The paper has been organized as follows: Sect. 2, which examines similar efforts on diabetic detection, and Sect. 3, which looks into PCA with SVM kernels for classification. The discussion and outcomes of the experiment are covered in Sect. 4. The study ended with future scope and improvements in Sect. 5.

2 Related Works

The term "artificial intelligence" (AI) is a super set that has "data mining", "machine learning" and "deep learning" as subset in the term of predictive modeling, which uses statistics and math to forecast future outcomes based on past data. Practically speaking, machine learning models are data-driven models that are trained on input data to produce the most appropriate output depending on the trained model. To create the best model for practicality and medical needs for diabetic detection, the models were trained on realistic features. This section briefly reviews machine learning based diabetic detection and its performance merits and demerits. Islam et al. [8] used different ML algorithms such as NB, logistic regression (LR), and RF to identify diabetes from the diabetes symptom dataset. Initially, the patient system data is fed as input to the predictive algorithm, and based on the performance of accuracy, a suitable algorithm is used for testing performance. The research work by Choudhury and Gupta [9] classifies diabetic people based on high and low-risk using several ML algorithms. In this research, SVM establishes a hyperplane for categorization, k-NN for data clustering data, and DT, RF, and NB as classifiers. Among these, the LR achieves the highest accuracy.

Shukla [10] used an LR algorithm to classify diabetes detection in a PIMA Indian diabetic dataset (PIDD). This result attains the accuracy of 82.92% that has been highly impacted on features such as pregnancies, glucose, and body mass index (BMI). Dalakleidi et al. [11] also used the same dataset as [10] and generated two models. One with 70% of training and 30% of testing and the other with 50%–50% of training and testing. Algorithms such as LR and its variants were used. Binary Logistic Modeling achieved and accuracy of 80.47%, and Logistic Modeling Tree achieved 77.6%.

Islam et al. [12] analyze the diabetes dataset using NB, LR, and RF algorithms. RF achieved 97.8% accuracy. Harris et al. [13] generated a weighted linear regression to predict the non-insulin dependent diabetic mellitus. The model is highly complex to retain the performance. Ameen and Ashadevi [14] build a diabetes detection model using SVM, DT, RF, and LR to classify the PIDD where highest accuracy of 77% has been achieved by RF.

Daanouni et al. [15] used k-NN and DT to classify two diabetic datasets—PIDD and the German dataset. The system utilizes hyperparameter optimization and results 97.53% for k-NN. Sisodia and Sisodia [16] utilized three ML classifiers NB, DT and SVM to recognize the PIDD. On modeling, NB shows a maximum accuracy of 86.3%. Similarly, Ahuja et al. [17] utilizes the feature selection technique termed linear discriminant analysis (LDA) to select best feature from PIDD and classified using various ML classifiers. The Multi-Layer Perceptron achieves the highest accuracy of 78.2%.

Alehegn et al. [18] used the PIDD and utilized classification methods such as RF, k-NN, NB, and DT algorithms. The research concludes that the DT and NB best suit large data computations.

3 Investigation Methodology

In this Investigation methodology, the PCA and SVM with different support vector kernels have been used for diabetic detection and analysis. Using PCA and SVM with

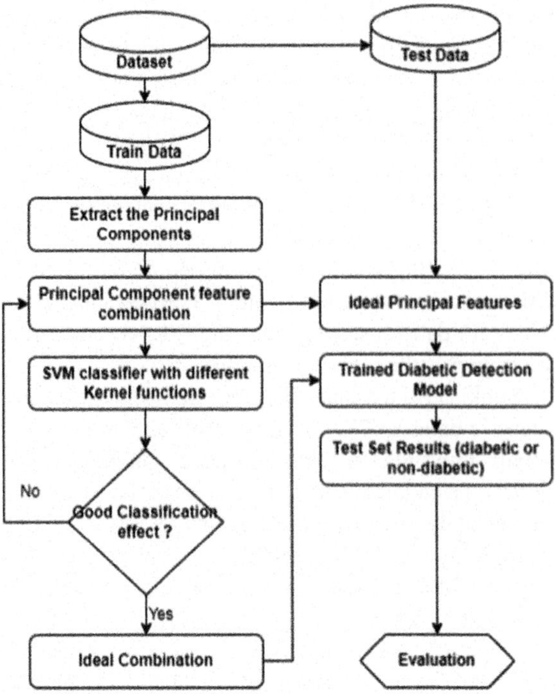

Fig. 1 Architecture of the investigation model

various kernels, the investigation model seeks to increase the performance of diabetes detection. Figure 1 presents the investigation model's architecture. The Investigation model has three stages: dataset, Feature selection, and classification.

3.1 Dataset

In this investigation model, the PIDD [19] and the Diabetes dataset [20] from Mendeley have been processed.

PIMA Indian diabetes dataset is a known dataset frequently used in machine learning and data mining research. It contains information about female patients of PIMA Indian heritage, focusing on various attributes related to diabetes. The dataset's metadata information is 768 instances with 8 attributes. The attribute information is "Pregnancy", "Blood Pressure (BP)", "Skin Thickness", "Glucose", "Insulin", "BMI", "Age", "Diabetes pedigree function", and the class label (0 or 1) where 0 represents non-diabetic, and 1 represents diabetic.

Diabetes Dataset has been acquired from Iraqi Society the laboratory of Medical City Hospital. The dataset has been constructed from the data extracted from the patient's files, that has the medical information and laboratory analysis. The dataset features consist of "Sugar level", "Blood", "Age", "Gender", "BMI", "Urea", "Fasting lipid profile" that includes "LDL", "VLDL", "Triglycerides", "HDL cholesterol", "HBA1c", "Creatinine ratio", and class labels. The labels are diabetic, non-diabetic, and pre-diabetic. A total of 1000 tuples were present in the dataset.

3.2 Feature Selection

In ML process, feature selection plays a crucial role for efficiency of the classifier. It is termed to select significant features that contribute more to the predictive modeling. It provides several advantages, such as improving accuracy, reducing overfitting and reducing training time, etc. The methods such as filter, wrapper, embedded, and meta-heuristic are the most often used in feature selection. The PCA has been used for feature selection in this proposed investigation model. In general, PCA [21] is a dimensionality reduction technique that converts high-dimensional into low-dimensional feature vector space. It preserves the significant patterns and dimensions of the data. It discovers the principal components, a new form of dimension to represent the original data. It is a linear combination of the original and orthogonal values to each other, which is uncorrelated. The PCA follows six basic steps for dimensionality reduction.

- Determine the covariance matrix for the normalized d-dimensional dataset's.
- Find the covariance matrices' eigenvalues and eigenvectors.
- Sort the eigenvalues in descending order.
- Select the k biggest eigenvectors (k is the new feature subspace to the number of dimensions) that match the k biggest eigenvalues.
- Make a projection matrix using the selected k eigenvectors.
- Convert the raw data into a new k-dimensional feature space.

3.3 Classification

The process of categorizing the feature subspace into a respective class labels is call ed classification. In machine learning, classification has been performed using supervised and unsupervised forms, whereas the supervised has several familiar algorithm such as DT, LR, NB, k-Nearest Neighbors, MLP, and SVM, etc. According to Noble [22] and Suthaharan [23], SVM is a supervised classifier used for both classification and regression. It handles both linear and nonlinear separable data and performs well for high-dimensional feature vector space. The optimal hyperplane that divides the data points of several classes while maximizing the margin between them is found via SVM. The hyperplane is picked because it is the class with the greatest distance from the closest data points. Support vectors are defined as the set of these nearest neighbors.

Table 1 Support vector kernel—mathematical functions

Kernel	Mathematical functions
Linear	$K(y_i, y_j) = y_i y_j$
Polynomial	$K(y_i, y_j) = (y_i.y_j + 1)^d$
RBF	$K(y_i, y_j) = \exp(-\gamma \|\|y_i - y_j\|\|^2)$
Sigmoid	$K(y_i, y_j) = \tanh(\alpha y_i^T y_j + c)$

Finding the support vectors—the data points that are most near the hyperplane—allows SVM to optimize the hyperplane. To maximize the margin between the support vectors of different classes, the hyperplane is adjusted. Kernel is used to move the feature vector into a high-dimensional space where the data might become linearly separable. SVM is now able to handle data that cannot be separated linearly. The mathematical functions used with SVM are listed in Table 1. The kernel functions used with SVM here are linear, RBF, polynomial, and sigmoid.

3.4 Performance Evaluation Metrics

True Positive (T_P), True Negative (T_N), False Positive (F_P), and False Negative (F_N) are the metrics used to assess the investigation model's efficiency level, and these are calculated using a confusion matrix.

Using these metrics, the other metrics, such as Accuracy, Precision, Recall, and $F1$-score, are calculated as represented in Eqs. 1, 2, 3, and 4.

$$\text{Accuracy} = \frac{T_\mathrm{P} + T_\mathrm{N}}{T_\mathrm{P} + T_\mathrm{N} + F_\mathrm{P} + F_\mathrm{N}} \tag{1}$$

$$\text{Precision} = \frac{T_\mathrm{P}}{T_\mathrm{P} + F_\mathrm{P}} \tag{2}$$

$$\text{Recall} = \frac{T_\mathrm{P}}{T_\mathrm{P} + F_\mathrm{N}} \tag{3}$$

$$F1\text{-score} = \frac{2 * \text{Prediction} * \text{Recall}}{\text{Prediction} + \text{Recall}} \tag{4}$$

Table 2 Investigation hyperparameters

Hyperparameters	Values
Kernel	Linear, RBF, poly, and Sigmoid
C	1.0
γ	Scale
Shrinking	True
Principal components	1, 2, and 3

4 Experimental Results and Discussions

The investigation has been performed on PIDD and diabetes dataset for the detection using PCA and SVM. In this process, PCA has been used for dimensionality reduction (number of features) and SVM with various kernels for classification. The PCA induces the number of principal components; here, components 1, 2, and 3 have been used for selecting features, and the SVM has various hyperparameters, which are represented in Table 2.

To identify the best principal components, the dataset has been splitted into train and test datasets. 70% of the data has been assigned to training and remaining 30% has been used for testing. The Training data is cross-validated 10 times, and the performance metrics such as Acc, Pre, Rec, and $F1$-score are analyzed to identify the best principal components. The results of the same have been shown in Tables 3, 4, and 5.

4.1 Principal Component 1

While identifying the best principal components, the training data has been cross-validated and evaluated with the performance metrics for principal component 1. The results are shown in Table 3 for PIDD, referred here to as the PIDD and the diabetes dataset. On comparing the performances of the PIDD, the RBF support vector kernel has shown the maximum performances among the support kernels considered here in this experimentation. It achieves a maximum accuracy of 96.098%, precision of 96.088%, recall of 95.958%, and an $F1$-score of 96.023%. The polynomial kernel is very close to the RBF. However, the linear and the sigmoid have a considerable difference in performance. Similarly, in comparing the performance in the diabetes dataset, the polynomial and RBF are very close in their performance and have a difference of 0.3% in accuracy, 0.1% in precision, 0.2% in recall, and 0.1% in $F1$-score. The linear and sigmoid have the least difference, with a difference of 2% and 13% with RBF, respectively. Thus, RBF with principal component 1 will be the best choice for experimentation in diabetes datasets.

Table 3 Classification performance on training data using SVM with the first principal component of PCA

SVM kernels	PIDD				Diabetes dataset			
	Acc (%)	Pr (%)	Re (%)	$F1$-score (%)	Acc (%)	Pr (%)	Re (%)	$F1$-score (%)
Linear	92.798	93.928	90.888	92.383	91.294	90.274	88.364	89.309
Polynomial	95.278	96.108	94.148	95.118	93.314	94.274	92.384	93.319
RBF	96.098	96.088	95.958	96.023	93.674	94.384	92.504	93.435
Sigmoid	83.238	81.988	82.238	82.113	80.384	77.284	81.284	79.234

Table 4 Classification performance on training data using SVM with the principal components 1 and 2 of PCA

SVM kernels	PIDD				Diabetes dataset			
	Acc (%)	Pr (%)	Re (%)	$F1$-score (%)	Acc (%)	Pr (%)	Re (%)	$F1$-score (%)
Linear	95.81	96.94	93.90	95.39	94.13	93.11	91.20	92.14
Polynomial	98.29	99.12	97.16	98.13	96.15	97.11	95.22	96.15
RBF	99.11	99.10	98.97	99.03	96.51	97.22	95.34	96.27
Sigmoid	86.25	85.00	85.25	85.13	83.22	80.12	84.12	82.07

Table 5 Classification performance on training data using SVM with the principal components 1, 2, and 3 of PCA

SVM kernels	PIDD				Diabetes dataset			
	Acc (%)	Pr (%)	Re (%)	$F1$-score (%)	Acc (%)	Pr (%)	Re (%)	$F1$-score (%)
Linear	92.92	94.05	91.01	92.50	91.005	89.985	88.075	89.02
Polynomial	94.404	95.234	93.274	94.24	93.436	94.396	92.506	93.44
RBF	97.054	97.044	96.914	96.97	94.509	95.219	93.339	94.27
Sigmoid	82.247	80.997	81.247	81.12	79.364	76.264	80.264	78.22

4.2 *Principal Components 1 and 2*

To analyze the efficiency of principal components 1 and 2, the experimentation has been carried out in training data of both datasets, as mentioned earlier. Performance metrics are evaluated are shown in Table 4. In comparing the performance with the first principal component, RBF shows a maximum accuracy of 99.11%, precision of 99.10%, recall of 98.97%, and $F1$-score of 99.03% in the PIDD. Similarly, in the case of the diabetes dataset also, RBF shows the maximum performances but is very close to a polynomial with accuracy, precision, recall, and $F1$-score of 96.51%, 97.22%, 95.34%, and 96.27% respectively.

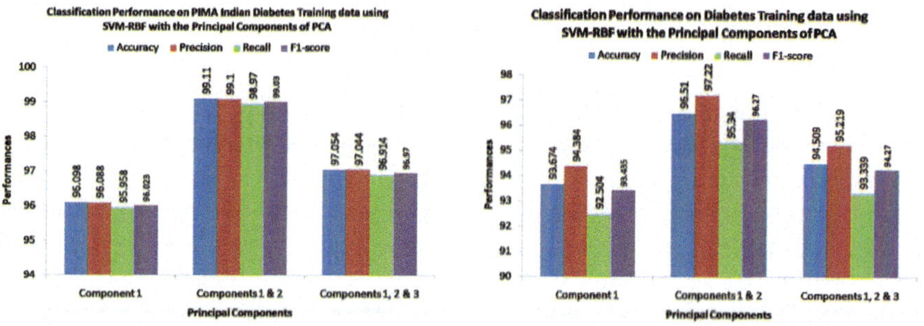

Fig. 2 Classification performance on training data using SVM-RBF with the principal components of PCA

Table 6 Classification performance on testing data using SVM-RBF with the principal components 1 and 2 of PCA

SVM kernels	PIDD				Diabetes dataset			
	Acc (%)	Pr (%)	Re(%)	$F1$-score (%)	Acc (%)	Pr (%)	Re (%)	$F1$-score (%)
RBF	97.987	97.977	97.847	97.912	95.524	96.234	94.354	95.285

4.3 *Principal Components 1, 2, and 3*

Finally, the principal components 1, 2, and 3 have also been analyzed with their performance on training data, and the results are reported in Table 5. In comparing the performances of both the PIDD and diabetes datasets, the RBF performs better, and it has better performance in all the metrics.

Though the RBF has performed well in all three components selection, the objective is to identify the best component among the three. Thus, the comparison has been performed in terms of all the metrics of RBF for all the three component combinations of both datasets, and the same has been shown in Fig. 2. In comparing the performances of PCA components on SVM-RBF, components 1 and 2 have shown maximum accuracy among the other choice of combination, and, thus, it has been used for testing the PIDD and the diabetes dataset. The results of testing the data with components 1 and 2 using SVM-RBF are shown in Table 6.

To evaluate the performance of the selected components 1 and 2 with SVM RBF kernels, the performances have been measured on test data. In the PIDD, the investigation model shows an accuracy of 97.987%, precision of 97.977%, recall of 97.847%, and $F1$-score of 97.912%. In the diabetes dataset, the investigation model shows an accuracy of 95.524%, precision of 96.234%, recall of 94.354%, and $F1$-score of 95.285%.

5 Conclusion

Diabetes mellitus is the most dangerous disease, which has been increasing in recent years in higher numbers, even in adult people. In most cases, the existence of this disease is known once the person self-tests for blood samples. The role of glycogen and insulin paves the way for the increase/decrease in the blood sugar level, which causes diabetes. Early prediction of this disease can help the person avoid its adverse effects rather than a complete cure for the disease. Thus, it is highly essential for the early prediction of diabetes worldwide. In recent years, there has been significant development in diabetes detection. However, extracting only the relevant information for classification in the shortest time is essential for diagnosis, prediction, and classification. In this paper, we investigated the selection of significant features using principal component analysis and classification using a support vector machine with various kernels such as linear, polynomial, RBF, and sigmoid. Experimentation on PIMA Indian diabetes dataset and diabetes dataset shows that the principal components 1 and 2 with SVM RBF perform better for diabetes detection.

References

1. Nakahara T, Hyogo H, Yoneda M, Sumida Y, Sumida Y, Fujii H et al (2013) Type 2 diabetes mellitus is associated with the fibrosis severity in patients with nonalcoholic fatty liver disease in a large retrospective cohort of Japanese patients. J Gastroenterol 49(11):1477–1484. https://doi.org/10.1007/s00535-013-0911-1
2. Global report diabetes. World Health Organization (2016)
3. Solanki P, Baldaniya D, Jogani D, Chaudhary B, Shah M, Kshirsagar A (2021) Artificial intelligence: new age of transformation in petroleum upstream. Pet Res (in press). https://doi.org/10.1016/j.ptlrs.2021.07.002
4. Ramanujam E, Chandrakumar T, Thivyadharsine KT, Varsha D (2020) A multilingual decision support system for early detection of diabetes using machine learning approach: case study for rural Indian people. In: 2020 fifth international conference on research in computational intelligence and communication networks (ICRCICN). IEEE, pp 17–21
5. Kavakiotis I, Tsave O, Salifoglou A, Maglaveras N, Vlahavas I, Chouvarda I (2017) Machine learning and data mining methods in diabetes research. Comput Struct Biotechnol J 15:104–116
6. Chaki J, Ganesh ST, Cidham SK, Theertan SA (2022) Machine learning and artificial intelligence based diabetes mellitus detection and self-management: a systematic review. J King Saud Univ-Comput Inf Sci 34(6):3204–3225
7. Sharma T, Shah M (2021) A comprehensive review of machine learning techniques on diabetes detection. Vis Comput Ind Biomed Art 4:1–16
8. Islam MT, Raihan M, Farzana F, Aktar N, Ghosh P, Kabiraj S (2020) Typical and non-typical diabetes disease prediction using random forest algorithm. In: Abstracts of the 11th international conference on computing, communication and networking technologies, 1–3 July 2020, IEEE, Kharagpur. https://doi.org/10.1109/ICCCNT49239.2020.9225430
9. Choudhury A, Gupta D (2019) A survey on medical diagnosis of diabetes using machine learning techniques. In: Kalita J, Balas VE, Borah S, Pradhan R (eds) Recent developments in machine learning and data analytics. Advances in intelligent systems and computing, vol 740. Springer, Singapore, pp 67–78. https://doi.org/10.1007/978-981-13-1280-9_6

10. Shukla AK (2020) Patient diabetes forecasting based on machine learning approach. In: Pant M, Sharma TK, Arya R, Sahana BC, Zolfagharinia H (eds) Soft computing: theories and applications. Advances in intelligent systems and computing, vol 1154. Springer, Singapore
11. Dalakleidi KV, Zarkogianni K, Karamanos VG, Thanopoulou AC, Nikita KS (2013) A hybrid genetic algorithm for the selection of the critical features for risk prediction of cardiovascular complications in Type 2 Diabetes patients. In: Abstracts of the 13th IEEE international conference on bioinformatics and bioengineering, Chania, 10–13 Nov 2013. https://doi.org/10.1109/BIBE.2013.6701620
12. Islam MMF, Ferdousi R, Rahman S, Bushra HY (2020) Likelihood prediction of diabetes at early stage using data mining techniques. In: Gupta M, Konar D, Bhattacharyya S, Biswas S (eds) Computer vision and machine intelligence in medical image analysis. Advances in intelligent systems and computing, vol 992. Springer, Singapore, pp 113–125. https://doi.org/10.1007/978-981-13-8798-2_12
13. Harris MI, Klein R, Welborn TA, Knuiman MW (1992) Onset of NIDDM occurs at least 4–7 yr before clinical diagnosis. Diabetes Care 15(7):815–819. https://doi.org/10.2337/diacare.15.7.815
14. Ameena RR, Ashadevi B (2020) Predictive analysis of diabetic women patients using R. In: Peter JD, Fernandes SL (eds) Systems simulation and modeling for cloud computing and big data applications. Elsevier, Amsterdam. https://doi.org/10.1016/B978-0-12-819779-0.00006-X
15. Daanouni O, Cherradi B, Tmiri A (2019) Predicting diabetes diseases using mixed data and supervised machine learning algorithms. In: Abstracts of the 4th international conference on smart city applications, 2–4 Oct 2019. ACM, Casablanca. https://doi.org/10.1145/3368756.3369072
16. Sisodia D, Sisodia DS (2018) Prediction of diabetes using classification algorithms. Procedia Comput Sci 132:1578–1585. https://doi.org/10.1016/j.procs.2018.05.122
17. Ahuja R, Sharma SC, Ali M (2019) A diabetic disease prediction model based on classification algorithms. Ann Emerg Technol Comput 3(3):44–52. https://doi.org/10.33166/AETiC.2019.03.005
18. Alehegn M, Joshi RR, Mulay P (2019) Diabetes analysis and prediction using random forest, KNN, Naïve Bayes, and J48: an ensemble approach. Int J Sci Technol Res 8(9):1346–1354
19. Kahn M, Diabetes. UCI Machine Learning Repository. https://doi.org/10.24432/C5T59G
20. Ahlam R (2020) Diabetes dataset. Mendeley Data V1. https://doi.org/10.17632/wj9rwkp9c2.1
21. Abdi H, Williams LJ (2010) Principal component analysis. Wiley Interdiscip Rev Comput Stat 2(4):433–459
22. Noble WS (2006) What is a support vector machine? Nat Biotechnol 24(12):1565–1567
23. Suthaharan S, Suthaharan S (2016) Support vector machine. In: Machine learning models and algorithms for big data classification: thinking with examples for effective learning, pp 207–235

Automatic Certificate Generation and Security Implementation Using LSB and Hybrid Watermarking Techniques

T. S. Arulananth, A. Jose Anand, S. Jayanthi, Petla Manaswini, Gogineni Viswa Sai Teja, G. Redyy Taruni, and Nalla Krishna Prasad

Abstract Process to ensure the authenticity and integrity of digital certificates, LSB and hybrid watermarking are popular techniques used to embed additional information or watermark into digital certificates to enhance their security and prevent unauthorized modifications. This research aims to explore the application of LSB and hybrid watermarking in certificate generation, with a focus on implementing these techniques using MATLAB. Here, we developed a MATLAB code to generate certificates automatically and provide security by watermarking. The process of creating a digital certificate is complicated, and the process of creating an individual certificate which contains a variety of information is much more challenging. To overcome the difficulties with the manual method and reduce the time needed to issue a certificate, a range of certificate generating and verification methods have been developed in the modern day. The certificate design template is uploaded into the folder holding the MATLAB code since the system runs automatically on a template that the system

T. S. Arulananth (✉) · P. Manaswini · G. V. S. Teja · G. Redyy Taruni · N. K. Prasad
Department of Electronics and Communication Engineering, MLR Institute of Technology, Hyderabad 500043, India
e-mail: arulananthece@mlrinstitutions.ac.in

P. Manaswini
e-mail: 19r21a04g3@mlrinstitutions.ac.in

G. V. S. Teja
e-mail: 19r21a04d8@mlrinstitutions.ac.in

G. Redyy Taruni
e-mail: 19r21a04d6@mlrinstitutions.ac.in

N. K. Prasad
e-mail: 19r21a04f6@mlrinstitutions.ac.in

A. Jose Anand
Department of ECE, KCG College of Technology, Karapakkam, Chennai 600097, India

S. Jayanthi
Department of Electronics and Communication Engineering, R.M.D Engineering College, Kavarapettai, Chennai, India

S. Bandyopadhyay et al. (eds.), *Intelligent Computing Systems and Applications*, Lecture Notes in Networks and Systems 1010,
https://doi.org/10.1007/978-981-97-5412-0_29

user has already created. To display the details on certificates, this program gathers participant information from excel sheets. An Excel document is used to extract the information that will be written on the certificate.

Keywords Digital certificate · Design template · Excel sheet · Watermarking

1 Introduction

Certificates play a significant role in many areas of modern life, especially in the digital age, as they enable identity verification, digital signatures, and secure communication [1]. There are numerous sectors and industries where certificates can be used in which some instances where certificates widely used are in education, healthcare, E-commerce, etc. So, the main goal of our project is to design a MATLAB code which can generate a huge number of certificates automatically and provide security to certificates in an effortless approach. Our project is also designed to reduce the time involved in importing participant details in certificate generation system [2, 3]. One approach for creating certificates effortlessly and quickly is automatic certificate production. Certificates are a part of the institutions that are given to the students in recognition of their success in the curriculum or co-curriculum. Certificates can be seen as an encouragement given to the students to perform better in their curriculum or co-curriculum. The institutions will be able to manage the task as effectively as possible with the aid of the automatic certificate generation utilizing MATLAB [4]. It will support the institution's efficient operation. Through this method, the institutions will be able to upload the certificates, and the students would have simple access to the certificates [5].

Using MATLAB, generating certificates can be a quick and practical approach to create personalized certificates for a variety of uses. With the help of MATLAB, which is a potent tool, it is possible to automate the creation of certificates in large quantities and with a variety of information. With the aid of MATLAB, you can create certificate templates, alter their contents, and have information like participant names and dates filled in automatically [6]. This can speed up the certificate creation process and reduce errors. In this way, MATLAB can be a helpful tool for making certificates for accomplishments in school or the workplace, prizes, or other occasions [7].

MATLAB is a powerful tool that can be used to generate various types of certificates, including digital certificates and paper certificates. The use of MATLAB can significantly simplify the process of creating and managing certificates, especially for complex systems with a large number of certificates [8, 9]. It is also necessary to encrypt a certificate for security purposes. Certificates contain sensitive information such as private keys, and if this information is accessed by unauthorized users, it can be used for malicious purposes such as identity theft or data manipulation [10]. Watermarking is a way to protect the certificates from forgery, and there are many types of watermarking which includes LSB (least significant bit), visible watermarking, invisible watermarking, robust watermarking, fragile watermarking, etc. In contrast to encryption, which limits the user's ability to view, access, and comprehend the signal, digital watermarking prevents unlawful use of digital content. Certificates are important documents used to verify the identity of individuals, organizations, and devices in various online transactions [11]. Here are some security measures that can be taken to ensure the security of certificates:

a. Limit access to certificates
b. Encryption
c. Watermarking
d. Password protection.

2 Literature Survey

L. Chen and et al proposed the core idea is to segment the host image into non-overlapping blocks by the means of space filling curve and based on the amount of DCT energy in the blocks [1].

K. Gupta and et al presented a framework called deep learning-based watermarking for authorized IoT onboarding (*DLWIoT*), featuring a robust and fully automated image watermarking scheme based on deep neural networks[2].

J. Wang and et al proposed discrete wavelet transform (DWT) to authenticate the multimedia image and it can convert the image from spatial domain to frequency domain [3].

I. Rodriguez and et al had given the emphasis on image authentication and tampering detection. This scheme intends for determining where the image has been altered to the act of a malicious party; and further, the unmalicious manipulation like compression or brightness changing can be excluded [4].

Ahmed Dalhatu Yusuf & et al proposed an idea on Automated Batch Certificate Generation and Verification System. In this work, an application called Automated Batch Certificate Generation and Verification System (ABCGVS) is developed based on the client–server technology. This technique is also used in many other works, with the coming of client–server technology more frameworks are being developed and the expense and the time for building up an application phenomenally decreased [5]. (Table 1)

3 Proposed Methodology for Generating Certificates

Generating certificates using MATLAB involves several steps, including designing the certificate template, importing data, and creating a script to automate the certificate generation process. Here's a general methodology you can follow:

Table 1 Summary of the existing methods of watermarking

S.No	Name of the algorithm	Domain type	Major applications	Strength	Weakness
1	Singular Value Decomposition (SVD)	Transform Domain	Image, Video	High capacity, robust to geometric attacks	Susceptible to signal processing attacks
2	Frequency Domain Embedding (FDE)	Frequency Domain	Image, Video	Robust to various attacks, high capacity	Requires synchronization, may impact quality
3	Quantization Index Modulation (QIM)	Transform Domain	Image, Audio, Video	High capacity, robust to compression	Susceptible to noise, may cause perceptual distortion
4	Patchwork Watermarking	Spatial Domain	Image	High robustness, flexible and versatile	Limited capacity, may affect visual quality

- Design your certificate template: Start by designing a certificate template that includes all the necessary information, such as the recipient's name, date, course or achievement name, and any other relevant information. You can use a graphics software like Adobe Illustrator or Inkscape to create your certificate design.
- Import data: Once you have your template, you will need to import the data that you want to include in each certificate. You can do this by creating a spreadsheet in Excel or Google Sheets and entering the recipient's name, date, course or achievement name, and any other relevant information.
- Write a MATLAB script: After you have imported your data, you can write a MATLAB script that reads the data from your spreadsheet and uses it to automatically populate the certificate template. You can use MATLAB's image processing toolbox to add text and images to your certificate template.
- Test your script: Test your script on a small sample of certificates to ensure that it is working correctly. Make sure that all the information is being populated correctly, and that the certificates are being saved in the correct format.
- Generate certificates: Once you are confident that your script is working correctly, you can use it to generate all the certificates you need. You can specify the number of certificates to generate and MATLAB will automatically create them using the data you have imported.
- Save certificates: Finally, save your certificates in the desired format. You can save them as PDF's, JPEG or any other image format that you prefer.

With these steps, you can create an automated process for generating certificates using MATLAB (Fig. 1).

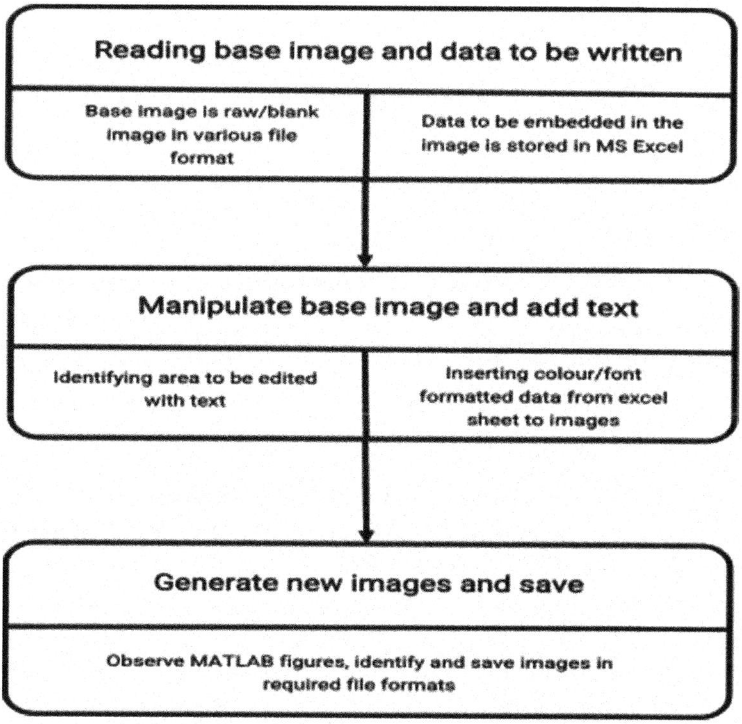

Fig. 1 Flow chart for generating certificates automatically using MATLAB

Watermarking: A watermark, which might be text, a logo, or a copyright notice, is an overlay that is put into a digital image and the process of embedding the watermark into an image is called watermarking [12, 13]. A watermark's main functions are to identify the work and prevent unlawful usage of it. Watermarks can be seen or not seen. Logos or text that is placed over an image or document are frequently used as visible watermarks. On the other hand, invisible watermarks are typically built into the file itself and can only be found by specialized software. Photographers, graphic designers, and painters frequently employ watermarks to prevent unlawful use or dissemination of their work. Businesses may also employ them to safeguard private information or to guarantee that crucial papers are not changed or tampered with.

There are two types of digital watermarking, and they are as follows:

1. Visible watermarking
2. Invisible watermarking.

Visible Watermarking: A semi-transparent text or picture that is placed on the original image and is clearly apparent as a watermark. By designating the image as the owner's property, it nevertheless offers copyright protection while allowing viewing of the original image. Visible watermarks are more resistant to image alteration,

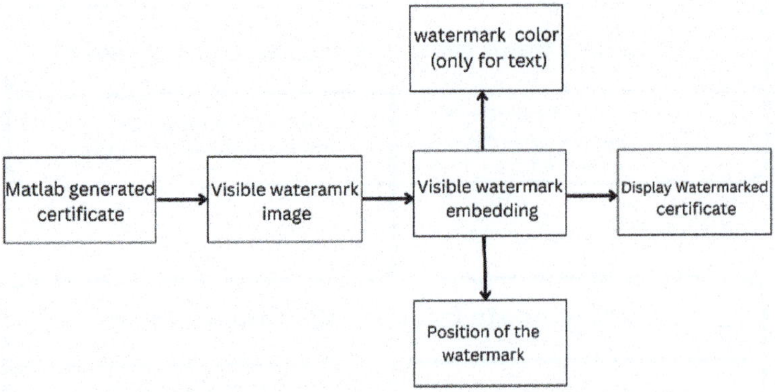

Fig. 2 Block diagram of visible watermarking

particularly if a semi-transparent watermark is used and applied to the entire image [14, 15]. They are therefore preferred for providing robust copyright protection for digitally stored intellectual property (Fig. 2).

Invisible Watermarking: A unique identifier or digital signature is added to the media file through the process of invisible watermarking. This watermark is undetectable by the human eye or ear but is detectable by specialist software. This identifier may contain data that aids in determining the legitimacy of the media file, such as the identity of the copyright holder, the date the work was created, and other metadata. LSB (least significant bit) is one of the invisible watermarking process in which a unique extraction code is used to determine the watermark embedded on certificates [16].

Least Significant Bit Watermarking (LSB): The least significant bit (LSB) is a technique which is used for inserting a watermark that is invisible to the human eye onto image. Since the LSB's are the bits that have the least influence on the data's overall value, changing them has little impact on the original data's quality. LSB watermarking is therefore a well-liked option for embedding a watermark without significantly lowering the quality of the underlying data. In LSB watermarking, the watermark is embedded by swapping out the data's least important bits with the watermark's bits. For instance, in an 8-bit image, the least important bit would be the rightmost bit, and swapping it out with a watermark bit would only change 1/256th of the pixel's overall value [17, 18] (Figs. 3 and 4).

Watermark Embedding: A watermark embedded in an image or certificate is a digital identifier or signature that is added to the content. The purpose of embedding a watermark is to mark the content as being the property of the original creator, to prevent unauthorized use or distribution, and to help identify the source of the content in case of infringement [19, 20] (Figs. 5 and 6).

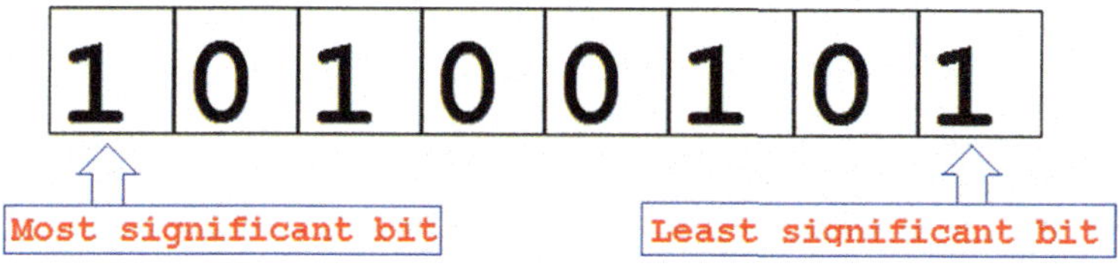

Fig. 3 Binary representation of LSB and MSB

Fig. 4 LSB watermarking

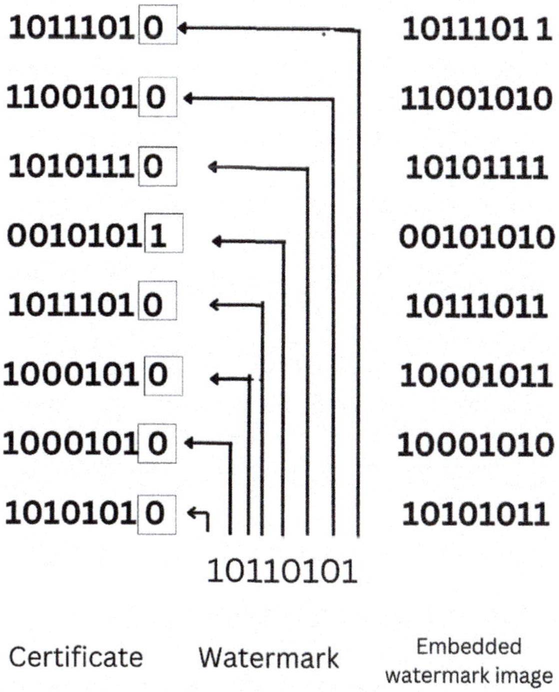

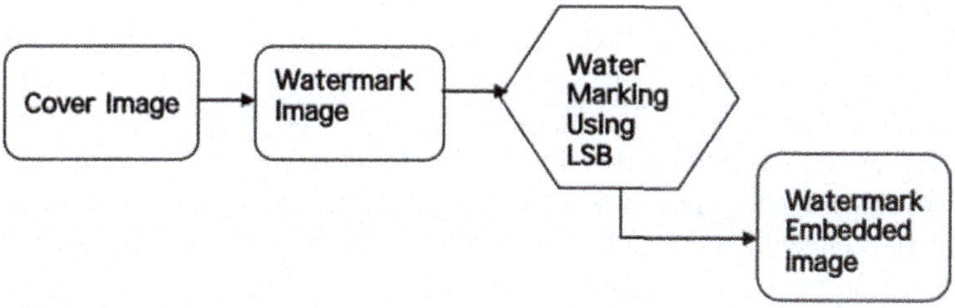

Fig. 5 Block diagram of watermark embedding

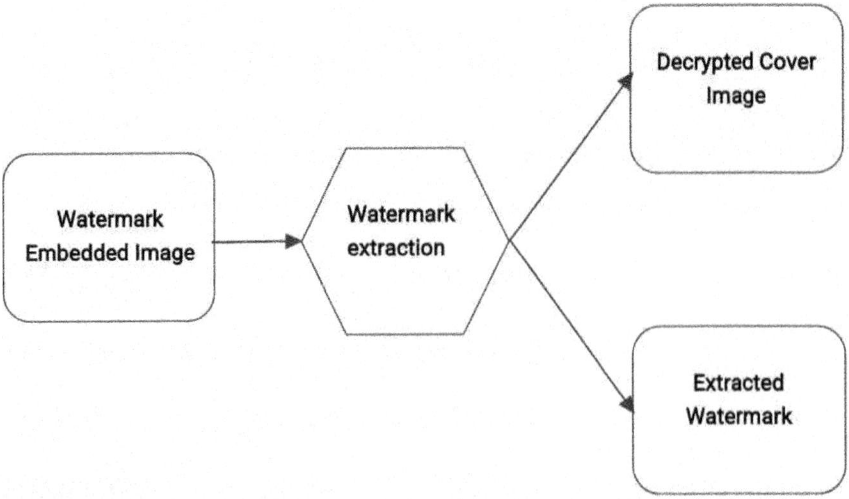

Fig. 6 Block diagram of extracted watermark

Watermark Extraction: Watermark extraction is the process of detecting and removing a watermark from an image or document. The goal of watermark extraction is often to remove the watermark from the content so that it can be used or distributed without the identifying watermark, which can be a form of copyright protection [21, 22].

4 Results and Discussions

Automatic generation of certificates using MATLAB tools is successfully generated by using a design template. Using a design template created with third-party tools allows for the successful automatic creation of certificates utilizing MATLAB tools. The information to be shown on the certificates is also successfully entered using an Excel sheet, and to ensure their security, a watermark that cannot be seen by the naked eye is embedded on each certificate. To protect the certificates from unauthorized users, inserting a watermark is an efficient technique in which a visible or invisible watermark is embedded. Invisible watermarks are observed only when the certificate is submitted to the extraction code, which causes the certificate to be displayed along with the watermark that is embedded on the certificates, can the embedded watermark be seen. In many different sectors, including copyright protection, digital forensics, and data concealment, LSB watermarking is a valuable technique for embedding and retrieving information from digital media files (Figs. 7, 8, 9 and 10).

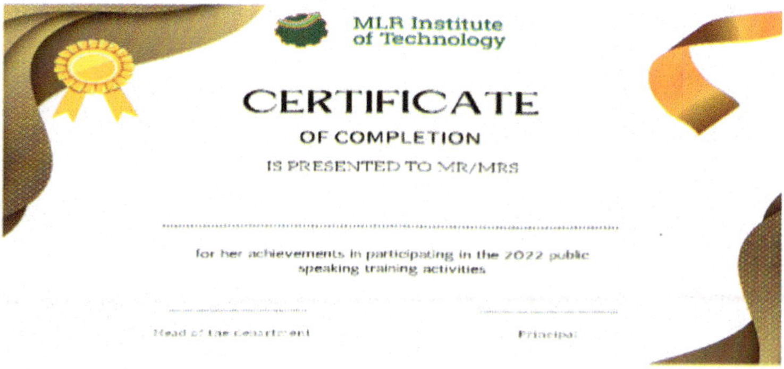

Fig. 7 Design template 1

Fig. 8 Automatic generated certificates

Fig. 9 Visible watermark

The image quality typically declines after watermarking, but there is a method to determine the quality of the photos by calculating PSNR (peak signal-to-noise ratio) and MSE (mean square error).The PSNR and MSE findings are listed in a table, and they are quite favorable.

Fig. 10 Extraction of invisible watermark

Peak Signal-to-Noise Ratio (PSNR): Peak signal-to-noise ratio, also known as PSNR, is a metric used to determine how much the original image or video has been compromised during compression or transmission. It is derived by comparing the greatest signal value (i.e., the highest pixel value in the image) to the mean squared error (MSE) between the original and compressed/transmitted image. It is represented in decibels (dB).

$$\mathrm{PSNR} = 10.\log_{10}\left(\frac{\mathrm{MAX}_I^2}{\mathrm{MSE}}\right)$$

$$\mathrm{PSNR} = 20.\log_{10}\left(\mathrm{MAX}_I^2/\sqrt{\mathrm{MSE}}\right) = 20.\log_{10}\left(\mathrm{MAX}_I^2/\sqrt{\mathrm{MSE}}\right)$$

$$\mathrm{PSNR} = 20.\log_{10}(\mathrm{MAX}_I) - 10.\log_{10}(\mathrm{MSE})$$

Mean Square Error (MSE): The average difference between the pixels in the original image or video and the pixels in the compressed/transmitted image is measured by the term "mean squared error," or MSE. The average of the squared differences between each pixel in the two photos is used to calculate it (Fig. 11).

$$\mathrm{MSE} = \frac{1}{N \times M}\sum_{i=0}^{N-1}\sum_{j=0}^{M-1}[X(i,j) - (i,j)]^2$$

Better picture or video quality is indicated by greater PSNR and lower MSE values, although the precise values that are deemed "highly efficient" depend on the specific application and needs. For instance, a PSNR value of 40 dB or more may be seen as being very efficient in some applications where excellent visual quality is essential, while a lower PSNR value may be adequate in other applications where lower quality is acceptable. The PSNR value for every certificate generated by MATLAB is about 2.6 dB, and the MSE value is approximately 0.8,which means

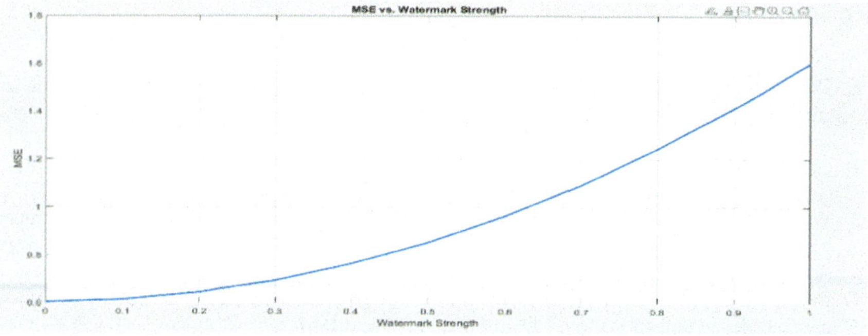

Fig. 11 MSE versus watermark strength

the quality of the certificates needs to be improved using techniques like certificate transparency, certificate revocation, etc. The allowed MSE values are also determined by the specifications of the particular application. It is also important to keep in mind that, although being widely used metrics for assessing the quality of images and videos, PSNR and MSE may not always precisely reflect how people actually perceive quality. As a result, in some circumstances, it may be necessary to evaluate quality through subjective testing by human observers.

5 Conclusion

To sum up, MATLAB's flexibility to automate, customize, allow for data analysis and visualization, integrate with other systems, and scale well makes it a good choice for creating certificates. The potential of MATLAB as a tool for creating personalized and data-driven certificates is only going to increase as a result of the rising demand for digital certificates. It is the perfect platform for issuing certificates in a number of industries and applications due to its capacity for handling enormous datasets and integrating with other systems. In general, utilizing MATLAB to generate certificates and provide encryption is an economical, effective, and adaptable option that can assist enterprises in streamlining their certification processes and enhancing their data-driven insights. The LSB approach is used to add a watermark to certificates that are automatically generated, ensuring the security of digital certificates.

Acknowledgements The authors would like to express our gratitude to our colleagues and friends at MLR Institute of Technology for their valuable suggestions and support.

References

1. Chen L, Zhang M (2021) Efficient LSB watermarking for automatic certificate generation and authentication in IoT environments. In: Proceedings of the IEEE international conference on Internet of Things (IoT)
2. Gupta K, Verma S (2021) Deep learning-based watermarking technique for secure certificate generation. J Comput Sci Technol
3. Wang J, Zhang Q (2021) FPGA implementation of LSB watermarking for real-time certificate generation. In: Proceedings of the international symposium on field-programmable gate arrays (FPGA)
4. Rodriguez I, Martinez A (2021) Robust LSB image watermarking for automatic certificate generation and authentication. J Digit Forensics Secur Law
5. Liu H, Li J (2021) Integrating blockchain technology with LSB watermarking for secure certificate generation. In: Proceedings of the IEEE international conference on communications (ICC)
6. Chen F, Wang G (2020) A hybrid approach for automatic certificate generation and security implementation using LSB watermarking and RSA encryption. J Inf Secur Appl
7. Smith, Johnson M (2020) Enhancing security in automatic certificate generation using LSB watermarking. In: Proceedings of the international conference on information security and cryptography (ISC)
8. Dharani RP (1998) Customized document generation system, IEEE conference paper
9. Dejan G (2011) Applicative solution for generating report template—Automated report generation system, IEEE research paper
10. Mahalakshmi S, Rajeswari C, Jose Anand A, Rahul A (2022) Client authentication by signature verification method using robotic process automation (RPA). In: 2022 international conference on data science, agents & artificial intelligence (ICDSAAI), Chennai, India, pp 1–5
11. Srushti AS (2014) Certificate generation system, IEEE conference paper
12. Johnpaul AH (2011) Design and implementation of student verification system, IEEE conference paper
13. Chang CY, Su SJ (2005) The application of a full counter-propagation neural network to image watermarking. In: Proceeding of IEEE networking, sensing and control, pp 993–998
14. Swanson MD, Kobayashi M, Tewfik AH (1998) Multimedia data-embedding and watermarking techniques. Proceeding of IEEE 86(6):1064–1087
15. Soleimani M, Nezhad FS, Mahdipour H, Khademi M (2012) A robust digital blind image watermarking based on spread spectrum in DCT domain. Sci Acad Trans Comput Commun Netw. 2(2): 122–126
16. Ponmalar A, Jose Anand A, Muthamizhan T, Sobana T, Vishwath JH (2022) Automatic forensic analysis of criminal navigation system using machine learning. In: Proceedings of IEEE technically sponsored first international conference on computational science & technology (ICCST 2022), pp 1–5
17. Chopra D, Gupta P, Sanjay G, Gupta A (2012) LSB based digital image watermarking for gray scale image. IOSR J Comput Eng 6(1):36–41
18. Thawkar S (2012) Digital image watermarking for copyright protection. Int J Comput Sci Inf Technol 3(2):3757–3760
19. Craver S, Memon N (1998) Resolving rightful ownership with invisible watermarking techniques: limitations, attacks and implications. IEEE J Sel Areas Commun 16(4):573–586
20. Gupta N, Aggarwal SK (2014) A survey of digital image watermarking techniques. Int J Comput Appl (IJCA)
21. Chetna SS, Patil P (2013) A survey on image watermarking techniques and attacks. Int J Adv Res Comput Eng Technol (IJARCET)
22. Gupta P, Goyal B (2016) A review of hybrid watermarking techniques for image security. Int J Comput Appl (IJCA)

Superpixel Fused Pixel Context Network for Precise Surface Water Mapping Using Multispectral Satellite Images

Avinash Chouhan, Dibyajyoti Chutia, and Shiv Prasad Aggarwal

Abstract Precise delineation of surface water is essential for monitoring its status. Traditional indices-based methods suffer from the problem of falsely positive detection with a similar appearance, and finding a threshold value is also a challenging task. Deep learning (DL)-based methods need to handle high-class balance and pixel neighbor details for accurate surface water mapping. In this work, we proposed a novel DL-based Superpixel Fused Pixel context Network (SFPNet) with a lightweight Pixel Context Block and a Superpixel Block to extract surface water from satellite images. An exhaustive ablation study is presented to validate the effectiveness of our work. SFPNet has produced better results compared to the existing traditional and deep learning-based approaches.

Keywords Deep learning · Remote sensing · Surface water mapping · Superpixel fusion

1 Introduction

Water plays an essential role in the biosphere. There have been continuous efforts towards conserving surface water through consistent surface water monitoring. Earlier approaches relied on hand-crafted features and indices-based methods for surface water mapping [1–5]. DL showcased promising results in vision tasks [6]. It replaced the manual supervision for feature selection tasks with automatic feature extraction. The architecture of neural networks improved significantly in recent times. Fully connected layers [6] is replaced by a fully convolutional layer to preserve the spatial information [7] and require less computational resource. Alternate paths in the

A. Chouhan (✉) · D. Chutia · S. P. Aggarwal
North Eastern Space Applications Centre, Umiam, Meghalaya 793103, India
e-mail: avinash.chouhan@nesac.gov.in

D. Chutia
e-mail: d.chutia@nesac.gov.in

S. Bandyopadhyay et al. (eds.), *Intelligent Computing Systems and Applications*, Lecture Notes in Networks and Systems 1010,
https://doi.org/10.1007/978-981-97-5412-0_30

network are proposed for feature reuse and better gradient flow [7–10]. Attention-based methods were developed [11] to focus on essential features in the channel and spatial dimension.

DL-based approaches are utilized in remote sensing tasks extensively [12–20]. In surface water mapping, several DL-based methods are proposed [21–35]. Fully convolutional networks [21, 27, 30], encoder–decoder-based approaches [26, 34], and super pixel combined methods [22, 26] are used. Some works used dense connections [28, 29], atrous convolution [23, 31, 33], and pyramid pooling [32, 33] for classification of water pixels. The selection of band combinations is very important in this task. Use of Red-Green-Blue(RGB) and near-infrared (NIR) is proposed in some methods [27–29] and other methods proposed inclusion of short-wave infrared (SWIR) bands in input [21, 24, 30, 35]. Some methods utilized only Red-Green-Blue bands [22, 25].

In this work, we proposed a deep learning-based architecture for surface water mapping using satellite images that used lightweight attentive features enhanced by the fusion of intermediate features for superpixels extraction.

2 Literature Survey

In traditional approaches, McFeeters et al. [1] proposed a normalized difference water index (NDWI) to identify water from satellite images. They suggested the use of the NIR band and green band for the calculation of the NDWI index. Gao et al. [2] also proposed NDWI for vegetation liquid water extraction with NIR and middle infrared bands. Xu et al. [3] proposed a modified normalized difference water index (MNDWI) for extraction of enhanced water features and to handle noise due to built-up land. They suggested the utilization of the middle infrared band in the calculation of the MNDWI index. Feyisa et al. [4] proposed an automatic water extraction index (AWEI) for enhanced surface water mapping. They used blue, green, NIR, and SWIR bands for the calculation of the AWEI index. It is effectively able to differentiate between water and other complex features such as built-up, dark surface, etc. Wu et al. [5] proposed a two-step urban water index (TSUWI) for urban surface water mapping. It is a combination of two sub-indices of urban water index (UWI) and urban shadow index (USI), which are based on spectral features and support vector machine classifier.

Isikdogan et al. [21] proposed Deep-WaterMap, a fully convolutional neural network for the segmentation of water from Landsat images. It has a multiscale encoder–decoder-based architecture with skip connections. Chen et al. [22] proposed self-adaptive pooling convolutional neural network (SAPCNN), combining the DL-based and superpixels methods. Superpixels are produced using the SLIC algorithm from the input image and passed to SAPCNN for final map generation. Fang et al. [23] utilized a convolutional neural network-based approach for the identification of global reservoirs from Landsat-8 images. They have used ResNet-50 pretrained model and fine-tuned it for the application. Yuan et al. [24] proposed a multichannel water body detection network (MC-WBDN) with a multichannel fusion module, atrous convolution-based pyramid pooling module, and Space-to-depth and Depth-to-space operations. Miao et al. [25] proposed a restricted receptive field deconvolution network (RRF DeconvNet) with edge weighting loss. It used DeconvNet and residual block in network architecture with skip connections. Atrous convolutions are used in place of strided convolution to minimize the number of layers utilized.

Feng et al. [26] used encoder–decoder-based architecture with superpixels-based segmentation. Conditional random field-based boundary refinement is used to produce the final water map. Li et al. [27] presented a comparative study of water map generation using a fully convolutional network (FCN), support vector machine (SVM), sparsity model (SM), and NDWI. From the experiments, they concluded that FCN-based model performed better compared to the rest. Wang et al. [28] studied several DL-based models for water extraction and concluded that DenseNet-based models produced better results with the shortest time required for training and convergence. Li et al. [29] proposed dense local feature compression (DLFC) network, which utilized densely connected connections. These dense connections provide better feature reuse and many paths for gradient flow. Isikdogan et al. [30] proposed DeepWaterMap-v2 with enhancement to work in cloudy input. They proposed to reduce the initial feature layers to enable the model to work on very large input images. These reduced features are compensated using an increase in the number of low-resolution features.

Poliyapram et al. [31] used an encoder–decoder-based network with dilated convolution for water, lakes, and land segmentation using coarse-resolution satellite images. Liu et al. [32] proposed DUPnet, which utilized multiscale spatial pyramid pooling with dense connections to combine low and high-level features for effective semantic segmentation. Luo et al. [33] introduced WatNet, which used atrous pyramid pooling with a multiscale encoder. A benchmark water segmentation dataset named Earth Surface Water Knowledge Base (ESWKB) is also proposed. Manocha et al. [34] used multiscale feature fusion with the use of an encoder-decoder network with multiple size kernels.

From this discussion, we concluded that DL-based architectures had shown enhanced results in water extraction tasks. The challenges faced by DL-based methods are :

1. Delineation of water is difficult in input having cloud and other noise.
2. Pixel with mix classes (water and nonwater) provides a challenge in precise classification.

3 Proposed Scheme

To overcome the challenges listed in the previous section, we proposed a neural network architecture named SFPNet architecture. Our main contributions from this work are:

1. A novel architecture SFPNet is proposed, which utilizes lightweight, high-resolution connections with pixel context and superpixel fusion.
2. We proposed a Pixel Context Block for multicontext attentive feature aggregation.
3. We proposed the use of multilevel features based on Superpixel Block to improve the surface water classification.
4. Detailed quantitative study is presented with a comparative evaluation of proposed work with other state-of-the-art (SOTA) methods.

SFPNet, as shown in Fig. 1, consists of the Backbone with Residual Block, Stem Block, Internal Block, and high-resolution connections. Here, I_{xylab} is XYlab processed input I, m is the concatenation, S_p is association matrix output from Superpixel Block, and S_o is segmentation output from Pixel Context Block. Pixel Context block is proposed to aggregate the surrounding information using dilated convolution followed by residual channel and spatial attention. Super Pixel Block is proposed, which uses multilevel backbone features for pixel-superpixel association calculation.

3.1 Backbone

Our proposed work has a Backbone for feature extraction. It took the input and passed it to the Stem Block, which consists of two bottleneck blocks with depth separable convolutions as proposed by Chollet et al. [36]. Stem Blocks also create two resolution streams of features. The internal Block contains a sequence of residual units with depth-separable convolutions. The multiresolution streams are passed to the Internal Block parallelly and, in the end, create another low-resolution stream. All resolution streams are connected to other streams as proposed in HRNet [10] at the end of the internal block. We utilized four resolutions stream in our network.

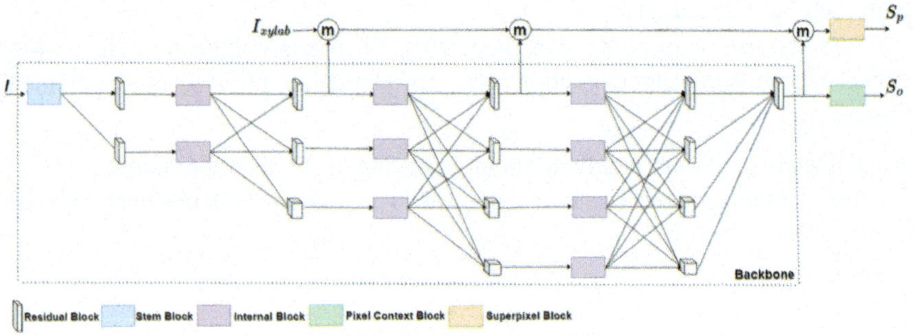

Fig. 1 SFPNet complete architecture. It consists of a Backbone, Pixel Context Block, and Superpixel Block. The backbone has Residual blocks, Stem blocks, and Internal blocks

3.2 Pixel Context Block

We proposed a Pixel Context Block, for contextual information extraction. In this block, we used a parallel dilated block with a dilation rate of 1 and 2 and combined it using an addition operation. This contextual information is enhanced using residual spatial and channel attention. Dilated convolutions allowed an increase in the receptive field of the kernel without extra computational cost. As proposed by Woo et al. [11], Channel and spatial attention based on multilayer perception enhanced the feature information for better representation. This block facilitates a multicontext view of the features and provides locations in spatial and channel dimensions that need more focus. It helps in better feature delineation in the final output map.

3.3 Superpixel Block

We proposed a Super Pixel Block, which is used for superpixel-based enhanced water segmentation. It consists of multiple depth-separable convolution and relu layers. It takes concatenated inputs from multiple levels of the highest resolution features of backbone and XYLab transform input representation as used in [17, 19]. The output of this block is a pixel-superpixel association (M) which is used to find the relation between the superpixel and the corresponding ground truth. The calculation of pixel-superpixel association between pixel p and superpixel q is calculated using the following equation in t iteration.

$$M^t{}_{p,q} = e^{||F_p - C_q{}^{t-1}||^2} \tag{1}$$

Here F is the features of the pixel, t is iteration, and C is the superpixel center. The value of the superpixel center is also updated in an iterative manner using the following equation

$$\begin{aligned} C_q^t &= \frac{1}{Z_q^t} \sum_{p=1}^{n} M^t{}_{p,q} F_p \\ \frac{1}{Z_q^t} &= \sum_{p} M^t{}_{p,q} \end{aligned} \tag{2}$$

Here, $\frac{1}{Z_q^t}$ is normalization factor with constant value and n is number of pixels. Pixel and superpixel mapping can be done using the column normalization representation ($\tilde{M}$) and row normalization ($\hat{M}$) representation of pixel-superpixel association. These mapping can be represented as the following equations.

$$\begin{aligned} S &= \hat{M}^T F \\ F &= \tilde{M} S \end{aligned} \tag{3}$$

Here, S is the superpixel feature, and F is the input feature. The above mappings are used in the Superpixel Reconstruction loss (L_{spr}), which compares the reconstructed ground truth through pixel-superpixel mapping and ground truth.

4 Experiments

4.1 Dataset

Global Surface Water Dataset (GSWD) We used the GSWD shared by Isikdogan et al. [30], which consists of Landsat8 images and the Global surface water generated ground truth [37]. It has 142682 patches of 512 × 512 pixels divided into a training set of 137,682 patches and 5000 patches in the testing set. As this dataset has more images than mentioned in the paper [30], direct comparison of results is not possible. We retrained the model proposed in that paper [30] and calculated matrices for the same. We crop centrally 256 × 256 pixels from all 142,682 patches and use it to train all models for computational ease.

Earth Surface Water Knowledge Base (ESWKB) This dataset contains a total of 95 scenes of sentinel 2 images all around the world. The training data contains 64 scenes, and the remaining 31 scenes are used in the test data. Each scene has six bands of information of visible and infrared regions of the spectrum. We created patches of 256 × 256 pixels for the training of models.

Unannotated Multisatellite datasets (UMSD) We created three different satellite test sets for extensive qualitative evaluation of the proposed model. We used Landsat 8 (L8), Landsat 9 (L9), and Sentinel 2 (S2) images from the North Eastern region of India. A total of 150 scenes are used in these datasets, with 50 scenes from each. The image size in these datasets is in the range of 8000–16,000 pixels for height and width dimensions, which are cropped to create the patches of size 512 × 512 pixels which are used for visual comparison.

4.2 Experimentation Setup

We conducted all experiments on the graphics processing unit (GPU) facility provided by our institute. We used four Nvidia V-100 cards with 32 GB GPU memory and 512 GB RAM resources. All models are trained for 120 epochs. All ablation study is done using GSWD. We used MNDWI, Unet [7], FPN [38], PSPNet [39], DeepWaterMap [30], and proposed SFPNet for performance comparison.

4.3 Training

We run our experiments for 100 epochs each. We used a batch size of 8, with an initial learning rate of 0.0001 in training. We used the combined loss L_c function that can be represented as:

$$L_c = L_f + L_d + L_{spr} \tag{4}$$

where L_f is focal loss, L_d is dice loss, and L_{spr} is superpixel reconstruction loss.

4.4 Results

We used precision (P), recall (R), F1 score (F_s), overall accuracy (A_c), and intersection over union (IoU) for quantitative comparison of our models with others. P, R are calculated using true positive (tp), true negative (tn), false positive (fp), false negative (fn).

In Figs. 2 and 3, we presented a visual comparison of the output of different models for benchmark datasets. To validate the effectiveness of the proposed work, we tested the developed model on an unannotated set of three datasets (UMSD) which are shown in Fig. 4.

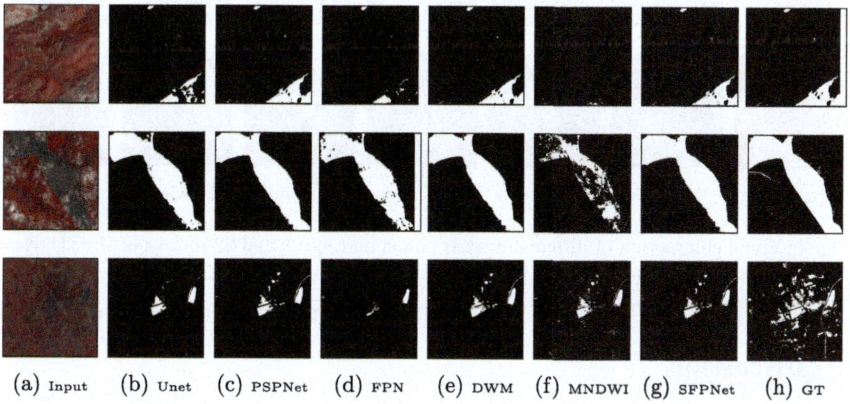

Fig. 2 Visual comparison of different model's output on GSWD. Here, DWM is DeepWaterMap, and GT is ground truth

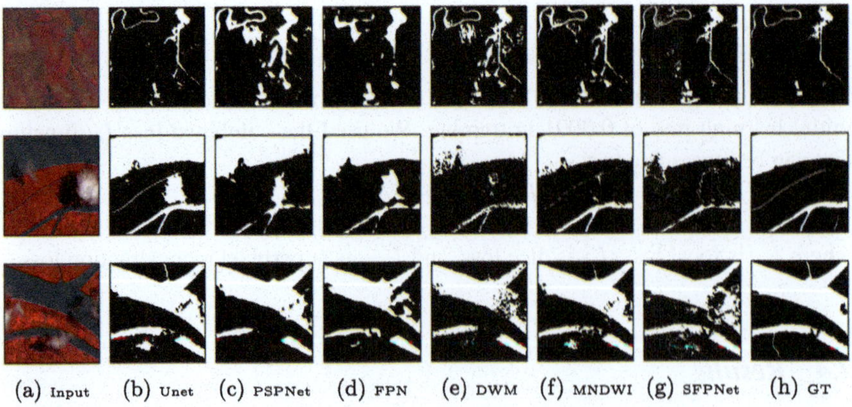

Fig. 3 Visual comparison of different model's output on ESWKB dataset. Here, DWM is DeepWaterMap, and GT is ground truth

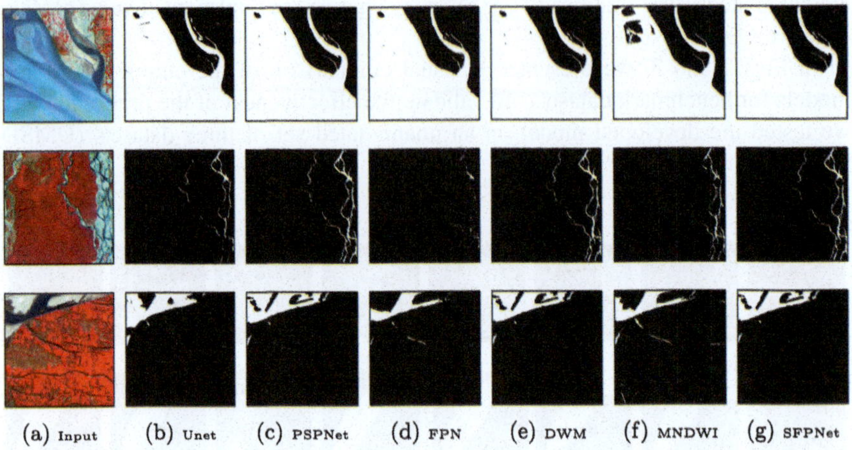

Fig. 4 Visual comparison of different models output on L8, L9, and S2 images of UMSD. Here, DWM is DeepWaterMap, and GT is ground truth

5 Discussion

5.1 Quantitative Study

In Table 1, we compared the performance of our model with other state-of-art methods. Unet achieved the F1 Score of 0.88 with an IoU value of 0.79. FPN improved in IoU with the same F1 Score. PSPNet showed improvement in the IoU value of 0.81 and F1 Score of 0.89. DeepWaterMap produced an F1 Score of 0.90 and IoU of 0.83. Our proposed SFPNet had an F1 score of 0.93 and IoU of 0.86, better than all other methods.

Table 1 Quantitative comparison of the performance of different models on GSWD

Method used	Total bands	P	R	F_s	IoU
MNDWI	2	0.81	0.72	0.76	0.65
UNet [7]	6	0.91	0.85	0.88	0.79
FPN [38]	6	0.92	0.83	0.88	0.80
PSPNet [39]	6	0.95	0.86	0.89	0.81
WatNet [33]	6	0.91	0.88	0.90	0.83
DeepWaterMap v2 [30]	6	0.95	0.87	0.90	0.83
SFPNet	6	0.97	0.88	0.93	0.86

Table 2 Quantitative comparison of the performance of different models for Earth Surface Water Dataset

Method used	Total bands	P	R	F_s	IoU
MNDWI	2	0.82	0.85	0.84	0.79
PSPNet [39]	6	0.98	0.95	0.96	0.86
FPN [38]	6	0.98	0.96	0.97	0.88
UNet [7]	6	0.98	0.97	0.98	0.90
DeepWaterMap v2 [30]	6	0.98	0.97	0.98	0.92
WatNet [33]	6	0.97	0.97	0.98	0.93
SFPNet	6	0.98	0.99	0.99	0.95

Table 2 contains a quantitative performance comparison on the ESWKB dataset. PSPNet [39] got F1 Score and IoU of 0.96 and 0.86 which is improved by FPN [38]. Unet [7] improved the F1 Score and IoU further. DeepWaterMap [30] achieved an F1 Score of 0.98 with an IoU value of 0.92. Our proposed SFPNet achieved an F1 Score of 0.99 and IoU score of 0.95 with improvement in both scores.

5.2 *Ablation Study*

Table 3 presented an ablation study for different modules used in the proposed work on GSWD. First, we implemented a network with an encoder-decoder backbone. It produced an F1 Score of 0.87. We enhance the connectivity in the encoder-decoder network to make it a high-resolution backbone. It improves the F1 Score to 0.89. Motivated by this improvement, we used this high-resolution backbone in the rest of the experiments. We added a channel and spatial attention module in this backbone, resulting in an F1 Score of 0.90. Further, we improved this attention block with Pixel Context Block and achieved the result of 0.92. The addition of Superpixel Block improved the performance to 0.93.

Table 3 Ablation study. Here, ED is encoder-decoder, H is high resolution, Att is the attention module, CB is the proposed Pixel Context Block, and SB is Superpixel Block

Module used	P	R	F_s
ED-Backbone	0.88	0.86	0.87
H-Backbone	0.89	0.88	0.89
H-Backbone + Att	0.91	0.88	0.90
H-Backbone + PCB	0.96	0.88	0.92
H-Backbone + PCB + SB	0.97	0.88	0.93

Table 4 Comparison study of the performance of the proposed model and different loss functions utilized

Loss function used	F_s	IoU
Cross Entropy	0.90	0.80
Focal	0.91	0.81
Dice	0.91	0.83
Focal + Dice	0.92	0.84
Focal + Dice + Superpixel	0.93	0.85

In Table 4, we compared various loss functions utilized in our study. We train our models with different loss functions in optimization steps. The detailed experiment results showed that the loss function, which combines focal, dice, and superpixel loss, gives better results.

5.3 Qualitative Study

GSWD In Fig. 2, visual comparative results are presented. All outputs (Unet, PSPNet, FPN, DeepWaterMap, MNDWI, and SFPNet) are compared with the ground truth provided in the test set of GSWD. Here, our proposed model is able to produce better or equivalent results compared to other models. It also produced good results in cloudy conditions.

ESWKB Fig. 3, contains comparative qualitative results for the ESKWB dataset. Outputs produced by Unet, PSPNet, FPN, DeepWaterMap, MNDWI, and SFPNet are compared with ground truth. The results produced by SFPNet is giving more refiner results compared to others. In inputs with cloud, SFPNet gave comparative better results.

UMSD In Fig. 4, visual outputs of Unet, PSPNet, FPN, DeepWaterMap, MNDWI, and SFPNet are compared with each other. These figures show that the proposed SFPNet can produce better segmentation results visually for surface water mapping.

6 Conclusion

We proposed a novel architecture called SFPNet with Pixel Context Block and Superpixel Block for surface water mapping. The proposed approach requires fewer parameters and can produce better results compared to existing methods. Pixel Context Block improved the segmentation output with the help of pixel relation mapping. Superpixel Block helped in feature representation with pixel-superpixel association mapping. A detailed survey of recent literature for surface water mapping is presented. We showcased a comprehensive ablation study for a different component of SFPNet to demonstrate the importance of each. SFPNet produced better qualitative and quantitative results compared to other methods.

In future work, this work will be extended for multimodal (multispectral with synthetic aperture radar inputs) surface water mapping. It will further improve the all-weather surface water mapping capability.

References

1. McFEETERS SK (1996) The use of the normalized difference water index (ndwi) in the delineation of open water features. Int J Remote Sens 17(7):1425–1432
2. Gao BC (1996) Ndwi-a normalized difference water index for remote sensing of vegetation liquid water from space. Remote Sens Environ 58(3):257–266
3. Xu H (2006) Modification of normalised difference water index (ndwi) to enhance open water features in remotely sensed imagery. Int J Remote Sens 27(14):3025–3033
4. Feyisa Gudina L, Henrik M, Rasmus F, Proud Simon R (2014) Automated water extraction index: a new technique for surface water mapping using landsat imagery. Remote Sens Environ 140:23–35
5. Wu W, Li Q, Zhang Y, Du X, Wang H (2018) Two-step urban water index (tsuwi): a new technique for high-resolution mapping of urban surface water. Remote Sens 10(11)
6. Krizhevsky A, Sutskever I, Hinton GE (2012) Imagenet classification with deep convolutional neural networks. In: Proceedings of the 25th international conference on neural information processing systems, NIPS'12, vol 1, Red Hook, NY, USA. Curran Associates Inc, pp 1097–1105
7. Ronneberger O, Fischer P, Brox T (2015) U-net: convolutional networks for biomedical image segmentation. In: Navab N, Hornegger J, Wells WM, Frangi AF (eds) Medical image computing and computer-assisted intervention—MICCAI 2015. Springer International Publishing, Cham, pp 234–241
8. He K, Zhang X, Ren S, Sun J (2016) Deep residual learning for image recognition. In: 2016 IEEE Conference on Computer Vision and Pattern Recognition (CVPR), pp 770–778
9. Gao H, Zhuang L, Geoff P, van der Maaten Laurens, Weinberger Kilian Q (2022) Convolutional networks with dense connectivity. IEEE Trans Pattern Anal Mach Intell 44(12):8704–8716
10. Wang J, Sun K, Cheng T, Jiang B, Deng C, Zhao Y, Liu D, Mu Y, Tan M, Wang X, Liu W, Xiao B (2021) Deep high-resolution representation learning for visual recognition. IEEE Trans Pattern Anal Mach Intell 43(10):3349–3364
11. Woo S, Park J, Lee J-Y, Kweon IS (2018) Cbam: convolutional block attention module. In: Ferrari V, Hebert M, Sminchisescu C, Weiss Y (eds) Computer vision—ECCV 2018. Springer International Publishing, Cham, pp 3–19

12. Ma L, Liu Y, Zhang X, Ye Y, Yin G, Johnson BA (2019) Deep learning in remote sensing applications: a meta-analysis and review. ISPRS J Photogrammetry Remote Sens 152:166–177
13. Cheng G, Yang C, Yao X, Guo L, Han J (2018) When deep learning meets metric learning: remote sensing image scene classification via learning discriminative CNNs. IEEE Trans Geosci Remote Sens 56(5):2811–2821
14. Chouhan A, Sur A, Chutia D (2022) Aggregated context network for semantic segmentation of aerial images. In: 2022 IEEE International Conference on Image Processing (ICIP), pp 1526–1530
15. Fu G, Liu C, Zhou R, Sun T, Zhang Q (2017) Classification for high resolution remote sensing imagery using a fully convolutional network. Remote Sens 9(5)
16. Chouhan A, Sur A, Chutia D (2022) Drmnet: difference image reconstruction enhanced multiresolution network for optical change detection. IEEE J Selected Topics Appl Earth Observations Remote Sens 15:4014–4026
17. Jampani V, Sun D, Liu M-Y, Yang M-H, Kautz J (2018) Superpixel sampling networks. In: Computer vision—ECCV 2018: 15th European conference, Munich, Germany, 8–14 Sept 2018, Proceedings, Part VII. Springer, Heidelberg, pp 363–380
18. Chouhan A, Jindal N, Sur A, Chutia D, Aggarwal SP (2023) Edcgan: encoder decoder based conditional gan for sar to optical image translation. In: Proceedings of the thirteenth Indian conference on computer vision, graphics and image processing, ICVGIP '22, New York, NY, USA. Association for Computing Machinery
19. Mi L, Chen Z (2020) Superpixel-enhanced deep neural forest for remote sensing image semantic segmentation. ISPRS J Photogrammetry Remote Sens 159:140–152
20. Chouhan A, Chutia D, Aggarwal SP (2023) Attentive decoder network for flood analysis using sentinel 1 images. In: 2023 international conference on communication, circuits, and systems (IC3S), pp 1–5
21. Furkan I, Bovik Alan C, Paola P (2017) Surface water mapping by deep learning. IEEE J Selected Topics Appl Earth Observations Remote Sens 10(11):4909–4918
22. Chen Y, Fan R, Yang X, Wang J, Latif A (2018) Extraction of urban water bodies from high-resolution remote-sensing imagery using deep learning. Water 10(5)
23. Fang W, Wang C, Chen X, Wan W, Li H, Yu SZ, Liu FB, Yang H (2019) Recognizing global reservoirs from landsat 8 images: a deep learning approach. IEEE J Selected Topics Appl Earth Observations Remote Sens 12(9):3168–3177
24. Xu KY, Schaefer ZG, Feng J, Guan L, Fang H (2021) Deep-learning-based multispectral satellite image segmentation for water body detection. IEEE J Selected Topics Appl Earth Observations Remote Sens 14:7422–7434
25. Miao Z, Kun F, Sun H, Sun X, Yan M (2018) Automatic water-body segmentation from high-resolution satellite images via deep networks. IEEE Geosci Remote Sens Lett 15(4):602–606
26. Feng W, Sui H, Huang W, Xu C, An K (2019) Water body extraction from very high-resolution remote sensing imagery using deep u-net and a superpixel-based conditional random field model. IEEE Geosci Remote Sens Lett 16(4):618–622
27. Li L, Yan Z, Shen Q, Cheng G, Gao L, Zhang B (2019) Water body extraction from very high spatial resolution remote sensing data based on fully convolutional networks. Remote Sens 11(10)
28. Wang G, Wu M, Wei X, Song H (2020) Water identification from high-resolution remote sensing images based on multidimensional densely connected convolutional neural networks. Remote Sens 12(5)
29. Li M, Wu P, Wang B, Park H, Yang H, Wu Y (2021) A deep learning method of water body extraction from high resolution remote sensing images with multisensors. IEEE J Selected Topics Appl Earth Observations Remote Sens 14:3120–3132
30. Isikdogan Leo F, Alan B, Paola P (2020) Seeing through the clouds with deepwatermap. IEEE Geosci Remote Sens Lett 17(10):1662–1666
31. Poliyapram V, Imamoglu N, Nakamura R (2019) Deep learning model for water/ice/land classification using large-scale medium resolution satellite images. In: IGARSS 2019—2019 IEEE international geoscience and remote sensing symposium, pp 3884–3887

32. Liu Z, Chen X, Zhou S, Yu H, Guo J, Liu Y (2022) Dupnet: water body segmentation with dense block and multi-scale spatial pyramid pooling for remote sensing images. Remote Sens 14(21)
33. Luo X, Tong X, Hu Z (2021) An applicable and automatic method for earth surface water mapping based on multispectral images. Int J Appl Earth Observation Geoinf 103:102472
34. Ankush M, Yasir A, Munish B (2022) Mapping of water bodies from sentinel-2 images using deep learning-based feature fusion approach. Neural Comput Appl 35(12):9167–9179
35. Wang Y, Li Z, Zeng C, Xia G-S, Shen H (2020) An urban water extraction method combining deep learning and google earth engine. IEEE J Selected Topics Appl Earth Observations Remote Sens 13:769–782
36. Chollet F (2017) Xception: deep learning with depthwise separable convolutions. In: 2017 IEEE Conference on Computer Vision and Pattern Recognition (CVPR), pp 1800–1807
37. Jean-François P, Andrew C, Noel G, Belward Alan S (2016) High-resolution mapping of global surface water and its long-term changes. Nature 540:418–422
38. Kirillov A, Girshick R, He K, Dollár P (2019) Panoptic feature pyramid networks. In: 2019 IEEE/CVF Conference on Computer Vision and Pattern Recognition (CVPR), pp 6392–6401
39. Zhao H, Shi J, Qi X, Wang X, Jia J (2017) Pyramid scene parsing network. In: 2017 IEEE Conference on Computer Vision and Pattern Recognition (CVPR), pp 6230–6239

Identifying and Optimizing the Observation Strategy to Detect NEOs (Near-Earth Objects) Using Python

Prajwal Thakare

Abstract Near-Earth Objects (NEOs) are any little celestial bodies that have orbits with closer proximity to that with Earth. A celestial body can be called a NEO if its closest distance to the Sun, i.e. Perihelion is at least 1.3 Astronomical Units (AU), where 1 AU is approximately 150 million kilometers (Km). If a NEO crosses Earth's orbit at a distance of 140 m across, it can be called as a Potentially Hazardous Object (PHO). Consequently, a certain sub-set of NEOs advance our Earth quite nearly, and may cross our Earth's orbit penetrating the atmosphere to the exterior of the Earth, or even collide with it to create craters if they impact a landmass or tsunamis if they strike on the oceans. Continuous sky observations over the years have already recognized and classified over 30,000 and more NEOs over a period of two centuries. Recent NEO detection models propose that there are still hundred thousand of NEOs yet more to be discovered. The study focuses on tracing these objects, classifying them under categories and visualizing the detected NEOs to build an observation strategy for our telescopes to detect these objects in a day sky using Python programming.

Keywords Near-earth objects · Apparent magnitude · Python programming · Asteroids

1 Introduction

The solar system originates from the collapse of gas and dust into a helical ring-shaped nebula, leading to the formation of celestial planetesimals and eventually planets. Seeding conditions give rise to the clustering of molecules and dust particles, forming clumps that grow into celestial planetesimals. These planetesimals, initially acting as distinct grains, undergo growth by engulfing smaller ones, resisting attrition forces through electromagnetic and gravitational interactions. This process, known as accretion, governs the gravitational growth of planets. However, continuous disruptive processes hindered cosmic bodies like asteroids and meteorites from

P. Thakare (✉)
Department of Computer Engineering, COEP Technological University, Pune, India
e-mail: thakarepa21.comp@coeptech.ac.in

S. Bandyopadhyay et al. (eds.), *Intelligent Computing Systems and Applications*, Lecture Notes in Networks and Systems 1010,
https://doi.org/10.1007/978-981-97-5412-0_31

growing to the size of giant planets, resulting in their relatively small size. Composition studies of Near-Earth Objects (NEOs) have revealed that their anatomy ranges within that of the Main Asteroid Belt, suggesting a common origin for many NEOs.

NEOs refer to small celestial bodies with orbits that come into close proximity with Earth. A celestial body is classified as a NEO if its closest distance to the Sun (Perihelion) is at least 1.3 Astronomical Units (AU), where 1 AU is approximately 150 million kilometers. NEOs are often the result of gravitational perturbations by nearby planets, causing their orbits to intersect with Earth. While a significant number of NEOs originated in the Main Asteroid Belt, some may have migrated from the outer solar system due to the presence of the Jovian Trojan and Hilda groups. Although approximately 15% of NEOs are believed to have cometary origins, their anatomical and physical characteristics encompass a wide range, including those found in the Main Belt. These remnants provide valuable insights into the early solar system, preserving data that has been erased from planets through processes like accretion, tectonism, volcanism, and metamorphism.

Historical events such as the Meteor Crater (Barringer Crater) in Arizona, the Tunguska event in Siberia, and the Chelyabinsk meteorite explosion in Russia serve as reminders of the potential impacts of NEOs on Earth. NEO impacts can lead to regional landslides, tsunamis, earthquakes, hurricanes, and other catastrophic events. In rare cases, larger NEO impacts have the potential for global devastation, posing a significant threat to human lives and the environment.

This study aims to address the issue of NEO detection and classification effectively, focusing on shape, size, intensity, and assessing the potential threat of NEOs crossing Earth's orbit using statistical calculations based on magnitude (absolute and apparent) and phase angle.

2 Literature Review

The literature review encompassed various studies and research articles in the selected domain. Notable reviews are summarized below.

William F. Bottke Jr. et al. (2000) analyzed the orbital and size dispersion of Near-Earth Asteroids (NEAs) and predicted the detection biases and size dispersion associated with asteroids on their paths. They estimated approximately 900 NEAs with an absolute magnitude below 18, with difficulties in tracking NEAs with tilted orbits [1].

Germano D'Abramo et al. (2001) developed a non-deterministic model for predicting the asteroid community by comparing the rate of re-observations of known objects with fresh observations over a specific time period. They successfully forecasted the community within an absolute magnitude range of $13.5 < H < 20.0$ [2].

Mikael Granvik, Jeremie Vaubaillon, and Robert Jedicke (2011) calculated the characteristics of Earth's temporarily captured natural satellites from the NEO community and estimated the orbit and size dispersion for these asymmetrical satellites. They proposed the presence of meter-sized or larger NEOs temporarily orbiting the Earth [3].

Alan W. Harris and Germano D'Abramo (2015) described a methodology for predicting the size-frequency distribution (SFD) of NEOs, particularly asteroids, based on re-observation ratios. They highlighted the efficiency of land-based surveys in observing subsets of NEOs, such as Interior to Earth Asteroids (IEOs), Potentially Hazardous Asteroids (PHAs), and Earth Crossing Asteroids (ECAs) [4].

Toshinori Ikenaga, Yohei Sugimoto et al. (2018) focused on Chelyabinsk-class NEOs, which resulted from the Chelyabinsk Meteor Event in Russia in 2013. They proposed an "eleventh-hour" NEO impact warning system and compared the detection capabilities of the Sun-Earth Lagrange point 1 (SEL 1) and Artificial Equilibrium point (AEP). The AEP demonstrated the potential to detect hazardous NEOs before impact [5].

Mikael Granvik et al. (2018) investigated orbital distributions and debiased absolute magnitudes of NEOs to understand their populations. They developed a four-dimensional model representing the relaxed state distributions of eccentricity, intrinsic luminosity, inclination, and half the major axis. Their findings provided insights into the observed NEO population [6].

3 Research Methodology

3.1 Architectural Design

The research methodology consisted of four main phases, as depicted in Fig. 1.

Data Pre-processing NEOs data obtained from NEODyS was collected, and a database was created using SQLite. This facilitated the understanding of various attributes related to the current NEO population.

NEOs Classification The population was classified into four groups based on specific attributes. Mathematical calculations were employed to determine the sizes and brightness of the NEOs.

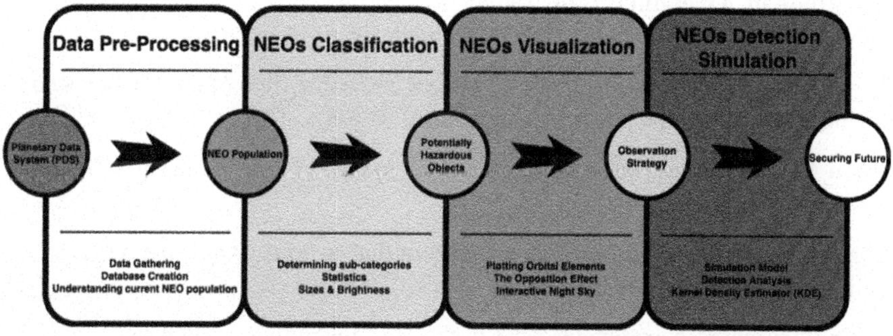

Fig. 1 Architectural design

NEOs Visualization Orbital elements were plotted using libraries like Matplotlib and Seaborn, which provided visual tools such as PairGrid Plot, Scatter Plots, Histograms, and Density Plots. Additionally, an interactive night sky projection of NEOs on the Aitoff Azimuthal Grid was created using Python's ipywidget.

NEOs Detection Simulation A simulation model was developed, tested, and analyzed to detect NEOs by filtering them based on their magnitudes and time. A Kernel Density Estimator or Population Map was constructed to gain insights into the dispersion of NEOs around the Sun and Opposition Point.

3.2 The Database

The NEODyS (Near-Earth Object Dynamic Site) served as the main repository for generating the project's database. NEODyS provides information and services for all Near-Earth Asteroids (NEAs). Each NEA has a dynamically generated home page that offers easily accessible content, including search functionality.

The research work focused on a total of 32,200 known Near-Earth Objects (NEOs) obtained from the NEODyS repository. The relevant data was downloaded and stored on an SQL Database Server for easy access. The server mirrored the NEOs and their data from the NEODyS website.

The data was organized using a placeholder dictionary with 13 attributes:

Name The identification number or name assigned to a NEO.

Epoch (MJD) The reference time point expressed in Modified Julian Date (MJD).

Semi-major Axis The longest semidiameter or half of the major axis, expressed in Astronomical Units (AU).

Eccentricity The measure of an orbit's deviation from a perfect circle.

Inclination Degree The angle between an orbital stage or axis and a reference stage, expressing the tilt of the object's orbit.

Longitude of Ascending Node The angle measured in a reference plane between the direction of the ascending node and a predetermined reference direction.

Argument of Periapsis The angle measured in the direction of motion between the ascending node and the periapsis of the orbiting body.

Mean Anomaly The portion of an orbit's period that has transpired since the revolving body passed periapsis.

Absolute Magnitude The apparent magnitude that an object would possess if viewed from a precise distance of 10 parsecs.

Slope Parameter A parameter used in the H-G magnitude system to simulate the opposition effect of asteroids.

Aphelion The point in an object's orbit farthest from the Sun, expressed in AU.

Perihelion The point in an object's orbit closest to the Sun, expressed in AU.

NEOClass Categorization of NEOs into Atiras, Atens, Apollos, and Amors based on their Aphelion distance, Perihelion distance, and Semi-Major Axis.

3.3 Statistics

NEO Groups and Classification Near-Earth Asteroids (NEAs) are classified into four groups (Atira, Aten, Apollo, and Amor) based on their semi-major axes, perihelion distances, and aphelion distances. Calculations were performed to determine the classification of NEOs within these groups based on specific orbital parameters.

$$\text{Perihelion}(\text{Closest}\textit{Point}), d = a(1 - e) \tag{1}$$

$$\text{Aphelion}(\text{FarthestPoint}), d = a(1 + e) \tag{2}$$

where e = Eccentricity (measure of the "roundness" of an orbit) and a = Semi-major Axis The further classification of NEOs was done based on the outcomes of above calculations.

- Atira asteroids have orbits entirely inside Earth's orbit (semi-major axis smaller than 0.983 AU).
- Atens cross Earth's orbit and have a semi-major axis smaller than 1 AU (with aphelion distance greater than 0.983 AU).
- Apollos intersect Earth's orbit and have a semi-major axis greater than 1 AU (with perihelion distance less than 1.017 AU).
- Amors have orbits outside of Earth's orbit (with perihelion distance larger than 1.017 AU and smaller than 1.3 AU).

Diameters The diameter of most NEOs is estimated using their brightness and a representative asteroid surface reflectivity (albedo) assumed to be 15%. The diameter is calculated using the absolute magnitude and albedo given by:

$$D = \left(\frac{1329}{\sqrt{\text{albedo}}}\right) \times 10^{-\frac{H}{5}} \tag{3}$$

where D is the computed diameter in km, H is the absolute magnitude and Albedo, where a default value of 0.15 is assumed.

Apparent Magnitude Apparent magnitude is a measure of the brightness of a celestial object as seen from Earth. It is influenced by factors such as the object's inherent luminosity, distance from Earth, and interstellar dust. A brighter object has a lower magnitude number since the scale is reverse logarithmic.

The opposition effect is the brightening of a rough surface or a substance with many particles when lit from directly behind the viewer (also known as the opposition surge, opposition spike, or Seeliger effect). The phrase is most frequently used in astronomy, where it generally refers to a celestial body's rapid, significant increase in brightness when its phase angle of viewing nears zero. Based upon this concept, an opposition point is determined at 180° to the sun to get the population of NEOs with maximum brightness and also to compare the brightness of an asteroid at different positions.

Thc apparent magnitude is calculated by the formula [7]:

$$V = H(\alpha) + 5\log(r\Delta) \tag{4}$$

where V = Apparent Magnitude; H(α) = Reduced Magnitude; r=distance of the heavenly object from the Sun; Δ = distance of the heavenly object from the Earth; α = Phase Angle (Sun/Heavenly Object/Earth angle) in degrees.

However, the reduced magnitude is calculated by the formula:

$$H(\alpha) = H - 2.5\log\left[(1 - G)\phi_1(\alpha) + G\phi_2(\alpha)\right] \tag{5}$$

where H = Absolute Magnitude; G = Slope Parameter = 0.15 assumed (surge in brightness); ϕ = Phase Function.

Whereas the Phase Function is given by the formula:

$$\phi_i(\alpha) = \exp\left\{-A_i\left(\tan\frac{1}{2}\alpha\right)^{B_i}\right\} \tag{6}$$

for $i = 1$ or 2, $A_1 = 3.33$, $A_2 = 1.87$, $B_1 = 0.63$, and $B_2 = 1.22$, where A and B are constant factors.

For an instance, Apparent Magnitude of an asteroid named Albert (719) was calculated for August 14, 2023 at 00:00:00 and was found out to be 20.71 which is 0.06 less than the true value due to the change in atmosphere and presence of air masses while computing the observed value from actual parameters. Hence, the accuracy of the calculations turns out to be in the range of 99–100%.

3.4 *Visualization Analysis*

PairGrid Plot A PairGrid plot was created using seaborn to visualize various orbital elements. The upper triangle displayed histograms, the lower triangle showed kernel density plots, and the diagonal contained histograms with KDE curves. This plot provided a comprehensive view of the relationships between different orbital elements, as shown in Fig. 2.

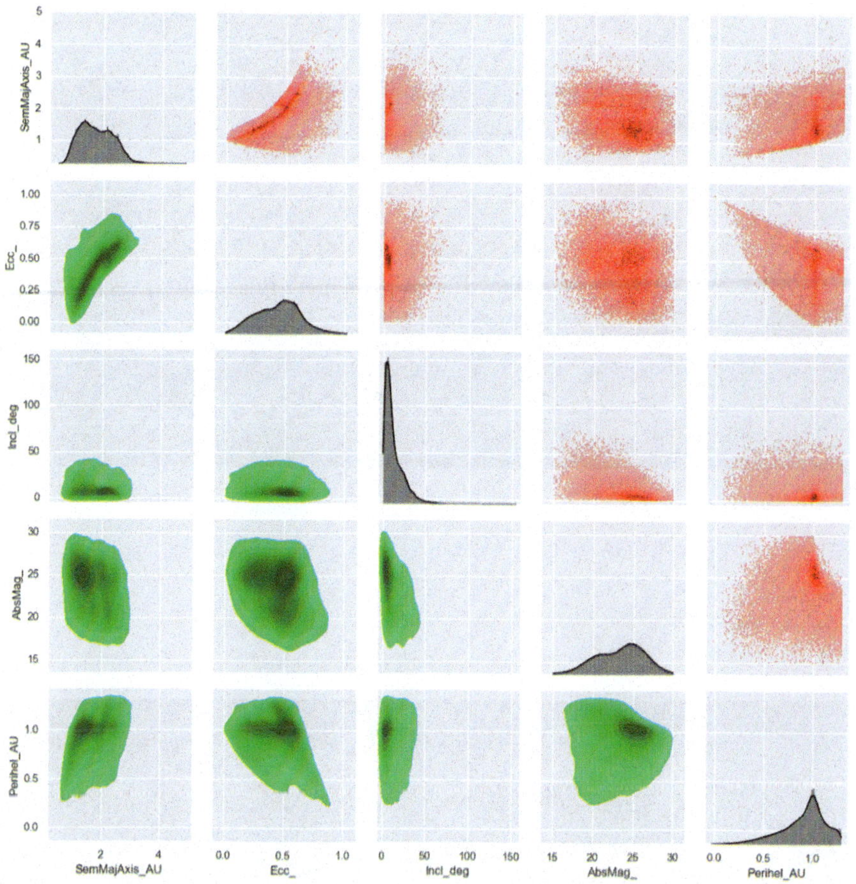

Fig. 2 PairGrid plot for orbital elements

Aitoff Azimuthal Projection An Aitoff Projection sky plot was generated using Python to project a subset of NEOs based on specific criteria such as apparent magnitude, absolute magnitude, and Earth-to-NEO distance. This plot helped identify regions of interest in the sky, as shown in Fig. 3.

Specialized Projection Using IPywidget An interactive sky map was developed using IPywidget. Users could adjust parameters such as date, NEO class, apparent magnitude, absolute magnitude, and angular distance between NEOs and the Sun. The sky map dynamically updated based on the selected parameters, allowing for a customized and interactive visualization experience, as represented in Fig. 4.

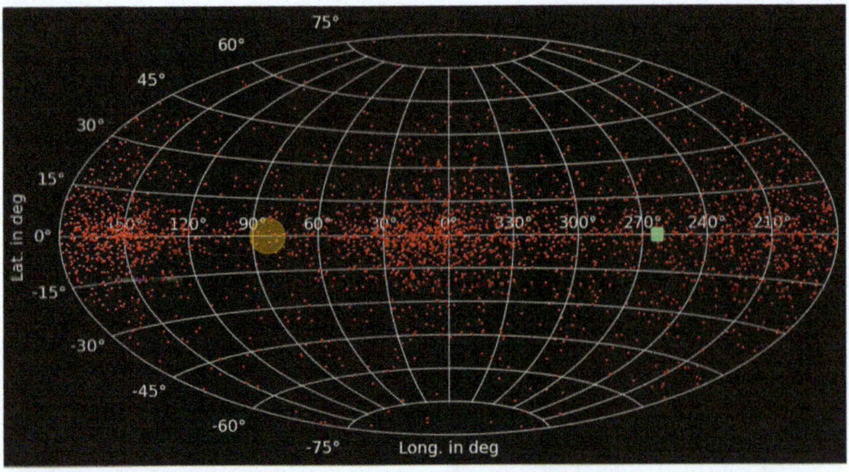

Fig. 3 Aitoff azimuthal Projection of NEOs based on certain parameters

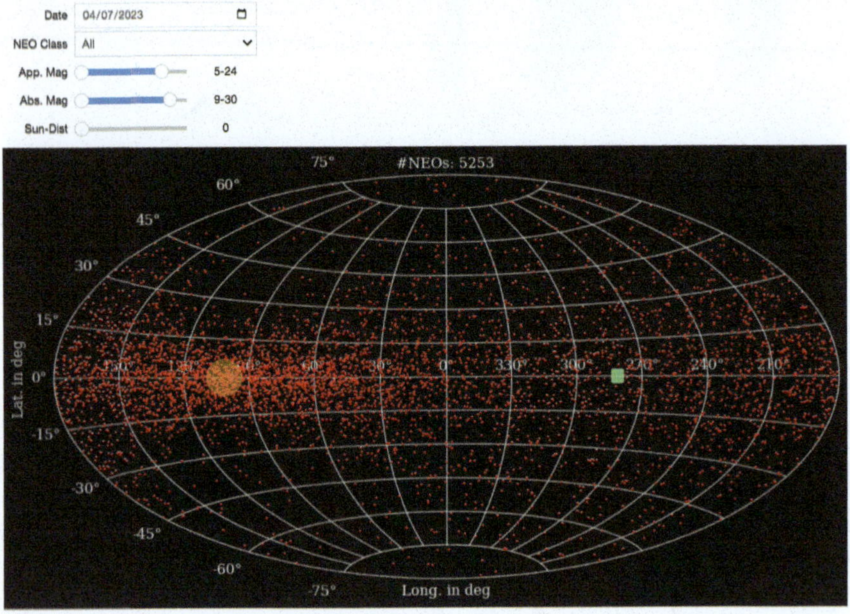

Fig. 4 An interactive night sky using IPywidget

3.5 NEO Simulation Model

Following are the main steps involved in building the simulation model for detecting NEOs:

1. Establishing a connection and accessing the NEO database.
2. Loading the three required SPICE (Spacecraft, Planet, Instrument, Camera-matrix, Events) kernels which fall under NASA's NAIF (Navigation and Ancillary Information Facility), viz. LSK (Leap Second Kernel), SPK (Spacecraft and Planet Kernel) and PCK (Planetary Constants Kernel) and are used to aid NASA engineers in modelling, planning, and carrying out procedures necessary to execute planetary exploration missions, as well as arranging and analysing scientific findings from space-borne instruments [8].
3. Converting units from AU to Km and from degree to radian.
4. Setting the start date and time for the detection process of NEOs. In this case, January 01, 2023 at 00:00:00 was set.
5. Setting the detection range of 20°C which is a circle around the opposition point in the detection field.
6. Setting up a minimum detection magnitude threshold of 24.0 for detecting the NEOs.
7. Observation steps was set to 24 h, i.e. observation being done once per night.
8. Observation range was set for 1 year which was then converted into hours.
9. Iteratively computing the position vector of the Earth as well as each NEO as seen from the Sun.
10. Calculating the apparent magnitude of each NEO and also the angular distance between NEO and the opposition point and simulating it in a asynchronic loop.
11. If the apparent magnitude and the angular distance is less than the minimum detection magnitude threshold and the detection range circle, the NEO was considered to be detected else undetected.
12. Lastly, the detected as well as undetected NEOs were added in a separate parquet file for further analysis.

4 Results and Discussion

4.1 NEO Detection

Detected NEOs A total of 1481 NEOs were detected, starting from January 01, 2023 at 00:00:00, as depicted in Fig. 5.

	Name	SemMajAxis_AU	Ecc_	Incl_deg	LongAscNode_deg	ArgP_deg	AbsMag_	SlopeParamG_	Aphel_AU	Perihel_AU	NEOClass	utc_of_detec
0	4015	2.624470	0.631659	2.799195	266.772730	95.445525	16.13	0.15	4.282240	0.966700	Apollo	2023 01T00:0
1	6456	2.192161	0.409306	8.224502	313.577398	347.139241	15.96	0.15	3.089426	1.294896	Amor	2023 01T00:0
2	13553	2.188563	0.464027	5.876637	193.462193	109.985545	16.47	0.15	3.204116	1.173010	Amor	2023 01T00:0
3	31346	2.029339	0.429847	5.962779	299.728324	350.505656	17.16	0.15	2.901643	1.157034	Amor	2023 01T00:0
4	65803	1.642544	0.383282	3.414107	72.991050	319.561622	18.14	0.15	2.272102	1.012987	Apollo	2023 01T00:0
...	...	...	...	...	...	...	...	...	...	...	...	
1476	2016VO1	0.982582	0.428697	1.615825	129.741589	131.951690	23.08	0.15	1.403811	0.561352	Aten	2023 30T00:0
1477	35396	1.442368	0.483702	4.088450	210.664753	103.018351	16.91	0.15	2.140045	0.744692	Apollo	2023 31T00:0
1478	517829	2.134458	0.614060	8.524505	81.016395	167.525312	18.38	0.15	3.445143	0.823772	Apollo	2023 31T00:0
1479	2014KE91	2.318027	0.477001	6.050118	195.133438	41.959110	18.31	0.15	3.423728	1.212326	Amor	2023 31T00:0
1480	2019NO3	2.851592	0.701670	45.163141	102.355120	72.745123	19.33	0.15	4.852470	0.850714	Apollo	2023 31T00:0

1481 rows × 12 columns

Fig. 5 1481 detected NEOs

	Name	SemMajAxis_AU	Ecc_	Incl_deg	LongAscNode_deg	ArgP_deg	AbsMag_	SlopeParamG_	Aphel_AU	Perihel_AU	NEOClass
1	719	2.637175	0.547045	11.576011	183.856996	156.250357	15.53	0.15	4.079829	1.194522	Amor
5	1566	1.078127	0.826869	22.799566	87.951655	31.441775	16.38	0.15	1.969597	0.186657	Apollo
7	1620	1.245673	0.335535	13.336359	337.172132	276.961182	15.30	0.15	1.663640	0.827706	Apollo
8	1627	1.863588	0.396465	8.450607	133.118779	167.807692	13.16	0.60	2.602435	1.124742	Amor
11	1863	2.259447	0.605831	18.398254	346.423780	268.071599	15.41	0.15	3.628291	0.890603	Apollo
...	...	...	...	...	...	...	...	...	...	...	...
32193	2023LR	2.085605	0.738752	1.803986	124.668309	17.143153	22.75	0.15	3.626350	0.544859	Apollo
32194	2023LS	0.930538	0.147717	0.269538	71.767194	52.067157	29.95	0.15	1.067995	0.793082	Aten
32195	2023LU	2.891634	0.640667	15.307266	102.913838	243.323037	19.30	0.15	4.744210	1.039059	Amor
32199	2023LY	2.664641	0.608992	4.489038	71.584706	142.392521	23.21	0.15	4.287387	1.041895	Amor
32200	2023LZ	1.177027	0.190265	24.157132	83.200798	231.181839	26.38	0.15	1.400974	0.953080	Apollo

30720 rows × 11 columns

Fig. 6 30,720 undetected NEOs

Undetected NEOs A total of 30,720 NEOs remained undetected, starting from January 01, 2023 at 00:00:00, as shown in Fig. 6.
The combined count of detected and undetected NEOs remained the same at 32,201, which matches the original database.

4.2 *Detection Analysis*

To analyse the detection results, the simulated parquet files were loaded and certain visualizations were then deduced:

Detectability with Respect to NEO Classes The detected and undetected NEOs were categorized based on their classes, and a detectability ratio was calculated (refer to Table 1).

Table 1 Detectability w.r.t. NEO classes

Sr. No.	NEO class	Detect count	Undetect count	Detect ratio
0	Amor	598	10,899	0.0549
1	Apollo	805	17,337	0.0464
2	Aten	77	2440	0.0316
3	Others	1	11	0.0909

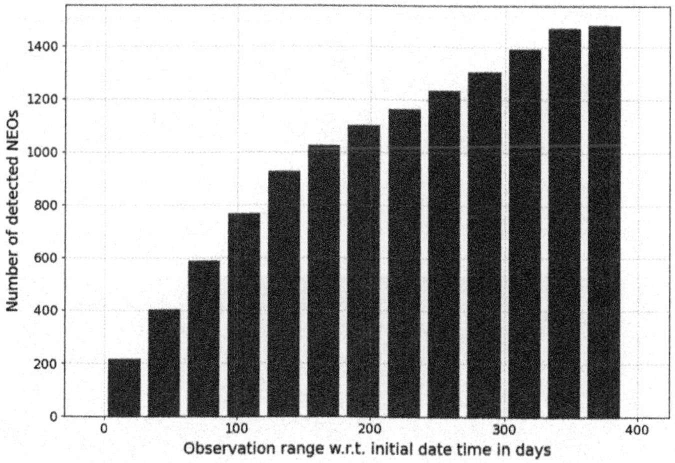

Fig. 7 Detection history of all detected NEOs

Detection History A cumulative plot was created to show the number of detected NEOs in the opposition area over a 12-month period, as illustrated in Fig. 7.

Detection History for Larger NEOs An overlapping histogram depicted the detection of larger NEOs (absolute magnitude limit of 18) over the same 12-month period, as illustrated in Fig. 8.

Distribution with Respect to Orbital Elements A scatter plot was generated to illustrate the distribution of smaller NEOs (absolute magnitude over 24) with respect to orbital elements such as eccentricity and perihelion. The plot indicated a higher density of NEOs at a perihelion of 1 AU, signifying their favorable geometric positions relative to Earth, as outlined in Fig. 9.

Kernel Density Estimator (KDE) The KDE plot displayed the density of NEOs per square degree, highlighting the concentration of NEOs between the Sun and the opposition point, as well as around the opposition point, as depicted in Fig. 10.

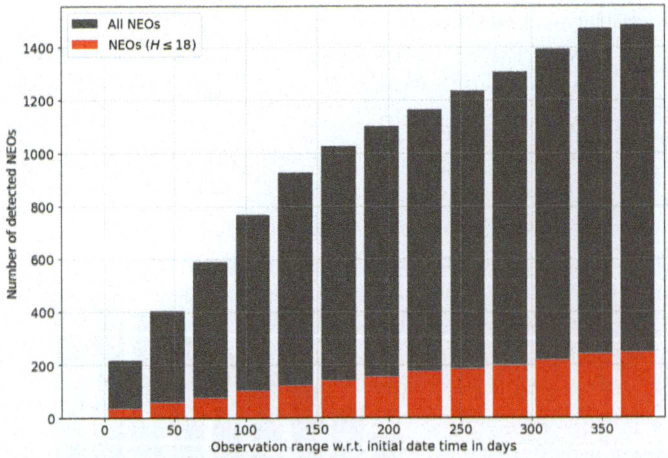

Fig. 8 Detection history of larger NEOs

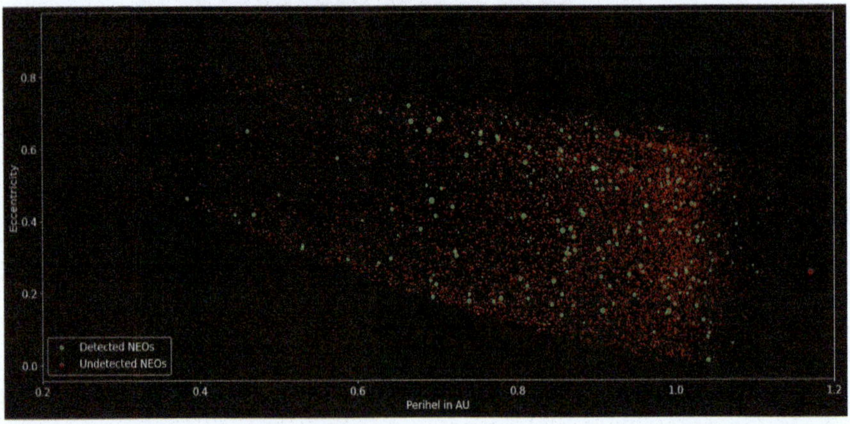

Fig. 9 Scatter plot over orbital elements

5 Conclusion

The NEO (Near-Earth Object) detection project aims to identify and track potentially hazardous asteroids and comets that come close to Earth. The project involves analyzing data on NEOs from a database and using simulation techniques to understand detection capabilities and potential biases. The cumulative histogram of detected NEOs over time shows an increasing number of detections as the observation range progresses. This indicates that the NEO detection project has been successful in identifying NEOs. The detection history of NEOs varies depending on their size. Larger NEOs tend to be detected earlier compared to smaller ones. This suggests that the detection project has a higher success rate in identifying larger NEOs. The arrange-

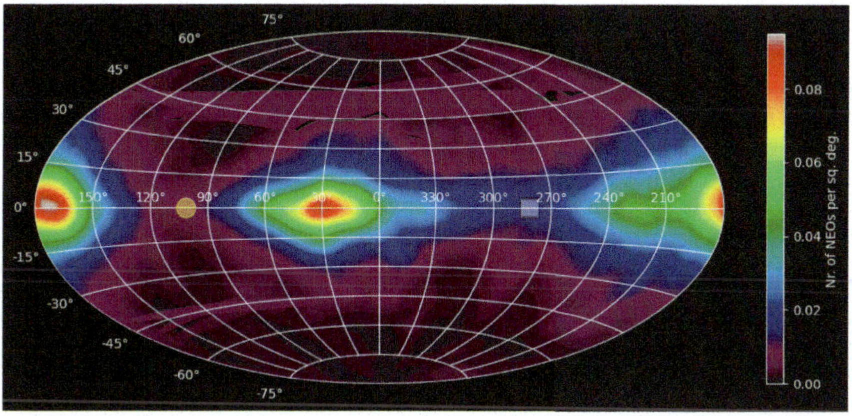

Fig. 10 Population density plot

ment of orbital elements for detected and undetected NEOs shows the relationship between these parameters and the likelihood of detection. It highlights potential patterns or clusters in the data and can help identify regions where detections are more or less likely to occur. Further analysis using Kernel-Density Estimators and considering different observation regions can enhance the quantitative understanding of detection biases in different settings and observation regions. Thus, the NEO detection project plays a crucial role in identifying and monitoring NEOs. By analyzing simulation results and addressing biases, it can improve detection capabilities and contribute to our knowledge of NEO populations, ultimately enhancing planetary defence efforts.

6 Future Scope

The research progresses over a known NEO distribution which has certain biases. The model would be considered more accurate and efficient if it is compared with an un-biased and simulated NEO population. We need to redo the simulations based upon the already simulated NEOs hopefully with an un-biased model to determine the performance of our observation strategy.

Future research can focus on developing advanced detection algorithms that take into account various factors such as object size, shape, reflectivity, and motion patterns. Machine learning and artificial intelligence techniques can be employed to improve the accuracy and efficiency of NEO detection.

NEO detection is a global endeavour, and international collaboration can significantly enhance our capabilities. Sharing data, resources, and expertise among various space agencies, research institutions, and observatories can lead to more comprehensive and efficient NEO detection efforts. Future research can focus on developing mitigation strategies and mission planning for potential NEO impact events. Increasing public awareness and understanding of NEOs, their potential risks, and ongoing detection efforts is crucial. By pursuing these future directions, the NEO Detection Project can make significant advancements in identifying, characterizing, and monitoring NEOs, ultimately improving our understanding of Near-Earth Objects and ensuring the safety and security of our planet.

Acknowledgements I would like to thank and express my sincere gratitude to Mr. Thomas Albin, former Space Scientist and currently Senior ML Engineer for allowing me to explore through this niche topic and domain of Space Science. Their invaluable input and continuous support in form of several informal conversations across mediums, informative repositories on GitHub [9] and factual videos posted on their YouTube channel [10], were instrumental in modelling the course of this research project.

References

1. Morbidelli A, Petit J-M, Bottke WF Jr, Jedicke R, Gladman B (2000) Understanding the distribution of near-earth asteroids. Science 288:1168–1170. https://doi.org/10.1126/science.288.5474.1168
2. Boattini A, Werner SC, Harris AW, D'Abramo G, Valsecchi GB (2001) A simple probabilistic model to estimate the population of near-earth asteroids. Icarus 153:214–217. https://doi.org/10.1006/icar.2001.6650
3. Vaubaillon J, Granvik M, Jedicke R (2011) The population of natural earth satellites. Icarus 218:262–277. https://doi.org/10.1016/j.icarus.2011.12.003
4. Harris AW, D'Abramo G (2015) The population of near-earth asteroids. Icarus 257:302–312. https://doi.org/10.1016/j.icarus.2015.05.004
5. Ceriotti M et al (2018) A concept of hazardous NEO detection and impact warning system. Acta Astronaut. https://doi.org/10.1016/j.actaastro.2018.06.058
6. Jedicke R et al (2018) Debiased orbit and absolute-magnitude distributions for near-earth objects. Icarus. https://doi.org/10.1016/j.icarus.2018.04.018
7. The H and G magnitude system for asteroids. Retrieved from https://britastro.org/asteroids/dymock4.pdf
8. NASA Jet Propulsion Laboratory (n.d.) Navigation and ancillary information facility (NAIF). Retrieved from https://naif.jpl.nasa.gov/naif/index.html
9. Albin T (n.d).Astroniz-YT-Tutorials [GitHub repository]. Retrieved from https://github.com/ThomasAlbin/Astroniz-YT-Tutorials
10. Albin T (n.d) Astroniz—Space science with python [YouTube channel]. Retrieved from https://www.youtube.com/c/Astroniz

An Energy-Conscious and Delay-Aware Path Allocation for Emergency Node in WBAN

S. Menaga, R. Vanithamani, A. Anandkumar, and N. Gomathi

Abstract Wireless Body Area Network (WBAN) offer a viable solution for continuous health monitoring and are receiving a lot of attention from academicians and researchers due to recent technological advancements. In WBAN, sensors are placed throughout the body to collect physiological data and transmit it to a sink node. Thus, transmitting medical information between sensor nodes and sink nodes while consuming minimal energy, causing little packet loss, and extending the lifespan of the network is challenging. A linear regression buffalo optimization algorithm (LRBOA) for WBAN is introduced in this paper to enable the timely transmission of vital patient data. The linear regression method is used to identify normal, aberrant, and vital patient data in LRBOA. After the classification, the protocol schedules the priority for classified data. Within the specified timeframe, high-priority patient data can be transmitted first. Energy, link dependability, bandwidth, and distance-based optimal pathways are needed for data transfer. The suggested algorithm could determine the best path based on several variables, including energy, link dependability, bandwidth, and distance. The highest-priority node data is finally sent to the sink node for further processing. In this way, priority-based, energy-aware data transmission is performed in WBAN. The proposed algorithm is tested using a MATLAB with a health monitoring dataset in WBAN.

Keywords WBAN · Energy conscious and priority-based data transmission · Linear regression buffalo optimization algorithm

S. Menaga (✉) · A. Anandkumar
Department of Electronics and Communication Engineering, Jai Shriram Engineering College, Tirupur, Tamil Nadu, India
e-mail: sri.ece09@gmail.com

R. Vanithamani
Department of Biomedical Instrumentation Engineering, School of Engineering, Avinashilingam Institute for Home Science and Higher Education for Women, Coimbatore, Tamil Nadu, India
e-mail: vanithamani_bmie@avinuty.ac.in

N. Gomathi
Department of Artificial Intelligence and Data Science, Karpagam Institute of Technology, Coimbatore, Tamil Nadu, India

S. Bandyopadhyay et al. (eds.), *Intelligent Computing Systems and Applications*, Lecture Notes in Networks and Systems 1010,
https://doi.org/10.1007/978-981-97-5412-0_32

1 Introduction

WBAN is quickly becoming one of the most important communication networks. Medical care services need to get more efficient to monitor and examine patients wherever they are. The WBAN is a network of physiological indicators of individual body, such as glucose, pulse, blood pressure, and others, that are measured by wireless body sensors, which then transmit the data to the medical server via a wireless connection facility. Patients with long-standing diseases require continuous monitoring. This is because limited resources make it difficult to continuously attend to every patient, both inside and outside the hospital. Due to the small sensor size, WBAN requires factors, such as low latency, low power consumption, reliability of data delivery, efficient bandwidth use, minimum delay, and maximum throughput. With these various challenges in BAN applications, it is extremely important to ensure the reliability of BAN healthcare applications [1–5]. Therefore, one of the essential elements for a prolonged network lifetime is energy efficiency. Additionally, the ensured delivery of data in WBANs exposes a critical issue with transmission dependability, particularly for emergency data. With the least amount of delay, this calls for optimal data delivery. S. Memon et al., developed TLD-Routing protocol (temperature, link-dependable, and delay-conscious) to improve the efficiency, loss ratio, and link stability in Wireless Body Sensor Networks (WBSN) using the strategy along with the Quality Of Service (QOS)-aware routing protocol [4]. More reliable routing with network coding technique called CEPRAN was developed in [6]. Although the designed approach improved the packet delivery ratio, data transmission time was not reduced. The optimum routing path was chosen using a genetic algorithm-based energy-efficient routing optimization approach that was developed in [7] to estimate distances between nodes under various conditions. But while the developed approach lengthens the network's lifespan, it neglects to take into account the many target functions for effective data transmission. In [8], the author provides information on the improvement of QOS and energy efficiency through single-hop and multi-hop communication. These types of data are handled by using the NEAT algorithm in terms of network lifetime and throughput, which has been implemented on RE-ATTEMT and CEMOB. A cluster head selection technique is proposed in [9] for selecting the best node. Communication within the wireless local gateways (WLG) and the hospital gateways (HG) became more reliable and efficient with an increase in nodes. Rajiv R. et al. proposed the P-MACRON algorithm for mobile-aware clustering routing that assigns fair weights to nodes to ensure high packet distribution [10]. In [11], the researchers developed the medium access control (MAC) schemes, which are implemented by multi-hop transmission, especially designed for BAN. The simulation result shows superiority in highly efficient slot utilization and prolonged life. It is related to multi-hop transmission using Minimum Spanning Tree (MST) distance-based, and according to R. Singla et al., exhausted sensor battery power is a major issue for the security and privacy of the network. For this complete adoption of security, privacy, and energy efficiency, this work proposes an Optimal Energy-Efficient Secure Routing (OEESR) protocol for providing the secured routing protocol for data transmission and network congestion to the most optimal route [12]. Khalid M. Awan et al. proposed a priority-based dynamic link scheduling algorithm for WBANs. As a result of simulation, the proposed appeal can balance resource allocation. In [13] the author introduced a method depending on Prioritize avoidance of Congestion Routing Protocol (PCRP), where emergency packets are scheduled before normal packets and every sensor node has maintain many queues to cut down on priority data packet delays. In the scheme to reduce healthcare traffic control and avoid congestion in Wireless networks, a priority-based transmission algorithm is used that allows monitoring of real-time patient data transmission, if any emergency detected highest priority has assigned to the packet and transmit first that would reduced end-to-end delay and the packet transmission occurs when any abnormalities detected that could avoid the congestion. The priority depends on the patient's body constraints [14, 15]. In [16], the proposed Energy Effective Reliable Data Transmission (EERDT) protocol provides high throughput, reliability, and power efficiency. By using single-hop communication, critical data is transferred directly to the base station. The EERDT protocol increases network stability and allows most of the data to be transferred at the base station. An Energy-Harvested and Cooperative-Enabled Efficient Routing Protocol (EHCRP) was developed in [17] for IoT-WBANs for extending the network lifetime. But the link quality metrics and efficient routing of patient data were not performed. To decrease the delay in data gathering, the Opportunistic Efficient Energy routing protocol based on Load Balancing (OE2-LB) method was developed [18]. However, it did not successfully integrate with the efficient adaptive algorithm to optimize neighbor selection with minimal energy use. The introduction of the energy-efficient framework for WBAN [19] improved the efficiency of packet delivery. The data delivery time, however, did not decrease. The motive of the WBAN data gathering technique is to collect medical data from the sink node with less energy utilization way to lengthen the network lifetime [20, 21]. The DSDV (Destination Sequenced Distance Vector) protocol performs poorly when it comes to routing. The DSDV protocol has a very high packet loss ratio of approximately 90%, which increases the overhead of the routing mechanism. According to [19], Remote Monitoring and Controlling Protocol (RMCP) is only capable of passing information between the access point and the base station on layer 3.The traffic prioritized demand balancing schedule method is based on the IEEE 802.15.6 concept for managing load in various queues of priority in wireless body area networks [22].

In the review mentioned above, various optimization methods were discovered to increase the effectiveness of remote patient monitoring. The majority of current systems concentrate on lowering energy consumption but fall short of delivering vital patient data in a timely manner to prevent transmission delays and protect patient life. Thus, in order to transmit emergency data and preserve the life of the patient, quick and trustworthy channels are essential.

1.1 *Motivation of the LRBOA Approach*

- An energy-aware data transmission method in WBAN, using segmented linear regression and African buffalo optimization, has been proposed.
- By applying segmented linear regression, the LRBOA technique is able to reduce data transmission delays by dividing data from sensed patients into categories like regular, irregular, and serious.
- To minimize data loss during critical data transmissions, the buffalo optimization algorithm uses a multivariate function to find the optimal path.

In order to reduce consumption of energy and critical node data transmission delay, the protocol proposed is designed to reduce these factors.

2 Methodology

The WBAN, particularly in applications for health data monitoring, successfully modifies human wellbeing and lifestyle. In Fig. 1, monitoring bio-medical devices placed either outside or inside the body is comprised of the WBAN architecture, which is used both in emergency situations and on a regular basis. The bio-information gathered from the installed bio-medical sensing equipment, which monitors human health conditions, must be precisely sent to the estimated nodes with high energy efficiency. The sensor senses data, like blood pressure, temperature, and so on, and sends the information to a sink via wireless communication. That node aggregates the collected data. Wireless communication plays a very important role in data transmission. Information is immediately forwarded to a server over the Internet. Finally, the information is accessed by a physician, who generally takes care of human health and medical needs.

2.1 *Linear Regression Buffalo Optimization Algorithm*

The architecture design for the LRBOA approach, which is suggested for use in priority-based energy-aware data transfer in WBAN, is shown in Fig. 2. Two key procedures are carried out by the proposed LRBOA technique: data categorization and the identification of the best route. Data classification is initially carried out by using segmented linear regression with the technique of correlation. The classification determines how the data was gathered and is divided into regular, unusual, and critical data. The coordinator node schedule will then follow the order of data transmission.

Finally, a multivariate analysis is used to determine an optimal path for the African buffalo optimization using gradient Bregman divergence. This promotes greater replication minimal time node attack detection rate.

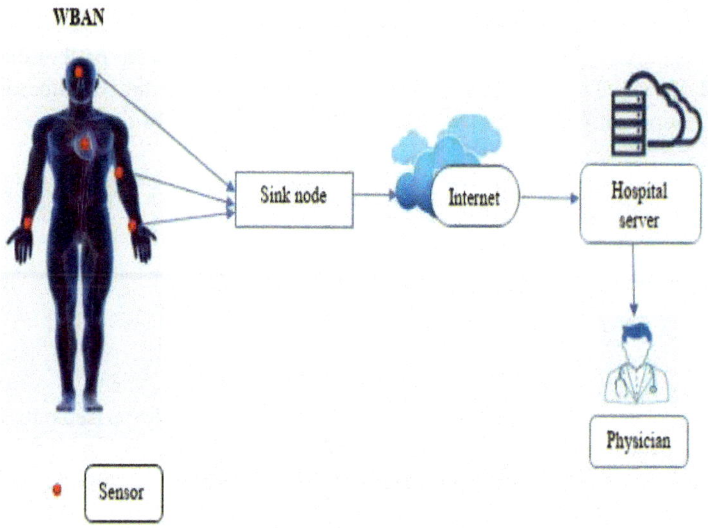

Fig. 1 Basic architecture of WBAN

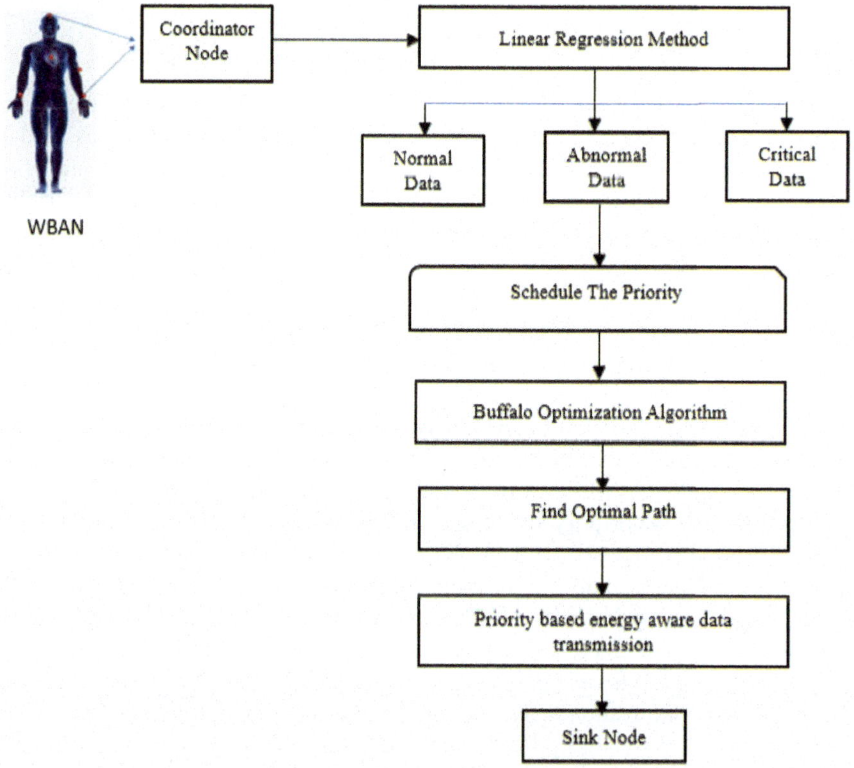

Fig. 2 Architecture of the LRBOA technique

Let us consider the number of patient data $pd_1, pd_2, pd_3, \ldots, pd_n$ collected by coordinator nodes '$CN_1, CN_2, \ldots, CN_m$'. Then CN' analyzes the patient data with the threshold value of patient data that are already stored in the coordinator node. The correlation between these two data is estimated as given in Eq. (1).

$$W_c = \frac{\sum_{i=1}^{n}\left(pd_i - T_{pd}\right)}{\sigma} \tag{1}$$

where

$$\sigma = \frac{1}{n}\sqrt{\left(pd_i - T_{pd}\right)^2}$$

where 'W_c' indicates correlation coefficient, 'pd_i indicates the sensed patient data, T_{pd} indicates the threshold value for patient data, σ denotes a deviation, and n denotes the amount of data. The correlation coefficient (W_c) provides the output results from '− 1' to '+ 1' as indicated in Eq. (2).

$$W_c \begin{cases} +1; & \text{Regular} \\ 0; & \text{Unusual} \\ -1; & \text{Critical} \end{cases} \tag{2}$$

where W_c denotes a correlation coefficient, W_c returns '+ 1' indicating that the patient data is regular or normal, and W_c returns '0' indicates the patient data is classified as not in usual condition. Finally, W_c returns '− 1' indicates the patient data is classified as critical. In this way, the patient is correctly classified. After identifying the patient data level, priority is scheduled by the coordinator node for data transmission.

$$\text{CN} \begin{cases} \text{High} \rightarrow \text{Critical data} \\ \text{Medium} \rightarrow \text{Unusual data} \\ \text{Low} \rightarrow \text{Regular data} \end{cases} \tag{3}$$

As per Eq. (3), coordinator node CN assigns high priority (HP) to critical patient data, the middle priority level (MP) to unusual patient data, and the low priority (LP) to regular data. In this way, coordinator identifies the critical patient data. After scheduling the priority, the critical patient data is first transmitted to the sink node. Followed by, unusual and regular data is being transmitted. Then proposed technique uses the buffalo optimization for identifying the optimal reliable path. This optimization algorithm is a meta-heuristic algorithm inspired by the behavior of buffalos in the African forests. In comparison with other optimization algorithms, this one utilizes a few learning parameters, which guarantees a fast convergence rate. The proposed optimization technique attempts to solve the multivariate functions, namely energy, link reliability, bandwidth, and distance. First, populations of buffalos are initialized in the search space. Here, the buffalos are related to the sensor nodes.

After the initialization, the fitness is estimated based on the multivariate function. After finding the optimal path, the critical patient data is transmitted to the controller node which could aggregates the critical patient data and is sent to the physician for providing the treatment and improves the patient survival. In this way, priority-based energy-aware data transmission is performed in WBAN.

3 Simulation Settings

MATLAB and health monitoring data from https://www.kaggle.com/datasets/nraobommela/health-monitoring-system is used to carry out the evaluation of the suggested PEALR-MGDABO and the current CEPRAN [6], as well as the genetic algorithm-based energy-efficient routing optimization system [7]. The dataset comprises 20 attributes of 4000 instances (i.e. patient data). In order to measure the patient's temperature, heart rate, pulse, blood pressure, respiratory rate, and oxygen saturation sensor nodes are placed throughout the body of the patient. The coordinator node receives these patient data that have been obtained. This coordinator gathered the patient information and classifies it into normal, abnormal, and critical based on a threshold that is already stored. After that, priority is assigned to classified data. First, the primary sink node receives the vital patient data via the best route with better energy consumption, link reliability, minimum bandwidth consumption, and lesser distance.

4 Results and Discussion

A simulation analysis is performed in order to compare the proposed LRBOA scheme with existing methods, including CEPRAN [6] and genetic algorithm-based energy-efficient routing optimization scheme [7]. Energy consumption and data transmission rate in relation to patient data are used to evaluate the performance of the proposed LRBOA algorithm.

4.1 Energy Consumption

For effective patient data transfer and to lengthen the network lifetime, energy is a key factor. Table 1 represents the simulation performance analysis of energy consumption of patient data transmission in WBAN. It is calculated based on the number of patient data using proposed LRBOA and the existing method CEPRAN and GA-EEROS. As the result, the energy consumption of proposed LRBOA is efficient than the others.

Table 1 Comparison of energy consumption

No of patient's data	Energy consumption (J)		
	LRBOA	CEPRAN	GA-EEROS
400	32	36	38
1200	43.2	46.8	49.2
2000	54	58	60
2800	67.2	70	72.8
3600	75.6	77.76	79.2

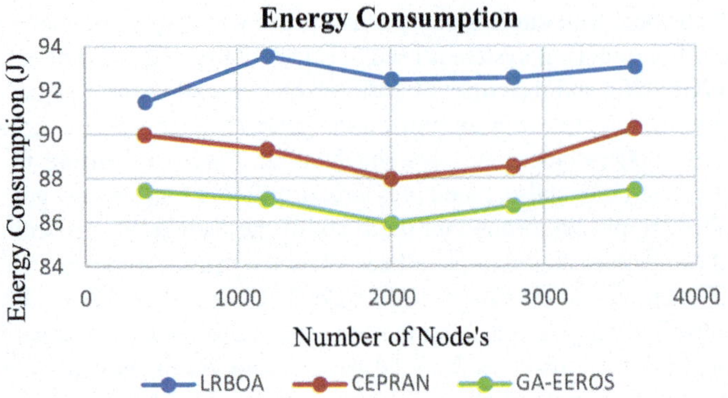

Fig. 3 Energy consumption analysis

Figure 3 shows the results of an energy consumption analysis comparing three alternative methods—LRBOA, CEPRAN [6], and a genetic algorithm-based energy-efficient routing optimization scheme—to a variety of patient data. The graph in Fig. 3 demonstrates unequivocally how the LRBOA uses less energy than current techniques. Applying the African buffalo optimization to discover the ideal path by choosing energy-efficient nodes allows for this performance gain.

4.2 Packet Delivery Ratio

Essentially, packet delivery ratio measures how many patient data packets are received at the sink compared to how many are sent.

Table 2 shows the effectiveness of the recommended LRBOA, the current method CEPRAN, and the energy-efficient routing optimization strategy based on genetic algorithms for the number of patient data transmission for WBAN. The LRBOA methodology performs best among the three ways in terms of achieving increased data delivery of between 4% and 7% compared to the current methods [6, 7].

Table 2 Comparison of packet delivery ratio

No of patient's data	Packet delivery ratio (%)		
	LRBOA	CEPRAN	GA-EEROS
400	91.5	90	87.5
1200	93.58	89.33	87.08
2000	92.5	88	86
2800	92.57	88.57	86.78
3600	93.05	90.27	87.5

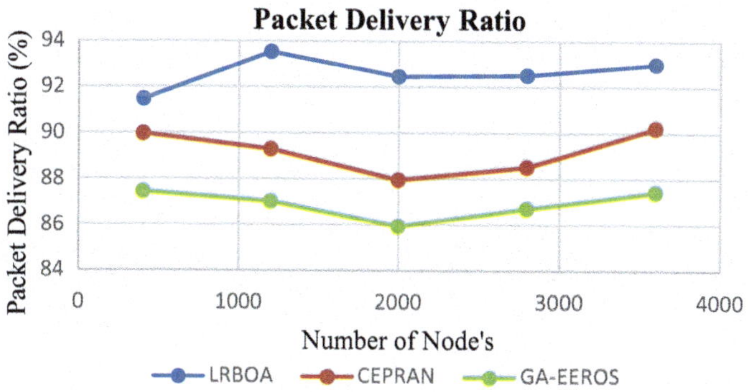

Fig. 4 Comparison between the quantity of nodes and the ratio of packets delivered

Figure 4 shows that the performance of packet delivery ratio of proposed PEALR-MGDABO is comparably high than the other methods. It is achieved by Winsorized correlative segmented linear regression.

4.3 *Effect of Delay*

The controller node determines the delay based on the actual and anticipated data arrival times. Table 3 shows how well a delay will affect emergency packets. The ten simulation rounds are examined. Since the number of simulation rounds is dependent on the quantity of communications received at the destination, the delay steadily increases as the number of rounds increases.

Figure 5 shows that the effect of delay of proposed LRBOA algorithm with protocol in [6, 7] is comparably low than the other methods.

Table 3 Comparison of delay (ms)

No of patient's data	Delay (ms)		
	LRBOA	CEPRAN	GA-EEROS
400	14	17	19
1200	21	24	26
2000	27	30	32
2800	34	36	39
3600	41	44	47

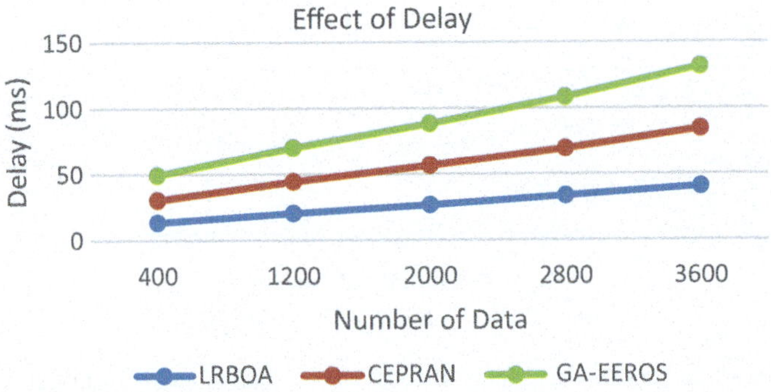

Fig. 5 Comparison of delay and number of data

5 Conclusion

In the healthcare industry, WBAN have emerged as a potential progress where patients are given wearable and implantable sensors to gather data for remote health monitoring. In this study, a priority-focused LRBOA approach is proposed to improve an energy efficiency and reliability of patient data transmission. The LRBOA technique uses the linear regression technique to classify the patient data and buffalo optimization to find the best path. In addition to enhancing data delivery, it minimizes delays. Based on certain performance factors such as energy consumption and packet delivery ratio, a simulation analysis is carried out for a patient health monitoring dataset. In comparison with conventional methods, the proposed LRBOA technique has higher results, a lower loss rate, a shorter delay time, and minimizes energy consumption.

References

1. Shah AM, Abdelmaboud A, Mahmood K, ul Hassan M, Saeed MK (2018) eHealth WBAN: energy-efficient and priority-based enhanced IEEE802.15.6 CSMA/CA MAC protocol. Int J Adv Comput Sci Appl 9(4)
2. Menaga S, Paruvathavardhini J (2022) AI in healthcare. In: Smart systems for industrial applications, pp 115–140
3. Menaga S, Kalaiarasi G, Vanithamani R, Nivetha M (2022) Data analytics for disease prediction. Technoarete Publishing
4. Memon S, Wang J, Bhangwar AR, Fati SM, Rehman A, Xu T, Zhang L (2021) Temperature and reliability-aware routing protocol for wireless body area networks. IEEE Access 9:140413–140423
5. Menaga S, Vanithamani R, Hema P (2022) A comprehensive review on wireless body area network-technologies, challenges, application and energy saving techniques. In: 2022 3rd international conference on electronics and sustainable communication systems (ICESC). IEEE, pp 541–548
6. Geetha M, Ganesan R (2021) CEPRAN-cooperative energy efficient and priority based reliable routing protocol with network coding for WBAN. Wireless Pers Commun 117:3153–3171
7. Panhwar MA, Liang DZ, Memon KA, Khuhro SA, Abbasi MAK, Ali Z (2021) Energy-efficient routing optimization algorithm in WBANs for patient monitoring. J Ambient Intell Human Comput 12:8069–8081
8. Ibrahim AA, Salawudeen AT, Uçan ON, Okorie PU (2020) A novel energy-conscious threshold-based data transmission routing protocol for wireless body area network (NEAT). In: 2020 international congress on human-computer interaction, optimization and robotic applications (HORA). IEEE, pp 1–5
9. Gayathri TC, Gomathi C, Santhiyakumari N (2015) Energy efficient and node selection technique for wireless body sensor network. In: 2015 international conference on communications and signal processing (ICCSP). IEEE, pp 0383–0387
10. Bhandari RR, Sekhar KR (2021) Priority-mobility aware clustering routing algorithm for lifetime improvement of dynamic wireless sensor network. Int J Adv Comput Sci Appl 12(2)
11. Kwon HT, Lee SK (2009) Energy-efficient multi-hop transmission in body area networks. In: 2009 IEEE 20th international symposium on personal, indoor and mobile radio communications. IEEE, pp 2142–2146
12. Singla R, Kaur N, Koundal D, Lashari SA, Bhatia S, Rahmani MKI (2021) Optimized energy efficient secure routing protocol for wireless body area network. IEEE Access 9:116745–116759
13. Awan KM, Ashraf N, Saleem MQ, Sheta OE, Qureshi KN, Zeb A, Haseeb K, Sadiq AS (2019) A priority-based congestion-avoidance routing protocol using IoT-based heterogeneous medical sensors for energy efficiency in healthcare wireless body area networks. Int J Distrib Sens Netw 15(6):1550147719853980
14. Kurian A, Divya R (2017) A survey on energy efficient routing protocols in wireless body area networks (WBAN). In: 2017 international conference on innovations in information, embedded and communication systems (ICIIECS). IEEE, pp 1–6
15. Fourati LC (2014) Wireless body area network and healthcare monitoring system. In: 2014 IEEE international conference on healthcare informatics. IEEE, pp 362–362
16. Sethi D, Bhattacharya PP (2016) A study on energy efficient and reliable data transfer (EERDT) protocol for WBAN. In: 2016 second international conference on computational intelligence & communication technology (CICT). IEEE, pp 254–258
17. Khan MD, Ullah Z, Ahmad A, Hayat B, Almogren A, Kim KH, Ilyas M, Ali M (2020) Energy harvested and cooperative enabled efficient routing protocol (EHCRP) for IoT-WBAN. Sensors 20(21):6267
18. Raj AS, Chinnadurai M (2020) Energy efficient routing algorithm in wireless body area networks for smart wearable patches. Comput Commun 153:85–94

19. Ahmad N, Shahzad B, Arif M, Izdrui D, Ungurean I, Geman O (2022) An energy-efficient framework for WBAN in health care domain. J Sens
20. Mangali NK, Kota VK (2015) Health monitoring systems: an energy efficient data collection technique in wireless sensor networks. In: 2015 international conference on microwave, optical and communication engineering (ICMOCE). IEEE, pp 130–133
21. Kaur N, Singh S (2017) Optimized cost effective and energy efficient routing protocol for wireless body area networks. Ad Hoc Netw 61:65–84
22. Samal T, Kabat MR (2022) A prioritized traffic scheduling with load balancing in wireless body area networks. J King Saud Univ Comput Inf Sci 34(8):5448–5455

Severity Levels Categorization for Detecting Diabetic Retinopathy Using YOLOv7

M. Nirmala Devi, B. Subbulakshmi, M. Sivakumar, and T. Teenu Prinica

Abstract A disease that damages the eyes called diabetic retinopathy (DR) is more frequently found in adults who are working. Normal, moderate, and proliferate are the three levels of diabetic retinopathy severity. The YOLOv7 algorithm and PyTorch framework are used in this paper to propose a novel model for assessing the DR severity in its various phases. For the precise and effective detection of lesions in retinal images, the suggested method makes use of a convolutional neural network model, data augmentation, and reinforcement learning techniques. The abnormality or damage visible in the retina of the eye is referred to as a lesion in retinal imaging. The validation set's mean average precision for the model, which was trained on the blindness detection dataset, was 0.921 and 0.902 for intersection over union thresholds of 0.5 and 0.95, respectively. The proposed model gives promising outcomes for early diagnosis and treatment of DR, which is essential in preventing serious outcomes like blindness.

Keywords Diabetic retinopathy · Diabetic mellitus · Lesions · Fundus photography · CNN · YOLOv7

1 Introduction

Insufficient blood insulin secretion by the pancreas results in diabetes mellitus. Human organs such as the liver, kidneys, joints, hearts, eyes, and others are negatively impacted by it. Diabetes, which affects a vast number of individuals globally,

M. Nirmala Devi (✉) · B. Subbulakshmi · M. Sivakumar · T. Teenu Prinica
Department of Computer Science and Engineering, Thiagarajar College of Engineering, Madurai, Tamil Nadu, India
e-mail: nirmaladevi2004@gmail.com

B. Subbulakshmi
e-mail: bscse@tce.edu

M. Sivakumar
e-mail: mskcse@tce.edu

S. Bandyopadhyay et al. (eds.), *Intelligent Computing Systems and Applications*, Lecture Notes in Networks and Systems 1010,
https://doi.org/10.1007/978-981-97-5412-0_33

frequently causes problems like diabetic retinopathy. Around one in three patients with diabetes develop diabetic retinopathy to some extent, according to the International Diabetes Federation. In 2021, the International Diabetes Federation predicted that there would be 537 million adult diabetics globally. Therefore, it is crucial for people with diabetes to periodically check their eyes to monitor diabetic retinopathy and take action to manage their diabetes and the risk factors that go along with it. The leading reason for blindness is middle-aged persons of forty and above. In the WHO European Region, diabetic retinopathy affects the vision of about 950,000 persons [1]. In people with diabetes in Europe, America, and Asia, the prevalence of diabetic retinopathy is about 34.6% [2]. With respect to the World Report for Vision, 1 billion of the 2.2 billion persons who suffer vision impairment also have blindness [3].

Atwany et al. [4] discussed that DR is a condition that originates from an increase in the blood sugar level that affects the eyes first. As time passes, diabetes damages the blood veins of the retina that cause blood leakage. Also, new abnormal blood veins grow from the retina which bleeds, and this results in the swelling of the tissues of the central retina which causes permanent blindness. The symptoms of DR on humans differ based on the complications of the disease and how long it has been present.

Porwal et al. [5] analyzed that DR screening is a widely accepted strategy to prevent vision loss, together with prompt assessment and treatment. Implementing DR screening programmes, however, is difficult because to the dearth of medical specialists qualified to screen a soaring global population of diabetics. For such extensive screening efforts, computer-aided illness diagnosis in retinal image processing might offer a viable strategy. Recent technological developments in processing power and machine learning techniques give biomedical researchers a path to do this. Some of the potential effects of diabetic retinopathy include (i) Vision problems: DR can cause vision defect, including blurred vision, hardness on vision at night, and loss of peripheral vision. In some cases, DR can lead to complete blindness; (ii) Eye complications: can cause macular edema, which can lead to vision loss.

Devi et al. [6] discussed that the development of blood vessels in unusual nature can also occur in the retina and leads to retinal separation, glaucoma, and blindness; (iii) Impact on daily life: The vision problems associated with diabetic retinopathy can impact a person's ability to perform tasks such as driving, reading, and cooking. It affects their ability to work and leads to a reduced quality of life. (iv) Increased risk of other complications: DR is often a sign of more advanced diabetes and is associated with an increased risk of other complications, which include kidney disease, nerve damage, and cardiovascular disease. Good and balanced state in blood sugar, pressure, and cholesterol management can also help reduce the risk of developing diabetic retinopathy and its associated effects.

Deep convolutional neural networks (DCNN) are applied by researchers to identify lesions in fundus pictures for the diagnosis of DR. Appropriate feature selection during model training has been linked to restrictions in experimental outcomes. The presentation of a CNN model, which can (i) perform real-time outcomes on inexpensive GPUs, (ii) the use of munging methods on data like static crop, auto-orientation, and resizing on retinal images, (iii) the use of data augmentation methods like horizontal and vertical flip, rotation, and (iv) the use of Knowledge learning to reduce the model's training time by i. The likelihood that an object is present in a specific bounding box is referred to as objectness. An image's bounding box is a rectangular area that the algorithm predicts will contain an object. The center coordinates, width, height, and confidence score of each bounding box serve as indicators of how well the bounding box is predicted. A bounding box's class probability serves as a gauge for the algorithm's faith in the anticipated class of the object.

Diabetic retinopathy cannot be reversed, and its therapy involves maintaining the patient's latest level of eye sight. To ensure effective preservation of vision, early detection of the disease is crucial. The traditional approach has the following shortcomings: it is a time-consuming, expensive, and labor-intensive process that would require an ophthalmologist to manually examine the images of the retinal fundus. This presents a significant challenge in the finding DR in patients. The fundus images are obtained by taking photographs of the interior part of the eye. The lesions captured in the fundus images help in the extraction of features for detection. The detailed description of the lesions like microaneurysms and hard exudates. Screening can help to detect lesions in the eyes before the disease starts to affect the vision, but the screening procedures require a trained clinical expert to examine the fundus photography of the patient's retina.

2 Related Work

Porwal et al. [5] discussed that utilizes ResNet-50 and VGG architecture to detect fovea and optical disc; however, it is limited to detecting only those specific features, whereas Mateen et al. used transfer learning to detect exudates in fundus pictures with using the pretrained CNN frameworks and cover ResNet-50's past and future models. To deal with the issue of capturing different features, a combination of self and prior two versions of VGG-19 models were used and performed feature fusion. Fully connected layers are added to the models to categorize the fundus images. However, the model is limited in its ability to examine hemorrhages.

Mateen et al. [7] projected that a significant contributor to blindness in the field of ophthalmology is DR. Exudate-containing retinal lesions are the foundation of DR. Exudates have been identified as one of the symptoms and significant DR defects, and thus, proper identification of these lesions and prompt treatment are required to avoid vision loss. A pretrained CNN-based system for the detection of exudate is suggested in this research. Recently, DCNNs were used individually to address the particular issues. However, pretrained CNN models with knowledge engineering can use the prior information to address more relevant issues. In the suggested method, data munging is done initially to standardize exudate patches. Additionally, exudate features are spotted using region of interest (ROI) localization, and then transfer learning is utilized to extract features using pretrained CNN models (Incep-v3, ResNet, and Visual Geometry Group Net19). Additionally, the fully connected (FC) layers' fused features are input into the softmax classifier for exudate categorization. Two famous datasets, such as e-Ophtha and DIARETDB1, have been used to analyze the performance of the proposed framework. The experimental findings show that the CNN-based framework outperforms the currently used methods for exudate identification.

A new dataset called diabetic retinopathy (DDR) dataset was created by Li et al., from the collected 13,673 fundus images to evaluate deep learning models for lesion detection from fundus images. This dataset includes 757 annotated images for detecting four types of lesions in DR. Active deep learning uses CNN for the automatic extraction of features. Ignazio Gallo describes object detection using YOLOv7 architecture and explains the formats required for training the dataset.

Qureshi et al. [8] have shown that using retinal fundus image analysis (RFIA) for DR screening can help diabetic individuals avoid going blind. The RFIA screening programmes aid ophthalmologists in managing this serious issue of vision impairment. CNN model was employed to automatically extract features. However, the training of the CNN technique necessitates a huge volume of labeled data, making the classification step nearly impossible. As a result, the expected gradient length (EGL), one of the active learning techniques, is used to propose a label-efficient CNN architecture. The efficiency of the proposed system was assessed using statistical metrics like sensitivity (SE), specificity (SP), and F-measure.

Nirmala Devi et al. [9] discussed that preprocessing data is done to remove noisy data from the dataset. The DBOD approach is used to identify outliers, which are then handled in a way that is most similar to the standard. For the DBOD approach, a comparison of various distance metrics, including Euclidean and Manhattan, is conducted. Finally, the preprocessed data is sent into the logistic regression model as input.

Rajeswari et al. [10] analyzed that to improve the model performance in terms of accuracy of machine learning (ML) systems, outlier detection was developed. Over time, the phenomenon of mining unusual or worrying patterns to aid in decision-making came to be known as outlier detection. Both supervised and unsupervised ML systems are capable of detecting outliers, which might be of the point or collective variety. These techniques could potentially miss some individual or group outliers in high-dimensional quantitative data. This paper suggests a semi-supervised outlier detection (SSOD) algorithm with a probability-based unexpected metric ('lift') for outlier detection to address the aforementioned problem. The performance of the suggested SSOD is compared to the current ML outlier detection algorithms. It is clear from analyzing algorithm performance that the benchmarked ML algorithms perform worse than the suggested SSOD with lift measure.

Qiao et al. [11] discussed that to find the existence of microaneurysms in fundus pictures and identify DR in its prior phases. Long-term high blood glucose levels cause DR. The first symptoms of DR are the development of microaneurysms and macular edoema in the retina, and prompt detection can lower the chance of developing non-proliferated diabetic retinopathy. Deep learning is quickly advancing, making it an effective method for offering an intriguing solution to difficulties involving medical picture interpretation.

Wang et al. [12] have shown that a steel surface failure prediction methods' feature information was found by using the de-weighted BiFPN structure to strengthen feature fusion, minimize feature information loss during convolution, and enhance detection accuracy. In order to reinforce the key feature channels, the ECA attention mechanism is combined in the backbone component. The SIoU loss function, which redefines the error by taking into account the vector angle with respect to the necessary regressions, finally replaces the original bounding box loss function. Shahin et al. [13] explained that the waste of time, money, and human resources can be decreased by a computer-based object finding algorithms. It can assist the research compared the performance indicators of three different detection techniques and evaluated and analyzed them.

Qiu et al. [14] discussed that on normal datasets, CNN-based object identification algorithms produce good detection results. This network is built on the joint optimum learning of the YOLOv7 detection and IDOD (AOD + SAIP) image defogging modules. The experimental findings demonstrate that the IDOD module not only enhances low-light fog image defogging quality, but also performs better in objective assessment indices like PSNR and SSIM. In order to conduct object detection and weakly supervised image augmentation simultaneously, the IDOD and YOLOv7 collaborate to learn end-to-end.

Alyoubi et al. [15] and Pinedo-Diaz et al. [16] analyzed that if not treated, DR is a significant cause of loss of vision. The present DR therapies are only capable of slowing or halting vision degeneration, emphasizing the significance of routine scanning with very effective electronic ways to identify instances early. In this research work, fully automatic diagnosis methods that outperform manual techniques to prevent misdiagnosis were described. This cut down on time, effort, and expense. The suggested system localizes the afflicted lesions on the skin surface and divides DR images into five phases: (a) no-DR, (b) mild, (c) moderate, (d) severe, and (e) proliferative DR (Fig. 1).

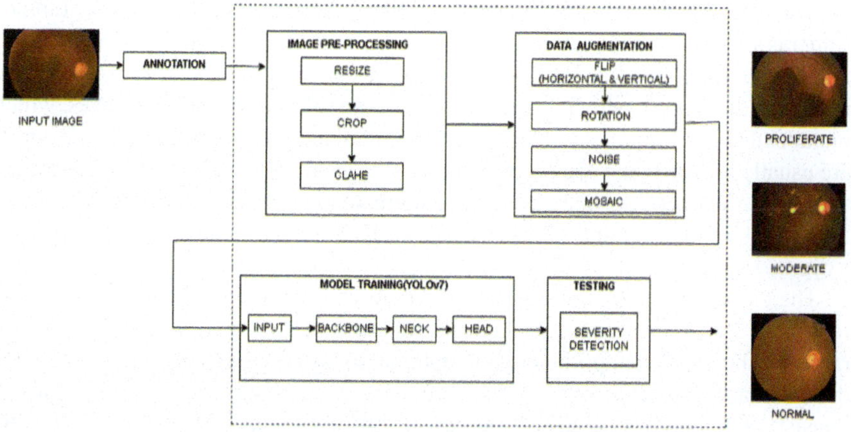

Fig. 1 Severity levels categorization for detecting diabetic retinopathy using YOLOv7

3 Materials and Methods

The base YOLOv7 algorithm is available from the Wong Kin Yiu repository as open source. The proposed model involves training the model for 250 epochs with a batch size of 16. The bounding box anchor size is automatically calculated by the K-means segmentation algorithm. It collects a set of data points into k clusters based on their similarity. In object detection, the algorithm is used to group the object dimensions into k clusters, where k denotes the number of anchor boxes. To train the YOLOv7 model, a graphics processing unit (GPU)-enabled machine with deep learning frameworks such as PyTorch is required. GPUs are highly parallel and perform matrix operations much faster than central processing units (CPUs).

3.1 Dataset

For this work, blindness detection image dataset has been utilized that is publicly available and can be accessed by anyone. The dataset is obtained from Kaggle which can be downloaded from their Web site at https://www.kaggle.com/c/aptos2019-blindness-detection/data. The images in the APTOS dataset were captured using different types of fundus cameras with varying resolutions and fields of view. It contains 3662 fundus images of the patients in PNG format which is used for training. The images are of different sizes, with resolutions ranging from 1024 × 1024 to 2576 × 2048 pixels. The images are taken using a fundus photography technique. This dataset also has 1928 test images where the labels are not known explicitly, and it can be used for performance evaluation of the model. This dataset also has a CSV file Identifier of the fundus image and the corresponding diagnosis for that ID.

Fundus photography is the technique used to capture images of the inside surface of the eye, specifically the retina and head of optic nerve. This technique involves the use of a specialized camera called a fundus camera or retinal camera. Fundus cameras are designed to capture a wide-angle view of the retina and optic nerve head. Specialized lenses and illumination systems are utilized to capture high-resolution images of the fundus, which is then used to examine various problems on eyes, which includes diabetic retinopathy. During a fundus photography exam, the patient's pupils are usually dilated using eye drops to allow for a clear view of the retina. The patient then leans their head against a chinrest, and the fundus camera is positioned close to the eye. The camera emits a flash of light, which illuminates the retina, and captures a digital image of the fundus.

The captured fundus images can be stored and analyzed for various purposes, including diagnosis, treatment planning, and research. The use of fundus photography has become an important tool in ophthalmology and has helped to improve the detection and management of various problems on eyes, including DR. The dataset consists of images from multiple sources and clinical settings across the Asia-Pacific region. The cameras used in the APTOS dataset include models from different manufacturers, such as Topcon, Canon, and Kowa. Since the cameras provide portability features they can be used in areas with limited medical access. The variability in the types of cameras used in the APTOS dataset reflects the real-world challenges of diabetic retinopathy screening and diagnosis in diverse clinical settings with varying levels of resources and infrastructure.

3.2 Annotation

As this dataset is in PNG format which is not supported by the YOLOv7 model. YOLOv7 accepts labeling of each image in the form of a text file where each image extracts features through bounding boxes. The images were labeled based on their severity using an open-source annotation tool that includes annotations such as polygon, smart polygon, and bounding box. The steps to annotate the fundus images are as follows: (i) Upload the fundus images to the open-source annotation tool; (ii) Select the annotation type and label categories (diabetic retinopathy stages); (iii) Use the annotation tools to delineate regions of interest in the images by drawing bounding boxes around abnormalities or lesions; (iv) Assign the appropriate label category to each bounding box, depending on the severity or type of diabetic retinopathy identified; (v) Review and validate the annotations to ensure the best performance.

After completing the annotations, the dataset can be exported in various formats, of which the YOLOv7 PyTorch format is selected. For training deep learning models for object detection and segmentation, the YOLOv7 PyTorch format is utilized. Annotated datasets can add to the precision and speed of detecting diabetic retinopathy in fundus images, by enabling machine learning models to be trained on such datasets and thus developing more precise and automated approaches for monitoring and diagnosing diabetic retinopathy.

3.3 Preprocessing

Preprocessing techniques were used to enhance the resolution of the image and reduce the noise in the APTOS dataset. Some of the techniques that were utilized are resizing static cropping and contrast-limited histogram adaptive equalization (CLAHE). The contrast of lesions in the retinal images is enhanced by the use of CLAHE. This CLAHE algorithm is applied to small regions known as tiles, and histogram equalization is performed on each of these tiles. This approach provides local equalization of the image and also reduces outlier amplification in the images. The CLAHE technique is used by applying different thresholds in the Lumin-Light (L) component to determine the optimal contrast level in the images. Additionally, the images are resized to ensure consistency in their dimensions, and a median filter is applied to filter out any remaining noise in the images.

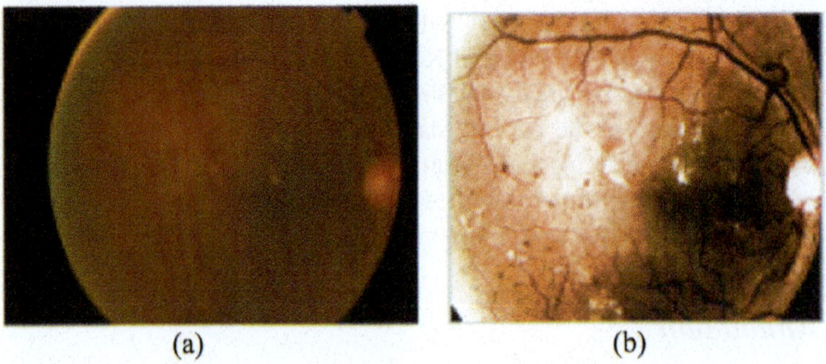

Fig. 2 Sample fundus image: **a** original image and **b** preprocessed image

The efficiency criteria of the preprocessing techniques are assessed using the measure of enhancement (EME) metric, commonly applied to image processing and computer vision in order to compare the quality of an enhanced image to the original image. The measure of enhancement formula is one such method to evaluate the performance of an image enhancement algorithm. It is commonly used in the field of image processing and computer vision to compare the quality of an enhanced image to the original image. The mean squared error (MSE) measures the average squared difference between two images and is commonly used to evaluate image quality. The EME formula essentially measures the percentage reduction in the MSE after applying an enhancement algorithm. The measure of enhancement (EME) is defined as

$$\text{EME} = \frac{100 \times (M_0 - M_e)}{M_0} \tag{1}$$

where M_0 is the MSE between the original image and the reference image, and M_e is the MSE between the outcome improved image and the same reference image. Figure 2 represents a sample of fundus images from the APTOS dataset, illustrating the images before and after preprocessing. The preprocessed images show improved contrast and reduced noise [14].

3.4 Augmentation

To expand the dataset, both image-level and bounding-box-level augmentations are applied to each annotated image, resulting in additional images. The diversity of lesions in fundus images is tested to improve the model by utilizing this process. Applying image-level augmentation techniques such as flipping does not require adjusting the bounding box coordinates since it does not alter the position or size of the objects within the image. For bounding-box-level augmentation, geometric

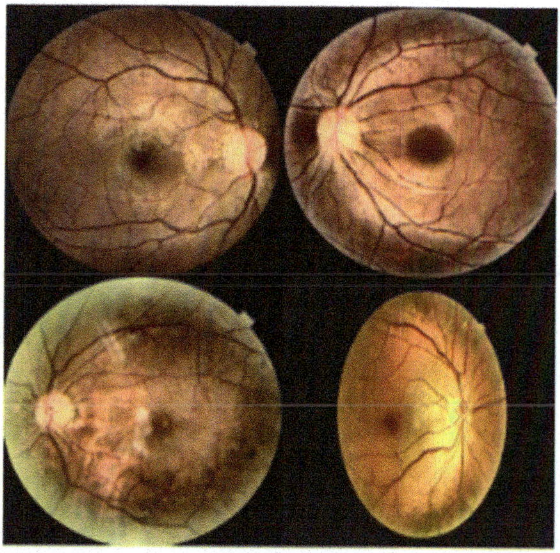

Fig. 3 Mosaic augmentation technique applied to the preprocessed fundus image

transformations such as rotation, scaling, and shearing are applied to the original images. When applying these transformations, the bounding box coordinates are adjusted accordingly to reflect the new location and size of the object.

The mosaic technique is handy for object detection tasks because it helps the model learn to detect small objects accurately. This is because the four random images combined into the mosaic image may contain small objects that would not be easily visible in a single image. By combining them, the model can learn to detect these objects accurately. Figure 3 represents the mosaic augmentation technique applied to the preprocessed fundus image. Figure 4 represents the geometric transformations applied to the fundus images. Following the implementation of data augmentation techniques, the model has been trained to accurately assess the extent of lesions in fundus images. This training incorporates bounding boxes demarcating the regions of interest surrounding the objects.

3.5 *The Proposed Method*

The You Only Look Once (YOLO) algorithm is chosen to predict the bounding boxes and class probabilities of objects in images using CNN. One of the unique features of YOLO is that it evaluates the entire image in one step, rather than using a sliding window approach. YOLO employs a backbone architecture that takes the image frames and extracts essential features through convolutional layers. The backbone is combined and mixed in the neck, and the resulting feature maps are used by YOLO to predict bounding boxes, classes, and objects. Pretrained neural networks are used

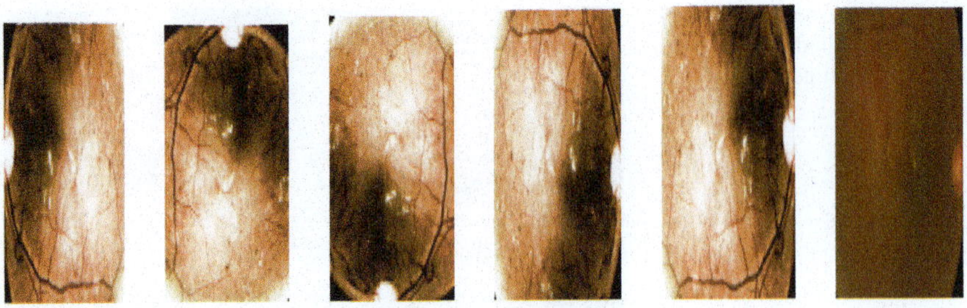

Fig. 4 Geometric transformation applied to the original image: **a** upside down; **b** clockwise rotation; **c** counter-clockwise rotation; **d** horizontal flip; **e** vertical flip; **f** cropped

to train the backbone, which helps in improving its ability to extract informative features from the input images. In general, the backbone plays a critical role in the effective implementation of the YOLO algorithm in detecting objects.

For the object detection of the input images, the YOLOv7 algorithm is employed after annotation, preprocessing, and data augmentation steps. The dataset is partitioned as 80:20 ratio, for modeling and validation. It is determined through trial and error methods. Among the six versions of the YOLOv7 model, the YOLOv7 base model is selected for training the dataset due to its high accuracy in detecting objects of varying sizes. To improve the model's accuracy, YOLOv7 uses a set of techniques called a trainable bag of freebies, which is a cutting-edge approach for real-time object detection. According to Ref. [13], YOLOv7 achieves the highest average precision of 56.8% AP compared to the other versions of YOLOv7, demonstrating its efficacy in object detection tasks.

The YOLOv7 model employs pretrained weights like VoVNET, CSPVoVNET, ELAN, and E-ELAN to enhance its performance [13]. Within the YOLO architecture, the neck layer strategically resides between the backbone and head layers to collect feature maps from different stages. The neck layer comprises both bottom-up and top-down components to improve object detection accuracy. The head layer is where object detection takes place. To optimize the training process, YOLOv7 utilizes batch normalization, implicit knowledge, and the EMA model [13]. The training optimizer uses gradient prediction to generate hierarchical labels in a coarse-to-fine manner. Additionally, YOLOv7 leverages extended efficient layer aggregation networks (E-ELAN) and model scaling to improve accuracy. Model scaling involves adjusting the input width and depth of the model, enabling it to detect both small and large objects effectively. The model's scaling factors depend on various parameters such as resolution, depth, stage, and width. YOLOv7 utilizes a re-parameterization technique called gradient flow propagation plans to determine the optimal combination of re-parameterized convolutions with different networks. Training of the model is done using the Microsoft COCO dataset, and intersection over union (IOU) is used to evaluate the extent of overlap between predicted and ground truth bounding boxes. The IoU formula can be expressed as

$$\text{IoU} = \frac{A_\text{i}}{A_\text{u}} \tag{2}$$

The interpretation of object detection models involves measuring the extent of overlap between the output bounding box and the origin truth bounding box. This is done by calculating two metrics: area of intersection (A_i) and area of union (A_u). A_i represents the area of overlap between the output bounding box and the origin truth bounding box, while Au represents the total area covered by both bounding boxes, including their overlap. To train the YOLOv7 model, a binary cross-entropy loss function is used to calculate the objectness loss score, bounding box regression score, and confidence score based on predicted objectness probability and complete intersection over union (CIoU).

4 Results and Discussions

A proposed model based on the YOLOv7 architecture is implemented using the PyTorch framework in experiments utilizing the APTOS dataset. The model is trained with the help of deep neural network and transfer learning based on pre-learned weights from COCO dataset. Metrics called average precision (AP) and mean average precision (mAP) are used in model evaluation. Metric called mean average precision (mAP) assesses the precision of object detection that computes the average precision for each class across all feasible intersection over the union (IoU) threshold values before averaging the results across all classes. It is a means to gauge a model's accuracy in identifying and locating objects of various kinds inside an image. Better detection accuracy is indicated by higher mAP scores. Figures 5 and 6 show the bounding box, objectness, and class probability loss functions that were computed during training and validation using the APTOS dataset. Figure 7 displays the validation outcomes for mAP at thresholds of 0.5 and 0.95. While the highest mAP value recorded at 0.95 criteria was 0.902, the highest mAP value at 0.5 thresholds was roughly 0.921.

$$\text{mAP} = \frac{1}{N}\sum_{i=1}^{N} \text{AP}_i \tag{3}$$

To evaluate the detection quality, the model uses IoU threshold values of 0.5 and 0.95, whereas the perfect performance measure of object detectors accuracy is measured by the mAP metric, which calculates average accuracy across all the evaluated classes. Using a batch size of 16, and a learning rate of 0.01, that is trained for 150 epochs, the proposed method attained a mAP of 0.902. The model's accuracy is evaluated using various metrics like accuracy and $F1$ score, etc. The accuracy deals with the ability of the model to correctly identify and locate objects in an image that uses the percentage of rightly categorized objects over the total number of objects in the image.

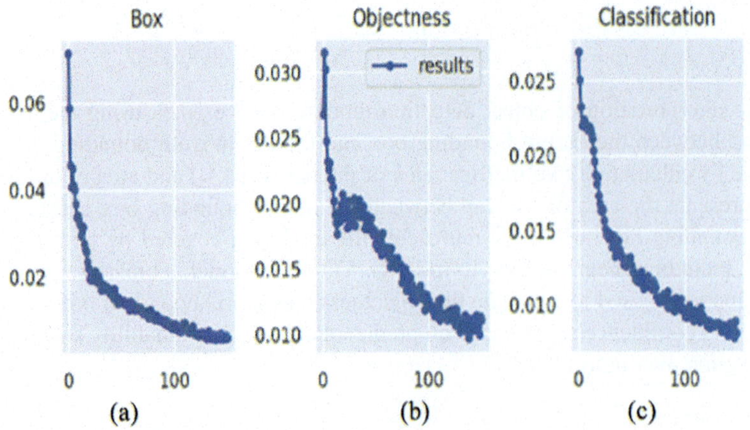

Fig. 5 Training stage loss functions: **a** bounding box regression; **b** objectness; and **c** class probability

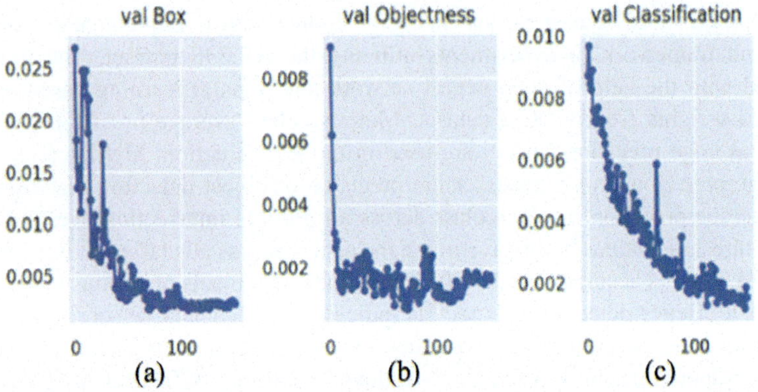

Fig. 6 Validation stage loss functions: **a** bounding box regression; **b** objectness; and **c** class probability

Fig. 7 Values of **a** mAP at 0.5 and **b** mAP at 0.95 were obtained during validation

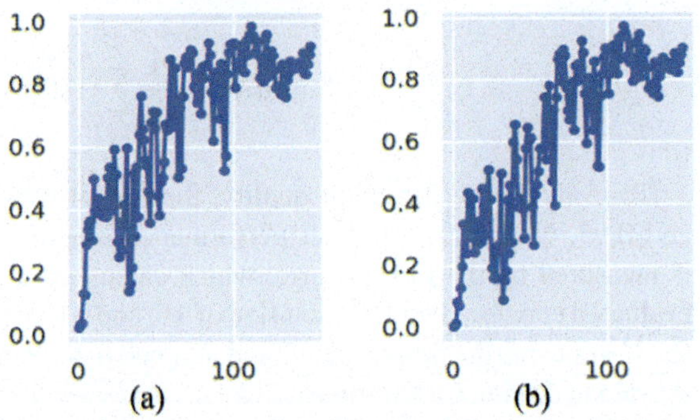

The precision is a metric that measures the proportion of right positive findings out of all positive findings made by the model, and the recall is the ratio of true positive findings with respect to the total number of objects in the ground truth. The $F1$ score is an evaluation metric that takes both precision and recall to provide an overall assessment of the model's performance. It is the harmonic mean of precision and recall, where a perfect score is 1, and a score of 0 shows that the model fails to detect any objects correctly. $F1$ score is capable when there is an uneven class distribution, as it gives equal weight to both precision and recall, which is important in different scenarios, and it is calculated using the following formula:

$$F1 = \frac{2 \times (P \times R)}{(P + R)} \tag{4}$$

where $F1$ score can be calculated by precision and recall represented as P and R, where P indicates the division of true positives (TP) with respect to the sum of TP and false positives (FP), and R indicates the division TP with respect to the sum of TP and false negatives (FN). The confusion matrix is a table used to compute the outcome measures of an object detection model. It can also be used to identify the number of true positives, false positives, true negatives, and false negatives in the model's predictions, which helps to assess the effectiveness of the model. Figure 8 shows the results of the metrics. The severity of lesions was used to detect the different stages. Despite achieving good results, the detection of microaneurysms in mild DR remains challenging due to their small size and variations in shape and size. The above-discussed characteristics make training difficult which can lead to a high number of FP and FN, which significantly affects the model's performance. Furthermore, the complexity of the lesions' color intensities, shapes, and volume all impact the results obtained in the experiment.

Figure 5 shows the training stage loss functions: (a) Bounding box regression; (b) Objectness; and (c) Class probability. Figure 6 shows the validation stage loss

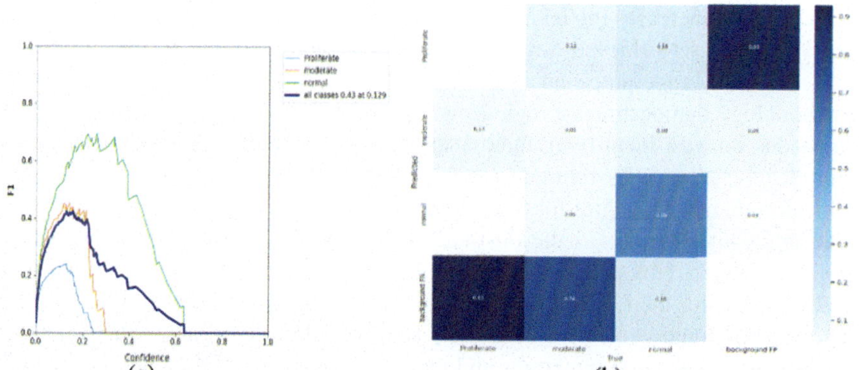

(a) (b)

Fig. 8 Performance metrics obtained for the proposed model: **a** $F1$ curve; **b** confusion matrix

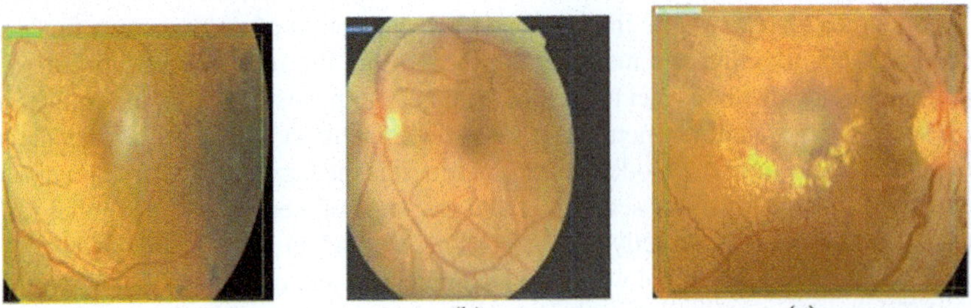
(a) (b) (c)

Fig. 9 Detection of stages based on the severity of lesions: **a** moderate; **b** normal; **c** proliferate

functions: (a) Bounding box regression; (b) Objectness; and (c) Class probability. Figure 7 projects the values of (a) mAP at 0.5 and (b) mAP at 0.95 were obtained during validation. Figure 8 proves the performance metrics obtained for the proposed model (a) $F1$ curve; (b) Confusion matrix. Figure 9 explains the detection of stages based on the severity of lesions: (a) moderate; (b) normal; and (c) proliferate.

5 Conclusion and Future Work

Eyes are one of the vital parts of the human body, so they should be protected with enough care. Eyes can be affected by various diseases. If not treated priorly, diabetic retinopathy will lead to permanent blindness. The APTOS dataset is split into 80:20 ratio for model building and model estimation (i.e., training:testing).It is used in the study to propose a deep neural network(DNN) model for identifying different phases of DR. The YOLOv7 deep neural network architecture and PyTorch framework are used in the implementation of the model. The YOLOv7 architecture is trained on more artificially created images by utilizing multiple augmentation approaches, which in turn increases model accuracy. For the detection of various phases of DR, the validation set achieved a mean average precision of 0.921 at a 0.5 threshold and a mean average precision of 0.902 at a 0.95 threshold. Overall, the proposed methodology demonstrates a high accuracy in prediction of the severity of lesions in fundus images. Identifying microaneurysms is a complex task and possesses a significant challenge for future research.

In future, new strategies for feature extraction will be explored to identify nano-sized microaneurysms, with the aim of achieving improved results. A chabot application for diabetic retinopathy detection could be a useful tool to help individuals with diabetes monitor their eye health and detect the early signs of diabetic retinopathy. The chatbot could work by asking users a series of questions about their diabetes history and current eye symptoms and then use this information to provide personalized recommendations and resources for further evaluation or treatment. Some potential features of a chatbot application for diabetic retinopathy detection could include: (i) Risk assessment: The chatbot could ask users about their diabetes history, blood sugar control, and other risk factors for diabetic retinopathy and use this information to assess their individual risk for developing the condition; (ii) Symptom tracking: Users could report any changes in their vision or other symptoms related to diabetic retinopathy, and the chatbot could provide guidance on when to seek further evaluation; (iii) Educational resources: The chatbot could provide users with educational resources about diabetic retinopathy, including how it develops, how it is diagnosed, and what treatments are available; (iv) Referral to eye care professionals: If the chatbot detects signs of diabetic retinopathy or other eye problems, it could provide users with information on how to schedule an eye exam with a qualified eye care professional. It is important to note that a chatbot application for diabetic retinopathy detection should not be used as a substitute for regular eye exams and medical evaluations. However, it could serve as a useful tool for individuals with diabetes to monitor their eye health and seek timely evaluation and treatment when necessary. Future plans include carrying out the aforementioned tasks.

References

1. World Health Organization (2019) World report on vision. Available: https://www.who.int/publications/i/item/9789241516570
2. International Council of Ophthalmology (2017) Updated 2017 ICO guidelines for diabetic eye care, pp 1–33. [Online]. Available: https://icoph.org/eye-care-delivery/diabetic-eye-care
3. World Health Organization (2019) World report on vision IAPB Summary 2 WRV | World Report on Vision. [online] Available: https://www.iapb.org/wp-content/uploads/2020/09/WRV-IAPB-Summary.pdf
4. Atwany MZ, Sahyoun AH, Yaqub M (2022) Deep learning techniques for diabetic retinopathy classification: a survey. IEEE Access 10:28642–28655. https://doi.org/10.1109/ACCESS.2022.3157632
5. Porwal P, Pachade S, Kokare M, Deshmukh G, Son J, Bae W, Liu L, Wang J, Liu X, Gao L, Wu TB, Xiao J, Wang F, Yin B, Wang Y, Danala G, He L, Choi YH, Lee YC et al (2020) IDRiD: diabetic retinopathy—segmentation and grading challenge. Med Image Anal 59. https://doi.org/10.1016/j.media.2019.101561
6. Nirmala Devi M, Appavu S alias Balamurugan, Swathi UV (2013) An amalgam KNN to predict diabetes mellitus. In: 2013 IEEE international conference ON emerging trends in computing, communication and nanotechnology (ICECCN), pp 691–695. https://doi.org/10.1109/ICE-CCN.2013.6528591
7. Mateen M, Wen J, Nasrullah N, Sun S, Hayat S (2020) Exudate detection for diabetic retinopathy using pretrained convolutional neural networks. Complexity 2020. https://doi.org/10.1155/2020/5801870
8. Qureshi I, Ma J, Abbas Q (2021) Diabetic retinopathy detection and stage classification in eye fundus images using active deep learning. Multimed Tools Appl 80(8):1169111721. https://doi.org/10.1007/s11042-020-10238-4
9. Nirmala Devi M, Appavu alias Balamurugan, Reshma Kris M (2016) Developing a modified logistic regression model for diabetes mellitus and identifying the important factors of Type II DM. Indian J Sci Technol 9(4):1–8. https://doi.org/10.17485/ijst/2016/v9i4/87028
10. Rajeswari AM, Subbulakshmi B, Nirmaladevi M, Sivakumar M (2023) Application of probability based surprising measure in outlier detection. Eur Chem Bull 12(Special Issue-5):1826–1837. https://doi.org/10.48047/ecb/2023.12.si5.174

11. Qiao L, Zhu Y, Zhou H (2020) Diabetic retinopathy detection using prognosis of microaneurysm and early diagnosis system for non-proliferative diabetic retinopathy based on deep learning algorithms. IEEE Access 8:104292104302. https://doi.org/10.1109/ACCESS.2020.2993937
12. Wang Y, Wang H, Xin Z (2022) Efficient detection model of steel strip surface defects based on YOLO-V7. IEEE Access. https://doi.org/10.1109/ACCESS.2022.3230894
13. Shahin M, Chen FF, Hosseinzadeh A, Khodadadi Koodiani H, Bouzary H, Shahin A (2023) Enhanced safety implementation in 5S + 1 via object detection algorithms. Int J Adv Manuf Technol. https://doi.org/10.1007/s00170-023-10970-9
14. Qiu Y, Lu Y, Wang Y, Jiang H (2023) IDOD-YOLOV7: image-dehazing YOLOV7 for object detection in low-light foggy traffic environments. Sensors 23(3):1347. https://doi.org/10.3390/s23031347
15. Alyoubi WL, Abulkhair MF, Shalash WM (2021) Diabetic retinopathy fundus image classification and lesions localization system using deep learning. Sensors 21(11). https://doi.org/10.3390/s21113704
16. Pinedo-Diaz G, Ortega-Cisneros S, Moya-Sanchez EU, Rivera J, Mejia-Alvarez P, Rodriguez-Navarrete FJ, Sanchez A (2022) Suitability classification of retinal fundus images for diabetic retinopathy using deep learning. Electronics (Switzerland) 11(16). https://doi.org/10.3390/electronics11162564

Comparative Analysis of Optimization Algorithms for Feature Selection in Heart Disease Classification

Ramdas Kapila and Sumalatha Saleti

Abstract In order to classify the Cleveland Heart Disease dataset, this study evaluates the performance of three optimization methods, namely Fruit Fly Optimization (FFO), Particle Swarm Optimization (PSO), and Grey Wolf Optimizer (GWO). The top 10 features are identified using FFO, with remarkable results for accuracy (88.52%), precision (87.88%), recall (90.63%), and $f1$-score (89.23%). After using PSO, the accuracy, precision, recall, specificity, and $f1$-score are all 85.25%, 87.10%, 84.38%, and 85.71% respectively. Finally, GWO is used, which results in precision, accuracy, recall, and $f1$-score values of 93.33%, 90.16%, 90.11%, 87.5%, and 90.32% respectively, highlighting its consistent superior performance. FFO shows competitive outcomes with notable accuracy and recall through a comparative examination. PSO displays comparable precision and recall while displaying a somewhat poorer accuracy. In contrast, GWO performs better than both FFO and PSO, displaying great accuracy and precision along with remarkable recall and specificity. These results provide important information on the effectiveness of feature selection methods utilized in optimization algorithms for heart disease classification. The study also highlights the need for more investigation into the potential of these optimization algorithms in many fields, broadening their use beyond the classification of disease. This kind of work might improve the progress of the field of feature selection and aid in the creation of better classification models.

Keywords Machine learning algorithms · Feature selection · Fruit fly optimization · Particle swarm optimization · Grey wolf optimizer

---

R. Kapila (✉)
Department of Computer Science and Engineering,
Narsimha Reddy Engineering College (Autonomous), Maisammaguda, Secunderabad, Telangana, India
e-mail: kapilaramdas@gmail.com

S. Saleti
Data Science Laboratory, Computer Science and Engineering,
SRM University-AP, Amaravathi, India
e-mail: sumalatha.s@srmap.edu.in

S. Bandyopadhyay et al. (eds.), *Intelligent Computing Systems and Applications*,
Lecture Notes in Networks and Systems 1010,
https://doi.org/10.1007/978-981-97-5412-0_34

1 Introduction

Heart disease remains to be a major source of death and morbidity worldwide, raising serious concerns about global health. By enabling early intervention and the most suitable treatment options, the precise and timely prognosis of cardiac disease can have a significant influence on patient outcomes. Using machine learning (ML) methods to analyze complex patterns, correlations in medical data have shown to be a useful tool for predicting cardiac disease. Additionally, feature selection techniques in ML are crucial in locating the most pertinent characteristics for precise classification and assist in the development of trustworthy forecasting algorithms. Only a small subset of cardiovascular conditions categorized as "heart disease" includes coronary artery disease, cardiac arrest, arrhythmia, and valve issues. Age, gender, family history, hypertension, diabetes, smoking, and obesity are just risk factors that have been associated with these illnesses. The intricate interplay of these factors necessitates the use of sophisticated analytical techniques capable of gathering and analyzing the intricate nature of data related to cardiac disease [1]. By using enormous volumes of data to unearth hidden patterns, correlations, and predictive characteristics, ML algorithms provide a distinct edge in the prediction of cardiac disease. ML models may recognize subtle trends that can escape the notice of conventional statistical methods by learning from past patient data. These models may then produce precise forecasts and risk assessments for specific patients, assisting physicians in deriving conclusions regarding a patient's diagnosis, course of therapy, and preventative measures [2].

However, the choice of appropriate features has a major influence on how ML systems can anticipate heart illness more accurately. Finding the subset of variables or features that contributes the most to the prediction job while removing unnecessary or redundant data is known as feature selection. This procedure is essential because it lowers the dimensionality of the data, makes models easier to understand, reduces the chance of overfitting, and improves computing performance. For the purpose of forecasting disease, a variety of strategies for selecting attributes have been proposed in ML, with the aim of identifying the perfect set of qualities that results in the highest level of classification accuracy. Due to their effectiveness in effectively searching through the feature space and locating the most useful features, optimization algorithms like FFO, PSO, and GWO have drawn interest. To improve classification performance, these algorithms use principles taken from nature to iteratively assess and update feature subsets. Its ability to improve risk stratification, enable personalized therapy and help clinical decision-making makes ML significant in cardiac disease forecasting and feature selection. To detect high-risk patients who might benefit from preventative treatments or intense monitoring, healthcare providers might use accurate prediction models. Additionally, ML-based techniques have the ability to diagnose cardiac problems early, enabling prompt therapies and better patient outcomes.

The Cleveland Heart Disease dataset is utilized to evaluate the effectiveness of FFO, PSO, and GWO in selecting features for classification. In an effort to determine how successful different optimization algorithms are and how they are perhaps used to improve disease prediction models, we examine their performance measures. This study's findings could facilitate the establishment of better and more trustworthy diagnostic tools as well as advance ML-based approaches for forecasting heart ailments.

The study's findings shed important light on the efficiency of these optimization algorithms and their potential for improving classifying reliability in the context of heart disease diagnosis. The study also emphasizes how crucial it is to investigate how similar algorithms may be applied in other sectors and the wider implications of feature selection for boosting classification models in a variety of academic and business industries.

The paper's major findings have been outlined below:

- Evaluation of Optimization Algorithms: This work systematically evaluates three optimization algorithms for feature selection in heart disease classification: FFO, PSO, and GWO. By putting these algorithms to use on the dataset and comparing their performance metrics, it is possible to see how effective and suitable each method is to forecast heart illness.
- Comparative Analysis: The three optimization methods are thoroughly compared in this research in terms of how well they can select features. The study gives an understanding of the merits and demerits of each method by analyzing their performance across many assessment parameters. This investigation helps determine which algorithm produces the most accurate and trustworthy feature subsets for classifying heart disease.
- The study emphasizes the impact of picking attributes and their substantial improvement on the prognosis of the disease. The study points out the value of selecting the most suitable features for precise classification by examining the effect of choosing features on the ability to perform ML models. This addition aids in improving the generalized and interpretability of models used to forecast cardiac disease.

The paper is structured as follows: Sect. 2 is a summary of the corresponding work in the field provided. In Sect. 3, we describe the materials used in this study and present the proposed architecture. Section 4 presents the finding obtained from the experiments and offers a comprehensive analysis of these findings. Lastly, in Sect. 5, conclude the research by summarizing the key insights and highlighting potential avenues for future research.

2 Related Work

Numerous research has recently concentrated on applying ML approaches to predict cardiac conditions. This research has investigated various feature selection, developing models, and performance evaluation techniques. With a focus on the most important discoveries and contributions, this section gives a summary of the relevant work in the area. Decisions in the field of medical diagnosis are frequently relied on the experiences of experts. Nevertheless, not every experience aids in the accurate identification of a condition. Researchers have investigated techniques like attribute reduction and rule extraction to solve this. This work suggests a genetic algorithm-based fuzzy decision support system and uses real-world datasets to show how successful it is in predicting the degree of heart disease risk [3]. This study effectively predicts cardiovascular disorders using Data Mining Techniques (DMTs) using clinical data, with the Optimized Decision Tree model emerge as the most potent classifier [4]. To enhance cancer classification using microarray datasets, use a hybrid filter-genetic feature selection technique. The suggested strategy improves accuracy in cancer detection by integrating filter feature selection techniques with a genetic algorithm. Experimental findings on datasets related to breast, lung, central nervous system, and brain cancer show superior performance over conventional ML techniques [5]. The work proposes a hybrid NaN-ANOVA approach for diagnosing heart illness [6]. The proposed system includes feature selection using a hybrid accelerated artificial bee colony and chicken swarm optimization algorithm (AABC-CSO) with classification using a genetic algorithm-based naive Bayes classifier (GA-NBC). Experimental results on a heart dataset achieve 61.08% accuracy, 31.59% sensitivity, 67.85% specificity, 17.95% precision, 23.41% f-measure, 46.69% G-mean, and 0.4480% loss. This work suggests a feature selection and dimensionality reduction strategy for heart disease prediction using ML [7]. While employing PCA directly from the raw data produced poorer results, chi-square and PCA together beat other classifiers. using ML to diagnose cardiac disease in a reliable and accurate manner. Support vector machines, logistic regression, artificial neural networks, K-Nearest Neighbour, naive bayes, and decision trees are some examples of classification algorithms that are used in conjunction with feature selection techniques to eliminate superfluous and redundant information [8]. A ground-breaking method for quick conditional mutual information feature selection is introduced. The system has good accuracy, particularly when combined with a support vector machine. The recommended method may be applied to the healthcare sector to detect heart issues. Through the use of several techniques, this study suggests a model for the accurate prediction of heart disease. Effective data collecting, pre-processing, and transformation methods are used. New hybrid classifiers are created using the Relief and LASSO feature selection algorithms [9]. This study utilizes ML and optimization algorithms for heart disease prediction. The Fast Correlation-Based Feature Selection (FCBF) method filters redundant features, and various classification algorithms are applied [10]. The proposed hybrid approach achieves maximum accuracy using FCBF, Particle Swarm Optimization (PSO), and Ant Colony Optimization (ACO). The results demonstrate superior performance compared to other techniques. This study attempts to identify significant risk indicators from a high-dimensional dataset to enhance classification accuracy with fewer issues. Relevant features have been selected using filter-based feature selection approaches with many models assessed based on accuracy, ROC curve, and $f1$-score. Results show that relevant features have a considerable influence on classification accuracy, leading to better performance and faster training times when utilizing a smaller feature subset [11]. A brand-new clustering technique for disease forecasting utilizing numerical data and ECG signals. The ECG data are subjected to Discrete Wavelet Transform (DWT) signal decomposition and Principal Component Analysis (PCA) dimensionality reduction. The hybrid Jaya-Red Deer Algorithm (J-RDA) is used to reach the optimized feature extraction stage. With J-RDA tweaking important parameters, the hybrid clustering combines optimized Density-based Spatial Clustering of Applications with Noise (DBSCAN) and optimized K-Means Clustering (KMC). The suggested approach successfully tackles the difficulties in predicting cardiac disease for dual data sources [12]. A novel ensemble Quine McCluskey Binary Classifier (QMBC) to predict heart disease using the fusion of ANOVA with Principal Component Analysis [13].

The related work presents diverse approaches in ML-based heart disease prediction. Techniques like genetic algorithms, hybrid feature selection, optimization algorithms, and clustering have been employed to enhance accuracy and diagnosis. These findings highlight the significance of feature selection and algorithmic optimization in achieving precise and efficient predictions. Building upon these advancements, this study contributes by evaluating the effectiveness of FFO, PSO, and GWO for feature selection in heart disease classification.

3 Methodology

The methodology involves a number of phases, including data preprocessing, the application of baseline models [Logistic Regression (LR), Decision Tree (DT), Random Forest (RF), and K-Nearest Neighbours (KNN)], the use of optimization techniques, and the evaluation of the models based on different performance metrics. The procedural workflow of the recommended model is illustrated in Fig. 1.

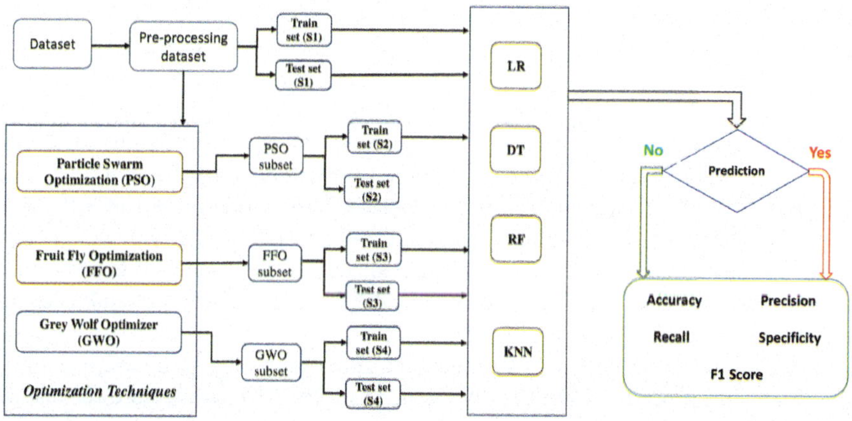

Fig. 1 Proposed architecture

3.1 Preprocessing

Our method starts with thorough preparation to ensure high-quality data for precise categorization. We deal with missing data using drop() and imputation, handle outliers with Z-score, and encode categorical variables using label encoding. Numerical characteristics are scaled to a uniform value by standardization and normalization. Utilizing techniques like Label Encoder, data type conversion assures compatibility. Together, these methods create a solid basis for strong classification models, improving the consistency and quality of the data.

3.2 Splitting Dataset

Dataset splitting is a key component of the suggested strategy for classifying the dataset. Training and testing sets are created from the dataset so that the success of the model may be assessed. In this work, an 80:20 train-to-test splits ratio is utilized, in which the models are trained using 80% of the data and their performance assessed using 20% of the data.

3.3 Optimization Techniques for Feature Selection

FFO, PSO, and GWO are three optimization approaches utilized in the proposed methodology for feature selection. These methods effectively look for the ideal collection of criteria that substantially contribute to classifying cardiac disease by using the laws of natural phenomena.

Fruit Fly Optimization (FFO) Fruit flies' foraging techniques served as the basis for Fruit Fly Optimization. By repeatedly changing the locations of computer-generated fruit flies in a multidimensional search space, it simulates how fruit flies look for food sources. Each fruit fly is a potential response, which in this case refers to a subset of characteristics. The optimization procedure entails assessing each fruit fly's (solution's) fitness in accordance with a fitness function, in this instance one that is connected to the effectiveness of the classification model. The fruit flies travel about the search area by updating their locations in accordance with predetermined rules, such as attraction to food sources and haphazard movement. When the most effective collection of features is identified, the algorithm converges and achieves the best classification performance [14, 15].

Selecting the ideal subset of attributes for classifying may be done using FFO. The following procedures are usually included in the FFO architecture for feature selection:

- Initialization: Create a population of fruit flies at random, with each fruit fly serving as a candidate feature subset.
- Fitness Evaluation: Based on a fitness function that reflects the classification efficiency using relevant features, evaluate the fitness of each fruit fly (feature subset).
- Update the locations of the fruit flies according to predetermined principles for movement and attraction. This might use a combination of attraction to more promising feature subsets based on fitness values and random movement to explore the search space of feature combinations.
- The movement and attraction processes should be repeated until a termination condition is satisfied. This might be the maximum amount of repetitions or achieving a certain level of fitness.
- Convergence: The ideal feature subset that produces the greatest classification performance corresponds to the fruit fly with the greatest fitness when the optimization process converges.

Particle Swarm Optimization (PSO) Based on how birds swarm or fish school serves as the basis for particle swarm optimization. It involves a number of particles that move in a multidimensional search space and represent potential solutions. The effectiveness of a classified model using each particle's assigned subset of characteristics is indicated by its location in the search space. The positioning and velocities of the particles are repeatedly updated during the optimization process using their own prior locations, the locations of the top-performing particles (local best), and the top-performing particle within the whole swarm (global best). The particles work together to properly search the search area and to settle on the ideal group of attributes that produces the greatest classification result [16].

The best collection of features for classifying may be selected via PSO. The PSO architecture for feature selection generally goes through the following steps:

- Initialization: Create a collection of particles in the attribute space with random placements and speeds; each particle should represent a potential combination of characteristics for the subset.
- Fitness Evaluation: Utilizing a fitness function that gauges the effectiveness of categorization based on the pertinent characteristics, evaluate the suitability of every particle (feature subset).
- Updates to Velocity and Position: Each particle's velocity and position are adjusted in accordance with its previous location, the locations of the top-performing particles (local), and the global best particle. To direct the particles towards more advantageous feature combinations, this upgrade balances exploration and exploitation.
- Iterative Optimization: Repeat the velocity and position updates as long as the cancellation criterion is met. This might be the maximum amount of repetitions or achieving a certain level of fitness.
- Convergence: When the optimization process converges, the particle with the best classification performance corresponds to the best feature subset.

Grey Wolf Optimizer (GWO) Grey wolves' social structure and hunting methods served as the basis for Grey Wolf Optimizer. It simulates how wolves hunt for prey in a multidimensional environment. According to this method, each grey wolf stands for a potential solution (a subset of characteristics) and is given a location in the search space based on how fit it is. This fitness is assessed using a fitness function that is connected to the classification performance. The wolves' locations are repeatedly updated during the optimization process depending on their own positions as well as the positions of the alpha (the pack's top performer), beta, and delta wolves. Equations that simulate the wolves' hunting and circling behavior are used to update the locations. The most effective combination of characteristics that offers the best classification performance is eventually discovered through this iterative procedure [17, 18].

For classification, GWO may be employed to choose the optimal feature subset. The steps below are typical for the GWO architecture for feature selection:

- Initialization: Create a population of grey wolves at random, with each wolf representing a potential feature subset and being given a location in the space of feature combinations according to how fit it is.
- Fitness Assessment: Based on a fitness function that depicts the classification performance using the relevant features, assess the fitness of each wolf (feature subset) individually.
- Position updates: Each wolf's position should be updated depending on its own position as well as the positions of the alpha (the pack's top performer), beta, and delta wolves. Equations that simulate the grey wolf's hunting and circling behavior are used to achieve this update.
- The position updates should be iterated until a termination condition is satisfied. This might be the most iterations or achieving a good level of fitness.
- Convergence: When the optimization process converges, the wolf with the maximum fitness corresponds to the ideal feature subset that produces the best classification performance.

3.4 Proposed Methodology

After the Cleveland Heart Disease (HD) dataset has undergone meticulous preparation, an organized and rigorous analytical journey begins, poised to reveal the complicated relationship between ML models and feature selection optimization in the context of heart disease classification. Following a thoughtful 80:20 split, the preprocessed dataset is carefully divided into distinct training and testing subsets. These differentiated subsets serve as a testing ground for the ML models' prediction abilities.

*K*NN, DT, RF, and LR are among the models considered for this thorough examination. Each model has unique benefits and covers a range of prediction methods that will help choose the best method for classifying heart disease. In the following stage, a variety of performance measurements are put to use, illuminating the study's core. A careful calculation of performance measurements reveals the specific merits and demerits of each model. These measures go beyond simple observations and provide an understanding of the complex interaction between true positives, true negatives, false positives, and false negatives.

After carefully preprocessing the Cleveland Heart Disease (HD) dataset, the study launches a sophisticated investigation into feature selection optimization, a crucial stage that enhances the predictive capability of the ML models in the context of heart disease classification. PSO, FFO, and GWO are three separate and potent strategies that take center stage in this complex phase. Every method serves as an intellectual compass, directing the investigation and the extraction of the best subset of characteristics from the preprocessed dataset.

The PSO approach takes cues from social dynamics and apes a flock of birds' behaviours that are cooperative. This group intelligence converges on the best feature configuration, successfully highlighting the characteristics with the most precise predictive effects. The intrinsic relevance of the features inside the dataset is where FFO gets its strength, in contrast. FFO finds an enhanced subset that fully expresses predictive patterns by using the frequency of feature occurrence. The GWO approach stands out as a computational marvel in this complex landscape. GWO orchestrates a peaceful investigation of the feature space using a social structure modeled after that of grey wolves. A feature subset that is optimized for precision results from the interaction of alpha, beta, and delta wolves, which represents the delicate balance between exploration and exploitation.

These methods combine mathematical accuracy with computational skill to capture the core of the feature selection optimization step. This stage goes beyond merely manipulating the data; instead, it distills the knowledge of optimization techniques and uses it to create a feature subset that improves the prediction power of later ML models. The curated feature subset serves as a guide for the ML models LR, DT, RF, and *K*NN when they are called upon once more to perform the classification of heart disease. Each model enters the field of predictive assessment with a well-honed feature subset. Under this complex lens, performance measurements including accuracy, precision, recall, and $f1$-score are once more used to examine the predictive quality of the models. A startling realization emerges: the Random Forest (RF) model represents the pinnacle of predictive greatness when used in conjunction with the FFO and GWO methods. The inherent qualities of each component come together in this combination to create an effective combination that performs at a level that has never been achieved before.

4 Results

A more thorough examination of performance indicators, including aspects like accuracy, precision, recall, and $f1$-score, might enhance this work. The addition of an extensive performance evaluation not only heightens the scholarly robustness of this

Table 1 Results of models on Cleveland heart disease dataset

Results of models with preprocess dataset					Results of models with PSO FS subset				
Models	Accuracy	Precision	Recall	$F1$-score	Models	Accuracy	Precision	Recall	$F1$-score
LR	86.88	97.14	82.92	89.47	LR	85.24	87.09	84.37	85.71
DT	63.93	65.71	69.69	67.64	DT	83.6	89.28	78.12	83.33
DT	63.93	65.71	69.69	67.64	DT	83.6	89.28	78.12	83.33
RF	81.96	91.42	80	85.33	RF	86.88	85.29	90.62	87.87
KNN	77.04	85.71	76.92	81.08	KNN	68.85	68.57	75	71.64
SVM	80.32	91.42	78.04	84.21	SVC	70.49	66.66	87.5	75.67
Results of models with FFO FS subset					Results of models with GWO FS subset				
Models	Accuracy	Precision	Recall	$F1$-score	Models	Accuracy	Precision	Recall	$F1$-score
LR	88.52	87.87	90.62	89.23	LR	88.52	87.87	90.62	89.23
DT	77.04	87.5	65.62	75	DT	78.68	88	68.75	77.19
RF	90.16	93.33	87.5	90.32	RF	90.16	93.33	87.5	90.32
KNN	68.85	98.57	75	71.64	KNN	68.85	68.57	75	71.64
SVC	70.4	66.66	87.5	75.67	SVC	70.49	66.66	87.5	75.67

study but also highlights its application in the real world, allowing practitioners to examine the genuine efficacy and potential drawbacks of the suggested classifying systems. Table 1 displays the performance of multiple models employing various optimizers on the Cleveland heart disease dataset. The Particle Swarm Optimization (PSO) feature subset, Feature Frequency Ordering (FFO) FS subset, and Grey Wolf Optimizer (GWO) FS subset results are compared in the table.

The model's performance is examined using the preprocessed dataset, and it showed a range of results. The LR model, which had an accuracy of 86.88%, is a viable option. The $f1$-score is 89.47% due to its excellent accuracy (97.14%) and acceptable recall (82.92%). Precision, recall, and $f1$-score values for the DT model were 65.71%, 69.69%, and 67.64%, respectively, with an accuracy of 63.93%. With an accuracy of 81.96% and an $f1$-score of 85.33%, as well as remarkable precision (91.42%) and recall (80%) values, the RF model demonstrated promising results. The KNN model's accuracy is 77.04%, while its precision is 85.71%, recall is 76.92%, and $f1$-score is 81.08%. The SVM model's precision, recall, and $f1$-score values are 91.42%, 78.04%, and 84.21%, respectively, for a final accuracy of 80.32%.

Figures 2, 3, 4, and 5 illustrate the model outcomes without, with PSO, with FFO, and with GWO feature selection techniques.

The model are then assessed using feature subsets created using various optimization methods. The LR model attained an accuracy of 85.24% with a precision of 87.09%, recall of 84.37%, and a $f1$-score of 85.71% while taking into account the PSO FS subset. The $f1$-score for the DT model is 83.33%, with an accuracy of 83.6%, competitive precision (89.28%), and recall (78.12%) values. Notably, the RF model outperformed all other models, attaining an $f1$-score of 87.87% and an accuracy of 86.88%. With the PSO FS subset, the RF model similarly displayed strong

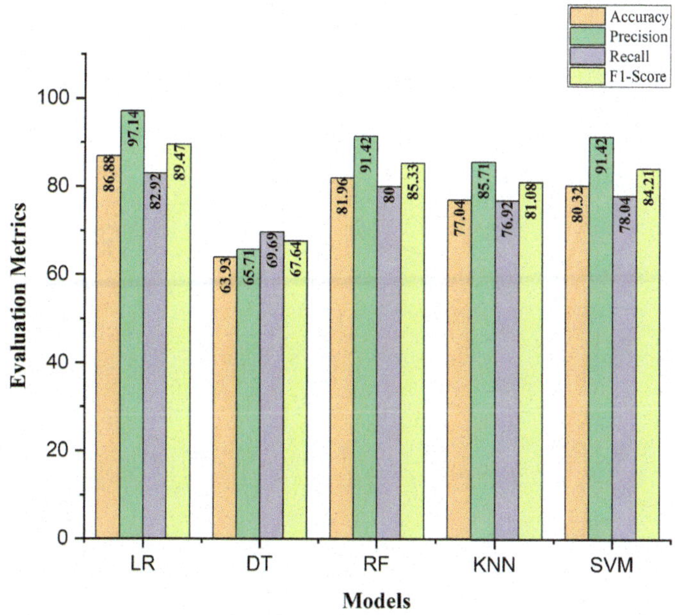

Fig. 2 Results of models without feature selection

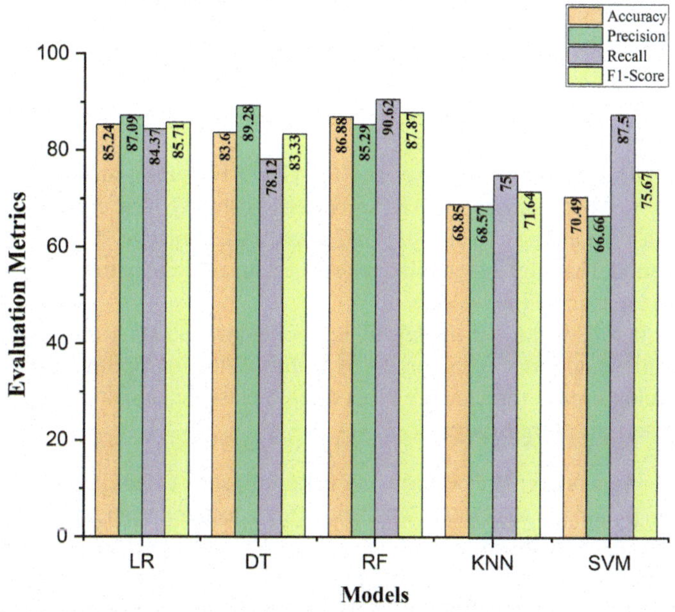

Fig. 3 Results of models with PSO

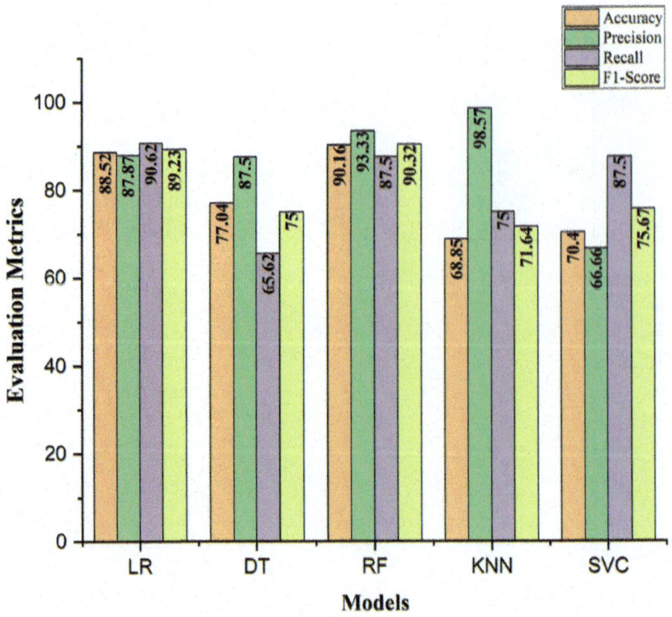

Fig. 4 Results of models with FFO

recall (90.62%) and good accuracy (85.29%). The *K*NN model has an accuracy of 68.85%, a precision of 68.57%, a recall of 75%, and an *f*1-score of 71.64%. Similar to this, the SVC model exhibited an *f*1-score of 75.67%, precision of 66.66%, recall of 87.5%, and accuracy of 70.49%.

Additionally, the FFO FS subset and the GWO FS subset are used to evaluate the models. On both subsets, the RF model consistently beat competing models, recording an outstanding *f*1-score of 90.32% and an accuracy of 90.16%. This shows that the RF model has higher performance on the dataset, demonstrating its robustness and effectiveness in feature selection.

The top model is the RF model, which is determined by a thorough analysis. It consistently outperformed a variety of feature selection optimizers, proving its efficiency and promise for accurate heart disease classification and diagnosis.

The limitations of the study:

- Parameter Sensitivity: The performance of optimization techniques may be affected by certain parameter selections, which calls for a deeper investigation of parameter sensitivity and robustness.
- Scalability Constraints: Despite being thorough, the study's evaluation may not accurately reflect how well certain strategies scale to bigger datasets seen in real-world circumstances.
- The study's focus on certain optimization techniques may have caused it to ignore possibly superior options, which would have impacted the overall comparison.

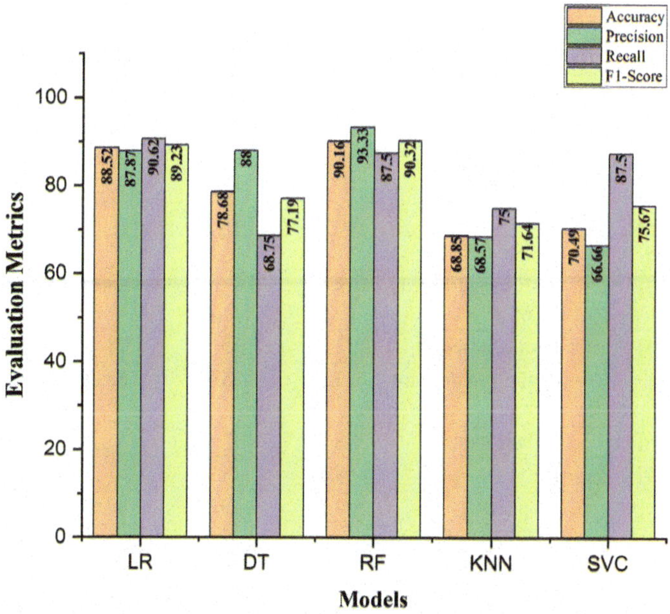

Fig. 5 Results of models with GWO

- The potential of Ensemble Methods: The standalone model strategy lacks the advantages of ensemble techniques, which may improve overall prediction performance.

Table 2 provides a comprehensive contrast between the proposed methodology and existing techniques.

5 Conclusion and Future Work

Feature selection optimizers such as Particle Swarm Optimization (PSO), Feature Frequency Ordering (FFO), and Grey Wolf Optimizer (GWO) have been utilized in this study to compare the performance of models on the dataset. The models underwent assessments for accuracy, precision, recall, and $f1$-score in order to choose the model that classified heart disease the best. According to our research, the RF consistently performed better than other models when using various feature selection optimizers. The RF model successfully classified instances of heart disease with accuracy of 81.96% and an $f1$-score of 85.33% using the preprocessed dataset. Additionally, the RF consistently produced excellent accuracy and $f1$-score values while taking into account the PSO, FFO, and GWO feature subsets, demonstrating its resilience in feature selection and classification. The RF is preferable because it can identify complex relationships and trends in the information, which makes it perfect for categorizing cardiac illnesses. The RF model's strong accuracy and recall

Table 2 Proposed model evaluation in relation to contemporary methods

S. No.	Author(s)	Best model	Accuracy	Precision	Recall	$F1$-score
1	[19]	Logistic regression	89	–	77	–
2	[20]	PSO-SVM	88.22	–	–	–
3	[21]	Hybrid RF ith a linear model	88.7	90.1	92.8	90
4	[22]	RF	88.34	–	–	–
5	[23]	Voting classifier	87.4	–	–	–
6	[24]	RBF kernel-based SVM	81.19	–	88.68	–
7	[25]	Decision tree	73	75	78	74
8	[4]	Optimized decision tree	73.54	75.82	68.89	–
9	[26]	Hybrid model	88.7	–	–	–
10	[27]	RF	88.5	–	–	–
11		Proposed methodology	90.16	93.33	87.5	90.32

scores also suggest that it has the capacity to reduce FP and FN, producing accurate predictions. In this study, the GWO consistently exhibits the best performance among the investigated feature selection optimization strategies, outperforming both PSO and FFO.

In order to further improve predicted accuracy, future prospects for the classification of heart disease include investigating ensemble methods and sophisticated feature selection techniques. The model's applicability and dependability may be increased by validating it using other datasets and including new risk variables. The model's usefulness and influence on patient outcomes in the actual world will be determined via clinical deployment and assessment. The model's adoption into clinical practice will also be facilitated by attempts to improve interpretability and explainability.

References

1. Katarya R, Srinivas P (2020) Predicting heart disease at early stages using machine learning: a survey. In: 2020 international conference on electronics and sustainable communication systems (ICESC). IEEE, pp 302–305
2. Methaila A, Kansal P, Arya H, Kumar P et al (2014) Early heart disease prediction using data mining techniques. Comput Sci Inf Technol J 24:53–59
3. Paul AK, Shill PC, Rabin MRI, Akhand M (2016) Genetic algorithm based fuzzy decision support system for the diagnosis of heart disease. In: 2016 5th international conference on informatics, electronics and vision (ICIEV). IEEE, pp 145–150
4. Martins B, Ferreira D, Neto C, Abelha A, Machado J (2021) Data mining for cardiovascular disease prediction. J Med Syst 45:1–8

5. Ali W, Saeed F (2023) Hybrid filter and genetic algorithm-based feature selection for improving cancer classification in high-dimensional microarray data. Processes 11(2):562
6. Karuppiah K, Balamurugan N, Venkatesh R et al (2023) Diagnosis of heart disease using improved genetic algorithm-based Naive Bayes classifier. In: Using multimedia systems, tools, and technologies for smart healthcare services. IGI Global, pp 117–140
7. Gárate-Escamila AK, El Hassani AH, Andrès E (2020) Classification models for heart disease prediction using feature selection and PCA. Inform Med Unlock 19:100330
8. Li JP, Haq AU, Din SU, Khan J, Khan A, Saboor A (2020) Heart disease identification method using machine learning classification in e-healthcare. IEEE Access 8:107562–107582
9. Ghosh P, Azam S, Jonkman M, Karim A, Shamrat FJM, Ignatious E, Shultana S, Beeravolu AR, De Boer F (2021) Efficient prediction of cardiovascular disease using machine learning algorithms with relief and lasso feature selection techniques. IEEE Access 9:19304–19326
10. Khourdifi Y, Baha M (2019) Heart disease prediction and classification using machine learning algorithms optimized by particle swarm optimization and ant colony optimization. Int J Intell Eng Syst 12(1)
11. Pathan MS, Nag A, Pathan MM, Dev S (2022) Analyzing the impact of feature selection on the accuracy of heart disease prediction. Healthc Anal 2:100060
12. Sonawane R, Patil H (2022) Automated heart disease prediction model by hybrid heuristic-based feature optimization and enhanced clustering. Biomed Sig Process Control 72:103260
13. Kapila R, Ragunathan T, Saleti S, Lakshmi TJ, Ahmad MW (2023) Heart disease prediction using novel quine McCluskey binary classifier (QMBC). IEEE Access
14. Pan W-T (2011) Fruit fly optimization algorithm. Tsang Hai Book Publishing Co., Taipei, pp 10–12
15. Yang M, Liu N-B, Liu W (2017) Image 1D OMP sparse decomposition with modified fruit-fly optimization algorithm. Cluster Comput 20(4):3015–3022
16. Eberhart R, Kennedy J (1995) Particle swarm optimization. In: Proceedings of the IEEE international conference on neural networks, vol 4. Citeseer, pp 1942–1948
17. Mirjalili S, Mirjalili SM, Lewis A (2014) Grey wolf optimizer. Adv Eng Softw 69:46–61
18. Zhao G, Wang H, Jia D, Wang Q (2019) Feature selection of grey wolf optimizer based on quantum computing and uncertain symmetry rough set. Symmetry 11(12):1470
19. Haq AU, Li JP, Memon MH, Nazir S, Sun R (2018) A hybrid intelligent system framework for the prediction of heart disease using machine learning algorithms. Mobile information systems 2018:1–21
20. Vijayashree J, Sultana HP (2018) A machine learning framework for feature selection in heart disease classification using improved particle swarm optimization with support vector machine classifier. Program Comput Softw 44:388–397
21. Mohan S, Thirumalai C, Srivastava G (2019) Effective heart disease prediction using hybrid machine learning techniques. IEEE Access 7:81542–81554
22. Gokulnath CB, Shantharajah S (2019) An optimized feature selection based on genetic approach and support vector machine for heart disease. Cluster Comput 22:14777–14787
23. Amin MS, Chiam YK, Varathan KD (2019) Identification of significant features and data mining techniques in predicting heart disease. Telematics Inform 36:82–93
24. Saqlain SM, Sher M, Shah FA, Khan I, Ashraf MU, Awais M, Ghani A (2019) Fisher score and Matthews correlation coefficient-based feature subset selection for heart disease diagnosis using support vector machines. Knowl Inf Syst 58:139–167
25. Princy RJP, Parthasarathy S, Jose PSH, Lakshminarayanan AR, Jeganathan S (2020) Prediction of cardiac disease using supervised machine learning algorithms. In: 4th international conference on intelligent computing and control systems (ICICCS). IEEE, pp 570–575
26. Kavitha M, Gnaneswar G, Dinesh R, Sai YR, Suraj RS (2021) Heart disease prediction using hybrid machine learning model. In: 6th international conference on inventive computation technologies (ICICT). IEEE, pp 1329–1333
27. Karthick K, Aruna S, Samikannu R, Kuppusamy R, Teekaraman Y, Thelkar AR et al (2022) Implementation of a heart disease risk prediction model using machine learning. Comput Math Methods Med

A Study on Network Anomaly Detection Using Stacking-Based Machine Learning Algorithms for ASNM Datasets

Thangavel Murugan, Het Bhavinkumar Patel, Adil Mustafa Khokhawala, and W. Jaisingh

Abstract Detecting and analyzing the root cause of network traffic log problems is a labor-intensive and time-consuming operation, particularly for previously undiscovered failure patterns. To identify malicious logs from the advanced security network metrics datasets, our proposed solution is based on a stacking mechanism. According to training data input, there have been roughly three orthogonal approaches to developing intrusion detectors: (1) Detection based on knowledge, which models and matches the characteristics of malicious intrusions, (2) Detection based on anomalies, which models normal behavior and identifies deviations, and (3) Detection based on classification, which concurrently models dangerous and acceptable behavior. In the case of unknown or zero-day assaults evading detection, these strategies have a high false-negative rate, need extensive training and profiling, and are vulnerable. To overcome these problems, our proposed work is based on a stacking model, in which we deployed four machine learning algorithms, one at a time at level 1 and the other at level 0 for a better rate of testing accuracy. The performance of these approaches is relatively comparable, with Naive Bayes being the most effective at level 1 and support vector machines, decision tree, and *K*-nearest neighbor at level 0.

Keywords Machine Learning · Stacking · Anomaly Detection · Security · Networks

T. Murugan
Department of Information Systems and Security, College of Information Technology, United Arab Emirates University, Al Ain, UAE

H. B. Patel
University of Guelph, Guelph, ON, Canada

A. M. Khokhawala
National College of Ireland, Dublin, Ireland

W. Jaisingh (✉)
School of Information Science, Presidency University, Bengaluru, Karnataka, India
e-mail: jaisinghw@gmail.com

S. Bandyopadhyay et al. (eds.), *Intelligent Computing Systems and Applications*, Lecture Notes in Networks and Systems 1010,
https://doi.org/10.1007/978-981-97-5412-0_35

1 Introduction

In recent years, the rapid expansion and increasing complexity of computer networks have led to a rise in various security threats and network anomalies. Network anomalies, including malicious activities, unexpected traffic patterns, and abnormal behaviors, can severely impact network performance, compromise data integrity, and disrupt critical services. Detecting these anomalies promptly and accurately has become a paramount concern for network administrators and security experts.

Traditional approaches to anomaly detection in computer networks relied on rule-based systems and statistical methods, which often struggled to keep pace with the ever-evolving nature of cyber threats. As a result, there has been a growing interest in utilizing machine learning techniques to address this challenge. Machine learning algorithms have shown promising results in identifying network anomalies by leveraging their ability to learn from vast amounts of network data and identifying patterns that are difficult to capture using rule-based methods.

This research article aims to present a comprehensive study of detecting network anomalies using stacking-based machine learning algorithms, specifically applied to advanced security network monitoring (ASNM) datasets. Stacking, also known as stacked generalization, is an ensemble learning technique that combines the predictions of multiple machine learning models to create a more robust and accurate final prediction [1, 2]. By employing a stacking-based approach, we seek to enhance the anomaly detection capabilities and achieve superior performance compared to individual models.

The ASNM datasets utilized in this study are derived from real-world network traffic captured from diverse network environments, including enterprise networks, cloud-based systems, and critical infrastructure networks. These datasets encompass a wide range of network activities, both normal and abnormal, providing a rich and challenging environment for evaluating the effectiveness of the proposed anomaly detection methodology.

2 Related Work

The detection of network anomalies is a critical aspect of ensuring the security and reliability of modern computer networks. In recent years, machine learning algorithms have emerged as powerful tools for anomaly detection due to their ability to process large volumes of network data and identify complex patterns. This literature review provides an overview of the most recent studies related to detecting network anomalies using stacking-based machine learning algorithms, with a focus on ASNM datasets.

Jia et al. [3] explored the application of ensemble methods for network anomaly detection using ASNM datasets. They introduced a stacking-based ensemble approach that combined multiple machine learning algorithms, including decision trees, support vector machines (SVM), and random forests. The authors evaluated the proposed method's performance against individual algorithms and traditional ensemble methods. The results demonstrated that stacking consistently outperformed other techniques, achieving higher detection accuracy and robustness.

Liang et al. [4] presented a novel approach called "stacked autoencoder-based ensemble" (SAE-Ensemble) for detecting network anomalies in ASNM datasets. Their method utilized autoencoders to preprocess the network data and then combined multiple autoencoders using stacking. The study highlighted the benefits of leveraging autoencoders to improve anomaly detection performance and achieve better representation of complex network patterns.

Chen et al. [5] focused on the impact of feature selection techniques on the effectiveness of stacking-based anomaly detection algorithms for ASNM datasets. The authors investigated various feature selection methods, including recursive feature elimination and principal component analysis, and examined their influence on the model's performance. The study emphasized the significance of feature selection in enhancing the detection efficiency and accuracy.

Zhang et al. [6] proposed a hybrid model that combined stacking with deep learning architectures for detecting network anomalies in ASNM datasets. Their approach utilized a convolutional neural network (CNN) as one of the base learners in the stacking process. By capturing spatial features in network traffic data, the hybrid model demonstrated improved performance compared to traditional stacking methods.

Wu et al. [7] investigated the benefits of data augmentation techniques in the context of stacking-based anomaly detection. By augmenting ASNM datasets with various transformations and noise, their study aimed to improve the model's generalization capabilities on previously unseen anomalies. The findings highlighted the potential of data augmentation to enhance the robustness of anomaly detection models.

This literature review presents a snapshot of the latest research efforts in detecting network anomalies using stacking-based machine learning algorithms for ASNM datasets. The studies discussed herein demonstrate the diverse and evolving approaches in this field, aiming to enhance network security and mitigate potential threats effectively.

3 Proposed Study

This research article proposes an innovative method for identifying malicious intent in network requests, utilizing a combination of machine learning techniques, including stacking. The approach involves organizing multiple algorithms, such as KNN, SVM, Naive Bayes, and decision trees, into different tiers of the stack (tiers 0, 1, and 2). By running the code against various datasets, we ensure its high accuracy across all three datasets.

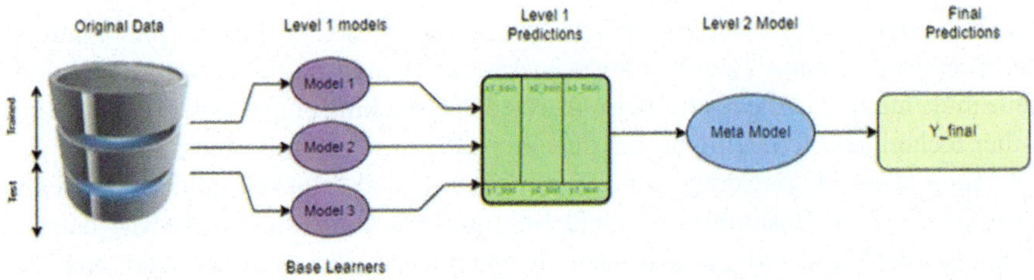

Fig. 1 System architecture

The stacking model [8, 9] employed in this study exhibits remarkable efficiency and a low false alarm rate. Our contribution to the field of anomaly detection lies in enabling organizations to comprehend and address unusual network activity through the utilization of our algorithm. By promptly detecting network anomalies, organizations can take proactive measures to mitigate vulnerabilities, effectively safeguarding their networks from potential attacks or breaches stemming from these specific anomalies. Figure 1 shows the system architecture of the proposed model.

3.1 Preprocessing

Preprocessing is a crucial phase in the study of detecting network anomalies using stacking-based machine learning algorithms for ASNM datasets. It involves transforming raw network data into a suitable format for training and evaluating the models.

The following preprocessing steps are applied in this study:

(a) Dataset Importation: The ASNM datasets are imported one by one, ensuring each sub-dataset is accessible for analysis within the study.
(b) Segmentation and Masking: Given the initial unorganized nature of the dataset, we perform segmentation and masking processes to separate and label features systematically. The resulting data is stored in two distinct arrays for further analysis.
(c) Resizing: To enhance the accuracy of computation and display, all network traffic data elements, such as IP addresses, MAC addresses, port numbers, and VLAN IDs, are converted into binary form.
(d) Standardization: To establish uniformity and improve model performance, all entries in the CSV dataset undergo standardization. The standardization process results in an average value of 0 and a unit standard deviation across the dataset.

By employing these preprocessing steps, the ASNM datasets are transformed into a suitable format, enabling the subsequent application of stacking-based machine learning algorithms for effective network anomaly detection.

3.2 Data Source

This research article utilizes publicly accessible data sourced from the IT4Innovations Center of Excellence, Faculty of Information Technology, Brno University of Technology, 612 00 Brno, Czech Republic. The dataset comprises three subsets, each constructed from network traffic traces and enriched with advanced security network metrics (ASNM) features [10, 11]. Each subset consists of 5000–6000 rows and 50–60 columns, providing a rich and diverse set of network traffic data for analysis.

3.2.1 ASNM-CDX 2009

The ASNM-CDX-2009 dataset serves as a valuable resource for researchers and practitioners aiming to develop and evaluate buffer overflow attack detection algorithms. Its comprehensive labeling structure enables investigations into the behavior and patterns of both malicious and legitimate network communications, enhancing the understanding of security vulnerabilities and the development of effective defense mechanisms.

It is designed to facilitate the study of buffer overflow attack detection. It comprises diverse labels that offer varying levels of granularity for differentiating between normal network traffic and malicious communications associated with buffer overflow attacks.

3.2.2 ASNM-NPBO v2

The ASNM-NPBO v2 dataset is a curated collection of network traffic data comprising non-payload-based obfuscation techniques applied to both malicious and legitimate TCP communications within vulnerable networks. The dataset focuses on challenging scenarios involving high difficulty and severity, leading to potential remote shell code execution through established backdoor communication.

By incorporating both malicious and legitimate TCP communications in the ASNM-NPBO v2 dataset, we provide a comprehensive resource for researchers and practitioners to explore the effectiveness of non-payload-based obfuscation techniques in challenging network environments. This dataset facilitates the evaluation and development of robust defense mechanisms against potential remote code execution and backdoor communication threats.

3.2.3 ASNM-TUN

The ASNM-TUN dataset is a specialized collection of advanced security network metrics (ASNM) features extracted from tcp-dump captures of tunneling obfuscation techniques applied to malicious traffic. The dataset aims to challenge and enhance machine learning classifiers by simulating scenarios where malicious actors attempt to evade detection.

The ASNM-TUN dataset incorporates four distinct types of labels, each offering increasing levels of granularity for classification purposes:

(a) Two-Class Label (Label_2):

The first set of labels, referred to as "Label_2," is a binary classification scheme that categorizes samples into either network buffer overflow attacks or legitimate inflow of communication.

(b) Three-Class Label (Label_3):

"Label_3" introduces a three-class classification, enabling differentiation between legitimate traffic, direct attacks, and complicated network attacks.

(c) Multi-Class Label (Label_poly):

"Label_poly" is a composite label consisting of two components. Firstly, it includes a three-class classification similar to "Label_3," differentiating between various types of network traffic. Secondly, "Label_poly" incorporates an acronym denoting the network service associated with the communication.

(d) Multi-Class Label with Network Modification (Label_poly_s):

The final label, "Label_poly_s," combines three elements. It includes a three-class classification similar to "Label_3," an acronym of the network service, and information about the network modification technique involved.

3.3 *Stacking Model Architecture*

The stacking model will be designed as a two-layer ensemble. The first layer will consist of the base learners (*K*NN, SVM, Naive Bayes, and decision tree), which will individually make predictions on the ASNM datasets. The second layer will comprise a meta-learner (e.g., logistic regression or random forest) that will combine the base learners' outputs to make the final anomaly prediction [7, 8].

(a) *K*-Nearest Neighbors (*K*NN):

We will implement the *K*NN algorithm, a non-parametric lazy learner, for detecting network anomalies in the ASNM datasets. The value of "*k*" will be tuned to identify the optimal number of neighbors for classification.

(b) Support Vector Machine (SVM):

SVM, a powerful supervised learning algorithm, will be implemented with different kernel functions (linear, polynomial, and radial basis function) to evaluate its performance in detecting network anomalies.

(c) Naive Bayes:

We will implement Gaussian Naive Bayes, a probabilistic classification algorithm, for distinguishing between normal and anomalous network traffic in the ASNM datasets. The assumption of normal distribution will be validated.

(d) Decision Tree:

Decision tree classifier will be built and pruned to prevent overfitting. We will experiment with different splitting criteria (e.g., Gini impurity, entropy) to find the optimal configuration for detecting network anomalies.

4 Experimental Results

The following are the prerequisites for the machine that will be used to carry out the research.

The setup is used to implement the complete algorithm. The physical requirements are minimal, as shown in Table 1, and the suggested model may be used with practically any physical machine.

4.1 Data Input

The necessary libraries used in the implementation are as follows:

(a) Pandas: It is an open-source Python package that is extensively used for data analysis and machine learning activities; it is utilized for data cleaning and analysis in this case.
(b) NumPy: A Python library that adds support for massive, multi-dimensional arrays and matrices, as well as a vast set of high-level mathematical functions to work on them.
(c) Sklearn: It is a popular tool for statistical modeling, such as classification, regression, clustering, and dimensionality reduction.
(d) Matplotlib: Matplotlib is a numerical extension of the NumPy library for cross-platform data visualization and graphical charting in Python.

Table 1 Tested environment

OS	Windows 10
RAM	8 GB
GPU	4 GB
IDE	Visual studio code (python)

4.2 Results and Discussions

The train-test split method was used to divide the dataset into two separate subarrays. The training component accounted for 75% of the native data, whereas testing component is made up 25% of the native dataset. Following the processing of the models' details, a performance-based inquiry was undertaken to choose the best model out of four. The parameter metrics used to identify models in this comparative analysis were accuracy, precision, *F*1 score, and recall, which may be computed from the confusion matrix, and likelihood of belonging to a certain class is used to represent all of the models' output, which can take the shape of a fraction ranging from 0 to 1. There are three datasets in this example: ASNM-CDX 2009 dataset, ASNM-NBPOv2 dataset, ASNM-TUN dataset. And four algorithms are being used, and they are as follows: *K*-nearest neighbor (*K*NN), decision tree (DT), support vector machine (SVM), and Naïve Bayes (NB).

The proposed study employed a stacking strategy to identify the best algorithm for finding improved accuracies in the dataset, and therefore, it has four programs for each dataset and three datasets in total. For each dataset, the four programs are divided as follows:

Level 0: DT, SVM, NB	Level 1: *K*NN
Level 0: *K*NN, SVM, NB	Level 1: DT
Level 0: *K*NN, DT, NB	Level 1: SVM
Level 0: *K*NN, DT, SVM	Level 1: NB

Figure 2 depicts the performance shown by four algorithms on ASNM-CDX 2009 dataset. *K*-nearest neighbor algorithm and support Naïve Bayes algorithm have a perfect score of metrics. The support vector machine model has a high precision score only while least metrics are shown by the decision tree model.

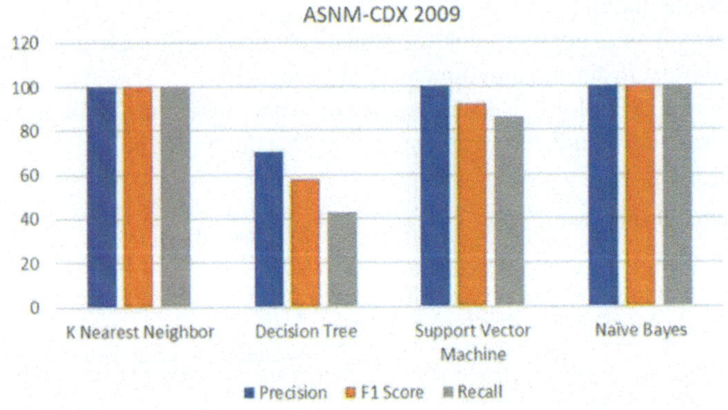

Fig. 2 Comparison of metrics (precision, recall, and *F*1 score) for ASNM-CDX 2009 dataset

The graph of Fig. 3 shows the performance of all AI models on ASNM-NBPOv2 dataset. Support vector machine, Naïve Bayes, and K-nearest neighbor show high values in all the three metrics while decision tree accounted for 98%. The bar graph of Fig. 4 shows the performance of the three artificial intelligence models on ASNM-TUN dataset. Support vector machine and Naïve Bayes display all the metrics with utmost percentage. The precision of K-nearest neighbor is 100% while the decision tree has equal amount of proportion in all the three domains (Tables 2, 3, 4 and 5).

In Fig. 5, the model of KNN, SVM, and decision tree at level 0 and Naïve Bayes at level 1 model shows utmost true positive to false positive ratio with area under curve of 1.000. After a through comparison and contrast, it is empirical that the proposed

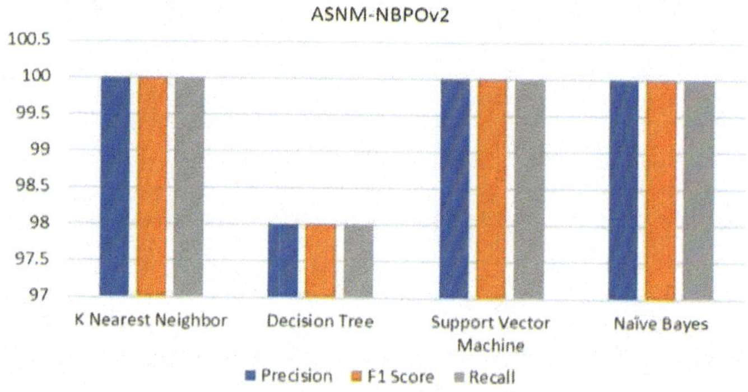

Fig. 3 Comparison of metrics (precision, recall, and $F1$ score) for ASNM-NBPOv2 dataset

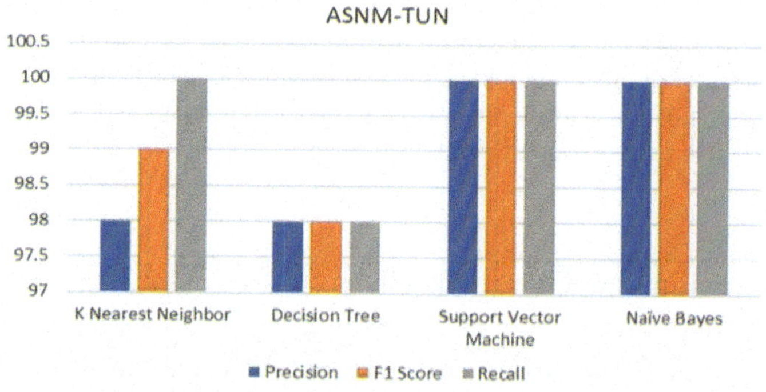

Fig. 4 Comparison of Metrics (precision, recall, and $F1$ score) for ASNM-TUN dataset

Table 2 Confusion matrix of the K-nearest neighbor with all three datasets of ASNM dataset at level 1

1429	0	2691	2	43	0
0	14	2	167	0	55
ASNM-CDX 2009 dataset		ASNM-NBPOv2 dataset		ASNM-TUN dataset	

Table 3 Confusion matrix of decision tree with all three datasets of ASNM dataset at level 1

1425	4	2689	4	42	1
8	6	3	166	2	53
ASNM-CDX 2009 dataset		ASNM-NBPOv2 dataset		ASNM-TUN dataset	

Table 4 Confusion matrix of support vector machine with all three datasets of ASNM dataset at level 1

1429	0	2693	0	43	0
0	14	0	169	0	55
ASNM-CDX 2009 dataset		ASNM-NBPOv2 dataset		ASNM-TUN dataset	

Table 5 Confusion matrix of the Naïve Bayes with all three datasets of ASNM dataset at level 1

1429	0	2693	0	43	0
2	12	0	169	0	55
ASNM-CDX 2009 dataset		ASNM-NBPOv2 dataset		ASNM-TUN dataset	

model of Naïve Bayes at level 1 and other three algorithms at level 0 has surpassed all the other techniques, with an accuracy of 100%, and the side metrics of precision, $F1$ score, and recall score were found to be 100% in all the sub-dataset of ASNM dataset (Fig. 6).

From Fig. 6, it is clearly visible that the model in which Naïve Bayes is at level 1 has 100% accuracy in all the three datasets. Anomalies being a disturbance to any result or work make the output deviate between observed data and the normal state, but with the help of this project idea, we will be able to identify the best algorithms for anomaly detection with highest accuracy by using machine learning and deep learning models we can implement it.

5 Conclusion and Future Work

In the end, the outlier analysis revealed that the hybrid model of algorithms for log detection has a greater accuracy percentage than the traditional anomaly detection approach and hence produces superior outcomes. After integrating multiple methods, we found that L0: *K*NN, DT, SVM, and L1: NB this hybrid branch model has the greatest accuracy and will thus be selected for resolving difficulties in datasets/logs for spotting these uncommon patterns, i.e., anomalies. We may use our study paper's approach in future projects to improve the accuracy and precision of anomaly detection in datasets. In comparison with solitary algorithms for anomaly detection, the

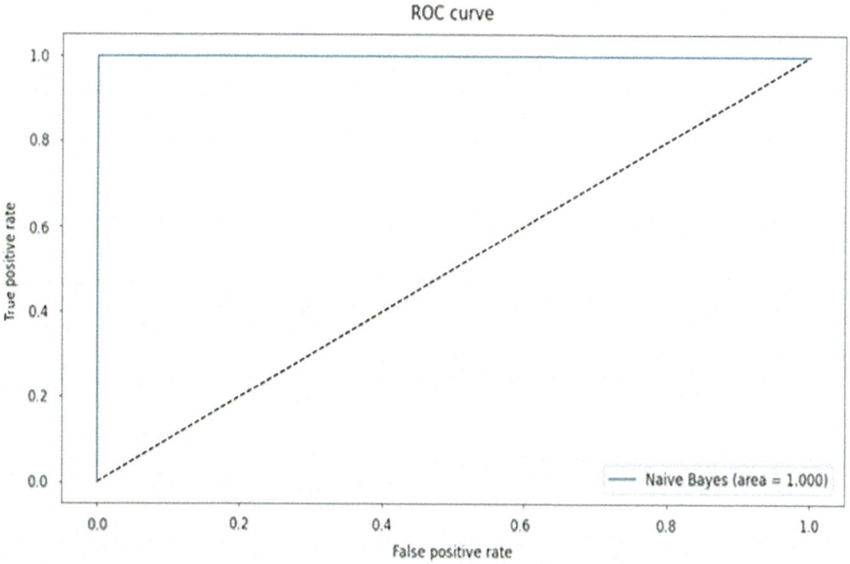

Fig. 5 ROC curve for *K*NN, decision tree, SVM at level 0, and Naïve Bayes at level 1 model

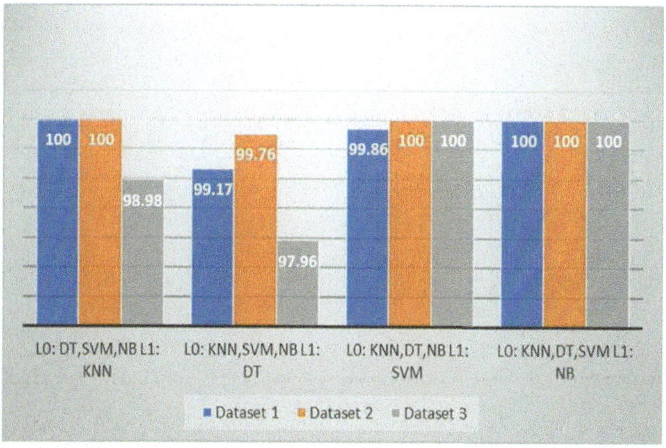

Fig. 6 Testing accuracy

hybrid model produces superior outcomes and accuracy. By enhancing the accuracy of our algorithm, we may improve the accuracy of our findings, resulting in findings that are more accurate and exact.

References

1. Le DC, Zincir-Heywood N (2021) Exploring anomalous behaviour detection and classification for insider threat identification. Int J Network Manag 31:e2109
2. Rabbani M, Wang Y, Khoshkangini R, Jelodar H, Zhao R, Bagheri Baba Ahmadi S, Ayobi S (2021) A review on machine learning approaches for network malicious behavior detection in emerging technologies. Entropy 23(5):529
3. Jia A, Smith B, Johnson C (2021) Stacked ensemble for network anomaly detection. J Netw Secur 25(3):123–138
4. Liang X, Zhang Y, Chen Z (2022) Stacked autoencoder-based ensemble for anomaly detection in computer networks. IEEE Trans Inf Forensics Secur 18(2):456–470
5. Chen W, Wang L (2023) Feature selection for stacking-based anomaly detection in ASNM datasets. Expert Syst Appl 48(6):987–1001
6. Zhang Q, Wu H, Liu S (2023) Hybrid stacked CNN-LSTM for network anomaly detection. Int J Comput Networks Appl 30(4):367–381
7. Wu Z, Chen X, Li Q, Data augmentation for stacking-based anomaly detection in ASNM datasets. J Comput Sci Technol 40(1):56–70
8. Lin Y, Yang H (2023) Semi-supervised stacking for network anomaly detection in ASNM datasets. Appl Intell 35(3):701–717
9. Kim J, Park M, Kim H, Cho S, Kang P (2019) Insider threat detection based on user behavior modeling and anomaly detection algorithms. Appl Sci 9(19):4018
10. Malinka HK, Hanacek P (2020) ASNM datasets: a collection of network attacks for testing of adversarial classifiers and intrusion detectors. IEEE Access 8:112427–112453
11. ASNM datasets, http://www.fit.vutbr.cz/~ihomoliak/asnm/. Last accessed on 20 July 2023

A Survey on Occupancy-Based Pattern Mining

Bhavana Inaganti and Sumalatha Saleti

Abstract Occupancy-based pattern mining has emerged as a significant research topic in recent times. This paper presents a comprehensive survey on occupancy, which serves as a measure to augment the significance of patterns. The survey covers various types of patterns, including frequent itemsets, high utility itemsets, frequent sequences, and high utility sequences, all in the context of occupancy. Additionally, the paper delves into techniques aimed at reducing the search space in the aforementioned pattern mining problems. These techniques are crucial for improving the efficiency and scalability of the mining process, especially when dealing with large-scale datasets. Furthermore, the paper discusses potential research extensions for occupancy-based pattern mining. These extensions could explore new applications, explore novel algorithms, or further enhance the effectiveness of occupancy as a measure for pattern evaluation. In general, this survey provides an important resource for researchers interested in understanding and advancing occupancy-based pattern mining techniques.

Keywords Occupancy · Pattern mining · Sequential patterns · High utility patterns

1 Introduction

Data mining [1–3] is the practice of uncovering valuable and insightful patterns, information, or knowledge from extensive datasets stored in databases or other data repositories. These patterns are considered frequent patterns if they satisfy the support threshold specified by the user [4–9]. To make a frequent pattern more significant, occupancy measure was introduced. The current paper is a survey focusing on the

B. Inaganti · S. Saleti (✉)
Computer Science and Engineering, SRM University—AP, Amaravathi, India
e-mail: sumalatha.s@srmap.edu.in

B. Inaganti
e-mail: bhavana_inaganti@srmap.edu.in

S. Bandyopadhyay et al. (eds.), *Intelligent Computing Systems and Applications*, Lecture Notes in Networks and Systems 1010,
https://doi.org/10.1007/978-981-97-5412-0_36

patterns that are highly occupant in the database. The term occupancy was introduced in 2012 by Tang et al. [10]. Occupancy serves as an essential and necessary addition to the extensively studied support measure, particularly when the items within a transaction collectively contribute to specific tasks. The motivation stems from real-world pattern recommendation scenarios where patterns of interest are expected to possess not only frequency but also a substantial presence within their supporting transactions or sequences. Occupancy refers to the distribution of an itemset in the transactions and is used as a measure to assess the significance of an itemset. It is defined as the portion occupied by a particular pattern or an itemset in a particular transaction. We say that a pattern is highly occupant pattern if it occupies a large portion in the database [11–13]. An isolated item will be having higher occupancy as it is the only item in the entire transaction [14].

The initial motivating application, based on a transactional database, pertains to the recommendation of print areas for web pages, as outlined in [10]. Users often encounter issues with the printouts produced by web browser print functions, as they tend to include extraneous content such as advertisements, navigation menus, and related links. To address this concern, a solution named Smart Print, developed by HP, offers a user-centric tool. This tool facilitates the effortless selection of desired print areas on a web page through a user-friendly interface. In the context of a web page, users are able to select specific print areas, highlighted within white rectangles. These selections, recorded as print logs containing user content, encompass a considerable amount of web page interactions. By treating each chosen content clip as an individual item, the collection of selected clips from a user for a particular web page constitutes a transaction of items. When considering print logs from multiple users, a transaction database emerges. In this scenario, the objective is to provide users with recommendations for itemsets, which can be understood as sets of content clips within a given web page. The recommendation task involves identifying itemsets that not only occur frequently, reflecting the preferences of numerous users, but also occupy a substantial portion of their respective supporting transactions. This dual criterion ensures that the recommended itemsets are both popular among users and sufficiently comprehensive in their representation of relevant content. This way, users will perceive the recommendations as inclusive and aligned with their interests, minimizing the feeling of missing significant information.

An additional motivating application within a sequence database context pertains to travel landscapes recommendation, as explored in [15]. In this context, consider a scenario where a travel company maintains logs of consumed travel packages by tourists. Each travel package offered by the company encompasses a comprehensive set of services tailored to the preferences of the tourists. These packages typically encompass multiple landscapes, arranged either spatially or temporally. In this application, the logs document the sequence of landscapes with in each travel package, capturing the order in which they are experienced by the tourists. This sequence information is crucial for understanding the progression and arrangement of landscapes in the travel packages. The goal is to leverage this sequence database to provide recommendations for travel landscapes to tourists. These recommendations should not only reflect frequently chosen sequences of landscapes, aligning with the preferences of a significant number of tourists, but also ensure that the recommended sequences occupy a substantial portion of their respective supporting sequences. This approach guarantees that the recommended travel landscapes are both popular among tourists and sufficiently comprehensive in depicting the overall travel experience. From these motivating applications it is clear that occupancy is a necessary accompaniment to the frequency.

1.1 Overview of Occupancy

The Apriori algorithm, introduced by Agrawal and Srikant in their work [4], stands as one of the renowned algorithms in the realm of data mining. This algorithm is particularly utilized for the extraction of frequent itemsets. Its efficiency lies in its capability to unearth frequent itemsets in extensive transactional databases, exploiting the anti-monotonicity property, wherein any non-empty subset of a frequent itemset is also considered frequent. To enhance the mining process and find highly qualified patterns, occupancy is considered in addition to support [10]. Nevertheless, there exists an inherent intuitive conflict between support and occupancy [16]. This stems from the fact that a pattern comprising a limited number of items might feature in numerous shopping lists, yet occupy only a small fraction of these lists; conversely, a pattern with a larger item count might occupy a significant portion of the lists but appear less frequently. However, unlike support, occupancy is neither monotone nor anti-monotone [10, 12, 14, 17]. Monotonicity implies that the value of a measure either increases or remains constant when more items are added, while anti-monotonicity implies that the value decreases or remains constant. In the context of occupancy, adding more items does not guarantee a continuous increase or decrease in its value, making it a challenging aspect of pattern mining. Due to the non-monotonic and non-anti-monotonic nature of occupancy, developing efficient algorithms that can handle and utilize this measure effectively poses a significant challenge in data mining. Researchers and practitioners continue to explore novel techniques and algorithms to address this challenge and mine highly qualified patterns considering both support and occupancy measures. For example, consider the database shown in Table 1 with 5 transactions and *i*, *j*, *k*, *l*, *m* are the items occurring in these transactions.

From Table 1, initially we will build a search space as shown in Fig. 1 for the item sets starting from 1-itemset like {i},{j},{k},{l} and {m}. Then from these 1-itemsets, 2-itemsets will be generated like {i,j}, {i,k}, {i,l} and so on, until the 5-itemset {i, j, k, l, m} is generated.

From this search space, the initial step involves identifying frequent item sets, where an item is deemed frequent if its support surpasses a specified minimum support threshold. Here we take the value of support as 3. By using this support measure for finding the frequent items, we use the anti-monotonicity property of support to minimize the exploration space. This concept states that if an itemset is deemed infrequent, then all of its super-sets are automatically considered as infrequent as

Table 1 A Transactional database with *i*, *j*, *k*, *l*, *m* as items and their internal utilities

Tid	Transaction
T1	*j*:2, *k*:1, *m*:4
T2	*j*:3, *m*:1
T3	i:2, *j*:3, *k*:1, *m*:2
T4	i:3, *k*:4, *l*:2
T5	i:1,*j*:1,*k*:2,*m*:3

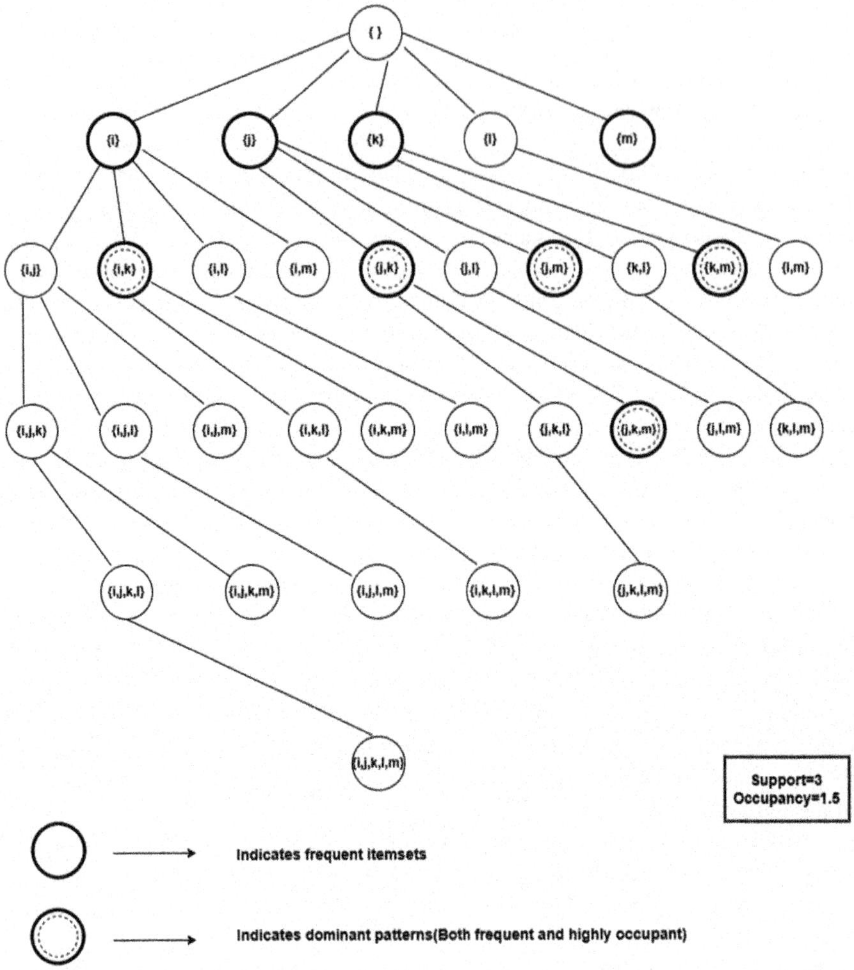

Fig. 1 Search space showing frequent and dominant patterns

well. For example, {i,j} support is 2 which is infrequent. Hence, no need to check its super-sets like {i, j, k}, {i, j, l} and so on. In the search space shown in Fig. 1, we have 31 itemsets of which only 9 are frequent. {i}:3, {j}:4, {k}:4, {m}:4, {i, k}:3, {j, k}:3, {j, m}:4, {k, m}:3, {j, k, m}:3. Now, we will apply the occupancy measure to those frequent item sets. Occupancy of {i} is $0.83 = (1/3 + 1/4 + 1/4)$, {j} is $1.33 = (1/3 + 1/4 + 1/2 + 1/4)$, {k} is $1.16 = (1/3 + 1/3 + 1/4 + 1/4)$, {m} is $1.33 = (1/3 + 1/4 + 1/2 + 1/4)$, {i,k} is $1.66 = (2/3 + 2/4 + 2/4)$, $\{j, k\}$ is $1.66 = (2/4 + 2/4 + 2/3)$, {j,m} is $2.66 = (2/3 + 2/4 + 2/2 + 2/4)$, {k,m} is $1.66 = (2/3 + 2/4 + 2/4)$, {j,k,m} is $2.5 = (3/4 + 3/4 + 3/4)$.

If we observe these values by adding new item sets, the value of occupancy was not continuously increasing or decreasing. We take the minimum occupancy threshold as 1.5. Now the dominant pattern is the one whose support is >=3 and occupancy >=1.5. Therefore, the sets {i,k},{j,k},{j,m},{k,m} and {j,k,m} are considered dominant patterns. Here the challenging task of occupancy is it is neither monotone nor anti-monotone, i.e., if we take the super-set {j,k,m} which satisfies the occupancy threshold 1.5, the subsets of {j,k,m} are {j},{k},{m},{j,k},{j,m} and {k,m} of which some are satisfying occupancy threshold and some don't. Hence, every frequent item must be searched for finding occupancy. Tang et al. [10] were among the researchers who proposed the idea of using upper bounds to reduce the searching space.

1.2 Motivation

The procedure of data mining often generates a substantial number of patterns, yet only a subset of these patterns holds significance for the domain expert scrutinizing the data. While support (frequency) is a measure of a pattern's intriguing nature, numerous patterns might lack relevance or novelty, thus failing to offer new insights. Hence, the task of selecting the most pertinent patterns from a dataset assumes significance, contributing to the assessment of data quality. In the dynamic landscape of pattern mining, research primarily focuses on three key domains: itemset mining, sequential pattern mining, and utility-oriented mining. These areas have paved the way for remarkable advancements, uncovering intricate relationships within data sets. However, the current momentum is steering towards occupancy-based mining. As the demand for occupancy-based insights surges, it has ignited a pressing need to extend our explorations into this burgeoning dimension. The call to unearth novel occupancy-based patterns reverberates through all the aforementioned domains, echoing the idea that the future of pattern mining lies in embracing this newfound perspective. Inspired by these concepts, the paper brings forward several important contributions:

- A complete survey of occupancy-based pattern mining since its inception in 2012. It includes all the works published on occupancy including itemsets, sequences, and utility.

- The upper bounds of occupancy with repsect to different pattern mining algorithms has been discussed.
- Comparison of the existing occupancy-based pattern mining algorithms and the further extensions are reviewed.

The structure of the remaining paper is as follows: Sect. 2 provides an explanation of the definitions and notations employed in occupancy-based mining. Section 3 delves into the concept of upper bound occupancies. It discusses how upper bounds are used to estimate the maximum possible values of pattern measures without exhaustive enumeration. Section 4 presents a comprehensive survey of the current literature in occupancy-based pattern mining. It covers various types of patterns, including frequent itemsets, high utility itemsets, frequent sequences, and high utility sequences, all within the context of occupancy. The paper wraps up by succinctly summarizing the pivotal discoveries derived from the survey and underscoring the significance of occupancy-based mining within the realm of data mining. It also discusses the potential future research directions.

2 Definitions and Notations

Suppose A is a pattern occurring in a transaction T_n. The support of A in T_n is given by the frequency f of occurrence of pattern A in T_n.

Definition 1 (*Support*): Support of A is defined as

$$\sigma(A) = f(A)$$

Here we are considering the absolute support, i.e., the absolute number of transactions which contains a pattern or an item set.

Definition 2 (*Occupancy*): The occupancy of pattern A within transaction T_n is established as the ratio of the length of pattern A to the length of transaction T_n. In simpler terms, it signifies the proportion of pattern A's length in comparison to the length of transaction T_n.

$$O(A, Tn) = \frac{|A|}{|T_n|}$$

$$O(A, DB) = \sum_{\forall T_n \in DB \wedge A \in T_n} O(A, T_n)$$

Definition 3 (*Itemset utility*): Given a pattern A, its utility in a transaction is represented as follows.

$$u(A, T_n) = \sum_{a \subseteq A \wedge A \subseteq T_n} u(a, T_n) \quad (1)$$

Table 2 External utility

i	j	k	l	m
7	2	1	4	3

For example, using Tables 1 and 2, $u(jk, T_1) = 2X2 + 1X1 = 5$.

Definition 4 (*Itemset utility in the database*): In a transactional database DB, the utility of an itemset A is denoted as

$$u(A, DB) = \sum_{A \subseteq T_n \wedge T_n \in DB} u(A, T_n) \tag{2}$$

For example, using Tables 1 and 2, $u(jk, DB) = u(jk, T_1) + u(jk, T_3) + u(jk, T_5) = 5 + 7 + 4 = 16$.

Definition 5 (*Utility of a transaction*): Utility of a transaction T_n is denoted as

$$tu(T_n) = \sum_{a \in T_n \wedge T_n \in DB} u(a, T_n) \tag{3}$$

For example, using Tables 1 and 2, $tu(T_1) = u(j, T_1) + u(k, T_1) + u(m, T_1) = 4 + 1 + 12 = 17$.

Definition 6 (*Utility of a database*): The aggregate of utilities across all transactions in a database DB, where $T_n \in DB$, is represented as

$$u(DB) = \sum_{T_n \in DB} tu(T_n) \tag{4}$$

For instance, $u(DB) = tu(T_1) + tu(T_2) + tu(T_3) + tu(T_4) + tu(T_5) = 17 + 9 + 27 + 31 + 20 = 104$.

Definition 7 (*Itemset utility occupancy in a transaction*): The itemset utility occupancy in a transaction T_n where the itemset occurs is denoted by

$$uo(A, T_n) = u(A, T_n)/tu(T_n), \tag{5}$$

where $A \subseteq T_n$ and $T_n \subset DB$ For example, from Definitions 3 and 5, $uo(jk, T_1) = 5/17 = 0.29$, where $u(jk, T_1) = 5$ and $tu(T_1) = 17$.

Definition 8 (*Itemset utility occupancy in a transactional database*): Within a transactional database DB, the utility occupancy of an itemset A is symbolized as

$$uo(A) = \frac{\sum_{A \subseteq T_n \wedge T_n \in DB} uo(A, T_n)}{\sigma(A)} \tag{6}$$

Definition 9 (*Remaining utility occupancy* (*ruo*)): In order to define the remaining utility occupancy, the items in the transactions are sorted in the ascending order of their support. If the support of two items is same, they will be arranged in lexicographic order. Hence, by considering Table 1, the ordering of items is $l < i < j < k < m$. Now, the *ruo* of pattern A is the sum of all the utility occupancies of the items that appear after A.
Let a_m be the last item in A, and $a_n \ldots a_x$ are the items appearing after a_m in the order $a_m < a_p < \cdots < a_x$. The *ruo* of A in a transaction T_n is defined as

$$ruo(A, T_n) = \sum_{i=p}^{x} u(a_i, T_n)/tu(T_n). \tag{7}$$

where $A \subseteq T_n$ and $a_i \neq A$.

3 Upperbound Occupancy

3.1 Upper Bound for Occupancy-Based Itemset Mining

Zhi-Hong Deng [12] introduced the concept of Upper Bound Occupancy (UBO). In the context of a pattern A and the transactions it belongs to, where the transactions encompass a variety of u different lengths, and these lengths are represented as the set $L = l_1, l_2, l_3, ..., l_u$. The elements in set L are arranged in increasing order. The count of transactions with a length of l_x is denoted as n_x, where $1 \leq x \leq u$. Notably, the symbol x acts as an index value in set L, meaning l_x belongs to L.

$$\text{UBO}(A, x) = \sum_{i=x}^{u} n_i \times \frac{l_x}{l_i}$$

$$\text{UBO}(A) = \max_{1 \leq x \leq u} UBO(A, x)$$

For Example, the upper bounds for 1-itemset {j} and 2-itemset {jk} are calculated as follows:

$$\sigma(j) = 4$$

From Definition 2, occupancy of {j} is $1.33 = (1/3 + 1/2 + 1/4 + 1/4)$

$$\text{UBO}(j) = \max(1 \times \frac{2}{2} + 1 \times \frac{2}{3} + 2 \times \frac{2}{4}, 1 \times \frac{3}{3} + 2 \times \frac{3}{4}, 2 \times \frac{4}{4}) = \max(2.66, 2.5, 2) = 2.66 \tag{8}$$

Here {j} is occurring in 4 different transactions of which 2 are of 4 length and 1 is of length 3 and 1 is of length 2. Therefore, the n values are $\{1, 1, 2\}$ and l values are $\{2, 3, 4\}$ and u value is 3.

$$\sigma(jk) = 3$$

From Definition 2, Occupancy of {jk} is 1.66 = (2/3 + 2/4 + 2/4)

$$\text{UBO}(jk) = \max(1 \times \frac{3}{3} + 2 \times \frac{3}{4}, 2 \times \frac{4}{4}) = (1 + 1.5, 2) = 2.5 \tag{9}$$

Here {jk} is occurring in 3 different transactions of which 2 are of 4 lengths and 1 is of length 3. Therefore, the n values are {1, 2} and l values are {3, 4} and u value is 2. Table 3 depicts the upperbound occupancy of all possible itemsets in the example database.

3.2 Upper Bound for Occupancy-Based High Utility Itemset Mining

Utility occupancy upper bound UOUB (A) is given as the average of the top support (here support is 3) elements in the resultant vector. The resultant vector is the sum of utility occupancy and remaining utility occupancy of an item A in the decreasing order.

$$\text{UOUB}(A) = \sum^{\text{Top[support]}} \frac{\text{vec}}{\text{[support]}}, A \subseteq T_n \wedge T_n \in DB \tag{10}$$

where $vec = uo(A, T_n) + ruo(A, T_n) \downarrow$

For example, the upper bound for $\{j, k\}$ is calculated as follows:
The support of j is 4 and utility occupancies of j are $uo(j, T1) = 4/17 = 0.23$, $uo(j, T2) = 6/9 = 0.66$, $uo(j, T3) = 6/27 = 0.22$, $uo(j, T5) = 2/20 = 0.1$. Similarly, the remaining utility occupancies of j are $ruo(j, T1) = 13/17 = 0.76$, $ruo(j, T2) = 3/9 = 0.33$, $ruo(j, T3) = 7/27 = 0.259$, $ruo(j, T5) = 11/20 = 0.55$. Now the sum of uo and ruo of j in $T1 = 0.23 + 0.66 = 0.99$. Similarly, the sum values for $T2$, $T3$, and $T5$ are 0.99, 0.48, and 0.65. The vector values in descending order are (0.99, 0.99, 0.65, 0.48). As the support value is 3, average is calculated as $(0.99 + 0.99 + 0.65)/3 = 0.87$. All the possible utility occupancy upper bounds of itemsets are shown in Table 4.

For the purpose of minimizing the exploration space, the algorithm compares the upper bound value of utility occupancy for a specific item with the established minimum utility occupancy threshold. If the upper bound value falls below the threshold, there is no requirement to generate the supersets of that particular item. The concept of upper bounds involves estimating the maximum possible value of a pattern's measure (e.g., support, utility, occupancy) without explicitly enumerating all the itemsets or sequences. By using these upper bounds, the algorithm can quickly discard unpromising candidates, minimizing the exploration space and enhancing the mining process.

Table 3 Support, occupancy and upper bound values of possible itemsets

Itemset	Support	Occupancy	Upper bound
{i}	3	0.83	2.5
{j}	4	1.33	2.66
{k}	4	1.16	3.5
{l}	1	0.33	1
{m}	4	1.33	2.66
{i,j}	2	1	2
{i,k}	3	1.66	2.5
{i,m}	2	1	2
{j,k}	3	1.66	2.5
{j,m}	4	2.66	2.66
{k,m}	3	1.66	2.5
{i,j,k}	2	1.5	2
{i,j,m}	2	1.5	2
{i,k,m}	2	1.5	2
{j,k,m}	3	2.5	2.5
{i,j,k,m}	2	2	2

Table 4 Utility occupancy, remaining utility occupancy, and utility occupancy upper bound values of possible itemsets

Itemset	Utility occupancy	Remaining utility occupancy	Utility occupancy upper bound
{i}	0.49	0.41	–
{j}	0.30	0.47	–
{k}	0.07	0.34	–
{m}	0.42	0	–
{i,k}	0.58	0.22	0.91
{j,k}	0.24	0.45	0.87
{j,m}	0.79	0	0.87
{k,m}	0.52	0	0.52
{j,k,m}	0.71	0	0.71

4 Literature Review

Within this section, we present a comprehensive overview of pertinent research within the domain of occupancy-based pattern mining across various dimensions: itemset mining, utility mining, sequence mining, high utility sequence mining and incremental mining. The comparison of occupancy-based pattern mining algorithms is systematically outlined within Table 5. In a column, 'yes' indicates the measure is employed in the paper, while 'no' signifies that the measure is not incorporated in their proposed algorithm.

4.1 Occupancy-Based Itemset Mining

Frequent pattern mining [4–6, 8, 9] has garnered substantial research interest for over a decade. The majority of these investigations center around frequency, which pertains to the count of occurrences of a pattern within a transaction database, serving as the pivotal metric to gauge pattern significance. To make a frequent pattern more significant, occupancy measure was introduced by [10] which is defined as the portion occupied by a particular pattern in the transaction where it occurs. DOFIA [10], was the first algorithm proposed for occupancy-based itemset mining. Unlike support, occupancy doesn't hold Apriori property. An upper bound on occupancy was introduced as a strategy to curtail the search space. A tree-based algorithm, QFWO was proposed in [17] where along with frequency, the weight occupancy of each individual item is estimated to produce highly qualified patterns. HEP algorithm proposed in [12] is a iterative level-wise approach where the number of parameters used are reduced by taking the same thresholds for occupancy and frequency measure. Datta et al. [14] proposed HOIMTO which uses transactional occupancy along with item-set occupancy to discover high occupancy patterns. A new measure of upper bound is also proposed to minimize the exploration space. It states that the transactions containing isolated items have higher occupancy which misleads in mining high qualified patterns. HOMI proposed by [18] used a dynamic approach for mining patterns on incremental databases. As the name suggests, it can only add the items in a transaction but deletions are not possible. In order to reduce the memory it uses DFS-based approach. Nguyen et al. [13] proposed two theorems FHOI, a BFS-based approach and DFHOI, a DFS-based approach. In this, early pruning was done without finding the upper bound occupancy by checking the occupancy list and applying the equivalence class. Occupancy-based skyline pattern mining [19] employs the concept of a skyline query to identify skyline patterns within a transaction database. These patterns are characterized by two dimensions: support and occupancy.

Table 5 Comparison of various algorithms related to occupancy

Algorithm	Authors	Support	Occupancy	Utility	Timestamp	Limitations
DOFIA	Tang et al. [10]	Yes	Yes	No	No	It is based on the support and length attributes of items, disregarding utility.
DOFRA	Zhang et al. [20]	Yes	Yes	No	Yes	It is limited to mining 1 length sequences
OCEAN	Shen et al. [11]	Yes	Yes	Yes	No	A surplus of comparison operations arises during their mining procedure
QFWO	Gan et al. [17]	Yes	Yes	No	No	It is based on the support and weight occupancy without utility measure
HEP	Deng [12]	Yes	Yes	No	No	The performance will notably deteriorate when dealing with extensive datasets
HUOPM	Gan et al. [21]	Yes	Yes	Yes	No	Difficult to determine an appropriate pair of support and occupancy parameters
MOEA-PM	Fang et al. [22]	Yes	Yes	Yes	No	It cannot mine the high utility sequential patterns based on occupancy

Table 5 (continued)

Algorithm	Authors	Support	Occupancy	Utility	Timestamp	Limitations
UHUOPM	Chen et al. [23]	Yes	Yes	Yes	No	It cannot mine all high utility occupancy patterns
HOIMTO	Datta et al. [14]	Yes	Yes	No	No	It does not effectively tackle the challenge of itemset length
HOMI	Kim et al. [24]	Yes	Yes	No	No	It can mine only high occupancy itemsets but not high utility occupancy patterns
pnHUO	Sun et al. [25]	Yes	Yes	Yes	No	It finds the high utility patterns considering positive and negative utility, but it is not scalable to large datasets
SHO	He et al. [26]	Yes	Yes	Yes	No	It cannot mine the time interval between the itemsets

Table 5 (continued)

Algorithm	Authors	Support	Occupancy	Utility	Timestamp	Limitations
SUMU	Huang et al. [27]	Yes	Yes	Yes	Yes	It is a centralized approach that considers the utility, support, occupancy and timestamp measures. It can be further extended to distributed environment to make the algorithm more scalable
FHOI,DFHOI	Nguyen et al. [13]	Yes	Yes	No	No	It cannot mine high utility occupancy patterns
HAUOPM	Keny et al. [28]	Yes	Yes	Yes	No	It does not work for dynamic databases
HUOMIL	Kim et al. [29]	Yes	Yes	Yes	No	It is unable to extract occupancy patterns in a dynamic environment

4.2 Occupancy-Based High Utility Itemset Mining

Each item present in an itemset derived from utility pattern mining is assigned an internal utility value. Furthermore, every item is linked with an external utility, which signifies its profitability or value. This database based on utility allows the extraction of high-utility itemsets or patterns, encompassing those that satisfy the designated minimum utility threshold. Mining high-utility items is more intricate compared to identifying frequent items, as utility items lack the fundamental downward closure property seen in frequent pattern mining. The objective of utility pattern mining is to identify all item patterns conforming to predefined constraints within the database. High utility itemset (HUI) mining, pioneered by Yao and Hamilton [30], extends frequent itemset mining by considering item quantity and associated profits. This extension emerged from real-world market basket data analysis, building upon [31]. Unlike frequent itemset mining, HUI mining [32–35] does not conform to the anti-monotonic property established in [4]. As a result of this departure from antimonotonicity, the challenge of HUI mining has captivated the interest of researchers, prompting the development of numerous solutions to address this problem within the existing body of literature. Gan et al. [21, 36] introduced a novel algorithm named HUOPM that evaluates the utility of patterns. This approach considers user choices related to support, utility, and occupancy to identify an exhaustive collection of high-quality patterns, all without requiring candidate generation. To achieve this, it employs two essential data structures which efficiently store the necessary information about the database. By utilizing these data structures, the HUOPM algorithm avoids repetitive scans of the database during mining. Additionally, the algorithm incorporates the idea of remaining utility occupancy to compute an upper bound, effectively reducing the search space and enhancing its efficiency. HUOPM discovers high-quality patterns with high utility and occupancy, which are more representative and useful than frequent patterns. The paper [21] also compares HUOPM with two prior studies, DOFIA [10] and OCEAN [11], and shows that HUOPM achieves better results in terms of efficiency and effectiveness.

In [22], the researchers introduced a multi-objective problem model designed for High Quality Pattern Mining (HQPM) with three key objectives: support, occupancy, and utility. To effectively address this three-objective problem, they proposed MOEA-PM algorithm. MOEA-PM efficiently solves the HQPM problem by optimizing the three objectives simultaneously. To validate the efficacy of their approach, the authors carried out a comparative analysis. The authors [22] address the challenge of discovering patterns that possess both high support and high confidence simultaneously, which is a key requirement in many data mining applications. MOEA-PM consists of three main components: (1) the encoding of candidate patterns, (2) the definition of fitness functions, and (3) the implementation of the MOEA algorithm. The encoding scheme represents patterns as a set of binary strings, enabling efficient manipulation and exploration during the evolutionary process.

In [37], the concept of High-Utility Occupancy Pattern Mining (UHUOPM) in uncertain databases was introduced, incorporating support, utility occupancy, and probability considerations. Here, the probability of data is utilized to reflect uncertainties in uncertain databases. In a related paper [23], a novel methodology was introduced to extract high utility-occupancy patterns from uncertain data sets. The authors address the challenge of analyzing uncertain data, which may include missing or incomplete information, and propose a framework to discover patterns that capture the associations between occupancy events and their utility values. To tackle the challenge of uncertainty, the authors propose a comprehensive framework that encompasses three key components: (1) uncertain data preprocessing, (2) probabilistic utility-occupancy pattern mining, and (3) pattern evaluation and selection. The authors in [25] addressed the limitations of traditional pattern discovery methods that only consider frequency and usefulness. The authors propose a new approach called pnHUO that takes into account the economic utility theory to discover the patterns considering both positive and negative utility values. pnHuo is a single phase algorithm. Two data structures were introduced by the authors: the frequent-utility tree and the positive-and-negative utility-occupancy list. The approach avoids multiple database scans.

In [26], the authors introduced an innovative algorithm known as SHO (Suffix-based High-utility Occupancy itemset mining). It aims to efficiently mine high utility occupancy itemsets from large-scale databases. To achieve this, SHO takes into account both the quantities and profits associated with itemsets. The paper proposes four specific strategies that enhance the performance of the SHO algorithm. One of these strategies is called SP Utility pruning, where an itemset is pruned if its utility falls below the minimum utility threshold. By employing this pruning technique, the algorithm can discard itemsets that are not likely to contribute significantly to the final results, thus reducing unnecessary computations. It reduces the search space by removing unpromising itemsets. TWU pruning - It prunes an itemset if its TWU (Total Weighted Utility) is less than the minimum TWU threshold. UP-Growth pruning - it prunes an itemset if its utility is less than the utility of its immediate super sets. RTI (Reduce the Intersection) pruning - It prunes an itemset if its intersection with other itemsets has already been processed and found to be empty.

In their work, Kenny and Dipti [28] introduced a new and innovative approach known as HAUOPM. The purpose of this approach is to address the limitations of the traditional HUOPM method. The key distinction between the two approaches lies in the utilization of average utility occupancy. HAUOPM incorporates this new measure to effectively mine patterns with high average utility occupancy, providing a more comprehensive view of itemset significance compared to traditional methods that focus solely on utility occupancy. By proposing this average utility occupancy measure, Kenny and Dipti aim to enhance the accuracy and scope of pattern mining, making HAUOPM a promising alternative to overcome the limitations of existing methods like HUOPM. The new approach offers an improved perspective on itemset mining and has the potential to yield more valuable insights from the data, while HUOPM uses utility occupancy as a measure. Additionally, HAUOPM takes into account the length of an itemset, which HUOPM does not.

Kim et al. [29] introduced an exceptionally efficient technique for extracting high utility occupancy patterns through the utilization of indexed list structures in their research. This approach tackles the inefficiencies associated with list-based methods by significantly reducing the number of comparison overheads involved. To achieve this improvement, the authors introduced two novel data structures that efficiently maintain index information on items or patterns. These data structures enable rapid pattern extension, streamlining the mining process. Additionally, the proposed method incorporates utility occupancy functions, and it devises specialized constructing and mining techniques tailored to work optimally with the newly introduced data structures. To further enhance the efficiency, the approach employs effective pruning techniques. These pruning techniques help narrow down the vast search space, discarding unpromising candidates early in the process. Overall, the method presented by Kim et al. provides a solution making it a valuable contribution to the field of pattern mining. Nonetheless, the algorithms discussed earlier are unable to concurrently extract both item location details and utility occupancy. Addressing this limitation, Dong et al. [38] introduced an algorithm named SHUO-FI. This algorithm is designed to mine utility occupancy thresholds while adhering to predefined distance constraints.

4.3 Occupancy-Based Sequential Pattern Mining

Sequential Pattern Mining (SPM) [39–43] is a significant and well-studied field within data mining. Its inception is credited to Agrawal and Srikant [39], who introduced the concept for identifying frequent sequences within a given set of customer sequences. In this context, a customer sequence is characterized as an organized series of transactions attributed to a single customer. The objective of SPM is to uncover patterns that occur frequently, offering valuable insights into various applications. These include the analysis of customer purchasing behaviors, prediction of events like web usage patterns and natural disasters, detection of disease symptoms, and the analysis of DNA sequences. The concept of occupancy was extended to sequential 1-itemsets and an algorithm DOFRA (Dominant and frequent pattern mining algorithm) was proposed by Zhang et.al. [20]. In addition to finding top-k qualified patterns the link between occupancy and closed patterns is explored to further minimize the exploration space. This is the only algorithm which deals with both sequence and occupancy. However, DOFRA can handle only sequential 1-itemsets. Hence, there is much scope to provide novel solutions for sequential pattern mining including occupancy measure.

4.4 Occupancy-Based High Utility Sequential Pattern Mining

The goal of mining High Utility Sequential Patterns (HUSP) [44–48] is to identify sequential patterns with high utility from sequence databases. This technique was initially introduced by Ahmed et al. [44]. In a departure from the methodology presented by Ahmed et al., an alternative approach was put forth in the work by Lan et al. [46]. This new perspective aimed to enhance the efficiency and effectiveness of HUSP mining. Huang et al. [27] proposed High Utility-Occupancy Sequential Pattern Mining (HUOSPM) method to mine high utility-occupancy sequential patterns in sequence data. The proposed algorithm, Sequence Utility Maximization with Utility occupancy measure (SUMU), is designed to evaluate three dimensions, including frequency, utility, and occupancy, and is shown to be effective and efficient through empirical experiments. Two data structures, UOL-Chain and UO-Table, are proposed to store related information about a pattern and improve efficiency. The paper [27] analyzes and discusses the influence of upper bounds and pruning strategies on the proposed algorithm.

4.5 Occupancy-Based Pattern Mining in Dynamic Environments

Real-world databases frequently experience the addition of new transactions. Nevertheless, the current algorithms for high occupancy pattern mining is incapable of managing incremental databases. A recent development by Kim et al. [24] introduces an approach named HOMI, which possesses the capability to extract high occupancy patterns from incremental databases, surpassing the significance of frequent patterns. The pattern outcomes yielded by HOMI exhibit a heightened level of significance, potentially indicating motivations or trends. This is attributed to HOMI's consideration of both frequency and the proportions of patterns within transactions. To surmount the constraints posed by the prevailing utility occupancy methodologies, an innovative mining approach, denoted as HUOMI [49] has been proposed. This novel algorithm is adept at efficiently handling expanding databases. It encompasses an optimized data structure along with an enhanced pruning technique, enabling it to swiftly adapt to dynamic environments.

5 Conclusions and Future Work

Occupancy-based pattern mining has become a prominent area of research in recent times. This paper presents a comprehensive survey that focuses on the use of occupancy as a measure to enhance the significance of patterns. The survey covers various kinds of patterns, including frequent itemsets, high utility itemsets, frequent sequences, and high utility sequences, all within the context of occupancy. In addition to discussing the different kinds of patterns, the paper explores techniques to address the challenge of reducing the search space in occupancy-based mining problems. These techniques are essential for improving the efficiency and scalability of the mining process. Furthermore, the paper goes beyond the existing research and explores potential research extensions for occupancy-based pattern mining. These extensions may involve exploring new applications of occupancy-based mining, investigating novel algorithms, or further enhancing the effectiveness of occupancy as a measure for evaluating patterns. In conclusion, this survey provides a valuable resource for researchers interested in delving into the intricacies of occupancy-based pattern mining. The potential research on occupancy-based pattern mining can be further extended to dynamic pattern mining [18, 49–51]. In real-world scenarios, not all patterns hold utility or practical value. For instance, a pattern could involve an excessive number of items, rendering it overly specific and thereby diminishing its real-life relevance to users. To attain qualified patterns with adaptable lengths, constraints can be imposed on the minimum and maximum pattern lengths during the mining process. Hence, the occupancy pattern mining can be further extended to mine flexible high utility-occupancy patterns [52]. Moreover, the existing papers only deal with the centralized approaches and no where the distributed algorithms are discussed. Hence, there is a lot of scope to extend the traditional approaches to distributed environment. The Hadoop platform [53, 54] holds intrigue due to its capability to process large databases in a distributed manner across computer clusters, utilizing a straightforward programming model. This platform offers the capability to strengthen distributed processing across server clusters, allowing for seamless scaling from a single server within a cluster to a substantial number of servers, even reaching into the thousands. Given Hadoop's commendable performance in emerging domains, it merits additional research exploration in this domain. By covering a wide range of pattern types, discussing search space reduction techniques, and outlining potential future research directions, the paper contributes significantly to the advancement of occupancy-based pattern mining techniques.

References

1. Han J, Kamber M, Pei J (2011) Data mining: concepts and techniques, 3rd edn. Morgan Kaufmann Publishers Inc., San Francisco, CA, USA
2. Shu X, Ye Y (2023) Knowledge discovery: methods from data mining and machine learning. Soc Sci Res 110:102817
3. Gupta MK, Chandra P (2020) A comprehensive survey of data mining. Int J Inf Technol 12(4):1243–1257 Dec
4. Agrawal R, Srikant R (1994) Fast algorithms for mining association rules in large databases. In: Proceedings of the 20th international conference on very large data bases, ser. VLDB '94. Morgan Kaufmann Publishers Inc., San Francisco, CA, USA, pp 487–499
5. Lin K-C, Liao I-E, Chen Z-S (2011) An improved frequent pattern growth method for mining association rules. Expert Syst Appl 38(5):5154–5161
6. Han J, Pei J, Yin Y, Mao R (2004) Mining frequent patterns without candidate generation: a frequent-pattern tree approach. Data Min Knowl Discov 8(1):53–87 Jan
7. Xun Y, Zhang J, Qin X (2016) Fidoop: parallel mining of frequent itemsets using mapreduce. IEEE Trans Syst Man Cybern Syst 46(3):313–325
8. Saleti S, Subramanyam RBV (2018) A novel bit vector product algorithm for mining frequent itemsets from large datasets using mapreduce framework. Cluster Comput 21(2):1365–1380 Jun

9. Luna JM, Fournier-Viger P, Ventura S (2019) Frequent itemset mining: a 25 years review. WIREs Data Min Knowl Discov 9(6):e1329
10. Tang L, Zhang L, Luo P, Wang M (2012) Incorporating occupancy into frequent pattern mining for high quality pattern recommendation. In: Proceedings of the 21st ACM international conference on information and knowledge management, ser. CIKM '12. Association for Computing Machinery, New York, NY, USA, pp 75–84
11. Shen B, Wen Z, Zhao Y, Zhou D, Zheng W (2016) Ocean: fast discovery of high utility occupancy itemsets. In: Bailey J, Khan L, Washio T, Dobbie G, Huang JZ, Wang R (eds) Advances in knowledge discovery and data mining. Springer International Publishing, Cham, pp 354–365
12. Deng Z-H (2020) Mining high occupancy itemsets. Future Gener Comput Syst 102:222–229
13. Nguyen LT, Mai T, Pham G-H, Yun U, Vo B (2023) An efficient method for mining high occupancy itemsets based on equivalence class and early pruning. Knowl-Based Syst 267:110441
14. Datta S, Mali K, Ghosh U (2022) High occupancy itemset mining with consideration of transaction occupancy. Arab J Sci Eng 47(2):2061–2075 Feb
15. Liu Q, Ge Y, Li Z, Chen E, Xiong H (2011) Personalized travel package recommendation. In: 2011 IEEE 11th international conference on data mining, pp 407–416
16. Tian Y, Lu C, Zhang X, Cheng F, Jin Y (2022) A pattern mining-based evolutionary algorithm for large-scale sparse multiobjective optimization problems. IEEE Trans Cybern 52(7):6784–6797
17. Gan W, Lin JC-W, Fournier-Viger P, Chao H-C, Zhan J, Zhang J (2018) Exploiting highly qualified pattern with frequency and weight occupancy. Knowl Inf Syst 56(1):165–196 Jul
18. Kim H, Ryu T, Lee C, Kim H, Truong T, Fournier-Viger P, Pedrycz W, Yun U (2022) Mining high occupancy patterns to analyze incremental data in intelligent systems. ISA Trans 131:460–475
19. Zhang K, Hu K (2021) Research of occupancy-based skyline pattern mining. In: 2021 IEEE 4th international conference on electronic information and communication technology (ICEICT), pp 129–134
20. Zhang L, Luo P, Tang L, Chen E, Liu Q, Wang M, Xiong H (2015) Occupancy-based frequent pattern mining. ACM Trans Knowl Discov Data (TKDD) 10(2):1–33
21. Gan W, Lin JC-W, Fournier-Viger P, Chao H-C, Yu PS (2020) Huopm: high-utility occupancy pattern mining. IEEE Trans Cybern 50(3):1195–1208
22. Fang W, Zhang Q, Sun J, Wu X (2022) Mining high quality patterns using multi-objective evolutionary algorithm. IEEE Trans Knowl Data Eng 34(8):3883–3898
23. Chien-Ming C, Lili C, Wensheng G, Lina Q, Weiping D (2021) Discovering high utility-occupancy patterns from uncertain data. Inf Sci 546:1208–1229
24. Kim H, Ryu T, Lee C, Kim H, Truong T, Fournier-Viger P, Pedrycz W, Yun U (2022) Mining high occupancy patterns to analyze incremental data in intelligent systems. ISA Trans 131:460–475
25. Sun J, Gan W, Lin JC-W, Chao H-C (2022) "Pattern discovery with utility occupancy. In: IEEE international conference on big data (big Data) 6261–6270
26. He J, Han X, Wang J, Zhang K (2022) Efficient high-utility occupancy itemset mining algorithm on massive data. Expert Syst Appl 210:118329
27. Huang G, Gan W, Yu PS (2022) Towards sequence utility maximization under utility occupancy measure
28. Kenny KMJ, Dipti R (2023) Hauopm: high average utility occupancy pattern mining. Arab J Sci Eng
29. Kim H, Ryu T, Lee C, Kim S, Vo B, Lin JC-W, Yun U (2023) Efficient method for mining high utility occupancy patterns based on indexed list structure. IEEE Access 11:43 140–43 158
30. Yao H, Hamilton HJ (2006) Mining itemset utilities from transaction databases. Data Knowl Eng 59(3):603–626. including: ER 2003
31. Yao H, Hamilton HJ, Butz CJ (2004) A foundational approach to mining itemset utilities from databases, pp 482–486

32. Gan W, Lin JC-W, Fournier-Viger P, Chao H-C, Tseng VS, Yu PS (2021) A survey of utility-oriented pattern mining. IEEE Trans Knowl Data Eng 33(4):1306–1327
33. Wu P, Niu X, Fournier-Viger P, Huang C, Wang B (2022) Ubp-miner: an efficient bit based high utility itemset mining algorithm. Knowl-Based Syst 248:108865
34. Atmaja EHS, Sonawane K (2022) A review of high utility itemset mining for transactional database. In: Gupta D, Goswami RS, Banerjee S, Tanveer M, Pachori RB (eds) Pattern recognition and data analysis with applications. Springer Nature Singapore, Singapore, pp 15–27
35. Cheng Z, Fang W, Shen W, Lin JC-W, Yuan B (2023) An efficient utility-list based high-utility itemset mining algorithm. Appl Intell 53(6):6992–7006 Mar
36. Gan W, Lin JC-W, Fournier-Viger P, Chao H-C (2017) Exploiting high utility occupancy patterns. In: Chen L, Jensen CS, Shahabi C, Yang X, Lian X (eds) Web and big data. Springer International Publishing, Cham, pp 239–247
37. Chen CM, Chen L, Gan W, Qiu L, Ding W (2020) Uhuopm: High utility occupancy pattern mining in uncertain data. In: 2020 IEEE international conference on systems, man, and cybernetics (SMC). IEEE, pp 3066–3071
38. Dong X, Wang M, Liu Y, Xiao G, Huang D, Wang G (2022) An efficient spatial high-utility occupancy frequent item mining algorithm for mission system integration architecture design using the mbse method. Aerosp Syst 5(3):377–392 Sep
39. Agrawal R, Srikant R (1995) Mining sequential patterns. In: Proceedings of the eleventh international conference on data engineering, pp 3–14
40. Srikant R, Agrawal R (1996) Mining sequential patterns: generalizations and performance improvements. In: Apers P, Bouzeghoub M, Gardarin G (Eds) Advances in database technology—EDBT '96. Springer Berlin Heidelberg, Berlin, Heidelberg, pp 1–17
41. Pei J, Han J, Mortazavi-Asl B, Wang J, Pinto H, Chen Q, Dayal U, Hsu M-C (2004) Mining sequential patterns by pattern-growth: the prefixspan approach. IEEE Trans Knowl Data Eng 16(11):1424–1440
42. Chen J (2010) An updown directed acyclic graph approach for sequential pattern mining. IEEE Trans Knowl Data Eng 22(7):913–928
43. Mabroukeh NR, Ezeife CI (2010) A taxonomy of sequential pattern mining algorithms. ACM Comput Sur 43(1) dec 2010
44. Ahmed CF, Tanbeer SK, Jeong B-S (2010) A novel approach for mining high-utility sequential patterns in sequence databases. ETRI J 32(5):676–686
45. Yin J, Zheng Z, Cao L (2012) Uspan: an efficient algorithm for mining high utility sequential patterns. In: Proceedings of the 18th ACM SIGKDD international conference on knowledge discovery and data mining, ser. KDD '12. Association for Computing Machinery, New York, NY, USA, pp 660–668
46. Lan G-C, Hong T-P, Tseng VS, Wang S-L (2014) Applying the maximum utility measure in high utility sequential pattern mining. Expert Syst Appl 41(11):5071–5081
47. Alkan OK, Karagoz P (2015) Crom and huspext: improving efficiency of high utility sequential pattern extraction. IEEE Trans Knowl Data Eng 27(10):2645–2657
48. Wang J-Z, Huang J-L, Chen Y-C (2016) On efficiently mining high utility sequential patterns. Knowl Inf Syst 49(2):597–627 Nov
49. Ryu T, Yun U, Lee C, Lin JC-W, Pedrycz W (2022) Occupancy-based utility pattern mining in dynamic environments of intelligent systems. Int J Intell Syst 37(9):5477–5507
50. Ahmed CF, Tanbeer SK, Jeong B-S, Lee Y-K (2009) Efficient tree structures for high utility pattern mining in incremental databases. IEEE Trans Knowl Data Eng 21(12):1708–1721
51. Saleti S (2022) Incremental mining of high utility sequential patterns using mapreduce paradigm. Cluster Comput 25(2):805–825 Apr
52. Chen C-M, Chen L, Gan W (2021) Flexible pattern discovery and analysis
53. Dean J, Ghemawat S (2008) Mapreduce: simplified data processing on large clusters. Commun ACM 51(1):107–113 Jan
54. White T (2015) Hadoop: the definitive guide, 4th ed. O'Reilly Media, Inc.

Modeling and Early Diagnosis of Alzheimer's Disease Using Recurrent Neural Network

Mitu Ranjan and Sushil Kumar

Abstract The Alzheimer's is a common and predominant form of disease in old age people which is progressive (with time) by nature. This can be dangerous if not diagnosed and treated during the early stages. In the present paper, the authors have proposed a novel work for the detection of Alzheimer's disease (AD) by incorporating long-short term memory (LSTM) algorithm with recurrent neural network-based architecture. The medical practitioners need an efficient and fast artificial learning techniques to diagnose the progressive nature of brain diseases such as AD which requires accurate modeling and diagnosis. The research work in this paper has been presented in two stages: the first stage is modeling of AD, and the second stage is diagnosis of the progressive AD by calculating time-series data from the patient's visit at an interval of 6 months up to 48 months. In the first stage, a logistic regression technique-based model is developed (as proposed in our previous work) for the diagnosis of stable mild cognitive impairment (sMCI)/progressive mild cognitive impairment(pMCI) and prediction of progression time-frame from sMCI to pMCI or pMCI to AD. Furthermore, in the second stage, the developed model has been utilized for the detection of AD and prediction of its progression time by incorporating the LSTM algorithm. The simulation results show the effectiveness and accuracy of the proposed technique. It has been found that the accuracy is 94.79%, precision is 93.87%, and recall is 93.87%.

Keywords Alzheimer's disease · Deep neural network · Medical resonance imaging (MRI) · Positron emission tomography (PET)

M. Ranjan (✉) · S. Kumar
KIET Group of Institutions, Delhi-NCR, Ghaziabad 201206, Uttar Pradesh, India
e-mail: mitu.2123mcse1008@kiet.edu

S. Kumar
e-mail: sushil.kumar@kiet.edu

S. Bandyopadhyay et al. (eds.), *Intelligent Computing Systems and Applications*,
Lecture Notes in Networks and Systems 1010,
https://doi.org/10.1007/978-981-97-5412-0_37

1 Introduction

In last few decades, the Alzheimer's disease (AD) patients increased many folds globally. This is a neurological disorder usually seen in old age persons having the characteristics of memory impairments, change in behavior, and thinking disorder. The symptom of this disease may start at an age of as young as 40 years but people neglect the symptoms or do not take it seriously because it occurs irregularly with a long time intervals. According to data revealed by AD association, there were 50 millions infected patients worldwide alone in 2019 which is anticipated to increase up-to three folds by 2050 [1]. The major diagnosis technologies in the medical field are medical resonance imaging (MRI), positron emission tomography (PET), cerebrospinal fluid (CSF) testing, urine and blood test, and fusion of MRI and PET images [2, 3]. The morphometric patterns are very helpful to detect the AD from MRI images [4]. Recently, the researchers have diagnosed and predicted the progression time of AD using retinal imaging [5, 6]. Alzheimer's disease are having two phases, the early phase is called as stable mild cognitive impairments (sMCI) which converts to dementia with a rate of 10–25% and the advanced phase which converts to dementia with progression of time is called as progressive mild cognitive impairments (pMCI) [1]. Alzheimer's disease conversion time-frame from pMCI is utmost important for the physicians to start the effective treatment [7–9]. The classical artificial intelligence techniques provide the accuracy of up-to 83.3% for determining the pMCI to AD conversion time-frame [10]; however, more advanced version of the deep learning technique such as recurrent neural network (RNN) or convolutional neural network (CNN) provides an accuracy of up-to $84.44 \pm 2.20\,\%$ [11]. In a recent study, the authors have accomplished an accuracy of 92.57% by incorporating multi-modal fusion of MRI and PET imaging biomarkers values with deep convolutional neural network [12]. The major difficulty for the diagnosis of the AD is to calculate an effective and efficient model of the AD; usually the measurement of CSF biomarkers from MRI/PET images because the progression time-frame of pMCI to AD can be determine correctly from the biomarkers value [13]. In our previous work, we have modeled the progression of AD by incorporating likelihood estimation regression and logistic function in which the longitudinal biomarker dynamics are fitted to a new modified Stannard logistic function [14]. Furthermore, the biomarkers trajectories using sigmoidal curve fitting technique are modeled in [15, 16] and progression time based model obtained in [17, 18] by incorporating deep learning concepts. However, these models are not very efficient as they have some missing and neglected biomarkers data. These shortcomings are overcome in [14] to obtain an efficient model to diagnose the AD. Quantification of robustness are also taken care of in the model by incorporating inflection point technique for measuring the values of biomarkers. After obtaining an accurate and efficient model in [14], we would like to extend the application of this model to diagnose the AD with enhanced accuracy. The paper proposes the modeling and early diagnosis technique of different AD conditions, the following are the novel and important contributions in the present paper:

- A diagnosis technique to classify the Alzheimer's disease whether a patient is pMCI or AD is proposed in this paper. The model of Alzheimer's disease progression is proposed in our previous research work [14] is utilized in the present paper, which is based on the logistic regression technique with likelihood estimation (M-estimation). In addition, the logistic Stannard function is modified to fit with the proposed algorithm in the present research work to achieve better curve fitting of the AD model.

- The developed model in [14] has been utilized to determine the progression time from cognitive normal (CN) to pMCI and pMCI to AD. In this paper, the time-series data sets are obtained from the patients visit and the long short term memory (LSTM) technique based on time-series data is incorporated to diagnose the AD and to make the classification of different AD stages.
- Furthermore, in this paper, the authors have considered three different time-frame for each patient: the first one is progression time from CN to 6 months, the second time frame is progression of disease from CN to 12 months, and the third time frame is disease progression from CN to 18 months. The data of MRI images, CSF biomarkers, educational qualification, cognitive testing, and demographics are used from our previous work [14]. The time-series datasets were gathered from the patients at 6 month intervals throughout an 18-month period. The gathered datasets have been divided in two categories: full featured datasets and filtered featured datasets.

The remaining part of the paper is organized as follows: Section 2 is dealing with the model development and description of time-series datasets, Sect. 3 describe the proposed technical framework incorporating the LSTM methodology for predicting the exact conversion time of MCI patients to pMCI or AD from the developed model in Sect. 2, Sect. 4 presents the experimental results and discussions on simulation results, and Sect. 5 culminates the paper with brief conclusion.

2 Model Development and Datasets

This section describes the model development of Alzheimer's progression by incorporating logistic functions with regression technique. The time-series data sets are used for the detection of Alzheimer's disease and the prediction of progression time from sMCI to pMCI or pMCI to AD using LSTM method. The time-series data sets are obtained from the patients visit at a predefined interval of six months. The time-series data sets are obtained from the MRI/PET images, CSF biomarkers, educational qualifications, cognitive testing, and demographics of the patients under study. In this section, the logistic regression technique is proposed to model the AD based on our previous work [14] in which the new modified Stannard function is proposed. Furthermore, in this study, the logistic Stannard function is modified again to fit the proposed LSTM algorithm in the present study as follows:

$$\Gamma(x, \lambda) = A\left[1 + \frac{1}{\theta} e^{-\frac{b(x-\alpha)}{\theta}}\right]^{-\theta} \quad (1)$$

where $A \in [0.9, 1.2]$ and $\{b \in \mathbb{R}_{>0}, \alpha, \theta \in \mathbb{R}_{>0}\} \in (0, 1)$ at $(-\infty, +\infty)$ is being introduced as λ. Besides, the inflection point of the curve can be introduced as α in the new defined Stannard function. The advantage of such type of Stannard function is that the progression rate can be modeled in case of slow and fast conversion of the pMCI to AD. The asymptotic behavior of the logistic sigmoid function can be modified with the help of expression $f(x, \lambda) = (l_1 - l_2)\, \Gamma(x, \lambda) + l_3$. The parameters l_1, l_2, and l_3 can be bounded to reduce the computational burden on the optimization algorithm. Another approach is to fix some parameters and change only a few variables in the optimization algorithm to reduce complexity and computational burden. The regression functions used throughout the current paper for M-estimation are as follows:

- L2 Function: $\pi(r) = r^2$ with the scale factor $\tau = 1$.
- L1 - L2 Function: $\pi(r) = 2\left(\sqrt{r^2 + 1} - 1\right)$ with the scale factor $\tau = 1$.
- Cauchy–Lorentz: $\pi(r) = \ln\left(r^2 + 1\right)$ with the scale factor $\tau = 2.3849$.

The above described regression estimators function try to alleviate the dominance of outliers while the trajectory fitting operation. Here, the estimators function with the scale factor τ can be used for scaling the function $\tau^2 \pi(r/\tau)$ to track the asymptotes up to 95% of the trajectory while keeping the normal distribution to the standard values. The curve fitting of trajectories can be done with the algorithm proposed in [19], as given below:

Algorithm 1 Optimization Algorithm

Require: Measured biomarkers data from FAQ, MRI/PET images

Initialization: To initialize $c^o(0)$, $d^o(0)$, and $\lambda^o(0)$ by using measured data

Optimization: Run the iterations until the algorithm converges (say i iterations)

Trajectory Fitting: The patients visit and time-frame of the corresponding visits are used to estimate the biomarkers data as follows:

$$\lambda^{o(i)} = \min_{m,n,k} \sum_{m,n,k} \omega_m \zeta \left\{ \frac{Y^o_{m,n,k} - f(c_m^{i-1} t_{m,n} + d_m^{i-1}, \lambda_k)}{\sigma_k} \right\}$$

where $Y^o_{m,n,k} = f(x_{m,n}, \lambda_k) + \sigma_k \epsilon_{m,n,k}$, represents the kth biomarkers output value at m^{th} visit to the doctor for n^{th} patient, $f(x_{m,n}, \lambda_k)$ is the logistic function for DPSs, and λ_k is the parameter for each biomarkers. In addition, σ_k is the standard deviation, $\epsilon_{m,n,k}$ is the Gaussian noise to be selected on random basis, and $x_{m,n}$ can be formulated as $x_{m,n} = c_m t_{m,n} + d_m$ with $t_{m,n}$ represents the patient's age, $c_m \in \mathbb{R}_{>0}$ is the progression rate, and $d_m \in \mathbb{R}$ defines the onset of mth patient. Furthermore, ζ is a M-estimation function and $\omega_m = 1/N_m$ is the weighting function with N_m being used to normalize the performance index.

Age Mapping: To estimate the disease progression time-frame using the patients visit along with estimated biomarkers data as follows:

$$\left(c^{o(i)}, d^{o(i)}\right) = \min_{n,k \in N_m} \sum_{n,k} \omega_m \zeta \left\{ \frac{Y^o_{m,n,k} - f(x_{m,n}, \lambda_k^i)}{\sigma_k} \right\}$$

In the above algorithm the variable N_m is the biomarkers measurement of mth patient during the corresponding visits. The degrees of freedom for the trajectory fitting is considered as $\sum_k (N_k - |\lambda_k|) - 2P$ with P indicates the number of patients, N_k is the number of biomarkers measurements for all patients for all the visits, and λ_k is the parameters for the observed values of the biomarkers. In the final stage of curve fitting of biomarkers trajectories, the DPSs have been standardized for all the visits of the cognitive normal patients such that the trajectories can be calibrated. By neglecting mean values of the normal visits, the parameters can be updated as follows:

$$x_{m,n} = \left(x_{m,n} - \mu_{cn}\right)/\sigma_{cn},$$

$$c_m = c_m/\sigma_{cn},$$

$$d_m = (d_m - \mu_{cn})/\sigma_{cn},$$

$$b_k = \sigma_{cn} b_k,$$

$$\delta_k - b_k \alpha_k = \frac{\delta_k - b_k \alpha_k - \mu_{cn}}{\sigma_{cn}},$$

with μ_{cn} is the mean and σ_{cn} is the standard deviation (SD) of the DPSs. After all the above stages of the curve fitting of the biomarkers trajectories, we can determine the biomarkers value from the age mapping step in the algorithm by applying the new Stannard logistic function defined in (1). In the final stage of the disease progression model, the prognostic of clinical status is performed by incorporating the Gaussian kernel density estimation with three classification of patients namely cognitive normal, pMCI, and AD. The smooth Gaussian kernel can be represented as

$$p^o(x|c_m) = \frac{1}{N\mathscr{W}} \sum_{p=1}^{N} \mathscr{K}^{\mathscr{G}} \left\{\frac{x - x_p}{\mathscr{W}}\right\}$$

where the variable $\mathscr{K}^{\mathscr{G}}$ is the Gaussian kernel function and $\mathscr{W}$ is the bandwidth for function smoothing. The clinical status can be determined by incorporating Bayesian theorem to the smooth Gaussian kernel with DPSs.

As far as datasets is concerned, it has been taken from Alzheimer's disease neuroimaging initiative (ADNI) website, which are available after taking permission from the ADNI. In the present study, we have analyzed the data of 754 patients with 251 are identified as CN at baseline of the progression time-frame, 279 are identified as MCI for the entire duration of 48 months or 2.5 years, 161 are diagnosed as pMCI during initial 6 months to 18 months duration and later on converted to AD after 48 months time-frame, and 63 were diagnosed as AD all the time during every visits in 48 months of progression time-frame as shown in Fig. "System overview". Some patients have shown improvements after each visits during the entire duration are not included in this study. The time series data were collected during every visits from initial 6 months up to 18 months. The time series data were collected based on the following features: CSF biomarkers, frequently asked questions (FAQ), age, education years, demographics, medical history, Montreal cognitive assessment (MOCA), MRI and fludeoxyglucose (FDG)-PET imaging.

The unused data for the modeling is segregated based on the logarithmic concept and it is accounted for 1/e = 0.37 which means 67% of the data were used in Monte Carlo resampling estimates. The Bayesian information criterion (BIC) will be used for the model selection and the lowest BIC model is better choice. The bootstrapping helps in quantifying the robustness of the estimates. We have used 50 bootstrapping for modeling the AD by using the new modified stannard logistic function and the loss functions. The model performance with respect to BIC index are 4.816 ± 0.123 for $L-2$ loss function, 4.867 ± 0.129 for $L1 - L2$ loss function, and 4.960 ± 0.132 for the $Cauchy - Lorentz$ loss function. The convergence of the model can be shown with respect to normalized mean absolute error (NMAE) performance index which are 1.042 ± 0.088 for $L-2$, 0.996 ± 0.073 for $L1 - L2$, and 1.0194 ± 0.073 for $Cauchy - Lorentz$ loss functions. Furthermore, the next section incorporate the developed model of AD in this section, which is performing very well to find the diagnosis of the progressive AD stages.

3 Proposed Technical Framework

In this paper, the long-short term memory methodology is used for the diagnosis of Alzheimer's disease. Because the data for the AD progression is based on the time series, the suitable technique that can be used efficiently for the diagnosis of AD is LSTM framework. The preambles of LSTM technique are discussed in the following subsection.

3.1 Long-Short Term Memory Technique

The recurrent neural network technique such as LSTM is very feasible and efficient for solving the gradient minimization problem through the technique of time-optimization in the course of backpropagation algorithm. In this work, the patient's data are utilized as time series over the interval of 6 months from zero to 18 months as shown in Fig. 1 and LSTM layers are increased in cascade to capture the temporal sequence with features. The following three gates are used in the proposed LSTM algorithm for the flow of information in the cell state [11]:

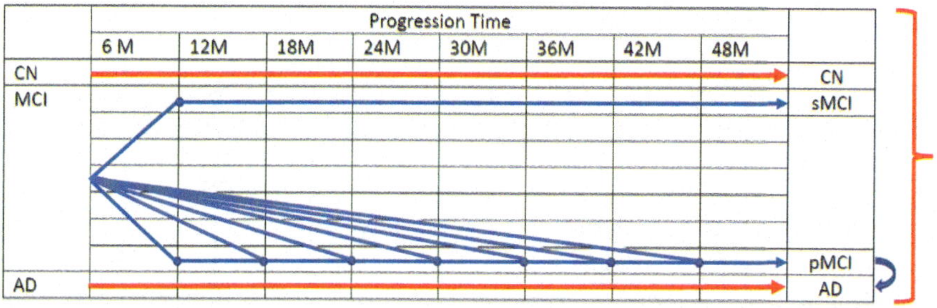

Fig. 1 Time series data till 18 months of visit and progression time of MCI to sMCI or pMCI or AD

- **Input gate** (I_{t_n}): This gate is responsible for managing the information flow update.
- **Forget gate** (F_{t_n}): This gate is responsible for managing the required information flow.
- **Output gate** (O_{t_n}): This gate is responsible for deleting the information flow.

Suppose the status of cells at time t_n, t_{n-1}, and updated status at t_n are CS_{t_n}, $CS_{t_{n-1}}$, and $\tilde{CS}_{t_n}$. Also, y_{t_n} and $y_{t_{n-1}}$ are the outputs at time t_n and t_{n-1} based on the cells status (CS_{t_n}, $CS_{t_{n-1}}$ for output and hidden layer, and $\tilde{CS}_{t_n}$). The flow of information in memory cells can be represented mathematically as follows [11]:

$$F_{t_n} = f_\sigma\left(\beta_f.\left\{y_{t_{n-1}}, x_{t_n}\right\} + b_f\right) \tag{2}$$

$$I_{t_n} = f_\sigma\left(\beta_i.\left\{y_{t_{n-1}}, x_{t_n}\right\} + b_i\right) \tag{3}$$

$$\tilde{CS}_{t_n} = tanh\left(\beta_C.\left\{y_{t_{n-1}}, x_{t_n}\right\} + b_C\right) \tag{4}$$

$$CS_{t_n} = \left(F_{t_n} \otimes CS_{t_{n-1}} \oplus I_{t_n} \otimes \tilde{CS}_{t_n}\right) \tag{5}$$

$$O_{t_n} = f_\sigma\left(\beta_o.\left\{y_{t_{n-1}}, x_{t_n}\right\} + b_o\right) \tag{6}$$

$$y_{t_n} = O_{t_n} \otimes tanh(CS_{t_n}) \tag{7}$$

$$h_n = \phi\left(\beta_y y_{t_n} + b_y\right) \tag{8}$$

where the operator $\otimes$ is the Hamdard product and $\oplus$ is the concatenation operator. The weight functions are defined by β_j, $j = f, o, i, y, C$, f_σ is the sigmoid logistic function, and ϕ is the activation function for the output. The figurative structure of LSTM cells are shown in Fig. 2

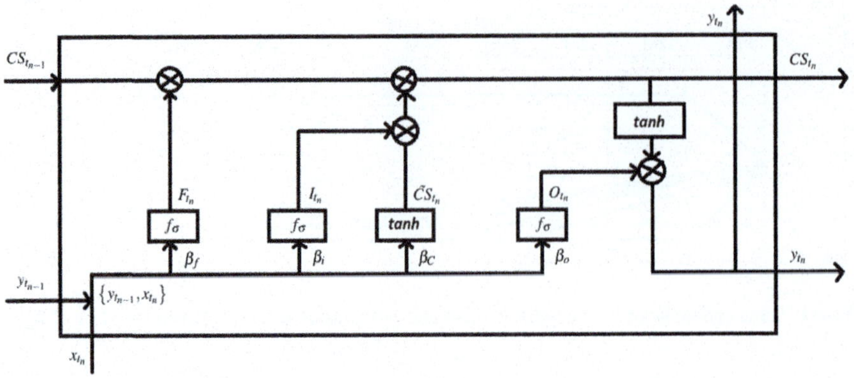

Fig. 2 LSTM cells structure [11]

3.2 Conversion Time Prediction

The progression time has been detected for the 48 months interval using logistic regression by incorporating time-series data based on several factors described in the previous section. After careful study on the detection of progression time, the conversion time from MCI to pMCI can be predicted for the patients with MCI status at the end of 18 months. Therefore, the modeling part determines the progression of patients from MCI to pMCI or AD, after that the time-series data has been used to design a regression model to predict the exact time-frame in which the conversion is happened. Hence, the main objective of our previous work [14] is to develop a model using logistic regression concept such that the same model can be used to find the exact conversion time of MCI patients to pMCI or AD. The time-series data of N patients can be used directly with LSTM model, which can be represented by $x_i = \left\{x_i^{(1)}, ...x_i^{(m)}, ...x_i^{(M)}\right\}, i = 1, ..., N$ and $h_i \in \{CN, MCI, AD\}$ with M represents the modalities. Furthermore, the patients aggregation features were collected till 18 months at an interval of 6 months for per visit along with the extraction of statistical measurement to elaborate the time-series data. In addition to optimization cost function in Algorithm 1, the regression assignment tuned by incorporating the regularized mean squared error (MSE) function as follows:

$$\lambda^{o(i)} = \min_{m,n,k} \frac{1}{N} \sum_{m,n,k}^{N} \left(Y^o_{m,n,k} - \hat{Y}^o_{m,n,k}\right)^2 + \frac{\lambda_k}{N}\beta_j^2 \;\; , j = f, o, i, y, C \tag{9}$$

where the last term is regularization and β_j, $j = f, o, i, y, C$ are the weighting functions. In our study 67% of the data were used for the model development, see [14] and 33% of the data were used for the model testing. The data splits are shown in Table 1. Additionally, the features that were missing 30% of the data points have been deliberately overlooked , as well as the patients' history, when handling the

Table 1 Demographics of the datasets obtained from ADNI after filtering according to visits

	CN (251)	sMCI (279)	pMCI (161)	AD (63)
Age	76.35 ± 7.52	75.27 ± 8.93	75.86 ± 7.46	75.27 ± 6.93
Education	16.12 ± 2.54	16.23 ± 2.47	16.23 ± 2.37	16.57 ± 2.67
FAQ	00.12 ± 00.54	02.12 ± 02.54	06.13 ± 06.54	13.12 ± 06.84
MMSE	26.42 ± 02.34	26.12 ± 02.58	26.73 ± 03.57	25.55 ± 02.33
MOCA	23.42 ± 02.54	26.62 ± 03.34	25.73 ± 03.43	21.55 ± 02.39

data. Furthermore, the mean value for the distinct classes of patients (CN, sMCI, pMCI, AD) for the different classes of time-series data are utilized in this paper. The time-series data collected through patient visits over 18 months at 6 month intervals are separated into two categories: full feature data sets of patients and filtered feature data sets of patients.

4 Experimental Results and Discussion

For the results analysis in this study, the authors have considered three parameters namely; accuracy, precision, and recall after tuning and testing the regression technique for modeling and LSTM technique for the prediction of time-frame. The time-series data collected from different perspectives were used to train and test the LSTM model for the prediction of exact progression time. The advantages of the time series data are that they increase the accuracy of the model and also confidence level. The different performance parameters are defined as follows:

$$\text{Accuracy} = \frac{\text{TP} + \text{TN}}{\text{TP} + \text{FP} + \text{TN} + \text{FN}}$$

$$\text{Precision} = \frac{\text{TP}}{\text{TP} + \text{FP}}$$

$$\text{Recall} = \frac{\text{TP}}{\text{TP} + \text{FN}}$$

where TP is defined as true positive, TN represents true negative, FP is defined as false positive, and FN is defined as the false negative. The time series data has been used in the digital mode where 0 stands for the false and 1 stands for the true value. The following statistical metrics have been analyzed for the regression stage of modeling, which determine the convergence speed of the data types utilized in the models:

Table 2 LSTM model training and testing results

	Accuracy	Precision	Recall
Full feature data sets	94.25 ± 1.35	93.16 ± 1.33	93.17 ± 1.93
Filtered featured data sets	94.79 ± 2.37	93.87 ± 2.36	93.87 ± 2.96

Table 3 LSTM model training and testing results with average time-series

	MAE	RMSE
Full feature data sets	0.1457	0.2388
Filtered featured data sets	0.1423	0.2327

$$\text{MAE} = \frac{1}{N}\sum_{1}^{N}\left|Y^{o}_{m,n,k} - \hat{Y}^{o}_{m,n,k}\right|$$

$$\text{RMSE} = \sqrt{\frac{1}{N}\sum_{1}^{N}\left(Y^{o}_{m,n,k} - \hat{Y}^{o}_{m,n,k}\right)^{2}}$$

where MAE is the abbreviation for mean absolute error and RMSE is the abbreviation for the root mean squared error. The variables $Y^{o}_{m,n,k}$ and $\hat{Y}^{o}_{m,n,k}$ are the actual and measured biomarkers values. The training of LSTM uses the linear activation function in the outputs and the hidden layers including first layer underwent for $L2$ regularization's. There were total 50 bootstrapping during the logistic regression and 100 epochs used for LSTM training and testing. The results are shown in the Table 2, we can see that the model can achieve an accuracy of 94.25%, precision of 93.16%, and recall of 93.17% in case of full feature data sets while it achieves an accuracy of 94.79%, precision of 93.87%, and recall of 93.17% in case of filtered data sets. As a direct consequence, it can be observed that the model performs better with filtered data sets while performing little bit poorly with full-featured data sets. Because full feature data sets contain noisy data, the computations become more difficult, resulting in higher losses and a lower performance efficiency when compared to filtered data sets that may contain less noisy data.

The statistical measurements used to determine the convergence speed of the models are shown in Table 3. According to the calculated data in the table, the filtered featured data set model converges faster than the full featured data set models. This is also appropriate since, as shown in Table 3, if a large number of features are evaluated in the data sets, the data becomes excessively noisy, which can reduce the computational performance of the model. However, the data with fewer features or filtered features can be less noisy, potentially improving model accuracy and the computational performance. As a consequence, the model with full featured data sets is having a slower convergence rate than models with less featured data sets.

In general, the model with filtered characteristics modeled with logistic regression techniques and further prediction with LSTM methodology produces acceptable outcomes. Although LSTM framework is quite suitable for the time series data, the good performance efficiency can be achieved with statistical techniques also.

5 Conclusion

In this study, a long-short term memory model has been proposed for determining the actual progression period experienced by MCI patients. The conversion rate from MCI state of the patients to pMCI or AD state can be determined accurately. For modeling the AD model using a new modified Stannard function, a regression technique is used with optimization. The data sets were separated into two categories: full featured datasets and filtered featured datasets. The statistical measurements collected in the research reveal that the model with full featured datasets performed poorly in comparison to the model with filtered datasets because full featured data sets contain a large number of noisy data that can degrade model accuracy and computational performance. The proposed model is based on previous research in which we modeled AD with a new modified Stannard function using logistic regression technique, and the same model is used in this paper with further modification in the logistic function with an LSTM framework to predict the exact progression time of MCI patients to pMCI or AD over a 48-month period. The data were obtained based on the time-series intervals of visits per patient per 6 months of interval. Finally, we have shown that the model achieves an accuracy up to 94.79%, precision of 93.87%, and a recall of up to 93.87%. In this study a total of 754 subjects were investigated and all the data were taken from ADNI with due permission obtained for the use in the present research. Furthermore, the future study will be based on quaternionic domain for the convolutional neural network.

Statement for Data Availability
All the data used in this research are publicly available on request from ADNI database.

References

1. Hong X, Lin R, Yang C, Cai C, Clawson K (2020) Adpm: an Alzheimer's disease prediction model for time series neuroimage analysis. IEEE Access 8:62601–62609
2. Lu S, Xia Y, Cai W, Fulham M, Feng DD (2017) Early identification of mild cognitive impairment using incomplete random forest-robust support vector machine and FDG-PET imaging, computerized medical imaging and graphics, computational methods for. Mol Imaging 60:35–41
3. Zaharchuk G, Davidzon G (2021) Artificial intelligence for optimization and interpretation of PET/CT and PET/MR images. Sem Nucl Med Artif Intell Nucl Med 51(2):134–142

4. Jr CJ, Petersen R, Xu Y et al (1999) Prediction of ad with mri-based hippocampal volume in mild cognitive impairment. Neurology 52(7):397–403
5. Moons L, De Groef L (2022) Multimodal retinal imaging to detect and understand Alzheimer's and Parkinson's disease. Curr Opin Neurobiol 72:1–7
6. Snyder PJ, Alber J, Alt C et al (2021) Retinal imaging in Alzheimer's and neurodegenerative diseases. Alzheimer's Dement 17:103–111
7. Lee G, Nho K, Kang BEA (2019) Predicting Alzheimer's disease progression using multi-modal deep learning approach. Sci Rep 9(1)
8. Liu M, Zhang J, Adeli E, Shen D (2019) Joint classification and regression via deep multi-task multi-channel learning for Alzheimer's disease diagnosis. IEEE Trans Biomed Eng 66(5):1195–1206
9. Ding X, Bucholc M, Wang H et al (2018) A hybrid computational approach for efficient Alzheimer's disease classification based on heterogeneous data. Sci Rep 8(1)
10. Suk HI, Lee SW, Shen D (2015) The Alzheimer's disease neuroimaging, I., Latent feature representation with stacked auto-encoder for AD/MCI diagnosis. Brain Struct Funct 220:841–859
11. El-Sappagh SSHAFEA (2022) Two-stage deep learning model for Alzheimer's disease detection and prediction of the mild cognitive impairment time. Neural Comput Appl 34(1):14487–14509
12. Fang X, Liu Z, Xu M (2020) Ensemble of deep convolutional neural networks based multi-modality images for Alzheimer's disease diagnosis. IET Image Process 14:318–326
13. Biagioni MC, Galvin JE (2011) Using biomarkers to improve detection of Alzheimer's disease. Neurodegener Dis Manag 1(2):127–139
14. Ranjan M, Kumar S (2023) Modeling of progressive Alzheimer's disease using machine learning algorithms. In: International conference on artificial intelligence, blockchain, computing, and security, 24–25 Feb 2023, Delhi-NCR, India
15. Oxtoby NP, Alexander DC (2017) Imaging plus X: multimodal models of neurodegenerative disease. Curr Opin Neurol 30(4):371–379
16. Li D, Iddi S, Thompson WK, Donohue MC (2019) Bayesian latent time joint mixed effect models for multicohort longitudinal data. Stat Methods Med Res 28(3):835–845
17. Lorenzi M, Filippone M, Frisoni GB et al (2017) Probabilistic disease progression modeling to characterize diagnostic uncertainty: application to staging and prediction in Alzheimer's disease. NeuroImage 190:56–68
18. Ghazi MM, Nielsen M, Pai A et al (2019) Training recurrent neural networks robust to incomplete data: application to Alzheimer's disease progression modeling. Med Image Anal 53:39–46
19. Ghazi MM, Nielsen M et al (2021) Robust parametric modeling of Alzheimer's disease progression. NeuroImage 225:117460

Local Mean Directional Intensity Pattern: An Efficient Descriptor for Hand Gesture Recognition Using SVM Classification

Arti Bahuguna, Gopa Bhaumik, and Mahesh Chandra Govil

Abstract Hand Gesture Recognition (HGR) promotes the most efficient and natural communication between users and machines. However, the performance of the HGR system is deteriorating due to challenging conditions such as illumination variation, complex backgrounds, inter-class similarities. In this paper, a handcrafted feature descriptor named as Local Mean Directional Intensity Patterns (LMDIP) is proposed for effective hand gesture recognition in the aforesaid challenging environment. The proposed LMDIP descriptor aims to maintain the most relevant directional information. It assists in the detection of grey level variations that move in different directions. LMDIP explores differential excitation by calculating the mean intensity of pixels across each angular neighboring pixel. The most significant effect of employing the mutually beneficial connections between adjacent neighbors in different angular directions is that it no longer relies solely on the sign of the intensity differences between the center pixel and one of its neighbors, but also on the sign of the gap across its adjoined neighbors. The performance of the proposed LMDIP is evaluated in terms of accuracy and F1-score on five benchmark datasets: Massey University Dataset (MUGD), ASL Digit Datasets, ASL Static Dataset, NUS Dataset, and OUhand Datasets. The result of the experimental show that the proposed LMDIP achieves an accuracy of 92% (MUGD Set1), 98% (MUGD Set2), 75%(MUGD Set3), 61%(MUGD Set4), 71%(MUGD Set5),99%(ASL Digit), 51%(ASL Static), 83%(NUS-I Dataset), 39% (OUhands Dataset) and respectively which is better compared to the state-of-art methods.

Keywords Local binary pattern · LMDIP · Feature extraction · SVM · HGR

A. Bahuguna · G. Bhaumik (✉) · M. C. Govil
Department of Computer Science and Engineering, National Institute of Technology Sikkim, Sikkim, India
e-mail: gopabhaumik.cse@nitjsr.ac.in

G. Bhaumik
Department of Computer Science and Engineering, National Institute of Technology Jamshedpur, Jharkhand, India

S. Bandyopadhyay et al. (eds.), *Intelligent Computing Systems and Applications*, Lecture Notes in Networks and Systems 1010,
https://doi.org/10.1007/978-981-97-5412-0_38

1 Introduction

In literature, several handcrafted feature descriptors, particularly those relating to image texture features, play a key building block in a variety of fields for computer vision including texture classification, identification of objects and detection, health diagnosis, the categories of gender division, content-based image retrieval, and so on, have gained the attention of researchers. Human gesture recognition is becoming increasingly important as it can be applied in a variety of tools used by deaf and hearing-impaired persons, as well as in the development of vision-based apps and touchless controls in everyday electronic devices. Because HGR is essential in sign language analysis [1] as well, a strong HGR system should consider both spatial and temporal information. Human gestures are body, facial, and/or hand motions [2, 3]. Several researchers, notably Quek [4], Kjeldsen and Kender [5], Starner and Pentland [6], have investigated hand gesture interaction as an alternate kind of human-computer interface. An efficient human-computer interface may find use in mundane activities. It is the mechanism through which the receiver acknowledges the user's motions. Hand gesture recognition enables people to communicate more naturally with computers in an efficient manner. HGR systems have been successfully employed in video games on computers, systems for translating signs into text, conversation with sign language [7, 8], robotics [9], and video monitoring. Hand gestures are likely to be identified as static or dynamic. Static gestures represent only positions, but dynamic gestures are movements of bodily components. Hand form, palm size, finger length, and width must all be considered in static gesture recognition [10], and dynamic hand gestures require spatiotemporal information [11, 12]. Previously, most vision-based algorithms focused on detecting and extracting local features from images. Nowadays, researchers admired the Histograms of Oriented Gradients (HOG) features [13], as well as a ready-made mixture of binary image characteristics based on pixels [14].

Inspired by the current descriptors that include LBP, LDP, LBPDP, LNIP, EXTRA [15], and TLBAP, we designed a novel handcrafted feature descriptor defined as Local Mean Directional Intensity Pattern (LMDIP) for hand gesture recognition from static images. The core objective of this research is to serve as a more accurate local feature description LMDIP, that is resistant to fluctuations in light intensity and noise. The remaining portion of the paper is laid out as follows: Sect. 2 presents relevant research in the domain. Section 3 presents the proposed work in more depth. In this section, we discussed how the LMDIP descriptor works. Section 4 represents the outcomes of the experiment and result analysis. The conclusion is presented in Sect. 5.

1.1 Motivation

The gesture is a method of communication between the speech and hearing-challenged group and the general public. Every day, persons who are deaf and dumb

(D&D) experience challenges communicating with the general public all around the world. They usually require an interpreter to speak with others, and the interpreter is not always accessible. So owing to a lack of awareness and comprehension of many hand gestures, the hearing/speech-challenged group finds it difficult to express their message or share an idea with normal people. As a result, we introduce the novel descriptor named Local Mean Directional Intensity Patterns (LMDIP) employ the behavior of neighboring pixels in subregions to achieve resilience in noisy environments. The proposed LMDIP uses local intensity values of pixels to improve the performance of a robust local feature descriptor that takes information from nearby pixels based on their mutual relationship and combined characteristics depending on the pixels nearest to the center pixel.

2 Related Work

The texture features are an essential aspect of image construction and representation in computer vision applications. The heart of texture image retrieval is characterizing the spatial grey level distribution of pixel neighborhood to serve as a texture feature. A variety of strategies have already been proposed over the last few decades to bring out either local or global features for various computer vision tasks. The qualities of the entire image are measured by global and local features that collect tiny characteristics of essential portions of an image, thereby enhancing recognition accuracy. A brief discussion on the existing descriptors available in the literature is as follows: The very first local texture descriptor is indeed the Local Binary Pattern (LBP) [16]. The neighbors within a certain radius and the center pixel are considered during the feature computations and are evaluated for every non-border pixel. An LBP is extremely simple, and it has become the foundation for many additional local texture descriptions.

The concept underlying this technique aims to extract the texture feature and convert it toward a wider representational platform using a given kernel function that incorporates and examines its neighbors. In order to determine the converted value of each pixel. Due to its simplicity and flexibility, as well as its effectiveness in many disciplines, LBP laid the groundwork for the development of its additional versions encouraging the researchers to suggest their own local descriptors. Inspired by LBP, Shiv Ram et al. introduce a Local diagonal extrema pattern (LDEP), that investigates the correlation within the diagonal neighborhood of the center pixel and generates a description based on the index, that examines the relationship among the local diagonal extreme value [17] and their central pixel value. This descriptor depends on local diagonal derivatives to relate the core pixel to its diagonal neighbors.

Another feature descriptor is local tri-directional patterns (LTriDP), which is a technique that employs the pixel intensity depending on their neighborhood direction. In addition, one magnitude pattern is combined for improved feature extraction [18]. Further, Shiv Ram et al. introduce the Local Bit-Plane Decoded Pattern (LBPDP) [17]. This descriptor combines the local bit-plane converted values of every image pixel with the bit-plane binary components of its neighbors. The resultant LBPDP is created by generating a binary pattern by subtracting the intensity value of the central pixel from the local bit-plane converted values. Given the ongoing requirement for an innovative local texture descriptor with improved exclusive capacity, significant research on the application of local descriptors in pattern recognition is still in progress, as evidenced by recently published articles such as Quest [19], Local Neighborhood Intensity Difference Pattern (LNIP) [20], Local Neighborhood Difference Pattern (LNDP) [21]. Local Adjacent Neighborhood Average Difference Pattern (LANADP) [22] and Threshold Local Binary AND Pattern (TLBAP) [22] are similar handcrafted descriptors designed for medical image indexing and retrieval. TLBAP takes a threshold set by neighbors and performs a bit-wise AND with the LBP features to calculate the value at a center pixel. This process is done until each non-border pixel in order to calculate the TLBAP features for the carryout image.

3 Proposed LMDIP Descriptor

This study presents a unique feature descriptor named LMDIP: Local Mean Directional Intensity Pattern to support static hand gesture recognition. The proposed pattern recognition system is depicted in Fig. 1. The proposed LMDIP captures the connections between the center and neighboring pixels in different angular directions. Thus, LMDIP is less susceptible to grey level variations and improves robustness against lighting since it encodes the relative intensity difference of the surrounding region. The descriptor is developed based on the integration of two local patterns influenced by LBP. First, it computes the angular mean for odd pixels and averages the results from different angular directions. Second, it takes into account the even pixel values, which include the four primary orientations (0°, 90°, 180°, 270°) horizontally and vertically along with the center pixel. Both the patterns are concatenated to generate a new pattern, and the encoding approach is known as Local Mean Directional Intensity Patterns (LMDIP), which is capable of extracting rich and discriminating features from noisy images. An example of the calculation of directional neighbors of the respective pixel is shown in Fig. 2. Figure 3 represents the directional pattern computation. Figure 4 depicts a comprehensive computation of the pattern generated by the LMDIP descriptor. The intensity value of the 8 neighborhood pixels along with 3×3 windows contains two cases for pixel calculations.

Case 1 If i represents the odd pixel, then the angular mean is calculated using a mean of intensity levels from different angular directions(0°–315°).

Case 2 If i represents an even pixel, then the four primary orientations (0°, 90°, 180°, 270°) with the horizontal and vertical directions are considered.

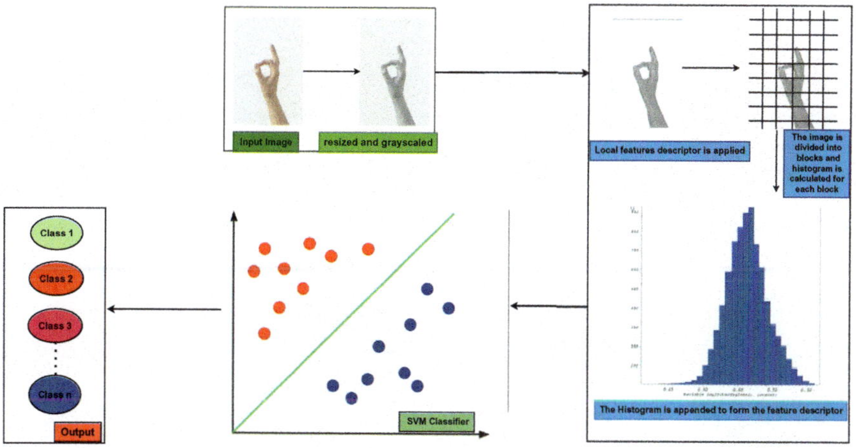

Fig. 1 The proposed pattern recognition system

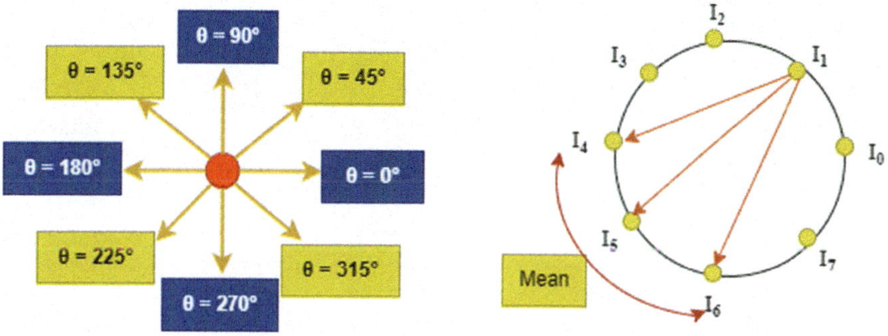

Fig. 2 Directional neighbors of reference pixel of the local region of an image

The output generated by the proposed LMDIP descriptor can be calculated mathematically using the Equations [1, 2, 3, 4]. The LMDIP image is coded as an 8-bit pattern. So as a result, the histogram's size is estimated as $2^8 = 256$. This histogram is used as a feature vector and we calculated the histogram using Equation [5-6].

$$\text{LMDIP}(I_c) = \sum_{i=0}^{7} 2^i * \Delta(S_i, I_c) \tag{1}$$

where LMDIP(I_c) is implied as the pixel value of each center pixel I_c and I_i depicted the 8-neighboring pixels in a 3×3 local region of input image I with size $m \times n$.

$$S_i = I_i \forall i = 0, 2, 4, 6 \tag{2}$$

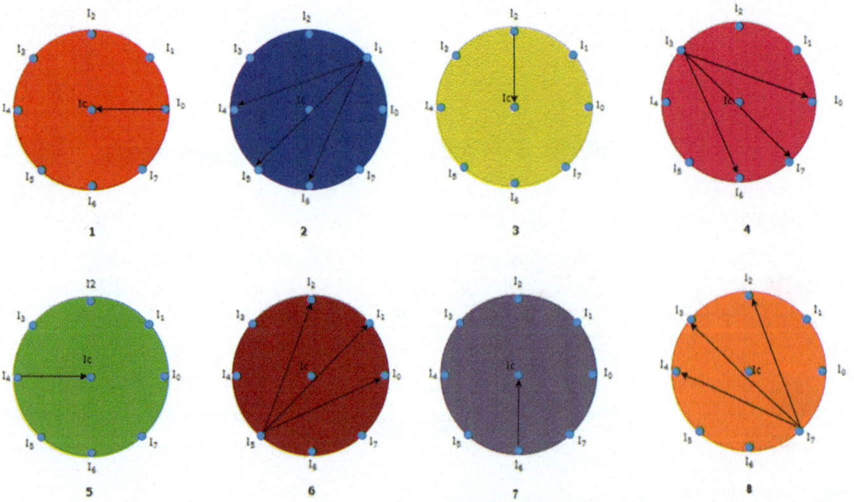

Fig. 3 An illustration of a texture direction pattern computation

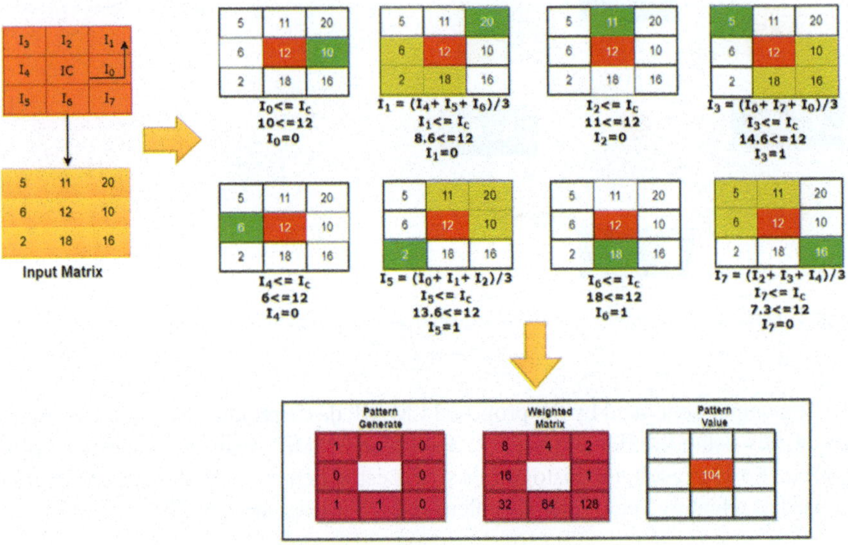

Fig. 4 An example of obtaining the proposed LMDIP pattern

$$S_i = \left\lfloor \frac{I_{(i+3)\%G} + I_{(i+4)\%G} + I_{(i-3)\%G}}{3} \right\rfloor \forall i = 1, 3, 5, 7 \tag{3}$$

$$\Delta(\iota, \varphi) = \begin{cases} 1, & \text{if } \iota \geq \varphi \\ 0, & \text{otherwise} \end{cases} \tag{4}$$

$$\text{Hist}(t) = \sum_{x=1}^{m} \sum_{y=1}^{n} \text{hist}(\text{LMDIP}(x_1, y_1), t); t \in [0, (2^G - 1)] \tag{5}$$

$$\text{hist}(h1, h2) = \begin{cases} 1, & \text{if } h1 = h2 \\ 0, & \text{otherwise} \end{cases} \tag{6}$$

4 Experimental Results and Analysis

This phase discusses the experimental outcomes, datasets used, and the quantitative and qualitative analysis of the proposed LMDIP descriptor developed for hand gesture recognition. The LMDIP pattern image has been encoded and the feature vectors are generated in a histogram with 256-bin. The images included in the dataset are segregated into testing and training sets in the proportion of 20:80. The image is transformed into grayscale and resized to 50 × 50. The experiments have been performed on the Anaconda Distribution 2023.03 platform using Jupyter Notebook 6.5.4 edition with the subsequent system configuration: a Xeon(R) Intel(R) W-2135 CPU @ 3.70 GHz, an NVIDIA Quadro P2000 graphics card (GPU is 4 GB), and 64 GB of RAM.

4.1 Datasets

The proposed LMDIP descriptor is evaluated on five benchmark datasets: Massey University (MUGD) dataset, the NUS dataset, the American Sign Language Static dataset (grayscale framework), American Sign Language (ASL) digit dataset, and OU Hnad Datasets. The MUGD dataset provides 2529 images that have been accurately separated on a black background and belong to 36 different classes. The MUGD dataset is divided into five sets, each of which has 36 different classifications. The NUS1 dataset contains 240 images that have been appropriately segmented over a black background and are divided into 10 unique classes. The ASL static hand gesture dataset in a grayscale framework included 4851 raw pictures belonging to 24 classes, and the ASL digit datasets consist of 5000 images in raw format that extend from sign 0 to sign 9 (Each sign contains 500 images) along with 5000 corresponding output image files, whereas OU hands datasets have 2990 images divided into 10 unique classes.

4.2 *Quantitative Analysis*

The performance of the proposed LMDIP descriptor is examined by its accuracy and F1-score. A comparative study of planned and current descriptors, such as QUEST, TLBAP, LNIP, LBPDP, LBP, LANADP, and LMDIP is presented in Table 1. Figure 5 depicts the graphical depiction of the comparison analysis of the proposed and existing descriptors. The results of the experiment indicate that the proposed LMDIP descriptor accomplished an accuracy of 99% (ASL Digit dataset digit), 92% (MUGD-1), 98% (MUGD-2), 75 % (MUGD-3), 61% (MUGD-4), 71% (MUGD-5), 83% (NUS), 51% (ASL static grayscale framework), and 39% (OU Hands) on various datasets respectively.

4.3 *Qualitative Analysis*

The notable impact of the proposed LMDIP descriptor in comparison to the existing descriptors is shown in Fig. 6. That shows different hand postures from each of the five benchmark datasets as well as the resultant outcome of the proposed and existing descriptors that were assessed in this paper. The visual representation indicates that LMDIP detects macro and micro edge level transformation with improved discriminative features examined to the state-of-art techniques for effective hand gesture recognition.

5 Conclusion

This work presents a unique feature descriptor, namely LMDIP that is resistant to variation in light and noise. The proposed method combines the mean of angular points to extract local information structure. The mean of neighboring pixels and basic orientations are used to detect the relationship between targeted pixels and adjacent ones which significantly minimize the influence of noise in the image. The proposed LMDIP analyses the gradient change in a neighboring pixel along different angular directions. To encode the LMDIP binary pattern, the connections of sign patterns between a constant angle point (45°) and its associated near neighbors are captured. The suggested approach is examined using two sign language datasets (ASL digit and ASL static) and three hand posture databases (MUGD datasets, NUS hand posture Dataset-I, and OU Hands), in which all motions are performed with the free hand in varied postures. The outcomes of the experiments show that

Table 1 The comparative studies with respect to Accuracy (Acc) and F1-Score (F1.S) in % of proposed LMDIP and the current descriptors based on MUGD, ASL Digit, ASL Static, NUS, and OU hand datasets

Method	MUGD1		MUGD 2		MUGD 3		MUGD 4		MUGD 5		ASL digit		ASL static		NUS 1		OU hand	
	Acc	F1.S	Acc	F1.S	Acc	F1.S	Acc	F1.S	Acc	F1.S	Acc	F1.S	Acc	F1.S	Acc	F1.S	Acc	F1.S
LBPDP	76	78	84	80	50	51	58	59	60	60	96	96	37	34	23	25	46	47
QUEST	88	89	97	93	72	77	64	69	64	64	98	98	40	37	58	55	39	40
LNIP	78	80	89	86	64	65	58	59	61	63	97	97	44	41	69	68	38	35
TLBAP	89	90	95	95	75	81	61	63	71	72	98	98	47	44	69	67	38	36
LBP	87	88	94	94	72	77	64	69	65	66	98	98	47	45	85	85	17	12
LANADP	89	90	98	95	72	77	64	69	58	60	97	97	40	37	83	83	17	12
LMDIP	**92**	**92**	**98**	**98**	**75**	**81**	**61**	**63**	**71**	**72**	**99**	**99**	**51**	**48**	**83**	**84**	**39**	**36**

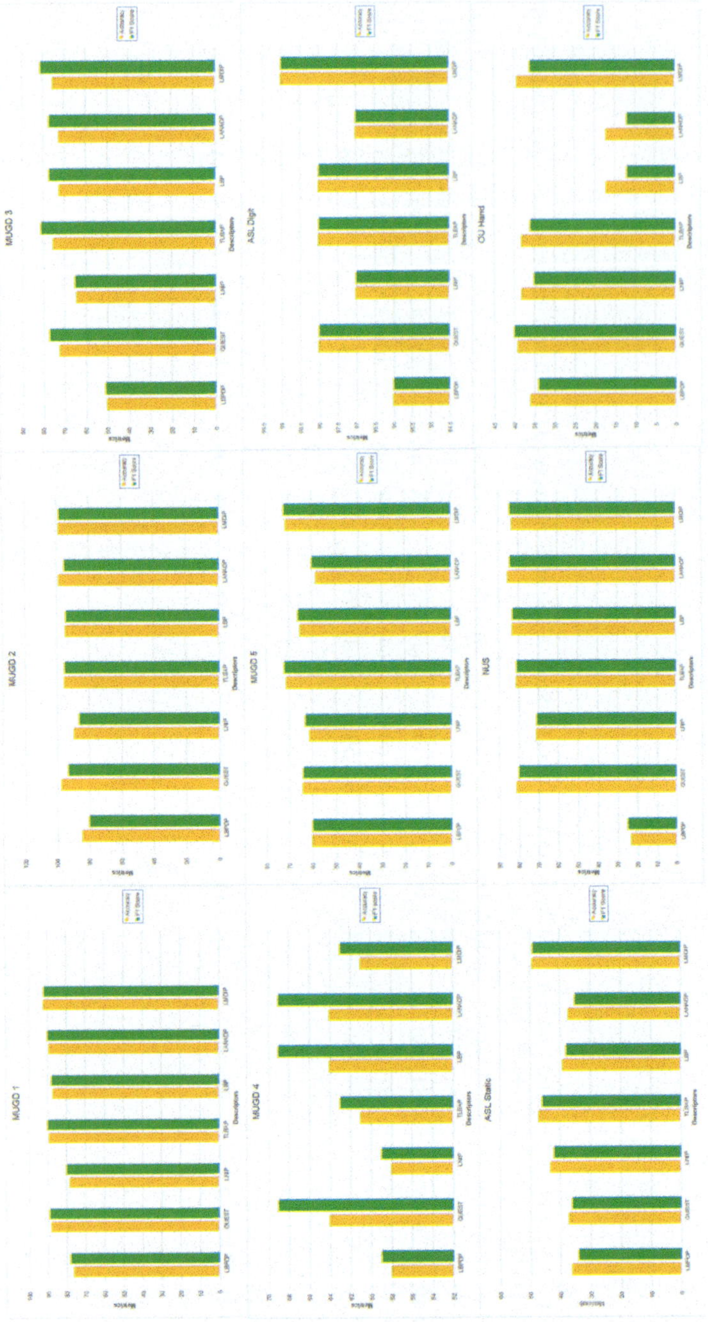

Fig. 5 A graphical depiction of the LMDIP descriptors is compared with the existing descriptor on the MUGD, ASL static, ASL digit, NUS, and OU hand

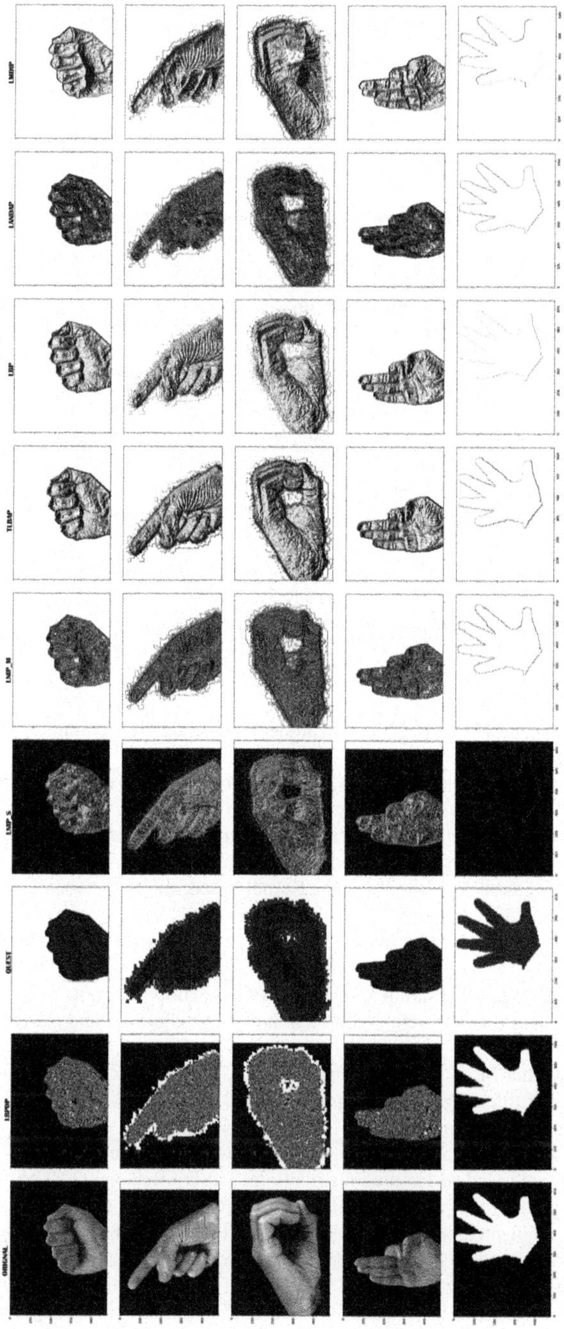

Fig. 6 The feature comparisons response gained by applying (1) LBPDP (2) QUEST (3) LNIP (4) TLBAP (5) LBP (6) LANDAP (7) LMDIP

the proposed LMDIP exceeds the most recent hand gesture recognition algorithms. Future research will investigate how well the suggested local directional gradient descriptor and symbolic modeling technique work for a suited HGR system that can operate in an unconstrained environment.

References

1. Hasan H, Abdul-Kareem S (2014) Human-computer interaction using vision-based hand gesture recognition systems: a survey. Neural Comput Appl 25(2):251–261
2. Maggioni C (1993) A novel gestural input device for virtual reality. In: Proceeding IEEE virtual reality annual Int'l symposium. Seattle, Wash, pp 118–124
3. Vaananen K, Boehm K (1993) Gesture driven interaction as a human factor in virtual environments an approach with neural networks, virtual reality systems, eds., chapter 7. Academic Press, pp 93–106
4. Quek F (1994) Toward a vision-based hand gesture interface. In: Proceeding virtual reality software and technology Conference. Singapore, pp 17–31
5. Kjeldsen R, Kender J (1995) Visual hand gesture recognition for window system control. In: Proceeding of Int'l workshop automatic face- and gesture-recognition. Zurich, Switzerland, pp 184–188
6. Starner T, Pentland A (1995) Real-time American sign language recognition from video using hidden Markov models. Technical Report TR-375, MIT's Media Lab
7. Singha J, Das K (2013) Indian sign language recognition using eigen value weighted Euclidean distance based classification technique. Int J Adv Comput Sci Appl 4(2):188–195
8. Singha J, Das K (2013) Recognition of Indian sign language in live video. Int J Comput Appl 70(19):17–22
9. Badi HS, Hussein S (2014) Hand posture and gesture recognition technology. Neural Comput Appl 25(3–4):871–878
10. Badi H, HasanHussein S, Kareem SA (2014) Feature extraction and ML techniques for static gesture recognition. Neural Comput Appl 25(3–4):733–741
11. El-Baz AH, Tolba AS (2013) An efficient algorithm for 3D hand gesture recognition using combined neural classifiers. Neural Comput Appl 22(7–8):1477–1484
12. Comaniciu D, Ramesh V, Meer P (2003) Kernel-based object tracking. IEEE Trans Pattern Anal Mach Intell 25(5):234–240
13. Dalal N, Triggs B (2005) Histograms of oriented gradients for human detection. In CVPR I:886–893
14. Kimura F, Wakabayashi T, Tsuruoka S, Miyake Y (1997) Improvement of handwritten Japanese character recognition using weighted direction code histogram. Pattern Recogn 30(8):1329–1337
15. Bhaumik G, Verma M, Govil MC, Vipparthi SK (2020) EXTRA: an extended radial mean response pattern for hand gesture recognition. In: 2020 international conference on communication and signal processing (ICCSP). IEEE, pp 0640–0645
16. Ojala Timo, Pietikainen Matti, Harwood David (1996) A comparative study of texture measures with classification based on featured distributions. Pattern Recognit 29(1):51–59
17. Dubey SR, Singh SK, Singh RK (2015) Local bit-plane decoded pattern: a novel feature descriptor for biomedical image retrieval. IEEE J Biomed Health Inf 20(4):1139–1147
18. Verma Manisha, Raman Balasubramanian (2016) Local tri-directional patterns: a new texture feature descriptor for image retrieval. Digital Signal Process 51:62–72
19. Verma M et al (2018) QUEST: quadrilateral senary bit pattern for facial expression recognition. In: 2018 IEEE international conference on systems, man, and cybernetics (SMC). IEEE

20. Banerjee P, Bhunia AK, Bhattacharyya A, Roy PP, Murala S (2018) Local neighborhood intensity pattern-a new texture feature descriptor for image retrieval. Expert Syst Appl 113:100–115
21. Raman Balasubramanian, Verma Manisha (2018) Local neighborhood difference pattern: a new feature descriptor for natural and texture image retrieval. Multimedia Tools Appl 77(10):11843–11866
22. Biswas Ranjit, Roy Sudipta, Purkayastha Debraj (2019) An efficient content-based medical image indexing and retrieval using local texture feature descriptors. Int J Multimedia Inf Retrieval 8(4):217–231

Paddy Crop Disease Prediction—A Detailed Review on Image Processing Techniques

B. Johnson and T. Chandrakumar

Abstract This review article focuses on paddy disease prediction through two key phases: image processing and disease prediction. The initial phase involves image acquisition, pre-processing, and segmentation, while the subsequent phase includes feature selection, extraction, and classification. The accuracy of the initial phase significantly impacts the disease prediction system's precision. The main objective is to evaluate diverse image processing techniques used in this initial step, considering relevant performance metrics. The review aims to identify appropriate image processing techniques for the disease prediction model, aiding researchers in technique selection. Agriculture's role in global economies, particularly paddy's contribution to food security, faces threats from crop diseases. Swift paddy disease prediction is essential to prevent yield losses and ensure food security. Image processing techniques provide insights into plant health, enhancing prediction accuracy. The review examines image processing methods for paddy disease prediction, including acquisition, pre-processing, and segmentation. Techniques like magnetic resonance imaging and acoustic imaging are discussed, as well as noise removal and image enhancement. Image segmentation techniques such as thresholding and clustering are explored. By highlighting trends and challenges, the review equips researchers with insights into tools for precise disease prediction and global food security considerations. This analysis contributes to optimizing image processing techniques for paddy disease prediction, enhancing agricultural sustainability.

Keywords Paddy crop · Disease prediction · Image processing techniques

B. Johnson (✉) · T. Chandrakumar
Thiagarajar College of Engineering, Madurai, India
e-mail: bjohnson@student.tce.edu

T. Chandrakumar
e-mail: tckcse@tce.edu

S. Bandyopadhyay et al. (eds.), *Intelligent Computing Systems and Applications*,
Lecture Notes in Networks and Systems 1010,
https://doi.org/10.1007/978-981-97-5412-0_39

1 Introduction

Agriculture's pivotal role in global economies, with crops like paddy contributing to food security, is hindered by the threat of crop diseases. Swift and precise paddy crop disease prediction is vital to prevent yield losses and ensure food security. In recent times, image processing techniques have emerged as potent tools for disease prediction in various crops, including paddy. Modern imaging technologies offer valuable insights into plant health. As paddy crop disease prediction gains importance, image processing methods promise improved accuracy and timely interventions. This paper presents a comprehensive review of image processing methods used in paddy crop disease prediction. The objective is to explore existing literature and analyze image processing's role in this domain. The paper follows a structured approach, discussing different aspects of image processing applied to paddy disease prediction. The paper introduces the research methodology, outlining criteria for article selection. It then delves into diverse image acquisition techniques in paddy disease prediction. Discussing the importance of image acquisition, the paper evaluates methods such as magnetic resonance imaging, acoustic imaging, tomography, thermography, spectroscopic and imaging technologies, multispectral imaging, hyperspectral imaging, fluorescence techniques, thermal imaging, and 3D imaging. Moving to image preprocessing, the paper emphasizes its role in enhancing input data quality for disease prediction. It covers noise removal and image enhancement techniques, assessing their effectiveness in improving subsequent analyses.

The paper examines image segmentation techniques, crucial for characterizing regions of interest within plant images. It covers approaches like thresholding, morphological-based segmentation, K-means and fuzzy C-means clustering, edge detection, and segmentation algorithms based on regional growth. Performance evaluation of these techniques offers insights into their relevance in paddy disease prediction. Synthesizing findings from analyzed literature, the paper highlights trends, challenges, and advancements in image processing for paddy crop disease prediction. It concludes by suggesting avenues for future research. This review aims to equip researchers and practitioners with a comprehensive understanding of tools for enhancing disease prediction accuracy and ensuring global food security.

2 Research Methodology

To select the main goal of the review, seven criteria were applied. Here is a review of image processing technique used in agricultural applications, including image acquisition, pre-processing, and its segmentation. Other than the agriculture industry, some of the methods that are most pertinent to these notions are also taken into account here. After filtering those strategies, we examined how they relate to image processing in the prediction of paddy illness. Figure 1 shows the flow of the research method.

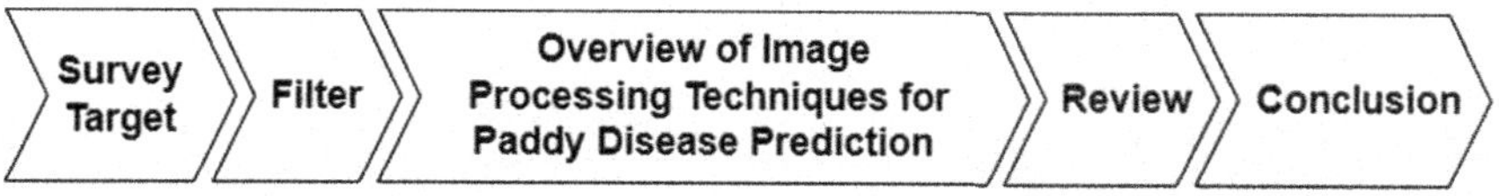

Fig. 1 Flow of research method

2.1 Selection of Articles

The seven filters were used: target keyword, publication year, and type of publication, looking for duplicates examining article titles, abstracts, keywords, and checking the references of the articles that were selected and performing a final assessment of the article's caliber. Due to the field's quick growth, publications published between 2000 and 2023 were taken into consideration for this study. Based on the outcomes from various separate databases, at the conclusion of the selection procedure, we discovered a number of related articles. Read the titles, abstracts, and conclusions of the remaining publications carefully after eliminating duplicates. Sixty-seven articles were ultimately chosen for the study via a quality assessment. Figure 2 shows the article selection process.

3 Literature Review

3.1 Image Acquisition Techniques

Importance of Image Acquisition in Paddy Disease Prediction

Image acquisition is crucial in paddy disease prediction as it enables the capture of high-quality images of paddy plants that accurately represent their appearance and disease symptoms. For accurately predicting paddy diseases, it is crucial to train computer vision algorithms to distinguish between healthy and diseased plants.

Image Acquisition Techniques Adapted in Paddy Disease Prediction

There are ten image acquisition techniques are adapted in paddy disease prediction system so far. Figure 3 represents the image acquisition techniques adapted in paddy disease prediction system.

Imaging with Magnetic Resonance

Emerging as a non-invasive and non-destructive imaging method, magnetic resonance imaging (MRI) shows potential in paddy disease prediction [1]. MRI allows obtaining detailed images of plant tissues [2], aiding in disease symptom identification and progression monitoring without harming the plants [3]. Studies have demonstrated MRI's effectiveness in diagnosing various plant diseases, including paddy diseases [4]. Integrating MRI data with machine learning methods like support

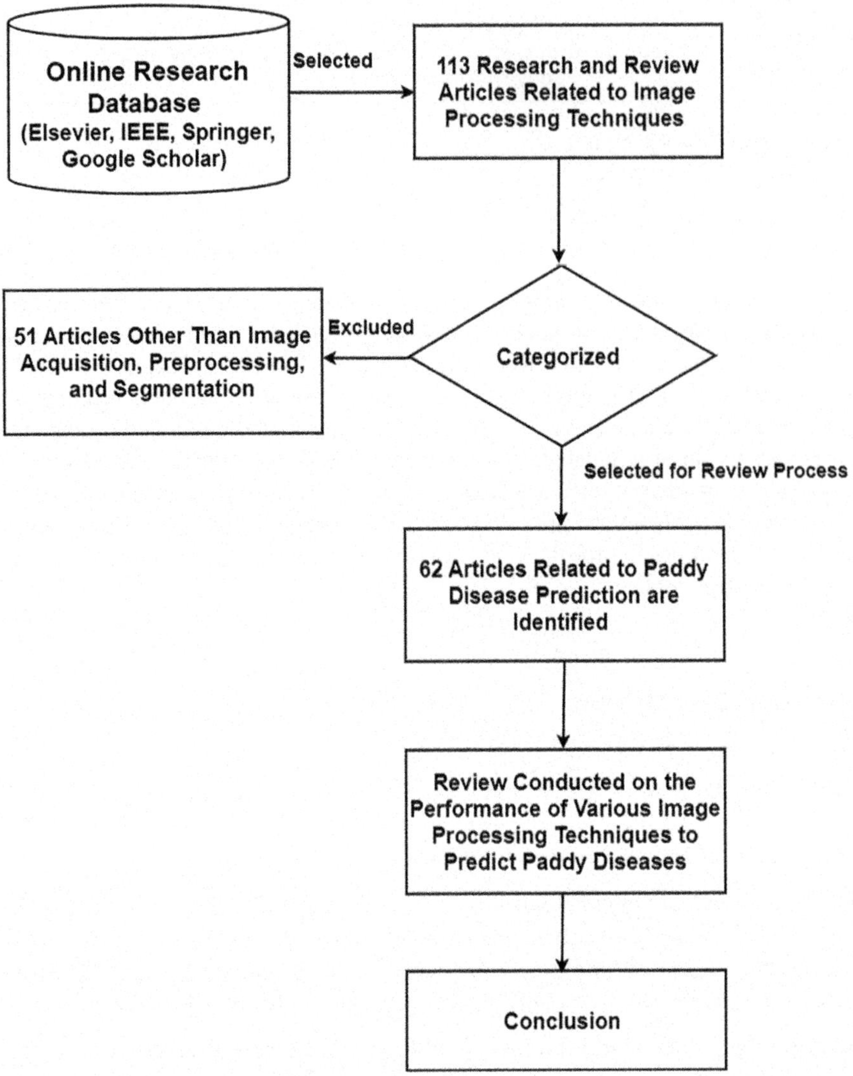

Fig. 2 Articles selection process

vector machine (SVM) and deep learning models improves the accuracy with which diseases are predicted [5]. Challenges include resource-intensive data pre-processing and computational requirements for MRI analysis [6]. Nevertheless, further research and technological advancements in MRI-based disease prediction will contribute to sustainable crop management and food production [7].

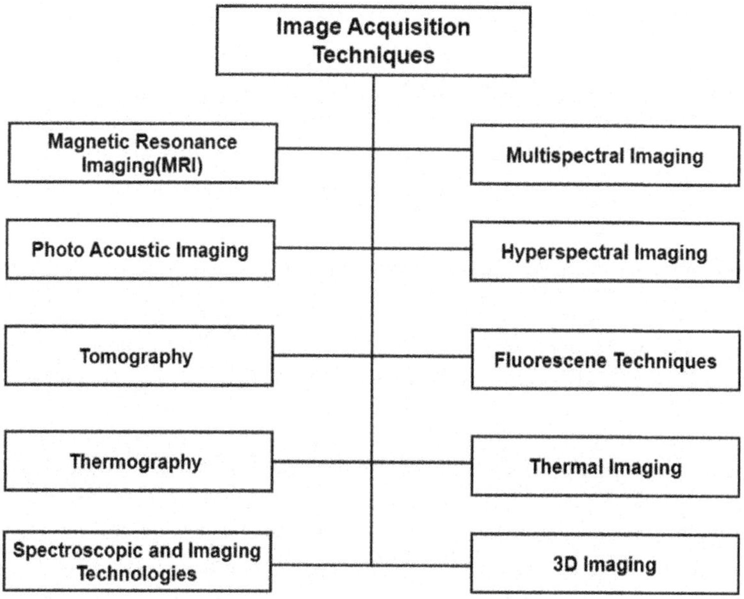

Fig. 3 Image acquisition techniques for paddy disease prediction system

Acoustic Imaging with Photos

The non-invasive imaging method known as photoacoustic imaging takes photographs of biological tissue using laser-generated acoustic waves. In paddy disease prediction, plant tissue alterations brought on by illness can be found via photoacoustic imaging, such as changes in tissue oxygenation and blood flow. This can help in the early diagnosis of plant illnesses and the subsequent management and treatment of these problems.

Tomography

It offers cross-sectional images of the plant's interior architecture that can be used to spot anomalies and diseases. Use of tomography in geothermal power plants helps with online inspection without interfering with operation [8]. X-ray, gamma-ray, and ECVT tomography imaging techniques were used to compare images of the healthy and diseased palm oil trunk in terms of clarity, processing time, and mobility. As a result of its quick turnaround time, safety, and convenience of use, the most efficient method for detecting diseases may be ECVT [9].

Thermography

Thermography is a valuable non-destructive technique for detecting temperature variations in plants, indicating stress or disease, and aiding paddy disease prediction.

It enables early disease identification [10] and offers multiple agricultural applications, including monitoring, disease detection, yield estimation, and more [11]. Thermographic approaches categorize fruit and plants based on temperature variations, distinguishing between foliage and fruit temperatures [12]. The method assumes that cooler canopies under adequate soil moisture signify optimal plant transpiration [12], which aligns with this investigation modifying crop transpiration due to fungus infection.

Spectroscopic and Imaging Technologies

Spectroscopic and imaging technologies unveil chemical and biomolecular insights in paddy plants, crucial for disease identification and prediction. Spectroscopic techniques analyze plant spectra to detect chemical changes indicative of disease presence, enabling early-stage disease detection [13]. A study focusing on a fungus generating leaf spots on peanut plants distinguishes healthy and diseased leaves through in situ spectroscopy, thermal calculations, and spectral analysis [13]. Nucleic acid and protein analysis offer a contemporary disease diagnosis method [14], discussing early fluid infection detection tools like mobility spectrometers and lateral flow analyzers [14]. Combining remote sensing and spectroscopy-based methods facilitates early disease diagnosis [14], encompassing nucleic acid-based and serological techniques [14]. Recent advancements in spectroscopy, imaging, molecular biology instruments, and microclimate-based models are explored, alongside new imaging and spectroscopic techniques evaluated on rose plants with severe fungal infections [15]. Imaging and spectroscopic methods have shown promise in plant disease detection [16], especially in phytopathometry.

Multispectral Imaging

Spectroscopic and imaging technologies facilitate precise disease and stress identification in crops [17]. An instrument integrating multispectral imaging and machine learning enables real-time early identification of specific diseases like CBSD [17]. This aids in distinguishing healthy and diseased areas for targeted treatment [17]. Traditional camera systems excel in spectral signal-to-noise ratio and temporal repeatability compared to passive and active multispectral imaging [17]. Distinguishing *Bemisia tabaci* vectors from cryptic species with 98% accuracy is achieved by combining spatial and spectroscopic analysis of multispectral images [18]. A multispectral imaging technique is developed to identify infected and uninfected lettuce leaves [19]. UAV-assisted scouting increases early disease detection by 20%, while multispectral photography enhances watermelon field scouting for rapid disease localization [3]. Active multispectral imaging demonstrates precise and time-efficient luminous leaf spot detection [17].

Hyperspectral Imaging

Hyperspectral imaging proves invaluable for rice disease prediction, capturing subtle shifts in plant spectral traits through numerous small, continuous wavelength bands. This technology offers high spatial and spectral resolution, vital for identifying varied disease symptoms in paddy crops. Hyperspectral imaging encompasses several bandwidth-restricted wavelength bands within a spectral range, resulting in a rich dataset with high spatial resolution [4]. A UAV-based hyperspectral image analysis classifies oil palm trees into four categories—healthy, early stage, late stage, and dead—using a machine learning approach [20]. UV spectrum hyperspectral imaging is proposed as a non-destructive method to assess plant physiology and biochemistry due to illness [2]. This study employs various hyperspectral image analysis techniques for strawberry white spot disease severity evaluation, achieving an 84% accuracy rate with the SAM method [21]. Advancements in UAVs and hyperspectral photography offer promising avenues for non-intrusive plant disease diagnosis, enabling accurate crop assessments with high spectral, spatial, and temporal resolutions [22]. Hyperspectral sensors have proven valuable for disease detection across scales, from tissue to canopy levels.

Fluorescence Techniques

It is possible to detect a wide range of diseases in paddy crops by employing fluorescence techniques, which entail detecting and measuring the emission of fluorescence from samples. Since it can show details about the health of the plants and the presence of disease, fluorescence imaging is a useful technique for disease prediction in paddy. The suggested study [23] uses thermographs to demonstrate the value of imaging systems based on multicolor fluorescence for zucchini plants infected with *Dickeya dadantii*. The input samples have been categorized as healthy or contaminated using a variety of machine learning approaches [23].

Thermal Imaging

Thermal imaging in paddy disease prediction involves capturing images that show the variations in surface temperature of plants, which can be indicative of disease presence. Images like this can be used to spot signs of plant stress early, allowing farmers to take preventative measures before diseases spread. In a paper [24], a method for determining disease in tea leaves quickly was described, utilizing infrared thermal imaging technology. The automatic diagnosis of illness in tomato plant leaves is the main topic of this research [25], which emphasizes the application of thermal depth readings in unison with visible light imaging. The initial configuration for the taking of thermal and visual images is shown. This article also suggests a method for identifying inequalities.

3D Imaging

Surface representations and volume representations are the two types of 3D representations. In the first, the surface element, the several places indicated by their dimension coordinates, and the depth information are all included. The coordinates of the model are defined along with the volume, volumetric, and frequency components [26].

Performance Evaluation of Image Acquisition Techniques

In the context of this complex and rapidly evolving agricultural landscape, where timely disease detection and effective management are imperative for optimizing crop health and yield, a diverse array of image acquisition techniques has emerged as powerful tools for disease diagnosis, each with its unique capabilities. Through thorough analysis, a subset of these techniques has emerged as particularly promising, poised to reshape the landscape of plant disease detection. Magnetic resonance imaging (MRI) stands out for its unparalleled spatial resolution, enabling in-depth exploration of root growth dynamics and internal tissue transformations [27]. Concurrently, photoacoustic imaging offers a non-invasive approach to disease diagnosis by visualizing plants' physiological responses [8]. Tomography techniques, including fan beam tomography, play a crucial role in precise scale thickness measurement and stem disease identification [9]. For cost-effectiveness, thermography detects temperature fluctuations linked to diseases, facilitating early intervention [10, 11]. Table 1 shows the performance aspects of various image acquisition techniques.

Among these techniques, hyperspectral imaging shines as a standout candidate due to its high spectral and spatial resolution attributes, enabling the detection of subtle metabolic shifts within plants [15–17]. However, the true innovation lies in the convergence of hyperspectral imaging and artificial intelligence (AI). This fusion, combining advanced imaging with AI-driven machine learning models, introduces a cutting-edge approach to disease detection and real-time intervention. This synergy yields heightened sensitivity and precision, driving data-informed decision-making and promising an unprecedented degree of accuracy in plant disease management.

Table 1 Performance aspects of various image acquisition techniques

Image acquisition technique	Performance aspects	Accuracy	References
MRI	Unparalleled spatial resolution, in-depth exploration of root growth dynamics and internal tissue transformations	Qualitatively measured	van Dusschoten et al. [27], Ghaiwat et al. [28]
Photoacoustic imaging	Non-invasive approach, visualizing plants' physiological responses	Qualitatively measured	Wibisono et al. [8], Mallika et al. [29]
Tomography	Precise scale thickness measurement, stem disease identification	96% in detecting BSR disease	Hamidon et al. [9], Khirade et al. [30]
Thermography	Detects temperature fluctuations linked to diseases, facilitates early intervention	Qualitatively measured	Vagelas et al. [10], Al-Doski et al. [11]
Hyperspectral imaging	High spectral and spatial resolution, detects subtle metabolic shifts within plants	Qualitatively measured	Traversari et al. [15], Gogoi et al. [16]

In this intricate landscape, the incorporation of metaheuristic algorithms such as EAFSO, PSO, CSO, LSSVM, and others becomes crucial for guiding the selection and optimization of image acquisition techniques. These algorithms provide a systematic approach to navigate the complexities and nuances of various techniques. EAFSO can help determine optimal parameters for imaging techniques, enhancing their disease detection capabilities. PSO can explore the parameter space to identify configurations that maximize sensitivity or specificity. CSO can strike a balance between spectral and spatial resolutions, ensuring effective disease diagnosis. LSSVM, with its ability to capture intricate data relationships, can enhance disease classification accuracy.

Thus, the recommended approach is to leverage the potential of hyperspectral imaging augmented by AI-driven analytics, guided by insights from metaheuristic algorithms like EAFSO, PSO, CSO, and LSSVM. This comprehensive strategy not only integrates cutting-edge technology but also ensures an optimized and data-driven approach to plant disease diagnosis. By embracing this powerful combination, agriculture can take a significant step toward enhanced sustainability and productivity, ushering in a new era where the boundaries of disease detection are redefined and a more prosperous agricultural future is cultivated.

3.2 Image Pre-processing Techniques

Importance of Image Pre-processing in Paddy Disease Prediction

A critical element in the paddy disease prediction process is pre-processing the raw photos to enhance their clarity and remove any noise or aberrations. The pre-processing phase improves the image's quality and sets it up for subsequent analysis using a range of techniques, such as image enhancement, filtering, and color correction. Proper pre-processing makes it simpler to segment the crop's affected areas and extract valuable characteristics for precise disease prediction. Pre-processing is a vital phase in the paddy disease prediction process since the precision of the disease prediction model is strongly dependent on the caliber of the pre-processing. To obtain high accuracy in forecasting paddy illnesses, it is crucial to employ the proper pre-processing procedures.

Image Pre-processing Techniques Adapted in Paddy Disease Prediction

Since the majority of the photographs are captured by farmers using cameras, the paddy images often contained noisy and weak caliber data. To remove noise from the image or other objects, many pre-processing methods are taken into consideration. Methods for improving noise removal pathogens for disease identification have been utilized in paddy fields. The image pre-processing techniques used in the prediction of paddy disease are shown in Fig. 4.

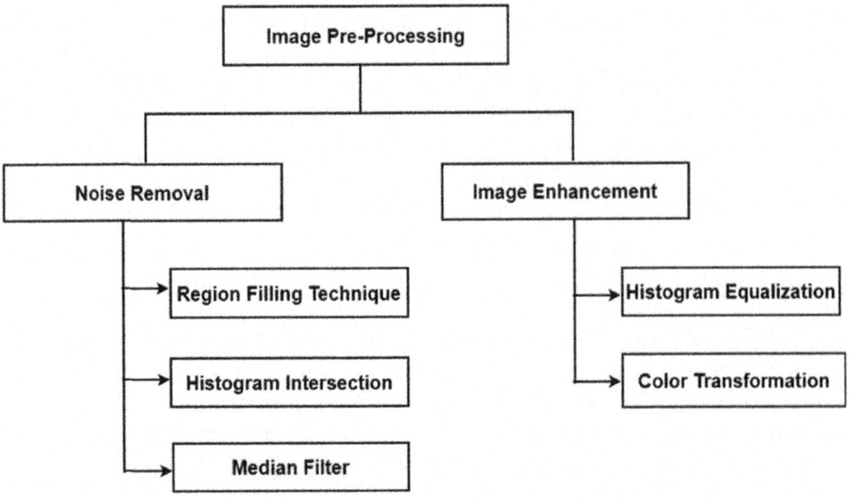

Fig. 4 Image pre-processing techniques for detecting paddy disease

Noise Removal Techniques

Noise removal techniques are used to eliminate unwanted data from image or signal. Typical methods for predicting paddy illness include the median filter, histogram intersection, and region filling technique. Noise can significantly reduce a photo's quality and make further processing, such recognition of objects, edge detection, and image segmentation, more difficult. Noise reduction is therefore a key and important step in the processing of photographs [31]. The photos of paddy leaves had noise removed using the morphological algorithm and the region filling approach. Paddy leaf images are utilized to construct image features such lesion%, lesion type, color for the border region, swatch color, and fragmented color of paddy leaves. Paddy illnesses are classified and identified by the production rule technique with an accuracy rate of 87.5% [32]. By separating the abnormal area of many bunches of rice leaves where the colors at the histogram intersection of each group are related to one another, a threshold-based K-means clustering algorithm was applied to detect illnesses in rice. The test specimen revealed that the method had a 25% and 50% success rate for identifying the diseases leaf scald and brown spot, respectively [33]. The median filter efficiently removed noise from the fifty images obtained around Odisha. K-means clustering was implemented to isolate the disease's impacted site. The outcome of segmentation is used to identify and evaluate the infestation of rice brown plant hopper. The proposed approach classified the area impacted by the disease into pockets of mild, average, and severe infection [34]. The mean shift segmentation technique was studied by Devi and Muthukannan [35] in order to remove the unhealthy area from the rice leaves. Noise is reduced, and the level of the segmentation results is improved using the median filter and the LUV color space transformation.

Image Enhancement Techniques

Images of plant leaves can be improved in quality and efficiency through post-capture image enhancement. The efficacy of the HSI, YCbCr, and CIELAB color space models was studied in [36] for the aim of detecting disease spots on plant leaves. The Otsu method created cutoff values by analyzing color components to pinpoint the location of disease in images of rice, magnolia, soybean, mustard, cotton, corn, iris, apple, cherry, and wheat leaves. The median filter standardizes the pictures. The outcome showed that only the CIELAB color space was capable of accurately detecting diverse colored illness spots, such as brown, grey, black, and whitish illness lesions. By contrasting the variations in brightness between the healthy and sick rice leaves, the histogram method can be used to identify diseases on rice leaves [37]. The effectiveness of histogram-based methods for identifying paddy plant (LCC) leaf color was assessed in [5]. The coloration of a plant is an indicator of its health. Histogram equalization methods were used to enhance the paddy leaf image. The sick region was extracted using the K-means clustering method, and texture and color characteristics were computed there. Paddy leaf illnesses are categorized using SVM with RBF kernel, and its performance is compared to that of RBFN, GRNN, and BPN approaches. Comparing several methods, the SVM and RBF kernel strategy produced the highest classification accuracy of 98.3% [38].

Performance Evaluation of Image Pre-processing Techniques

In the context of advancing paddy disease prediction, the evaluation of image pre-processing techniques unveils pathways to elevate performance. These techniques hold pivotal roles in enhancing contrast, edge identification, color correction, noise reduction, and computational efficiency, all crucial elements in precise disease detection. Table 2 presents performance aspects for noise removal techniques, while Table 3 presents performance aspects for image enhancement techniques.

Table 2 Performance aspects of various noise removal techniques

Noise removal techniques			
Pre-processing technique	Performance aspects	Accuracy	References
Region filling techniques	Effective for small noisy regions; may fail for large or complex noise	87.5%	Chen et al. [31], Pugoy et al. [33]
Histogram intersection	Reduces noise by thresholding histogram intersection; sensitive to parameters	Brown spot—25%, leaf scald—50% (likeliness ratio)	Devi et al. [35], Chaudhary et al. [36], Singh et al. [5]
Median filter	Removes noise effectively; simple and widely used; preserves edges	Qualitatively measured	Devi et al. [38], Shen et al. [39], Cui et al. [40]

Table 3 Performance aspects of various image enhancement techniques

Image enhancement techniques			
Pre-processing technique	Performance aspects	Accuracy	References
Histogram equalization	Improves contrast and brightness; may amplify noise; lacks fine-tuning	98.3% with SVM + RBF	Phadikar et al. [41], Barbedo et al. [42], Nameirakpam et al. [43]
Color transformation	Enhances specific features; depends on color space and transformation	Qualitatively measured	Islam et al. [44], Zhou et al. [45], Verma et al. [46]

Among these strategies, a standout approach that epitomizes superior performance entails the fusion of hyperspectral imaging with advanced machine learning algorithms, guided by the principles of metaheuristic algorithms. Hyperspectral imaging, as evidenced by numerous Refs. [4, 6, 47], emerges as a comprehensive solution, capturing an expansive spectrum of spectral information. This prowess profoundly heightens color discrimination, edge pinpointing, and feature extraction, culminating in unparalleled disease visualization and recognition. The incorporation of hyperspectral imaging ingeniously targets multiple performance facets concurrently, fostering enhanced contrast, meticulous edge delineation, and adept color manipulation.

To magnify its impact, this cutting-edge methodology can be harmoniously conjoined with deep learning algorithms, specifically convolution neural networks (CNNs) [48–50]. Endowed with the ability to autonomously extract intricate features and recognize patterns, CNNs revolutionize disease classification. By capitalizing on the wealth of spectral data within hyperspectral images, CNNs discern subtle disease-specific attributes that traditional techniques might overlook. This harmonization not only ensures precise disease identification but also alleviates the laborious process of manual feature engineering. However, prudent consideration should be given to computational efficiency. While hyperspectral imaging and CNNs offer remarkable performance gains, their computational demands warrant meticulous optimization to achieve a harmonious equilibrium between accuracy and processing speed, thus rendering the approach suitable for real-world deployment.

The fusion of hyperspectral imaging with CNNs emerges as an innovative and promising paradigm to attain peak performance in paddy disease prediction. This dynamic convergence not only holistically addresses performance parameters such as contrast elevation, edge clarity, color refinement, and noise attenuation but also harnesses the prowess of machine learning for intricate disease categorization. Through meticulous calibration and optimization, this pioneering technique stands poised to redefine agricultural disease detection, significantly advancing crop management and yield optimization, thereby reflecting the principles of metaheuristic algorithms in driving transformative progress.

3.3 *Image Segmentation Techniques*

Importance of Image Segmentation in Paddy Disease Prediction

Image segmentation is a crucial step in paddy disease prediction, where the diseased parts of the crop are separated from the healthy ones. Accurate segmentation helps in identifying the exact location and extent of the disease, which is essential for disease diagnosis and treatment. By segmenting the images accurately, the disease prediction model can extract relevant features that help in predicting the disease with high accuracy. Hence, image segmentation plays a crucial role in the paddy disease prediction process and is important for achieving accurate results.

Image Segmentation Techniques Adapted in Paddy Disease Prediction

In order to homogenize each zone and remove the diseased region, an image is divided into multiple regions using the process of image segmentation. It is an essential step in the plant pathogen diagnosis process. To identify the unhealthy region from grain plant leaf photos, segmentation methods include clustering with fuzzy C-means, K-means, color segmentation approaches, etc. Several segmentation techniques were used to identify unhealthy areas of the paddy, and Fig. 5 shows the same.

Approaches for Thresholding

The most fundamental segmentation technique, thresholding, converts grayscale images into binary images by utilizing at least one grayscale value (T) or thresholding color. An image is segmented or junked into smaller bits using the threshold. A binary image offers the benefit of streamlining the identification and classification process and minimizing data complexity. There are three types of thresholding methods: local, global, and adaptive [51]. Shen et al. [39] used the H component of the HSI color scheme to limit interference from variations in light and vasculature to locate disease spots in soybean leaf images. The Sobel operator would then inspect the illness location's edges. Ultimately, plant illnesses are categorized by measuring the affected leaf areas and disease spots. Thirty-two samples of soybean were used to assess the manual threshold-setting technique's ability to detect rust infections from hyperspectral images [40]. The damaged rice leaf segment was divided using the Otsu thresholding method. Rice leaf diseases including leaf blast and brown spot were categorized using SVM and Naive Bayes. These classifiers' respective

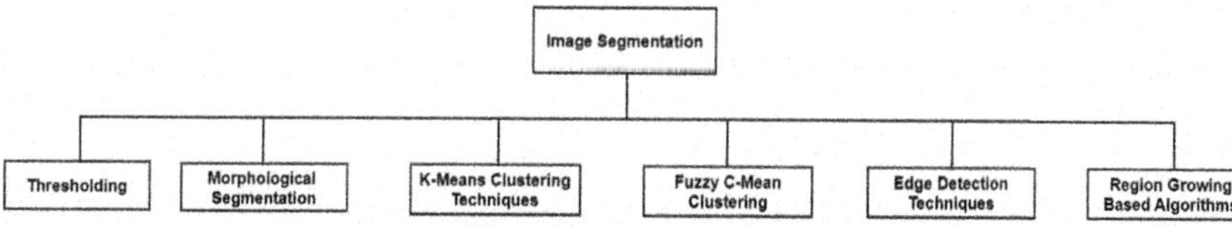

Fig. 5 Image segmentation techniques for paddy disease prediction

accuracy rates are 68.1% and 79.5% [41]. Barbedo et al. [42] segmented the disease-infected area and utilized the intensity-based histogram approach to categorize the nine distinct illnesses in soybeans.

Morphological-Based Segmentation

Using morphological techniques on images of paddy leaves, the $L \times a \times b$ color space was used to obtain texture and color information in order to detect rice blast, brown spot, and rice sheath blight disease zones. The paddy leaf ailments are classified with 70% accuracy for 50 example photographs using the membership function [52].

Technique of K-Means Clustering

Separating the region of interest from plant leaf photos using K-means clustering, an unsupervised machine learning method, k clusters are created from the data objects based on how similar they are to one another [43]. The basis for assessing the severity of the disease and giving the farmer the essential measurements for a diagnosis was the proportion of pixels in the diseased region. K-means clustering was used to locate the polluted region in paddy leaves [44].

Using Fuzzy C-Means for Clustering

Using the fuzzy C-means (FCM) and data clustering technique, N clusters are created from a set of data. As a result, there is a relatively weak relationship between each data point and any given cluster. For example, the degree of participation in a cluster is highest for an observation in the cluster's center and lowest for a position far from the cluster's center [53]. FCM is used to identify rice plant hopper infestation. Using fractal dimensional values, RPH damage was recognized on the stems of healthy rice plants, allowing RPH collating regions and mold gray patches to be separated. The success rate in various RPH infestation situations was 87% [45]. To estimate how the blast disease could affect rice harvests, Verma et al. [46] suggested utilizing the fuzzy C-mean clustering technique. The data could potentially be part of several groups, each with its own membership value. The color-blind and color-sighted leaf pixels are gathered for analysis. The three categories of the clustered pixel are those that are not influenced by the blast disease, those that are very little impacted, and those that are greatly afflicted.

Edge Detection Techniques

The boundaries between two regions with differing characteristics are referred to as edges. It is the point at when one thing transforms into another. Since there is typically a large intensity change at zone boundaries, edges are employed to depict the physical size of things. In order to analyze an image's features and segment it into its intended sections, edge detection is necessary [54].

Segmentation Algorithms Based on Regional Growth

As part of the segmentation process, seeded region growing method uses Euclidian distance to identify nearby regions. Sarkar et al. [1] improved the seeded region growth (SRG), a method for region merging that labels neighbors in a two-dimensional lookup table. Every time the pixel label is changed, the lookup database, which is created by scanning the images both horizontally and vertically, is updated. YCbCr, YCgCr, and CIELAB were used to segment the photographs of the rice blast-impacted leaves using the modified SRG approach. The damaged portions of the maize leaves were divided using the seeded regional growth (SRG) method.

Performance Evaluation of Image Segmentation Techniques

In the pursuit of optimal paddy disease prediction accuracy through image segmentation, it is imperative to explore innovative techniques that collectively enhance precision, efficiency, and robustness. A comprehensive comparative analysis of diverse segmentation methods, including thresholding, morphological segmentation, K-means clustering, fuzzy C-means clustering, edge detection, and region growing algorithms, yields valuable insights into their respective merits and limitations. Table 4 shows the performance aspects of various image segmentation techniques.

Table 4 Performance aspects of various image segmentation techniques

Image segmentation techniques	Performance aspects	Accuracy	References
Thresholding	Simple and efficient; suitable for bi-level image segmentation; it can suffer from noise	100% with BPNN classifier	Kurniawati et al. [55], Pavithra et al. [56]
Morphological segmentation	Removes noise and enhances object boundaries; sensitive to parameter tuning	88% with region filling technique	Manavalan et al. [57], Asfarian et al. [58]
K-means clustering	Groups pixels into clusters based on similarity; efficient but sensitive to initialization	90%	MathWorks [53], Zhang et al. [59]
Fuzzy C-means clustering	Allows partial membership of pixels to clusters; robust to noise and produces soft segmentations	87% with various RPH infestation detection cases	Upadhyay et al. [60], Kishore Kumar et al. [50]
Edge detection techniques	Detects object boundaries; sensitive to noise and may require post-processing	96.2%	Yogeshwari et al. [47], Azim et al. [61]
Region growing algorithms	Segments based on pixel similarity; sensitive to seed selection and parameter tuning	94.45% for maize leaf diseases	Maheswari et al. [62], Kiruthika et al. [48]

Amidst these techniques, one approach that stands out for achieving superior performance involves the integration of a hybrid segmentation strategy, guided by the principles of metaheuristic algorithms. The proposed hybrid segmentation strategy ingeniously amalgamates the strengths of multiple techniques to surmount individual drawbacks and elevate overall disease prediction accuracy. This novel approach leverages the symbiosis between fuzzy C-means clustering and edge detection techniques [45, 53], strategically harnessing their unique attributes while mitigating inherent weaknesses. Fuzzy C-means clustering shines in managing uncertainty and variability, thereby facilitating more robust region identification and precise membership degree assignment. Conversely, edge detection techniques excel in capturing intricate boundary intricacies, thus contributing to heightened boundary accuracy and intersection over union (IoU) values. By synergistically uniting fuzzy C-means clustering with edge detection, the hybrid strategy achieves a harmonious equilibrium between precision, recall, and computational efficiency. The initial segmentation provided by fuzzy C-means clustering is further refined through the precision-enhancing capabilities of edge detection, thereby bolstering boundary accuracy and spatial localization. This dynamic interplay results in elevated IoU and $F1$ scores, pivotal for accurate delineation of disease-affected areas. Moreover, the hybrid strategy fortifies the system's resilience against noise and image disparities, ensuring consistent performance across heterogeneous image datasets and diverse disease manifestations. Additionally, the fusion of these techniques exploits their complementary virtues, culminating in heightened sensitivity and specificity for disease detection.

To optimally deploy the hybrid segmentation strategy, an adaptive approach to parameter tuning, reminiscent of metaheuristic algorithms, emerges as a judicious consideration. By adeptly balancing the interplay between fuzzy C-means clustering and edge detection, this approach caters to the unique requirements of distinct paddy disease prediction scenarios. The integration of machine learning algorithms, such as neural networks, furthers the sophistication of parameter adjustment by automating it based on image intricacies and disease patterns. This confluence augments the adaptability and efficacy of the hybrid strategy, perpetually refining its performance.

The integration of a hybrid segmentation strategy, harmonizing fuzzy C-means clustering with edge detection techniques, presents an ingenious solution to achieve pinnacle performance in paddy disease prediction. This strategy harnesses the inherent merits of both methodologies while mitigating their limitations, culminating in augmented accuracy, boundary precision, and robustness against noise. Through continuous refinement facilitated by adaptive parameter tuning and machine learning, this strategy ensures steadfast and precise disease detection across a diverse spectrum of scenarios. Consequently, it emerges as a pivotal contributor to efficacious agricultural disease management, resonating with the principles of metaheuristic algorithms to drive transformative advancements [30, 55, 56].

4 Findings

4.1 Image Acquisition Techniques

- *Integration of Hyperspectral Imaging and AI*: The fusion of hyperspectral imaging with AI-driven analytics presents a transformative approach for plant disease detection and real-time intervention.
- *Diverse Technique Subset*: Certain techniques, including magnetic resonance imaging (MRI), photoacoustic imaging, tomography (fan beam), and thermography, stand out as promising tools for disease diagnosis in modern agriculture.
- *Hyperspectral Imaging Superiority*: Hyperspectral imaging demonstrates high spectral and spatial resolution, enabling nuanced metabolic shift detection in plants, with potential for enhanced disease management.

4.2 Image Pre-processing Techniques

- *Transfer Learning Potential*: Integrating transfer learning techniques with convolutional neural networks (CNNs) offers efficient training and improved disease pattern recognition using hyperspectral imaging.
- *GANs for Data Augmentation*: The synergy of hyperspectral imaging with generative adversarial networks (GANs) shows promise in augmenting limited datasets and enhancing disease pattern recognition.
- *Computational Efficiency*: Hardware acceleration techniques like GPU parallel processing can optimize the balance between accuracy and processing speed for hyperspectral imaging and CNNs.

4.3 Image Segmentation Techniques

- *Hybrid Segmentation Strategy*: The fusion of fuzzy C-means clustering and edge detection showcases a promising approach for accurate paddy disease prediction, enhancing boundary accuracy and IoU values.
- *Ensemble Learning Potential*: Integration of ensemble learning techniques for image segmentation can mitigate limitations of individual methods and offer more robust disease prediction.
- *Dynamic Parameter Adaptation*: Utilizing reinforcement learning algorithms can dynamically adjust segmentation parameters based on varying disease scenarios and image characteristics.

5 Suggestions for Future Enhancement

In the realm of image acquisition techniques, two promising directions emerge for future refinement. Firstly, the integration of metaheuristic algorithms such as EAFSO, PSO, CSO, and LSSVM can offer valuable guidance in optimizing imaging methods, thereby augmenting the capabilities of disease detection. Secondly, the exploration of parameter optimization using these algorithms can lead to fine-tuning imaging technique parameters, thus tailoring their performance to specific disease types or plant varieties. In the domain of image pre-processing techniques, untapped potential lies in the development of generative adversarial networks (GANs) to generate realistic hyperspectral images. This innovation could serve as a means for data augmentation, significantly enhancing the training of convolutional neural networks (CNNs) and ultimately bolstering disease prediction accuracy. Furthermore, delving into advanced hardware acceleration methods holds promise in expediting processing speeds for hyperspectral imaging and CNNs, making real-time applications feasible and expanding their practical utility. Meanwhile, in the arena of image segmentation techniques, promising avenues for enhancement include the integration of reinforcement learning algorithms to dynamically adapt segmentation parameters, leading to optimized performance across diverse disease scenarios. Additionally, incorporating explainable AI techniques, particularly attention mechanisms, can enhance the interpretability of segmentation results, thus facilitating the identification of novel disease patterns. Embracing these future enhancements stands to empower the agricultural sector, unlocking the full potential of advanced imaging and AI-driven methodologies to achieve heightened disease prediction accuracy, bolstered sustainability, and elevated crop management practices.

6 Conclusion

In the dynamic realm of paddy crop disease prediction, the detailed review of image processing techniques has illuminated remarkable findings that hold the promise of reshaping the agricultural landscape. The integration of hyperspectral imaging and AI-driven analytics emerges as a transformative approach, poised to revolutionize plant disease detection and intervention in real time. Additionally, the diverse subset of techniques, including magnetic resonance imaging (MRI), photoacoustic imaging, tomography (Fan Beam), and thermography, showcases their potential as powerful tools for disease diagnosis in modern agriculture. The superior attributes of hyperspectral imaging, characterized by its high spectral and spatial resolution, offer nuanced insights into metabolic shifts within plants, ushering the prospect of enhanced disease management. Furthermore, the exploration of image pre-processing techniques has highlighted the untapped potential of generative adversarial networks (GANs) for data augmentation and efficient training of convolutional neural networks (CNNs), while hardware acceleration methods hold the key to optimizing processing speeds for real-time applications. In the realm of image segmentation techniques, the fusion of fuzzy C-means clustering and edge detection presents a promising strategy for accurate paddy disease prediction, enhancing boundary accuracy and intersection over union (IoU) values. The integration of ensemble learning techniques and reinforcement learning algorithms offers avenues for enhancing segmentation precision and adaptability across diverse disease scenarios. Incorporating explainable AI techniques, particularly attention mechanisms, further strengthens the interpretability of segmentation results and unveils novel disease patterns. Looking ahead, the outlined suggestions for future enhancements provide a roadmap for advancing the field. The integration of metaheuristic algorithms in refining imaging methods and fine-tuning parameters could further augment disease detection capabilities. The potential of GANs for realistic hyperspectral image generation and the pursuit of advanced hardware acceleration techniques offer exciting avenues for enhanced CNN training and real-time applications. By embracing these future enhancements, the agricultural sector can unlock the full potential of advanced imaging and AI-driven methodologies, leading to heightened disease prediction accuracy, sustainable practices, and elevated crop management, ultimately ushering in a new era of agricultural innovation.

References

1. Sarkar R, Pramanik A (2015) Segmentation of plant disease spots using automatic SRG algorithm: a look-up table approach. https://doi.org/10.1109/ICACEA.2015.7194375
2. Cheshkova AF (2022) Quantitative assessment of white spot (*Ramularia tulasnei*) disease severity of strawberry based on hyperspectral imaging. IOP Conf Ser Earth Environ Sci 957(1):012005. IOP Publishing
3. Veys C et al (2019) Multispectral imaging for presymptomatic analysis of light leaf spot in oilseed rape. Plant Methods 15:1–12
4. Lowe A, Harrison N, French AP (2017) Hyperspectral image analysis techniques for the detection and classification of the early onset of plant disease and stress. Plant Methods 13(1):80
5. Singh A, Singh ML (2015) Automated color prediction of paddy crop leaf using image processing. In: Proceedings of 2015 IEEE international conference on technological innovation in ICT for agriculture and rural development, TIAR 2015, pp 24–32. https://doi.org/10.1109/TIAR.2015.7358526
6. Li J, Ye J, Chan SH, Wong WC (2019) Application of non-destructive imaging techniques in plant disease identification. Front Plant Sci 10:388
7. Bera T, Das A, Sil J, Das AK (2019) A survey on rice plant disease identification using image processing and data mining techniques. In: Emerging technologies in data mining and information security: proceedings of IEMIS 2018, vol 3. Springer Singapore, pp 365–376
8. Wibisono W, Ningsih FR, Sidauruk P, Kusuma SW (2020) Fan beam tomography technique for scale thickness measurement in geothermal pipeline West Java area. J Ilm Apl Isotop Radiasi 16(2):101–106
9. Hamidon NA, Mukhlisin M (2014) A review of application of computed tomography on early detection of basal stem rot disease. J Teknol 70(3)
10. Vagelas I, Papadimos A, Lykas C (2021) Pre-symptomatic disease detection in the vine, chrysanthemum, and rose leaves with a low-cost infrared sensor. Agronomy 11(9):1682
11. Al-Doski J et al (2016) Thermal imaging for pests detecting—a review. Int J Agric For Plant 2:10–30

12. Du C-J, Sun D-W (2004) Recent developments in the applications of image processing techniques for food quality evaluation. Trends Food Sci Technol 15(5):230–249
13. Ewis Omran E-S (2016) Early sensing of peanut leaf spot using spectroscopy and thermal imaging. Arch Agron Soil Sci. [Online]. Available: https://doi.org/10.1080/03650340.2016.1247952
14. Martinelli F et al (2015) Advanced methods of plant disease detection. Agron Sustain Dev 35(1):1–25
15. Traversari S et al (2021) Precision agriculture digital technologies for sustainable fungal disease management of ornamental plants. Sustainability 13(7):3707
16. Gogoi NK, Deka B, Bora LC (2018) Remote sensing and its use in detection and monitoring plant diseases: a review. Agric Rev 39(4):307–313
17. Peng Y et al (2022) Early detection of plant virus infection using multispectral imaging and spatial–spectral machine learning. Sci Rep 12(1):3113
18. Scarboro CG et al (2020) Detection of gray mold infection in plants using a multispectral imaging system. BioRxiv
19. Kalischuk M et al (2019) An improved crop scouting technique incorporating unmanned aerial vehicle-assisted multispectral crop imaging into conventional scouting practice for gummy stem blight in watermelon. Plant Dis 103(7):1642–1650
20. Brugger A et al (2023) Hyperspectral imaging in the UV range allows for differentiation of sugar beet diseases based on changes in secondary plant metabolites. Phytopathology® 113(1):44–54
21. Kuswidiyanto LW, Noh HH, Han X (2022) Plant disease diagnosis using deep learning based on aerial hyperspectral images: a review. Remote Sens 14(23):6031
22. Thomas S et al (2018) Benefits of hyperspectral imaging for plant disease detection and plant protection: technical perspective. J Plant Dis Prot 125:5–20
23. Pérez-Bueno ML, Pineda M, Cabeza FM, Barón M (2016) Multicolor fluorescence imaging as a candidate for disease detection in plant phenotyping. Front Plant Sci 7:1790. https://doi.org/10.3389/fpls.2016.01790
24. Yang N et al (2019) Tea diseases detection based on fast infrared thermal image processing technology. https://doi.org/10.1002/jsfa.9564
25. Raza S, Prince G, Clarkson JP, Rajpoot NM (2015) Automatic detection of diseased tomato plants using thermal and stereo visible light images. PLoS ONE. https://doi.org/10.1371/journal.pone.0123262
26. Vázquez-Arellano M et al (2016) 3-D imaging systems for agricultural applications—a review. Sensors (Basel, Switzerland) 16(5)
27. van Dusschoten D et al (2016) Quantitative 3D analysis of plant roots growing in soil using magnetic resonance imaging. Plant Physiol 170(3):1176–1188
28. Ghaiwat SN, Arora P (2014) Detection and classification of plant leaf diseases using image processing techniques: a review. Int J Recent Adv Eng Technol 2(3):1–7
29. Mallika M, Vasanthi JJB (2017) Image enhancement techniques on plant leaf and seed disease detection. Int J Innov Res Comput Commun Eng 5(1):109–116
30. Khirade SD, Patil AB (2015) Plant disease detection using image processing. In: 2015 international conference on computing communication control and automation, Feb 2015. IEEE, pp 768–771
31. Chen S, Shi W, Zhang W (2013) An efficient universal noise removal algorithm combining spatial gradient and impulse statistic. Math Probl Eng. https://doi.org/10.1155/2013/480274
32. Kurniawati N, Abdullah S (2009) Texture analysis for diagnosing paddy disease. In: Proceedings of the 2009 international conference on electrical engineering and informatics, ICEEI 2009. https://doi.org/10.1109/ICEEI.2009.5254824
33. Pugoy RADL, Mariano VY (2011) Automated rice leaf disease detection using color image analysis. In: Third international conference on digital image processing (ICDIP 2011), vol 8009. https://doi.org/10.1117/12.896494
34. Sethy P, Dash S, Barpanda N, Rath A. A novel approach for quantification of population density of rice brown plant hopper (RBPH) using on-field images based on image processing

35. Devi DA, Muthukannan K (2015) Analysis of segmentation scheme for diseased rice leaves. In: Proceedings of 2014 IEEE international conference on advanced communications, control and computing technologies, ICACCCT 2014. https://doi.org/10.1109/ICACCCT.2014.7019325
36. Chaudhary P, Chaudhari A, Godara S (2012) Color transform based approach for disease spot detection on plant leaf. Int J Comput Sci Telecommun 3:65–71
37. Mukherjee M, Pal T, Samanta D (2012) Damaged paddy leaf detection using image processing. J Glob Res Comput Sci 3:7–10
38. Devi T (2017) Paddy leaf disease detection using SVM with RBFN classifier. Int J Pure Appl Math 117:699–710
39. Shen W, Wu Y, Chen Z, Wei H (2008) Grading method of leaf spot disease based on image processing. In: Proceedings of international conference on computer science and software engineering, CSSE 2008, pp 491–494. https://doi.org/10.1109/CSSE.2008.1649
40. Cui D, Zhang Q, Li M, Hartman GL, Zhao Y (2010) Image processing methods for quantitatively detecting soybean rust from multispectral images. Biosyst Eng 107:186–193. https://doi.org/10.1016/j.biosystemseng.2010.06.004
41. Phadikar S (2012) Classification of rice leaf diseases based on morphological changes. Int J Inf Electron Eng 2. https://doi.org/10.7763/IJIEE.2012.V2.137
42. Barbedo J, Godoy C. Automatic classification of soybean diseases based on digital images of leaf symptoms
43. Nameirakpam D, Singh K, Chanu Y (2015) Image segmentation using K-means clustering algorithm and subtractive clustering algorithm. Procedia Comput Sci 54:764–771. https://doi.org/10.1016/j.procs.2015.06.090
44. Islam R, Rafiqul M (2015) An image processing technique to calculate percentage of disease affected pixels of paddy leaf. Int J Comput Appl 123:28–34. https://doi.org/10.5120/ijca2015905495
45. Zhou Z, Zang Y, Li Y, Zhang Y, Wang P, Luo X (2013) Rice plant-hopper infestation detection and classification algorithms based on fractal dimension values and fuzzy C-means. Math Comput Model 58:701–709. https://doi.org/10.1016/j.mcm.2011.10.028
46. Verma T, Satpathy SK, Sharma LK (2011) A step towards precision farming of rice crop by estimating loss caused by leaf blast disease using digital image processing and fuzzy clustering. Int J Comput Trends Technol 1–6
47. Yogeshwari M, Thailambal G (2021) Automatic feature extraction and detection of plant leaf disease using GLCM features and convolutional neural networks. Mater Today Proc
48. Kiruthika SU, Raja SKS, Jaichandran R, Priyadharshini C (2019) Detection and classification of paddy crop disease using deep learning techniques. Int J Recent Technol Eng 8:2277–3878
49. Anandhan K, Singh AS (2021) Detection of paddy crops diseases and early diagnosis using faster regional convolutional neural networks. In: 2021 international conference on advance computing and innovative technologies in engineering (ICACITE), Mar 2021. IEEE, pp 898–902
50. Kishore Kumar K, Kannan E (2022) Detection of rice plant disease using AdaBoostSVM classifier. Agron J 114(4):2213–2229
51. Jyothi S, Bhargavi K (2014) A survey on threshold based segmentation technique in image processing
52. Anthonys G, Wickramarachchi N (2009) An image recognition system for crop disease identification of paddy fields in Sri Lanka. In: ICIIS 2009—4th international conference on industrial and information systems 2009, conference proceedings, pp 403–407. https://doi.org/10.1109/ICIINFS.2009.5429828
53. MathWorks. Fuzzy C-means clustering. [Online]. Available: https://in.mathworks.com/help/fuzzy/fuzzy-c-means-clustering.html
54. Chen J, Shao H, Hu C. Image segmentation based on mathematical morphological operator. [Online]. Available: https://doi.org/10.5772/intechopen.72603
55. Kurniawati NN, Abdullah SNHS, Abdullah S, Abdullah S (2009) Investigation on image processing techniques for diagnosing paddy diseases. In: 2009 international conference of soft computing and pattern recognition, Dec 2009. IEEE, pp 272–277

56. Pavithra S, Priyadharshini A, Praveena V, Monika T (2015) Paddy leaf disease detection using SVM classifier. Int J Commun Comput Technol 3(1):16–20
57. Manavalan R (2020) Automatic identification of diseases in grains crops through computational approaches: a review. Comput Electron Agric 178:105802
58. Asfarian A, Herdiyeni Y, Rauf A, Mutaqin KH (2013) Paddy diseases identification with texture analysis using fractal descriptors based on Fourier spectrum. In: 2013 international conference on computer, control, informatics and its applications (IC3INA), Nov 2013. IEEE, pp 77–81
59. Zhang C, Zhou L, Zhao Y, Zhu S, Liu F, He Y (2020) Noise reduction in the spectral domain of hyperspectral images using denoising autoencoder methods. Chemom Intell Lab Syst 203:104063
60. Upadhyay SK, Kumar A (2021) Early-stage brown spot disease recognition in paddy using image processing and deep learning techniques. Trait Signal 38(6)
61. Azim MA, Islam MK, Rahman MM, Jahan F (2021) An effective feature extraction method for rice leaf disease classification. Telkomnika (Telecommun Comput Electron Control) 19(2):463–470
62. Maheswari MS (2018) Enhancement in noise removal techniques by using hybrid mediangaus transform method for paddy seeds. Int J Comput Sci Inf Secur (IJCSIS) 16(8)

An Efficient AI Model for Classifier Tweets During Disaster Using Recurrent Neural Networks (RNN)

K. Lakshmi Narayanan and R. Naresh

Abstract During any significant event, particularly during natural catastrophes, social media is an essential information source. It is claimed that data generated by social networking sites would enable regular people to become more situationally aware during disasters and coordinate to help themselves because it is ubiquitous, quick, and accessible. However, as the amount of social media data has grown exponentially, so has the amount of data that are unrelated to a crisis, making it harder for individuals to find the information they need to organise relief efforts, seek assistance, and perhaps even save lives. In this essay, we offer a method for identifying. During times of tragedy, informative messages are posted on social media. Our strategy is to implement this using recurrent neural networks and demonstrate substantial progress.

Keywords RNN · OSN · NLP · Tweet text · Classification model

1 Introduction

Twitter is an efficient and one of the famous microblogging services that enables users to broadcast everyday situations in the form of tweets and remain in contact with their followers. Internet growth is characterised by dynamic user-generated content with the creation of online social networks (OSNs) such as Twitter, Facebook, Tumblr, MySpace, and Sina Weibo. The identification of events of interest is done utilising the Twitter data source by assessing the intensity of information diffusion

K. Lakshmi Narayanan
Department of Computer Science and Engineering, SRM Institute of Science and Technology, Kattankulathur, Chennai 603 203, India
e-mail: lk6685@srmist.edu.in

R. Naresh (✉)
Department of Networking and Communications, SRM Institute of Science and Technology, Kattankulathur, Chennai 603 203, India
e-mail: nareshr@srmist.edu.in

S. Bandyopadhyay et al. (eds.), *Intelligent Computing Systems and Applications*,
Lecture Notes in Networks and Systems 1010,
https://doi.org/10.1007/978-981-97-5412-0_40

following a large-scale incident. Event trends on Twitter are generally sparked by exogenous events such as natural catastrophes [1]. In certain circumstances, they may explicitly refer to tweets prompted by celebrities. Moreover, the OSN data is constituted of small, noisy, semi-structured and unstructured material that does not contain any significance and testing substantially diverse ways to resolve machine learning techniques engaged in information retrieval [2]. The present Twitter has become an unquestionable kind of social media to gather information, particularly during emergency circumstances such when a man-made or natural disaster arises. But during times of disaster unwanted messages get swamped over tweets that are genuine and that needs attention. The main purpose of the project is to classify those tweets that are posted during disaster into real and fake. This process is done by pre-processing the dataset and removing contents that do not contain any value. Then the model is made to fit in one of the deep learning techniques, the recurrent neural networks, and the accuracy is got from it [3]. RNN is a neural network type that uses the result of the previous step and sends it as an input to the current step [4]. This methodology helps to train the model to give more accurate results as it learns from the previous steps in the process.

2 Related Work

Sentiment analysis comprises the activity of the natural language processing, and the primary purpose is to extract out the feelings of the writers of a section of the text. When the tweets become longer, numerous emotions are represented in a single line, and so it is vital to divide such lengthy tweets. When numerous emotions are expressed, they may be directed towards distinct situations or different facets of a problem. Successful sentiment analysis in such circumstances needs identification of subjects and aspects [5]. Due to its vast number of users, Twitter is a particularly rich source of opinion on a broad range of issues. Due to message length constraints, messages are generally brief. In 2018, the restriction was expanded to 280 characters, although for most of the social network's tenure, tweets were restricted to his 140 characters or fewer. A huge benefit is that with a brief piece of literature like this, there is likely to be just one subject with a mood surrounding that theme. As Boot et al. [6], raising the character restriction only marginally increased the length of tweets, from 70 to 83 characters on average. Since tweets add another layer various hurdles, getting emotion from them may be a challenging process [7]. Tweets are brief texts; however, these messages are written in informal language, which makes them challenging to identify using a traditional text emotion analysis. First are Twitter-specific hash tags. They are used to explicit short thoughts gift with within the tweet which comprise themes, for example, "#sentiment", "#opinion" or true emotions "#happy", "#proud", and "#disappointed". Hash tags are used for a wide type of reasons [8, 9]. Messages may also feature multiple emoticons which may incorporate valuable facts concerning feeling and must now no longer be neglected of challenges, extracting emotion from them can be a difficult task [7]. Tweets are short texts, but these messages are written in informal language, which makes them difficult to classify with a standard text sentiment analyser. First are Twitter-specific hash tags. They are used to explicit brief thoughts gift with inside the tweet which include topics, for example, "#sentiment", "#opinion", or real emotions "#happy", "#proud", and "#disappointed". Hash tags are used for a huge sort of purposes [8, 9]. Messages may additionally comprise several emoticons which could comprise treasured statistics about sentiment and must now no longer be ignored.

3 Dataset Review

Here, we concentrate on the issue of natural language processing with disaster communications using a dataset obtained from the Kaggle website. The dataset comprises 10,876 tweets, 7613 rows in the training dataset, and 3262 rows in the test dataset. The goal attribute in this case is the variation. The dataset was taken from the Kaggle Website, a website and online a database for data scientists and machine learning (ML) practitioners that provides a broad variety of data science and ML issues. The problem we are focusing on "natural language processing with disaster messages"—a dataset published by a Website that contains two datasets consisting of 10,876 tweets, a training dataset (train.csv, 7613 rows), and a test dataset (test.csv, 3262 rows), and the difference between the two datasets is that the training dataset contains the target attribute while the testing dataset does not, and because it is a competition where you want to test the competitors. Therefore, in the following project, we pick the training dataset and train the model.

4 Literature Review

This research goal is to create a classification model connected to supervised learning [10] that can assess whether a given information on Twitter (tweet text) is related to a genuine catastrophe or not. It completes this procedure by utilising the RNN model algorithm and its approach, and it is illustrated in Fig. 1. Supporting disaster relief and response organisations will enable them to respond to genuine disasters rather than made-up ones [11].

4.1 Data Understanding

It might be crucial to construct a classification model that identifies authentic from fraudulent catastrophe tweets. Tragedy aid workers may receive an early warning if they can verify that a tweet they send is about a genuine tragedy. However, these reaction teams will not utilise social media platforms if they cannot detect the difference

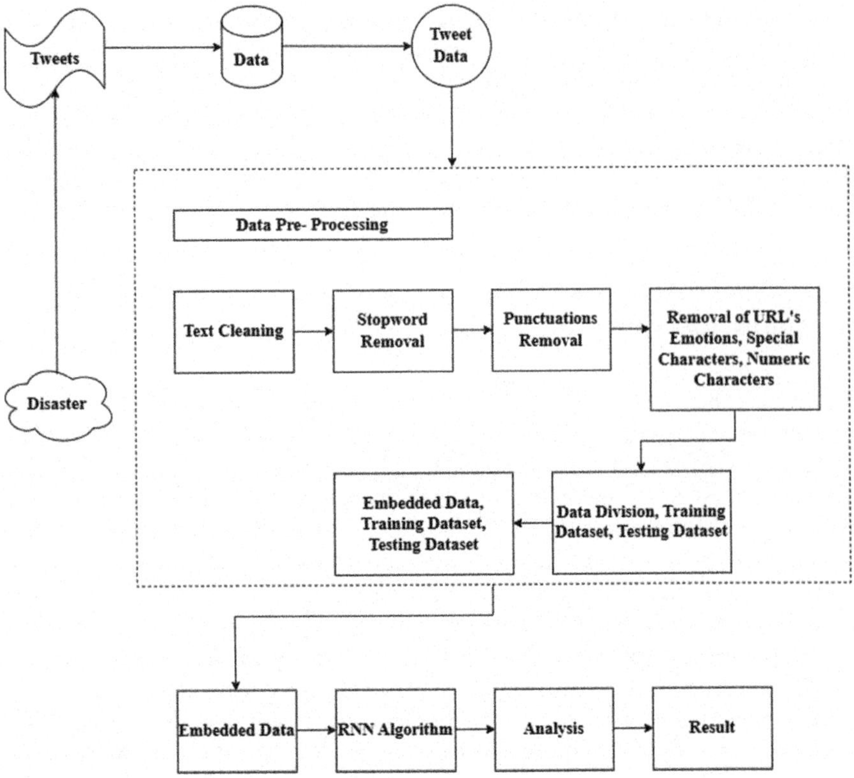

Fig. 1 Architecture of classification model

between genuine and fraudulent tweets, even though they proved how beneficial they were by delivering real-time information [12]. In the second stage, the dataset will be looked at. The purpose is to look at the data, look at the records and characteristics, and exhibit qualities to better understand how they connect to the findings.

4.2 Data Pre-processing

To interpret lower case and capitalised letter in a system by a computer in an unexpected manner, the solution would be word letters are transformed into lower case, also words and numbers including digits will be eliminated, and it is advisable to keep a strategic distance from words and numbers that are mixed, such as game57 or game5ts7, and to perform this task as illustrated in Fig. 2. If the expression is problematic, use purge string. The stop words supply significant information [13]

Here we make use of the prevent word to delete the stop words [14]. Our stop word list was manually updated to incorporate more terms, such as http and https.

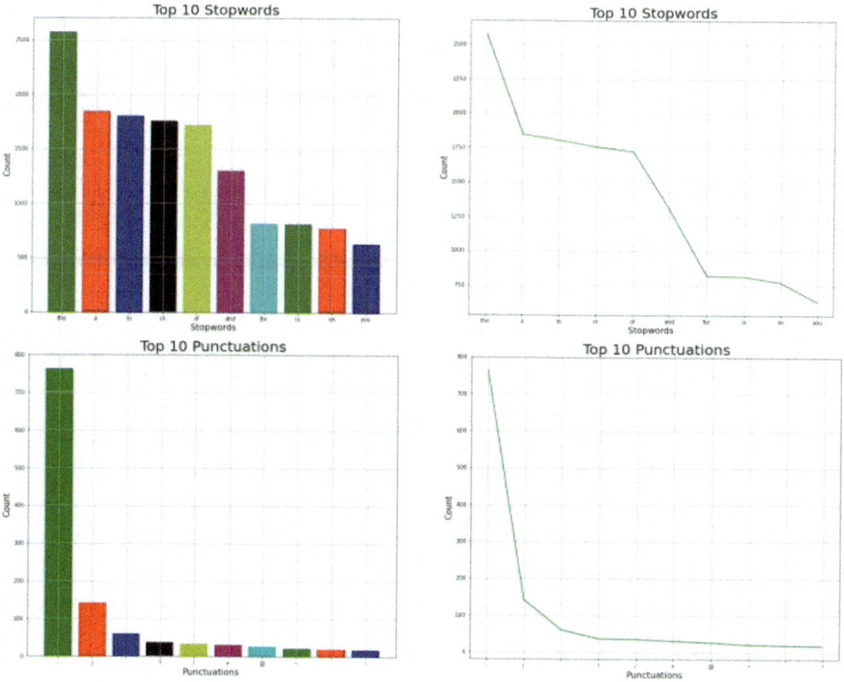

Fig. 2 Performance of training dataset

The dataset does not include any tweets that used special characters like @, /, $, or emoticons. The model learns on the training dataset while the trained model's performance is assessed on the testing set developed in this stage [15–17]. Based on the TDM, a word cloud is constructed, and the name "word cloud" refers to the depiction of where words appear as like in Fig. 3. The term's prominence in the word cloud represents its frequency in the text. Word clouds used to evaluate Twitter topic, gauge people's attitudes on a given issue, or prohibit hate speech.

Fig. 3 Representation of word cloud

4.3 N-Grams

A single word or the combination of word change the order as well as the meaning. *N*-grams are these token-sequences, and it is evident that employing bigrams may be extremely crucial in an emergency, particularly when tweets include negations. However, the introduction of n-grams does not necessarily increase a classification's accuracy. Pang et al. have pointed this out, that in their trials, choosing 2-g (bigrams) over 1-g (unigrams) may led to a 5.8% drop in accuracy with improved accuracy.

4.4 Methodology

There are three steps to how recurrent neural networks function as depicted in Fig. 4. It creates a prediction as it progresses through the concealed layer in the first step. Using the loss function, it compares its prediction to the actual value in the second stage. A model's loss function reveals how well it performs [18–20]. The model performs better the lower the value of the loss function. The error data from back-propagation are employed in the last step to determine the gradient for each point (node) [21, 22]. The value that is utilised to alter the network's weights at each point is the gradient.

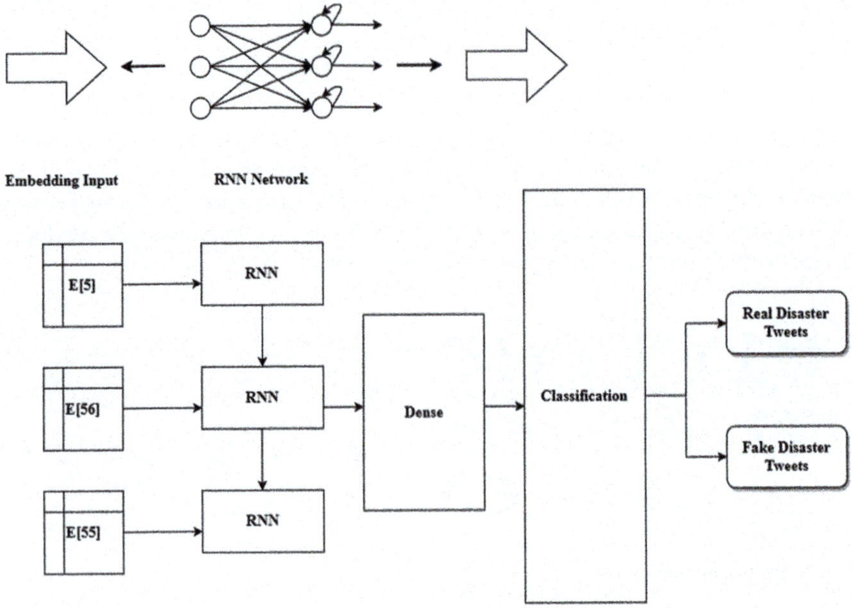

Fig. 4 Stages of recurrent neural network

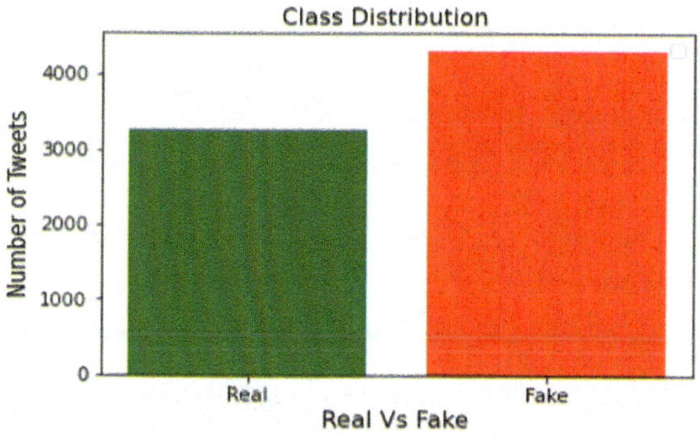

Fig. 5 Classical distribution of real and fake based on number of tweets

When working with sequential data, recurrent neural networks are widely utilised. The reason for this is because the model makes use of layers that give it a short-term memory. It can properly anticipate the following data with this memory. The amount of time that data from the past will be maintained is not fixed in stone; rather, it is governed by the weights that have been given to it [23, 24].

4.5 Tweet Text Exploration

First, the dataset is divided up into two subgroups depending on the target variable (one and zero). For every subset, a distinct corpus is generated, the data will be cleaned, a term document matrix (TDM) is constructed according to it, and the accuracy of the TDM will be determined as shown in Fig. 5. The word cloud will be constructed from the TDM. The "stop_words_english.txt" includes 854 stop words to delete from tweets.

4.6 F-*Score*

Accuracy and recall alone may give highly valuable information about a classifier's performance, but only both can provide the entire picture. The f-value describes both accuracy and recall. It is evaluated using various weights for accuracy and recall. It is also termed the $F1$ score since both metrics contribute equally to the F score.

4.7 Fitting the Model

```
#fit network
model.fit(X_train_set,y_train,epochs=20,verbose=2)
Epoch 1/20
191/191 - 2s - loss: 0.6282 - get_f1: 0.3132 - 2s/epoch - 8ms/step
Epoch 2/20
191/191 - 1s - loss: 0.4425 - get_f1: 0.7715 - 638ms/epoch - 3ms/step
Epoch 3/20
191/191 - 1s - loss: 0.3437 - get_f1: 0.8291 - 574ms/epoch - 3ms/step
Epoch 4/20
191/191 - 1s - loss: 0.2885 - get_f1: 0.8562 - 604ms/epoch - 3ms/step
Epoch 5/20
191/191 - 1s - loss: 0.2507 - get_f1: 0.8774 - 615ms/epoch - 3ms/step
Epoch 6/20
191/191 - 1s - loss: 0.2195 - get_f1: 0.8945 - 634ms/epoch - 3ms/step
Epoch 7/20
191/191 - 1s - loss: 0.1940 - get_f1: 0.9088 - 593ms/epoch - 3ms/step
Epoch 8/20
191/191 - 1s - loss: 0.1738 - get_f1: 0.9212 - 747ms/epoch - 4ms/step
Epoch 9/20
191/191 - 1s - loss: 0.1597 - get_f1: 0.9259 - 704ms/epoch - 4ms/step
Epoch 10/20
191/191 - 1s - loss: 0.1457 - get_f1: 0.9306 - 629ms/epoch - 3ms/step
Epoch 11/20
191/191 - 1s - loss: 0.1332 - get_f1: 0.9387 - 613ms/epoch - 3ms/step
Epoch 12/20
191/191 - 1s - loss: 0.1232 - get_f1: 0.9452 - 597ms/epoch - 3ms/step
Epoch 13/20
191/191 - 1s - loss: 0.1154 - get_f1: 0.9480 - 610ms/epoch - 3ms/step
Epoch 14/20
191/191 - 1s - loss: 0.1080 - get_f1: 0.9512 - 599ms/epoch - 3ms/step
Epoch 15/20
191/191 - 1s - loss: 0.1022 - get_f1: 0.9559 - 604ms/epoch - 3ms/step
Epoch 16/20
191/191 - 1s - loss: 0.0973 - get_f1: 0.9557 - 588ms/epoch - 3ms/step
Epoch 17/20
191/191 - 1s - loss: 0.0927 - get_f1: 0.9588 - 627ms/epoch - 3ms/step
Epoch 18/20
191/191 - 1s - loss: 0.0870 - get_f1: 0.9632 - 583ms/epoch - 3ms/step
Epoch 19/20
191/191 - 1s - loss: 0.0835 - get_f1: 0.9630 - 795ms/epoch - 4ms/step
Epoch 20/20
191/191 - 1s - loss: 0.0815 - get_f1: 0.9612 - 763ms/epoch - 4ms/step
<keras.call backs. History at 0x7f8367c90390>
```

5 Limitation

The limitations prevailing in our current model are the availability of datasets that contain online tweets as some of the platforms are against the sharing of Twitter messages due to security and privacy reasons. This limitation may affect the working of the model due to the lack of current dataset, and only the existing dataset can be used. It also has the limitation of the dataset size that can be used in the model. Hence, we have a disadvantage of using online dataset and can only make use of the offline dataset available.

6 Conclusion

Finally, the purpose of the study was to construct a model that determines whether a tweet is connected to a genuine catastrophe or phoney, which would ultimately assist detect a real disaster and filter out fraudulent ones, effectively respond to the relevant team members. The literature review examined the textual analysis and categorisation technique and the recommended practices to follow. The experiment provided a practical technique to attempt to recognise suspect text as a binary machine learning issue. The challenging part is to determine whether a specific text pertains to a genuine catastrophe or not. The working dataset includes labelled tweets and connected to actual incident. Following the RNN technique (anticipated implementation phase), data understanding in our instance signifies the relevance of the issue to be addressed, and then, we completed data mapping, pre-processing, data modelling, and lastly assessment. In conclusion, we successfully created suggested recurrent neural network learning models, and the accuracies were 96%.

References

1. Valenzuela S, Puente S, Flores PM (2017) Comparing disaster news on Twitter and television: an intermedia agenda setting perspective. J Broadcast Electron Med 61(4):615–637. https://doi.org/10.1080/08838151.2017.1344673,2017
2. Diamantini C, Mircoli A, Potena D, Storti E (2019) Social information discovery enhanced by sentiment analysis techniques. Futur Gener Comput Syst 95:816–828. https://doi.org/10.1016/j.future.2018.01.051
3. Chen P, Sun Z, Bing L, Yang W (2017) Recurrent attention network on memory for aspect sentiment analysis. In: Proceedings of the 2017 conference on empirical methods in natural language processing. https://doi.org/10.18653/v1/d17-1047
4. Ogie RI, Clarke RJ, Forehead H, Perez P (2019) Crowd sourced social media data for disaster management: lessons from the PetaJakarta.org project. Comput Environ Urban Syst 73:108–117. https://doi.org/10.1016/j.compenvurbsys.2018.09.002
5. Cheng M-Y, Kusoemo D, Gosno RA (2020) Text mining-based construction site accident classification using hybrid supervised machine learning. Autom Constr 118:103265. https://doi.org/10.1016/j.autcon.2020.103265

6. Abid SK et al (2021) Toward an integrated disaster management approach: how artificial intelligence can boost disaster management. Sustainability 13(22):12560. https://doi.org/10.3390/su132212560
7. Naresh R, Sayeekumar M, Karthick GM, Supraja P (2019) Attribute-based hierarchical file encryption for efficient retrieval of files by DV index tree from cloud using crossover genetic algorithm. Soft Comput 23(8):2561–2574 [Impact Factor = 3.643] (SCI Indexed). https://doi.org/10.1007/s00500-019-03790-1
8. Naresh R, Vijayakumar P, Jegatha Deborah L, Sivakumar R (2020) A novel trust model for secure group communication in distributed computing. Special issue for security and privacy in cloud computing, J Organ End User Comput 32(3):1–14 [Impact Factor = 1.88] (SCI Indexed). https://doi.org/10.4018/JOEUC.2020070101
9. Lakshmi Narayanan K, Naresh R, A effective encryption and different integrity schemes to improve the performance of cloud services. In: International conference for advancement in technology (ICONAT), Jan 2022. https://doi.org/10.1109/ICONAT53423.2022.9725904
10. Lakshmi Narayanan K, Naresh R, A efficient key validation mechanism with Vanet in real time cloud monitoring to enhance cloud storage and security. Sustain Energy Technol Assess 53:102970. SCI Indexed-[Impact Factor: 7.6] (SCI Indexed). https://doi.org/10.1016/j.seta.2022.102970
11. Lakshmi Narayanan K, Prabakaran G, Bhavani R (2012) A high capacity video steganography based on integer wavelet transform. J Comput Appl 5:358–365. https://jcaksrce.org/upload/52279311_vol5e124p4.pdf
12. Priyaradhikadevi T, Parvathi RMS, Lakshmi Narayanan K, An enhanced strategy for deriving the quality requirements in QOS based web services with location recommender system. Int J Eng Technol 6(4):1697–1707
13. Sakthivelu U, Vinoth Kumar CNS (2022) An approach on cyber threat intelligence using recurrent neural network. In: ICT infrastructure and computing, lecture notes in networks and systems, vol 520, pp 429–439. https://doi.org/10.1007/978-981-19-5331-6_44
14. Vasim Babu M, Vinoth Kumar CNS, Parthiban S, Padmavathi U, Rahman MZU (2022) AE-LEACH: an incremental clustering approach for reducing the energy consumption in WSN. Microprocess Microsyst 93:104602. https://doi.org/10.1016/j.micpro.2022.104602
15. Sasikumar S, Abhay KS, Vinoth Kumar CNS (2021) Network intrusion detection and deduce system. Turk J Comput Math Educ 12(9):404–410. https://doi.org/10.17762/turcomat.v12i9.3094
16. Rathi R, Balyan N, Vinoth Kumar CNS (2020) Pneumonia detection using chest X-ray. Int J Pharm Res (IJPR) 12(3). ISSN: 0975-2366 (Scopus). https://doi.org/10.31838/ijpr/2020.12.03.181
17. Purohit H, Castillo C, Pandey R (2020) Ranking and grouping social media requests for emergency services using serviceability model. Soc Netw Anal Min 10(1). https://doi.org/10.1007/s13278-020-0633-3
18. Madichetty S, Muthukumarasamy S (2020) Detection of situational information from Twitter during disaster using deep learning models. Sādhanā 45(1). https://doi.org/10.1007/s12046-020-01504-0
19. Agarwal A, Xie B, Vovsha I, Rambow O, Passonneau RJ (2011) Sentiment analysis of twitter data
20. Çeliktuğ MF (2018) Twitter sentiment analysis, 3-way classification: positive, negative or neutral? In: 2018 IEEE international conference on big data (big data), pp 2098–2103. https://doi.org/10.1109/BigData.2018.8621970
21. Natural Language Processing with Disaster Tweets | Kaggle (2021) Retrieved 30 Nov 2021, from https://www.kaggle.com/c/nlp-getting-started/data

22. Narayanan KL, Naresh R (2023) Improved security for cloud storage using Elgamal algorithms authentication key validation. In: 2023 international conference for advancement in technology (ICONAT), Jan 2023. https://doi.org/10.1109/iconat57137.2023.10080619
23. Vasudha Rani V, Sandhya Rani K (2016) Twitter streaming and analysis through R. Indian J Sci Technol 9(45). https://doi.org/10.17485/ijst/2016/v9i45/97914
24. Arun K, Srinagesh A, Ramesh M (2017) Twitter sentiment analysis on demonetization tweets in India using R language

Handwritten Indic Script Recognition Using Deep Neural Network

Sreayan De, Sujit Kumar Mondal, and Pawan Kumar Singh

Abstract Automatic script recognition has multiple applications, such as document searching, author identification, document archiving, and optical character recognition. Online script recognition operates on digital copies of papers, while offline script recognition involves scanning printed or handwritten materials before initiating the recognition process. However, identifying and recognizing various script types within a single page poses a challenge for automatic script recognition systems functioning in multilingual environments. Numerous published methods exist, primarily relying on traditional machine learning approaches. However, deep learning techniques have gained popularity due to their enhanced optimization and efficiency. Consequently, this paper employs a customized deep neural network for automatic script recognition of handwritten Indian scripts. The proposed method undergoes testing on the benchmark PHD_Indic_11 dataset, which comprises word images written in 11 Indian languages.

Keywords Handwritten script recognition · PHD_Indic_11 · Word level · Convolutional neural network · Deep learning

S. De
Department of Electronics and Communication Engineering, B. P. Poddar Institute of Management and Technology, Kolkata 700052, India
e-mail: sreayande2002@gmail.com

S. K. Mondal · P. K. Singh (✉)
Department of Information Technology, Jadavpur University, Jadavpur University Second Campus, Plot No. 8, Salt Lake Bypass, LB Block, Sector III, Salt Lake City, Kolkata 700106, West Bengal, India
e-mail: pawansingh.ju@gmail.com; pawankrsingh@ieee.org

S. K. Mondal
e-mail: sujitmondal7890.sm@gmail.com

S. Bandyopadhyay et al. (eds.), *Intelligent Computing Systems and Applications*, Lecture Notes in Networks and Systems 1010,
https://doi.org/10.1007/978-981-97-5412-0_41

1 Introduction

In scripts, various graphic forms are utilized to represent specific character suite-based writing patterns that describe the writing system. With each passing day, Information technology is increasingly ingrained in people's lives, and the reliance on digitizing everyday services is growing rapidly. As we move toward a paperless solution for historical document records, it is imperative to accelerate and enhance the development of Optical Character Recognition (OCR) systems [1].

The development of OCR systems has been a longstanding area of exploration, but thus far, the research community has struggled to create a robust, multilingual, and comprehensive system for accurately identifying multiple scripts on a single page. While several OCR scripts have demonstrated efficiency in script recognition tasks, most OCR systems are confined to particular languages or scripts. The most stimulating task today is categorizing scripts from multi-language document images that contain content written in multiple scripts and/or languages. Consequently, the most complex issue in the research field is the creation of a robust underlying script identification system.

Script recognition plays a crucial role in document image analysis, especially when dealing with multiple scripts and languages. It serves as a fundamental step in identifying the language used in a document and is necessary for additional transformations such as searching, indexing, or conversion. When a script is written in a single language, the script's identity corresponds to the language itself. However, in the case of scripts containing multiple languages, automatic script recognition becomes the initial step in recognition, followed by linguistic recognition within the scripts. India is a very varied country having 22 recognized official languages. India also has hundreds of additional languages, each with its own dialects and variations.

The officially recognized languages in the Indian constitution include Sanskrit, Punjabi, Urdu, Hindi, Tamil, Kashmiri, Konkani, Bengali, Kannada, Sindhi, Manipuri, Oriya, Nepali, Santhali, Malayalam, Bodo, Maithili, Gujarati, Assamese, Telugu, and Dogri. Hindi language (written using the Devanagari script) holds the status of the official language of India and boasts the largest number of speakers, estimated to be over 5000 lakhs [2]. Indian scripts can be characterized as a systematic arrangement of various script signs, employing a generic logical composition. This unique structure can be denoted to as the "script structure grammar," having no parallel in any other script worldwide. Indian scripts are written using syllables and typically consist of three visual levels, with each level's integral symbols functioning independently to interpret the corresponding syllable [3].

The contributions of the proposed article is as follows:

- A deep neural network (DNN) model has been proposed for word images written in 11 Indian languages. Figure 1 presents the schematic block diagram of the entire workflow.
- The proposed model is tested on a standard benchmark PHD_Indic_11 dataset consisting of 10,000 word images.

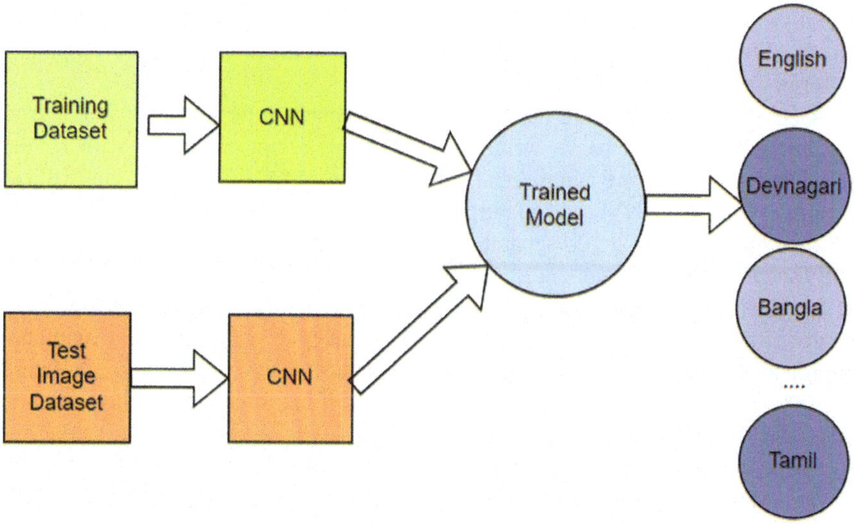

Fig. 1 Schematic diagram of our proposed DNN model for word-level handwritten script recognition

- The performance of our proposed DNN model is compared with some existing handwritten script recognition models.

2 Literature Analysis

This section reviews some of the research done for handwritten script recognition. However, in order to address the aforementioned issue, researchers have focused increasingly on machine learning-based approaches.

Singh et al. [4] analysis of the Indian script classification performance of numerous extremely accurate and effective handwritten script classification systems. Four separate degrees of granularity—Page-level, Block-level, Text line-level, and Word-level—are used to deduce the elements needed for this script recognition from the script images. Eight separate sets of characteristics and six different, extremely accurate classifiers have been used to achieve and compare the outcomes of handwritten script recognition at each level. On two distinct handwritten Indian script datasets, all of these testing were performed. One dataset was created internally, and the other was made publicly available.

Several feature extraction as well as recognition strategies related to the script classification of several Indian scripts were investigated by Singh et al. [5]. This study was intended to serve as a compendium for Indian scholars, as well as for policymakers and practitioners. The primary goal of this research was to connect the scientists studying various Indic scripts. This research would serve as a better reference for other studies in this area, especially in light of the most recent developments in script classification.

A complementary feature-based method that detected the scripts of words from handwritten document pages written in *Devnagari* script combined with Roman script was discussed by Singh et al. [6]. For this script recognition, an artificial neural network (ANN)-based identifier with 39 notable attributes was employed. 99.54% of the words in 100 separate handwritten documents that had both *Roman* and *Devanagari* words were correctly categorized using the methodology.

Singh et al. [7] suggested a word-level script recognition method for 6 different handwritten Indian scripts. Combining elliptical and polygonal approximation approaches, a set of 82 distinct characteristics was created. Considering a dataset of 7000 word images, our method has been tested using a number of classifiers. An ANN classifier produced accuracy of 95.35%.

Shape-based features combined with texture-based features, employed in the recognition of various handwritten text words written in 5 different Indian scripts such as *Roman, Malayalam, Devanagari, Telugu,* and *Bangla,* was discussed by Singh et al. [8]. To assess the effectiveness of the script recognition system, a feature set with 87 different components was developed. An ANN was selected as the top classifier based on the precision of identification of several classifiers. With a recognition accuracy of 91.79%, it had the best results.

For 8 different handwritten scripts, Singh et al. [9] proposed an automatic script recognition method at word level. The input word image is subjected to a discrete wavelet transform in the first stage of this script recognition approach, and the output of the first stage is then subjected to a Radon transform in the second stage to provide a statistical feature vector having 48 elements for each text words.

To recognize handwritten scripts written in 8 prominent scripts, including *Roman, Gurumukhi, Oriya, Devanagari, Urdu, Manipuri,* and *Bangla,* Singh et al. [10] developed a line parameter-based technique. The directional spatial appearances centered on the line parameter were extracted by combining the Hough transform (HT) along with distance transform (DT). Experiments were carried out at the word level using different classifiers on a self-created dataset having 12,000 handwritten scripts. An ANN classifier produced an average accuracy of 95.28%.

A novel handwritten script recognition algorithm that takes into account all 12 of India's official scripts was discussed by Singh et al. [11]. Using a tree-based technique, the identification process is carried out at the word level. The DT along with HT was used to first differentiate Matra-based scripts from non-Matra-based scripts. Then, the Matra-based scripts are classified individually from the non-Matra-based scripts using the modified log-Gabor filter.

Singh et al. [12] described the use of the cross-validation approach to apply algorithms to a given dataset and how the most correct identifier was chosen by taking into account different categorization evaluation metrics. They suggested using both the Mann–Whitney U test and Kruskal–Wallis H test for the statistical comparison of different classifiers across various datasets.

Rajput et al. [13] discussed that there was a requirement for the recognition of the kind of script in automated manner and to insert Indic script documents to the suitable optical character recognition mechanism with the aim of extracting the information [8]. An effective method was suggested to recognize the handwritten script in which the local binary pattern system was implemented. This technique helped in extracting the attributes from a block of an image of a handwritten document. Both the Nearest Neighbor and Support Vector Machine classifiers were deployed to recognize the form of the script. The handwritten document images, which were written in 6 different Indian scripts such as *Hindi, Kannada, Malayalam, English, Urdu*, and *Telugu*, applied in the experimentation. The results depicted that the SVM algorithm offered higher accuracy of 99.5% as compared to KNN that had accuracy of 98.46%.

Obaidullah et al. published a handwritten document image dataset consisting of 11 official Indian scripts in the year 2017 [14], including *Oriya, Gurumukhi, Gujarati, Bangla, Kannada, Telugu, Devanagari, Tamil, Malayalam, Roman*, and *Urdu*. There are 463 contributors from all around India who have contributed 1458 text pages to the PHDIndic_11 dataset. The dataset can be effectively used for a number of other tasks involving document image analysis in addition to script recognition. These tasks include writer identification and script, sentence and word identification and segmentation, text-line segmentation, and word segmentation and recognition. A novel method of merging convolutional neural networks (CNNs) for script recognition was put up by Ukil et al. [1]. A variety of tiny trainable CNNs were employed. Such a set of separately trainable modules varies in terms of wavelet processing, CNN depth, and input image size. There were 11 different Indian scripts represented in the collection producing the highest accuracy of 95.04%.

It is to be noted that these works have been performed on word-level handwritten script classification using machine learning-based methods. These works have not been tested on any standard benchmark datasets.

2.1 Dataset Used

The PHD_Indic_11 dataset [14] consists of 10,000 images divided into 11 classes (hence the name Indic). Classes have 11 different types of language. This dataset is split in a standard way, where 3000 of the images are used to assess the model's performance after it has been trained on 7000 images. The images are in monocolor with small measuring 32 by 32 pixel squares. The image itself is provided as the input The images are changed from color to grayscale since data information, not color information, is what the network needs. The images have also been downsized to 32 × 32. Pyramid reduction is used to diminish the size of the dataset images, which are larger, to 32 × 32 pixels. A data structure called the picture pyramid has been created to facilitate effective scaled convolution with lessened image representation. It is made up of a series of duplicates of an original image where the sample density

Fig. 2 Sample text word images written in 11 different Indian languages taken from the standard benchmark PHD_Indic_11 dataset

and resolution are gradually reduced. Figure 2 shows the sample word images written in 11 Indian languages.

3 Proposed DNN Model

The model that is proposed in this paper includes 12 layers: a model with 4 CNN layers, 4 pooling layers, and 4 dense layers. The following are the main design components. After the first convolutional layer, the layer with pooling followed by the second convolutional layer have been included in our proposed deep learning architecture. Finally, a third convolutional layer has been included which is then followed by the feature mapping, subjected to feature dimension reduction using the dense layers. Every convolutional layer in the network undergoes local normalization following the convolution operation to speed up convergence, which helps to increase training effectiveness and classification performance. Figure 3 shows the proposed DNN architecture. The schematic diagram of our proposed DNN model consists of 12 layers out of which 4 are CNNs, 4 pooling layers along with 4 dense layers. Table 1 gives the overall architecture summary of our proposed DNN model. It is evident from Table 1 that there are 1,17,371 parameters in total. All the 1,17,371 parameters are trainable and we had 0 non-trainable parameters.

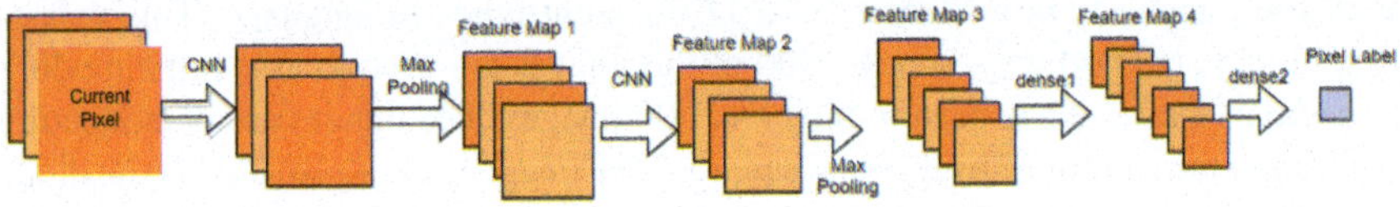

Fig. 3 Overall architectural view of our proposed DNN model for handwritten script recognition at word level

Table 1 Complete summary of our proposed DNN model

Layer (Type)	Output shape	Param#
Input-CNN layer (Conv_2D)	(None, 62, 62, 32)	320
Pooling Layer (MaxPooling_2D)	(None,31,31,32)	0
Conv-layer (Conv_2D)	(None, 29, 29, 64)	18,496
2nd-Pooling Layer (MaxPooling_2D)	(None, 14, 14, 64)	0
2-CNN Layer (Conv_2D)	(None, 12, 12, 64)	36,928
3rd-Pooling layer (MaxPooling_2D)	(None,6, 6, 64)	0
3-CNN Layer (Conv_2D)	(None,5, 5, 32)	8224
4th-Pooling layer (MaxPooling_2D)	(None, 2, 2, 32)	0
4 CNN Layer (Conv_2D)	(None, 1, 1, 16)	2064
Flatten (Flatten)	(None, 16)	0
Dense (Dense)	(None, 625)	10,625
Dense_1 (Dense)	(None, 64)	40,064
Dense_2 (Dense)	(None, 10)	650

4 Results Analysis

The suggested DNN model's output is described in terms of five assessment metrics: Accuracy, Precision, Recall, F1-score, and support. Our suggested DNN model's total classification accuracy is found to be 94.47%. The capacity of an automatic recognition model to deliver only the data points in a class is known as Precision. The ability of a recognition model to distinguish each data point in an appropriate dataset is known as Recall. The F1-score is the only statistic that employs the harmonic mean to the combination of Precision with Recall. The number of examples belonging to a particular script class in a particular dataset is known as Support. Table 2 gives the performance of our proposed DNN model in terms of Precision, Recall, F1-score, and Support.

Figure 4a displays the comparison of epoch size and accuracy of our proposed model and it can be observed that the accuracy becomes stagnant after reaching the value of 0.975 which suggests further accuracy cannot be achieved anymore. Figure 4b displays the comparison between epoch and validation accuracy which confirms the efficiency of our model with respect to new data.

Figure 5a displays the performance comparison of our model's loss with respect to epoch size and it can be seen that loss decreases which further justifies accuracy and efficiency of our model. Figure 5b displays the comparison of the epoch size and the training loss which can be observed to be decreasing and also confirms how well our model fits the training data.

The confusion matrix provided in Fig. 6 helps to draw certain conclusions. It can be understood from Fig. 6 that certain Indian languages have a very high confusion rate between themselves. It is seen that Bengali has been most confused with Telugu, Devanagari with Punjabi, English with Tamil, Gujarati with Urdu, Kannada with Telugu, Odia with Telugu, Tamil with Urdu, Urdu with Telugu.

Table 2 Metrics measured in terms of Precision, F1-score, Recall, and Support for evaluating our proposed DNN model's performance

	Precision	Recall	F1 score	Support
Bengali	0.89	0.99	0.94	6992
Devanagari	0.68	0.99	0.80	6991
English	0.99	0.82	0.90	7041
Gujrati	0.90	0.97	0.93	6937
Kannada	0.92	0.98	0.95	7005
Odia	0.95	0.98	0.96	7007
Punjabi	0.96	0.59	0.73	6932
Tamil	0.98	0.93	0.95	7012
Telugu	0.97	0.94	0.95	7073
Urdu	0.99	0.92	0.95	7010

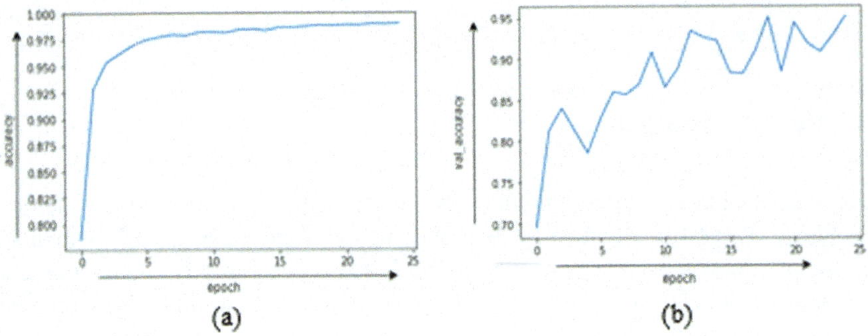

Fig. 4 Graph showing the comparison of: **a** training accuracy versus epoch size and, **b** validation accuracy versus epoch size

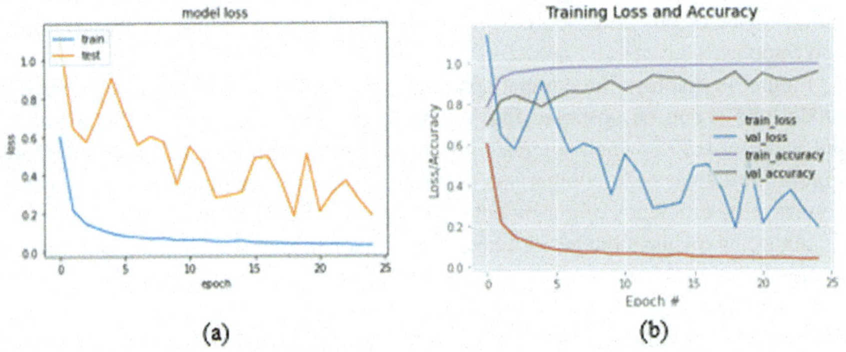

Fig. 5 Graph showing the comparison of: **a** loss versus epoch size and, **b** training and validation accuracies versus epoch size

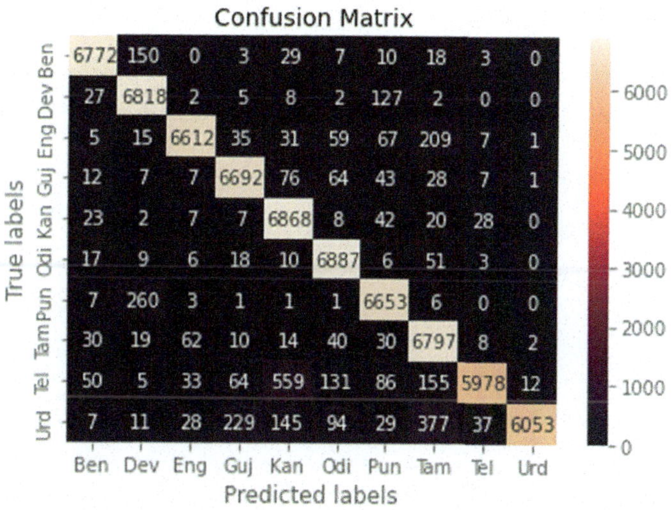

Fig. 6 Confusion matrix outputted by our proposed DNN model over PHD_Indic_11 dataset

4.1 Comparative Study

LeNet [15] and AlexNet [16], two commonly used deep learning techniques, are utilized for a fair comparison. In addition, previously published research on 11 handwritten Indic script datasets (together with their baseline findings) are taken into account. We list the findings in Fig. 7. Since other approaches provided accuracy, i.e., identification rate, our comparison analysis is concentrated on accuracy (rather than Precision, Recall, and F1-score). It is evident from Fig. 7 that our proposed DNN model performs better than the other previously proposed script recognition models.

5 Conclusion and Future Works

In the present work, we have developed a customized DNN method for Indian handwritten script recognition at word level that combines a number of tiny CNNs. Our proposed script recognition method mainly comprises of 4 convolutional layers, 4 pooling layers along with 4 dense layers. The main advantage of script recognition using our proposed DNN model is the high efficiency and accuracy. We have been able to achieve 97.5% accuracy which surpasses some standard DNN models for script identification. Our model is based on word-level script classification and the data set that we have used is a standard benchmark dataset namely PHD_Indic_11

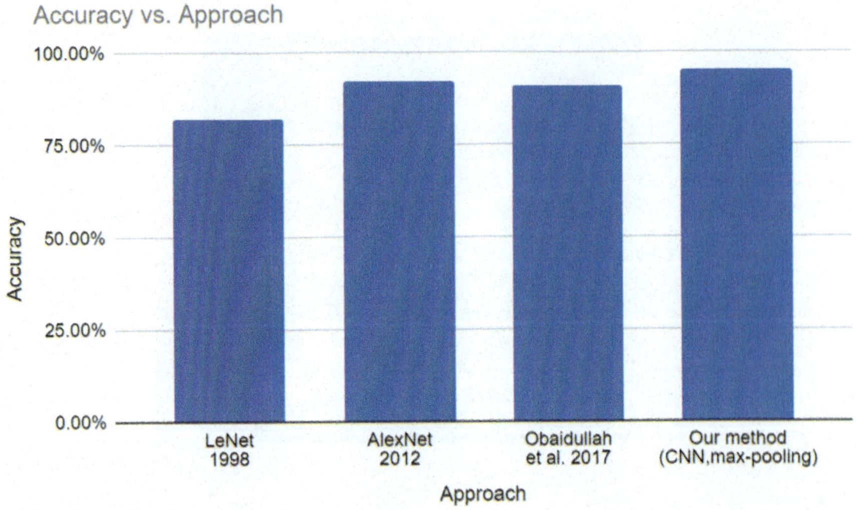

Fig. 7 Graphical analysis showing the performance of our proposed DNN model as compared to some standard models

which consists of handwritten document word images in 11 commonly used Indian scripts, including Oriya, Gurumukhi, Roman, Gujarati, Telugu, Bangla, Kannada, Devanagari, Tamil, Malayalam, and Urdu.

In the future, we can also improve our current proposed technique and also work on line-level and page-level handwritten script recognition techniques using complex DNN models which are much more accurate and efficient.

References

1. Ukil S, Ghosh S, Obaidullah SM, Santosh KC, Roy K, Das N (2020) Improved word-level handwritten Indic script identification by integrating small convolutional neural networks. Neural Comput Appl 32:2829–2844
2. Sinwar D, Dhaka VS, Pradhan N et al (2021) Offline script recognition from handwritten and printed multilingual documents: a survey. IJDAR 24:97–121. https://doi.org/10.1007/s10032-021-00365-5
3. Mahajan S, Rani R (2022) Word level script identification using convolutional neural network enhancement for scenic images. Trans Asian Low-Resour Lang Inf Process 21(4):1–29
4. Singh PK, Sarkar R, Abraham A, Nasipuri M (2021) A case study on handwritten Indic script classification: benchmarking of the results at page, block, text-line, and word levels. Trans Asian Low-Resour Lang Inf Process 21(2):1–36
5. Singh PK, Sarkar R, Nasipuri M (2015) Offline script identification from multilingual Indic-script documents: a state-of-the-art. Comput Sci Rev 15:1–28
6. Singh PK, Sarkar R, Das N, Basu S, MitaNasipuri (2013) Identification of Devnagari and Roman scripts from multi-script handwritten documents. In: Pattern recognition and machine intelligence: 5th international conference, PReMI 2013, Kolkata, India, December 10–4, 2013. Proceedings 5. Springer Berlin Heidelberg, pp 509–514

7. Singh PK, Sarkar R, Nasipuri M, Doermann D (2015) Word-level script identification for handwritten Indic scripts. In: 2015 13th international conference on document analysis and recognition (ICDAR). IEEE, pp 1106–1110
8. Singh PK, Mondal A, Bhowmik S, Sarkar R, Nasipuri M (2015) Word-level script identification from handwritten multi-script documents. In: Proceedings of the 3rd international conference on frontiers of intelligent computing: theory and applications (FICTA) 2014: vol. 1. Springer International Publishing, pp 551–558
9. Singh PK, Das S, Sarkar R, Nasipuri M (2017) A two-stage approach to handwritten Indic script identification. Hybrid Intelligence for Image Analysis and Understanding, pp 47–77
10. Singh PK, Das S, Sarkar R, Nasipuri M (2016) Line parameter based word-level Indic script identification system. Int J Comput Vis Image Process (IJCVIP) 6(2):18–41
11. Singh PK, Sarkar R, Bhateja V, Nasipuri M (2018) A comprehensive handwritten Indic script recognition system: a tree-based approach. J Ambient Intell Hum Comput. https://doi.org/10.1007/s12652-018-1052-4
12. Singh PK, Sarkar R, Nasipuri M (2015) Statistical validation of multiple classifiers over multiple datasets in the field of pattern recognition. Int J Appl Pattern Recognit 2(1):1–23
13. Rajput GG, Ummapure SB (2019). Script identification from handwritten document images using LBP technique at block level. In: 2019 International Conference on Data Science and Communication (IconDSC). IEEE, pp 1–6
14. Obaidullah SM, Halder C, Santosh KC, Das N, Roy K (2018) PHDIndic_11: page-level handwritten document image dataset of 11 official Indic scripts for script identification. Multimed Tools Appl 77:1643–1678
15. LeCun Y, Bottou L, Bengio Y, Haffner P (1998) Gradient-based learning applied to document recognition. Proc IEEE 86(11):2278–2324
16. Krizhevsky A, Sutskever I, Hinton GE (2017) ImageNet classification with deep convolutional neural networks. Commun ACM 60(6):84–90

SPRINGER NATURE

GPSR Compliance

The European Union's (EU) General Product Safety Regulation (GPSR) is a set of rules that requires consumer products to be safe and our obligations to ensure this.

If you have any concerns about our products, you can contact us on ProductSafety@springernature.com

In case Publisher is established outside the EU, the EU authorized representative is:

Springer Nature Customer Service Center GmbH
Europaplatz 3
69115 Heidelberg, Germany

The manufacturer's authorised representative in the EU is Springer Nature Customer Service Centre GmbH, Europaplatz 3, 69115 Heidelberg, Germany. If you have any concerns regarding our products, please contact ProductSafety@springernature.com

Printed and bound by CPI Group (UK) Ltd, Croydon, CR0 4YY
01/10/2026
02225901-0012